CliffsStudySolver™

Chemistry

By Charles Henrickson

Wiley Publishing, Inc.

Published by:
Wiley Publishing, Inc.
111 River Street
Hoboken, NJ 07030-5774
www.wiley.com

Published by Wiley, Hoboken, NJ
Published simultaneously in Canada

Library of Congress Cataloging-in-Publication data is available from the publisher upon request.

ISBN: 0-7645-7419-1

Printed in the United States of America

10 9 8 7 6 5 4 3 2 1

1B/RU/QT/QV/IN

For general information on our other products and services or to obtain technical support, please contact our Customer Care Department within the U.S. at 800-762-2974, outside the U.S. at 317-572-3993, or fax 317-572-4002.

Wiley also publishes its books in a variety of electronic formats. Some content that appears in print may not be available in electronic books. For more information about Wiley products, please visit our web site at www.wiley.com.

WILEY

About the Author

Charles Henrickson received his Ph.D. in chemistry from the University of Iowa with additional study at the University of Illinois, Urbana and the University of California, Berkeley. He has authored or coauthored several widely used learning aids and laboratory manuals in chemistry as well as a textbook. Dr. Henrickson is a professor of chemistry, emeritus at Western Kentucky University in Bowling Green, Kentucky.

Publisher's Acknowledgments

Editorial

Project Editor: Marcia L. Larkin

Senior Acquisitions Editor: Greg Tubach

Technical Editor: John Moore

Editorial Assistant: Amanda Harbin

Composition

Project Coordinator: Ryan Steffen

Indexer: Ty Koontz

Proofreader: Laura L. Bowman

Wiley Publishing, Inc. Composition Services

Table of Contents

Pretest

Pretest Questions

Circle the letter of the best answer to each question.

1. In scientific notation, 0.00450 would be written:

   A. 4.50×10^3

   B. 4.5×10^{-3}

   C. 4.50×10^{-3}

2. Round off 1,766,439 to two significant figures and express the answer in scientific notation.

   A. 1.7×10^6

   B. 1.8×10^6

   C. 1.77×10^6

3. 15.0 mm = _____ m, and 5.50 dL = _____ mL

   A. 150 m and 55.0 mL

   B. 0.15 m and 0.0550 mL

   C. 0.015 m and 550 mL

4. 0.250 L of an organic liquid had a mass of 194 grams. What is the density of this liquid in g/mL?

   A. 1.29 g/mL

   B. 0.485 g/mL

   C. 0.776 g/mL

5. 54.4°C = _____ °F

   A. 98.0°F

   B. 130°F

   C. 72.2°F

6. What amount of heat is needed to raise the temperature of 50.0 g of lead metal from 45.0°C to 50.0°C. The specific heat of lead is 0.128 J/g-°C.

   A. 32.0 J

   B. 1,950 J

   C. 320 J.

7. Which of the following is a physical change?

A. Sodium and chlorine forming table salt.
B. Table salt melting at 801°C.
C. Making solid soap from solid animal fat.

8. The chemical symbols for iron, silicon, sulfur, and calcium are:

A. Fe, Si, S, and Cl
B. Ir, Si, Su, and Ca
C. Fe, Si, S, and Ca

9. Which of the following is an ionic compound?

A. SO_2
B. PCl_3
C. K_2O

10. Of the three subatomic particles, which one is not found in the nucleus of an atom?

A. proton
B. neutron
C. electron

11. The composition of the $^{23}_{11}N$ atom is:

A. 11 protons, 11 neutrons, 23 electrons
B. 11 protons, 12 neutrons, 11 electrons
C. 23 protons, 11 neutrons, 23 electrons

12. Which pair represents isotopes of an element?

A. $^{10}_{5}X$ and $^{10}_{6}X$
B. $^{10}_{5}X$ and $^{11}_{6}X$
C. $^{10}_{5}X$ and $^{11}_{5}X$

13. How many electrons, protons, and neutrons are in one $^{56}_{26}Fe^{3+}$ ion?

A. 29 electrons, 26 protons, and 30 neutrons
B. 26 electrons, 29 protons, and 30 neutrons
C. 23 electrons, 26 protons, and 30 neutrons

14. Which of the following is not a base?

A. $NH_3(aq)$
B. $Ca(OH)_2$
C. NO_2

15. Which completion is false? One mole of carbon dioxide, CO_2 . . .

A. contains the same number of molecules as one mole of water, H_2O.

B. has the same mass as one mole of water, H_2O.

C. is 6.022×10^{23} molecules of CO_2.

16. How many atoms of iron are in 100. g of iron? The molar mass of iron is 55.85 g.

A. 1.09×10^{26} atoms

B. 1.08×10^{24} atoms

C. 2.69×10^{23} atoms

17. What is the molar mass of sodium sulfate, Na_2SO_4? The molar masses of the constituent elements are: Na = 23.0 g; S = 32.1 g; O = 16.0 g.

A. 71.1 g

B. 142.1 g

C. 213.3 g

18. How many molecules are in 100. g of nitrogen dioxide, NO_2? The molar masses of nitrogen and oxygen are: N = 14.0 g; O = 16.0 g.

A. 6.0×10^{25} molecules

B. 100 molecules

C. 1.31×10^{24} molecules

19. What is the mass-percent of nitrogen in nitrogen dioxide, NO_2? The molar masses of nitrogen and oxygen are: N = 14.0 g; O = 16.0 g.

A. 30.4% N

B. 46.7% N

C. 87.5% N

20. How many grams of aluminum are in 34.8 g of aluminum oxide, Al_2O_3? The molar masses of the elements are: Al = 27.0 g; O = 16.0 g.

A. 13.9 g Al

B. 21.9 g Al

C. 18.4 g Al

21. A sample of an organic compound was analyzed and found to contain 6.49 g carbon, 1.08 g hydrogen, and 5.76 g oxygen. What is the empirical formula of the compound? The molar masses for the elements are: C = 12.0 g; O = 16.0 g; H = 1.01 g.

A. C_6HO_5

B. CH_4O

C. $C_3H_6O_2$

22. When correctly balanced, the number written in front of $Mg(s)$ would be:

$Mg(s) + Fe_2(SO_4)_3(aq) \rightarrow MgSO_4(aq) + Fe(s)$

A. 1
B. 2
C. 3

23. How much heat is involved in the synthesis of 5.0 moles of ammonia, $NH_3(g)$?

$N_2(g) + 3\ H_2(g) \rightarrow 2\ NH_3(g) \quad \Delta H = -91.8\ kJ$

A. 459 kJ
B. 36.7 kJ
C. 230 kJ

24. Which is a correct interpretation of the following balanced chemical equation?

$S(s) + 3\ F_2(g) \rightarrow SF_6(g)$

A. 32.1 g of S reacts with 3 moles of F_2 to form 6.022×10^{23} molecules of SF_6.
B. 1.0 molar mass of S and 3 molecules of F_2 react to form 1 molecule of SF_6.
C. 100 grams of S react with 300 grams of F_2 to form 100 g of SF_6.

25. How many moles of CO_2 will be produced if 4.5 moles of C_2H_4 are consumed?

$C_2H_4(g) + 3\ O_2(g) \rightarrow 2\ CO_2(g) + 2\ H_2O(g)$

A. 4.5 moles
B. 2.25 moles
C. 9.0 moles

26. How many moles of SO_2 will form if 180. g of FeS_2 is burned? The molar mass of FeS_2 is 120. g.

$4\ FeS_2(s) + 11\ O_2(g) \rightarrow 2\ Fe_2O_3(s) + 8\ SO_2(g)$

A. 12 moles
B. 3 moles
C. 6 moles

27. How many grams of KO_2 must be used to prepare 100. g of O_2? The molar mass of KO_2 is 71.1 g and that of O_2 is 32.0 g.

$4\ KO_2(s) + 2\ H_2O(l) \rightarrow 4\ KOH(aq) + 3\ O_2(g)$

A. 889 g
B. 296 g
C. 94.8 g

28. How many subshells and orbitals are in the $n = 3$ principal shell?

A. 3 subshells and 3 orbitals

B. 3 subshells and 9 orbitals

C. 2 subshells and 8 orbitals

29. If two electrons are in the same orbital, then

A. they must spin in the same direction.

B. they must spin in opposite directions.

C. the electrons lose their spin.

30. Which of the following is the correct ground state electronic configuration for the 15 electrons of phosphorus, P?

A. $1s^2 2s^2 2p^3 3s^2 3p^6$

B. $1s^2 2s^2 2p^6 3p^5$

C. $1s^2 2s^2 2p^6 3s^2 3p^3$

31. Which of the following would be a likely electronic configuration of an alkali metal, an element in Group IA?

A. $1s^2 2s^2 2p^6 3s^2 3p^6 4s^1$

B. $1s^2 2s^2 2p^6 3s^2 3p^1$

C. $1s^2 2s^2 2p^6 3s^2 3p^6 4s^2 3d^1$

32. Which list of 4 elements correctly aligns them left to right in order of decreasing atomic radius?

A. O > P > K > Mg

B. K > Mg > P > O

C. Mg > K > P > O

33. What is the expected formula for the ionic compound formed from Ca and S?

A. Ca_2S

B. CaS

C. CaS_2

34. How many nonbonding pairs of electrons are in the Lewis structure of Cl_2O?

A. 4

B. 6

C. 8

35. Which term describes the three-dimensional shape of the SO_3 molecule?

A. trigonal planar
B. trigonal pyramidal
C. tetrahedral

36. Which gas law relates the volume of a gas and its temperature?

A. Boyle's law
B. Charles' law
C. Dalton's law

37. How many moles of hydrogen gas occupy 112 L at STP?

A. 5.00 moles
B. 7.00 moles
C. 12.0 moles

38. What is the volume occupied by 2.0 moles of nitrogen gas at 300°C and 1.25 atm?

A. 39 L
B. 57 L
C. 75 L

39. Which of the following is not an intermolecular force?

A. ionic bonding
B. hydrogen bonding
C. dipole-dipole forces

40. What mass of a 5.00 %(m/m) copper(II) sulfate solution would contain 19.4 g of $CuSO_4$?

A. 388 g
B. 397 g
C. 369 g

41. What mass of NaOH is needed to prepare 250. mL of a 3.50 M solution of NaOH in water? The molar mass of NaOH is 40.0 g.

A. 35.0 g
B. 56.0 g
C. 10.0 g

42. It required 36.45 mL of 0.115 M NaOH to just exactly neutralize the acid in 25.00 mL of $H_2SO_4(aq)$. What was the molarity of the sulfuric acid? The balanced equation is:

$2\ NaOH(aq) + H_2SO_4(aq) \rightarrow Na_2SO_4(aq) + H_2O(l)$

A. 0.168 M

B. 0.0838 M

C. 0.0394 M

43. An Arrhenius acid . . .

A. is a proton acceptor.

B. produces hydronium ions in water.

C. reacts with metals to produce oxygen.

44. What is the principal difference between a weak acid and a strong acid?

A. Weak acids are more dilute than strong acids.

B. Weak acids ionize much less than strong acids in solution.

C. Weak acids are nonelectrolytes; strong acids are strong electrolytes.

45. The net-ionic equation for the neutralization of hydrochloric acid with sodium hydroxide is:

A. $H^+(aq) + OH^-(aq) \rightarrow H_2O(l)$

B. $HCl(aq) + OH^-(aq) \rightarrow H_2O(l) + Cl^-(aq)$

C. $H^+(aq) + NaOH(aq) \rightarrow Na^+(aq) + H_2O(l)$

Pretest Answers

1. **C.**

   If you missed 1, go to "Writing Numbers in Scientific Notation," page 11.

2. **B.**

   If you missed 2, go to "Significant Figures and Rounding Off Numbers," page 13.

3. **C.**

   If you missed 3, go to "Metric-Metric Conversions," page 22.

4. **C.**

   If you missed 4, go to "Density," page 26.

5. **B.**

   If you missed 5, go to "Measuring Temperature," page 28.

6. **A.**

   If you missed 6, go to "Energy and the Measurement of Heat," page 30.

7. **B.**

   If you missed 7, go to "Properties and Changes," page 41.

8. **C.**

   If you missed 8, go to "The Periodic Table of the Elements," page 44.

9. **C.**

   If you missed 9, go to "Compounds," page 48.

10. **C.**

    If you missed 10, go to "Electrons, Protons, and Neutrons," page 64.

11. **B.**

    If you missed 11, go to "The Atomic Number and Mass Number," page 65.

12. **C.**

    If you missed 12, go to "Isotopes of the Elements," page 68.

13. **C.**

    If you missed 13, go to "Ions from Atoms," page 73.

14. **C.**

    If you missed 14, go to "The Common Acids and Bases," page 98.

15. **B.**

    If you missed 15, go to "The Mole and Avogadro's Number," page 110.

16. **B.**

    If you missed 16, go to "Elements: Moles and Counting Atoms," page 114.

17. **B.**

    If you missed 17, go to "Determining the Molar Mass of Compounds," page 118.

18. **C.**

    If you missed 18, go to "Compounds: Moles and Counting Molecules," page 123.

19. **A.**

If you missed 19, go to "Calculating the Percent Composition of Compounds," page 135.

20. **C.**

If you missed 20, go to "Calculating the Percent Composition of Compounds," page 135.

21. **C.**

If you missed 21, go to "Calculating Empirical Formulas," page 143.

22. **C.**

If you missed 22, go to "Balancing Chemical Equations," page 160.

23. **C.**

If you missed 23, go to "Chemical Reactions and Heat," page 177.

24. **A.**

If you missed 24, go to "The Meaning of the Balanced Equation," page 191.

25. **C.**

If you missed 25, go to "Mole-to-Mole Conversions," page 193.

26. **B.**

If you missed 26, go to "Mole-to-Mass and Mass-to-Mole Conversions," page 197.

27. **B.**

If you missed 27, go to "Mass-to-Mass Conversions," page 201.

28. **B.**

If you missed 28, go to "The Quantum Mechanical Atom," page 227.

29. **B.**

If you missed 29, go to "Electron Spin," page 231.

30. **C.**

If you missed 30, go to "The Electronic Configuration of Atoms," page 234.

31. **A.**

If you missed 31, go to "Electronic Configuration and the Periodic Table," page 239.

32. **B.**

If you missed 32, go to "Properties of Atoms and the Periodic Table," page 246.

33. **B.**

If you missed 33, go to "Ionic Bonding—Ionic Compounds," page 259.

34. **C.**

If you missed 34, go to "Lewis Structures," page 274.

35. **A.**

If you missed 35, go to "VSEPR Theory—Predicting Molecular Shapes," page 280.

36. **B.**

37. **A.**

If you missed 36 or 37, go to "The Gas Laws," page 297.

38. **C.**

If you missed 38, go to "The Ideal Gas Law," page 307.

39. **A.**

If you missed 39, go to "The Intermolecular Forces of Attraction," page 334.

40. **A.**

41. **A.**

If you missed 40 or 41, go to "Solution Concentrations," page 362.

42. **B.**

If you missed 42, go to "Solution Stoichiometry," page 370.

43. **B.**

If you missed 43, go to "The Arrhenius Concept of Acids and Bases," page 394.

44. **B.**

If you missed 44, go to "The Strength of Acids and Bases," page 398.

45. **A.**

If you missed 45, go to "Neutralization and Net-Ionic Equations," page 402.

Chapter 1

Measurement and Units of Measurement

At the core of any science is measurement. Being able to measure volumes, pressures, masses, and temperatures as well as the ability to count atoms and molecules allows chemists to understand nature more precisely. Modern science uses the International System of Units (SI) that was adopted worldwide in 1960. The metric system of measurement, which is consistent with the International System, is widely used in chemistry and is the principal system used in this book.

Chemists often have to work with numbers that are very, very small or very, very large. It is more convenient to express numbers of this kind in scientific notation, so that is the first topic to look at in this chapter.

Writing Numbers in Scientific Notation

It is likely that you have already seen numbers expressed in scientific notation on your calculator. With only 8 or 9 spaces to display numbers, calculators must resort to scientific notation to show very small or very large numbers. In scientific notation a number is expressed in this form

$$a \times 10^p$$

where "a" is a number between 1 and 10 (often a decimal number) and "p" is a positive or negative whole number written as an exponent on 10, often called the *power* of 10. The average distance from the earth to the sun is 93,000,000 miles, a very large number. In scientific notation this would be 9.3×10^7 miles. The power of 7 equals the number of places a decimal point would be moved from the right end of 93,000,000 *to the left* to get an "a" value between 1 and 10 (9.3). The 10^7 term equals 10,000,000 ($10 \times 10 \times 10 \times 10 \times 10 \times 10 \times 10$) and when multiplied by 9.3 would restate the original number in conventional form.

$$9.3 \times 10^7 \text{ miles} = 9.3 \times 10{,}000{,}000 \text{ miles} = 93{,}000{,}000 \text{ miles}$$

Likewise, the year 1492 would be written 1.492×10^3 in scientific notation.

Small numbers, those less than 1, are handled in a similar way, except the decimal point has to be moved to the *right* to get an "a" value between 1 and 10, and the "p" exponent is a *negative number*. For example, an atom of gold has a diameter of 0.000000342 meter. The decimal must be moved 7 places *to the right* to get 3.42, and the number is stated as 3.42×10^{-7} meter in scientific notation. The 10^{-7} term equals 1 divided by 10,000,000.

$$3.42 \times 10^{-7} \text{ meter} = 3.42 \times 1/10{,}000{,}000 \text{ meter} = 0.000000342 \text{ meter}$$

Similarly, the number 0.000045 would be written 4.5×10^{-5}.

One last thing: If you are given a number between 1 and 10 and need to write it in scientific notation, the power on 10 would be zero. In scientific notation, the number 8 would be written 8×10^{0}. In mathematics, 10^{0} equals 1.

Example Problems

1. Express these numbers in scientific notation.

   (a) 22,500,000

   Answer: 2.25×10^{7}

   This is a large number, so the decimal is moved 7 places to the left to form 2.25, and 7 is written as a positive exponent on 10.

   (b) 0.0006

   Answer: 6×10^{-4}

   This is a small number, so the decimal must be moved 4 places to the right to form 6, and 4 is written as a negative exponent on 10.

   (c) 602,200,000,000,000,000,000,000

   Answer: 6.022×10^{23}

   This is a very large number, so the decimal is moved 23 places to the left to form 6.022, and 23 is written as a positive exponent on 10.

2. Express these numbers in convention notation.

   (a) 6.35×10^{5}

   Answer: 635,000

   The power of 10 is 5, a positive number, so the decimal point is moved 5 places to the right.

   (b) 2.4×10^{-3}

   Answer: 0.0024

   The power of 10 is −3, a negative number, so the decimal point is moved 3 places to the left.

Work Problems

1. Express these numbers in scientific notation.

   (a) 1945 (b) 0.00000255 (c) 388000000000 (d) 0.023

2. Express these numbers in conventional form: (a) 7.55×10^{-4} (b) 8.80×10^{2}

Worked Solutions

1. **(a) 1.945×10^3.** Because this is a large number, the decimal is moved 3 places to the left to form 1.945. The exponent on 10 is 3.

   (b) 2.55×10^{-6}. This is a small number, so the decimal is moved 6 places to the right to form 2.55. The exponent on 10 is −6.

   (c) 3.88×10^{11}. This is a large number, so the decimal is moved 11 places to the left to form 3.88. The exponent on 10 is 11.

   (d) 2.3×10^{-2}. Because this is a small number, the decimal is moved 2 places to the right to form 2.3. The exponent on 10 is −2.

2. **(a) 0.000755.** The power of 10 is −4, a negative number, so the decimal point is moved 4 places to the left.

   (b) 880. The power of 10 is 2, a positive number, so the decimal point is moved 2 places to the right.

Significant Figures and Rounding Off Numbers

It is not possible to measure anything exactly; there will always be some amount of uncertainty. In many cases, the tool used to do the measurement causes the uncertainty. An inexpensive laboratory balance, for example, measures the mass of a gold ring to be 2.83 grams, while a more expensive analytical balance measures the mass to a greater accuracy, 2.8275 grams. The greater accuracy of the analytical balance is reflected in the larger number of digits in the numerical value of the mass. In either number, 2.83 or 2.8275, the right-most digit is the only digit that is not known with certainty. The mass of the ring is *closer* to 2.83 grams than to 2.82 or 2.84 grams on the first balance (2.83±0.01), and *closer* to 2.8275 grams than to 2.8274 or 2.8276 grams on the second (2.8275±0.0001). In both cases, all the digits are certain except the last one. The number of digits shown in a measured value (the certain digits + the one uncertain digit) indicates the accuracy of that value. These digits are referred to as significant digits or, more commonly, **significant figures (sig. figs.).**

Counting Significant Figures

You need to know how to count the number of significant figures in a number, because they affect the way answers are stated in calculations. Zeros can be a problem. A zero may or may not be significant depending on how it is used. To handle this "zero problem," follow this set of six rules:

Rule 1. All nonzero digits (1, 2, 3, 4, 5, 6, 7, 8, and 9) are always significant and must be counted.

Rule 2. A zero standing alone to the left of a decimal point is not significant. For example, in 0.63 and 0.0055, the 0 to the left of the decimal only helps you see the decimal point. It has no other use.

Rule 3. For a number less than 1, any zeros between the decimal point and the first nonzero digit are not significant. These zeros are simply placing the decimal point. The zeros in bold type in 0.**00**457 and 0.**0000**864 are not significant. Both numbers have three significant figures: 457 in the first number and 864 in the second.

Rule 4. A zero between two nonzero digits is significant. In 2.0056 and 0.0040558, both numbers have 5 significant figures. Because the second number is less than 1, only the 4, 0, 5, 5, and 8 are significant.

Rule 5. If the number has a decimal point, any zeros at the end of the number are significant. Both of these numbers have 4 significant figures: 4.500 and 0.01380.

Rule 6. If the number does not have a decimal point, like 1,500, the zeros at the end of the number may or may not be significant. If they are significant, place a decimal after the last zero, as in 1,500., or the number could be written in scientific notation, 1.500×10^3. Otherwise, 1,500 means the value is $1{,}500 \pm 100$ with 2 significant figures, the 1 and 5. For any number written in scientific notation, all digits in the first part of the number are significant. Both 1,500. and 1.500×10^3 show 4 significant figures.

Example Problems

Count the number of significant figures in each of these numbers.

1. 2.054

Answer: 4

The zero between two nonzero digits is counted (Rule 4) along with the 2, 5, and 4.

2. 0.00399

Answer: 3

The zeros between the decimal and the 3 are not counted (Rule 3), so only the 3, 9, and 9 are significant.

3. 0.99800

Answer: 5

The two zeros following the 8 are counted (Rule 5), so all five digits are significant.

4. 6.014×10^{-3}

Answer: 4

The zero is counted (Rule 4), so all four digits are significant.

5. 6,500

Answer: 2

At most, 2 significant figures. 6,500 without a decimal indicates $6{,}500 \pm 100$ (Rule 6).

Work Problems

Count the number of significant figures in these numbers.

(a) 93.082 (b) 0.00059 (c) 4.520 (d) 1.0×10^6 (e) 120,000.

Worked Solution

1. **(a) 5.** All digits are significant; the zero is counted (Rule 4).

   (b) 2. The 5 and 9 are significant; the zeros place the decimal (Rule 3).

   (c) 4. All digits are significant; the zero is counted (Rule 5).

   (d) 2. All digits in the non-exponential part of a number written in scientific notation are significant (Rule 1 and Rule 5).

   (e) 6. The decimal tells us all digits are significant (Rule 6).

Rounding Off Numbers

In calculations, we often obtain answers that have more digits than can be justified considering significant figures. This is a major problem when using calculators. Removing the digits that are not significant from the answer is called **rounding off.** Here are three rules to guide you:

Rule 1. If the next digit after those you want to retain is 4 or less, drop that digit and all that follow and keep the digits that remain.

Rounding to three figures: 3.253 → 3.25 | ~~3~~ → 3.25

Rule 2. If the next digit after those you want to retain is 5 or greater, drop that digit and all that follow, and then increase the last retained digit by 1. This is sometimes called rounding up.

Rounding to three figures: 6.3466 → 6.34 | ~~66~~ → 6.35

Rule 3. If the number to be rounded is less than 1 with zeros between the decimal point and the first nonzero digit (like 0.0004638), consider only the numerals that follow the zeros when counting digits.

Rounding to three figures: 0.0004638 → 0.000463 | ~~8~~ → 0.000464

Example Problems

1. Round off 45,317 to 2 digits and express the answer in scientific notation.

   Answer: 4.5×10^4

   Separating the first 2 digits (45 | ~~317~~) shows the next digit to be 3. It and the following digits are dropped (Rule 1), leaving 45,000.

2. Round off 368 to 2 digits and express the answer in scientific notation.

Answer: 3.7×10^2

The digit following the first 2 digits is greater than 5 (36 | ~~8~~). It is dropped, and 6 is increased to 7.

3. Round off 0.00941 to 1 digit and express the answer in scientific notation.

Answer: 9×10^{-3}

Consider only the digits following the zeros. Separating 1 digit from the rest (0.009 | ~~41~~) shows the next digit, 4, and all that follow can be dropped.

Work Problems

Round off each number to the indicated number of significant figures, shown in parentheses, and express the answer in scientific notation.

(a) 55,583 (4) (b) 38,953 (2) (c) 0.007665 (2)

Worked Solutions

(a) 5.558×10^4

55,58 | ~~3~~; Rule 1 applies; drop the 3, keep the rest.

(b) 3.9×10^4

38 | ~~953~~; Rule 2 applies; 38 is rounded up to 39 since the next digit is greater than 4.

(c) 7.7×10^{-3}

0.0076 | ~~65~~; Both Rules 2 and 3 apply, and 0.0076 is rounded up to 0.0077.

Significant Figures in Calculations

Keeping track of the number of significant figures in calculations depends on the kind of calculation you are doing.

Multiplication and division: The product or quotient can have no more significant figures than the number with the smallest number of significant figures used in the calculation.

Addition and subtraction: The sum or difference can have no more places after the decimal than there are in the number with the smallest number of digits after the decimal.

In addition and subtraction, those places after the decimal that are of unknown value (?) negate the numerals in those same places in the other numbers used in the calculation.

Example: 3.45762 ← five places after the decimal are known

\+ 4.32??? ← only 2 places after the decimal are known

Here is how these two numbers are added. 3.45762 is rounded off to two places after the decimal: 3.45762 → 3.45 | ~~762~~ → 3.46

It is then added to the other number.

$$\begin{array}{r} 3.46 \\ +\ \underline{4.32} \\ 7.78 \end{array}$$

Example Problems

1. 1,949 ÷ 6.33 =

   Answer: 308

   There are 4 significant figures in 1,949 and 3 in 6.33, so the answer can have no more than 3 significant figures.

2. 34.238442 + 9.35 =

   Answer: 43.59

   The answer can only have 2 digits after the decimal. 34.238442 is rounded up to 34.24 and added to 9.35.

3. $19.42 - (8.5 \times 10^{-3}) =$

   Answer: 19.41

   Start by writing 8.5×10^{-3} in conventional form, 0.0085. Only 2 digits after the decimal are allowed. 0.0085 is rounded up to 0.01 and then subtracted from 19.42.

Work Problems

1. 64.33 × 2.1416 =
2. 64.362 − 4.2 =
3. $19 - (3.3 \times 10^{-2}) =$

Worked Solutions

1. **138.0**

   The 4 significant figures in 64.33 limit the answer to 4 figures.

2. **60.2**

   Answer is limited to 1 digit after the decimal. 64.362 is rounded up to 64.4, and 4.2 is subtracted from it.

3. **19**

   $3.3 \times 10^{-2} = 0.033$. There are no digits after the decimal in 19, so subtracting 0.033 has no affect on 19.

Calculators and Significant Figures

You use a calculator to do your math chores, but calculators know nothing about significant figures. Calculators blindly grind out all the digits they can show, so it is up to you to get rid of the unnecessary digits and round off the number. If you divide 1.0 by 3.0 (both have 2 significant figures), the calculator gives 0.333333333. You need to round off this answer to 2 significant figures: 0.33 | ~~3333333~~ becomes 0.33, the correct answer with 2 significant figures.

Example Problems

Use your calculator to do each calculation and state the answer in scientific notation with the correct number of significant figures.

1. $600.3 \div 0.22 =$

   Answer: 2.7×10^3

   Use 2 significant figures. The calculator display, 27 | ~~28.636364~~, becomes 2,700 when rounded to 2 significant figures. The decimal is moved 3 places left to write the answer in scientific notation.

2. $3.1416 \times 6.52 =$

   Answer: 2.05×10^1

   Use 3 significant figures. 2.04 | ~~83232~~ $\times 10^1$ is rounded up to the correct answer.

Work Problems

1. $0.443 \div 9 =$

2. $33.0 \div 10.466 =$

Worked Solutions

Use your calculator to do each calculation and state the answer in scientific notation with the correct number of significant figures.

1. **5×10^{-2}**

   Use 1 significant figure. 0.04 | ~~92222222~~ is rounded up to 0.05, and the decimal moved 2 places right.

2. **3.15×10^{0}**

   Use 3 significant figures. 3.15 | ~~306707~~ could be stated correctly as 3.15, forgoing scientific notation.

The Metric System

The metric system is a system of measurement using units based on the decimal system. Today, in English, it is formally called the International System, abbreviated SI from the original French, *Système International*. The base units of the modern metric system used in general chemistry are given in the following table. From these, you can derive all other units of measure.

The Base Units of the Metric System

Quantity	*Base Unit*	*Symbol*
length	meter	m
mass	kilogram	kg
time	second	s
temperature	Kelvin	K
amount of substance	mole	mol

In this chapter, we are concerned only with the base units for length, mass, and temperature and those derived from them.

Length

The base unit of length is the **meter,** a length a little longer than the English yard; 39.37 inches to be exact. It isn't a convenient length for describing the length of things that are very small, like molecules, or for larger measures, like the distances between cities. To increase the usefulness of the meter, its length can be subdivided or multiplied by the use of the metric prefixes, which indicate different powers of 10. The most common metric prefixes, their abbreviations, and mathematical meaning of each are listed in the following table.

The Metric Prefixes

Prefix	*Abbreviation*	*Numerical Value*	*Equivalent Value in Power of Ten*
giga	G	1,000,000,000	10^9
mega	M	1,000,000	10^6
kilo	k	1,000	10^3
base unit	—	1	10^0
deci	d	0.1	10^{-1}
centi	c	0.01	10^{-2}
milli	m	0.001	10^{-3}
micro	μ*	0.000001	10^{-6}
nano	n	0.000000001	10^{-9}
pico	p	0.000000000001	10^{-12}

**μ is the Greek letter mu.*

The prefixes are used to modify the base unit, such as the meter (m), to make larger or smaller units that are more appropriate for a particular use. The most often used units of length in the metric system are shown in the following table.

The Metric Units of Length

Unit of Length	*Abbreviation*	*Meter Equivalent*	*Power Equivalent*
kilometer	km	1,000 m	1×10^3 m
meter (base)	m	1 m	1 m
decimeter	dm	0.1 m	1×10^{-1} m
centimeter	cm	0.01 m	1×10^{-2} m
millimeter	mm	0.001 m	1×10^{-3} m
micrometer	μm	0.000001 m	1×10^{-6} m
nanometer	nm	0.000000001 m	1×10^{-9} m

All the relationships between the unit and the meter in the preceding table are exact by definition; for example, 1 kilometer is *exactly* 1,000 meters *by definition.*

Relating each to 1 meter:

$$1 \text{ m} = 0.001 \text{ km} = 1 \times 10^{-3} \text{ km}$$
$$1 \text{ m} = 10 \text{ dm} = 1 \times 10^{1} \text{ m}$$
$$1 \text{ m} = 100 \text{ cm} = 1 \times 10^{2} \text{ cm}$$
$$1 \text{ m} = 1{,}000 \text{ mm} = 1 \times 10^{3} \text{ mm}$$
$$1 \text{ m} = 1{,}000{,}000\ \mu\text{m} = 1 \times 10^{6}\ \mu\text{m}$$

Because the inch-centimeter relationship is defined to be exact, all English-metric unit comparisons of length are also exact when used with the number of digits shown below. This exactness is only true for length comparisons.

1 inch (in) = 2.54 cm (exactly)
1 mile (mi) = 1.609 km (exactly)
1 m = 39.37 in (exactly)

Mass

The base unit of mass is the **kilogram** (kg), which, because of the *kilo* prefix (kilo = 1,000), indicates that 1 kilogram is 1,000 grams. Although the base is the kilogram, the more commonly used unit is the gram. The units of mass in the metric system are given in the following table.

The Metric Units of Mass

Unit of Mass	*Abbreviation*	*In Terms of Grams*	*Power Equivalent*
kilogram (base)	kg	1,000 g	1×10^{3} g
gram	g	1 g	1 g
decigram	dg	0.1 g	1×10^{-1} g
centigram	cg	0.01 g	1×10^{-2} g
milligram	mg	0.001 g	1×10^{-3} g
microgram	µg	0.000001 g	1×10^{-6} g

Relating each to 1 gram:

1 g = 0.001 kg
1 g = 10 dg
1 g = 100 cg
1 g = 1,000 mg
1 g = 1,000,000 µg

Comparing English and metric units of mass:

1 kg = 2.20 pounds (lbs) (3 sig. figs.)
1 lb = 454 g (3 sig. figs.)

Volume

There is no base unit for volume in the SI system since volume is derived from the base unit of length (volume = length × length × length). The derived unit of volume in the SI system is the cubic meter, m^3, a volume that is too large for most laboratory work, so a smaller volume has been adopted, the liter (L); note the uppercase "L." The liter is a little larger than the English quart and is exactly 1 cubic decimeter, dm^3. The most often used units of volume are given in the following table.

The Metric Units of Volume

Unit of Volume	*Abbreviation*	*In Terms of the Liter*	*Power Equivalent*
kiloliter	kL	1,000 L	1×10^3 L
liter	L	1 L	1 L
deciliter	dL	0.1 L	1×10^{-1} L
centiliter	cL	0.01 L	1×10^{-2} L
milliliter	mL	0.001 L	1×10^{-3} L
microliter	μL	0.000001 L	1×10^{-6} L

Relating each to 1 liter:

1 L = 0.001 kL
1 L = 10 dL
1 L = 100 cL
1 L = 1,000 mL
1 L = 1,000,000 μL

The most common unit of volume used in laboratory work is the milliliter (mL), which is the volume of exactly 1 cubic centimeter, cm^3, or cc. The units of cc, cm^3, and mL are interchangeable.

$$1\ \text{mL} = 1\ \text{cm}^3 = 1\ \text{cc (exactly)}$$

Comparing English and metric units of volume:

1 L = 1.06 quart (qt) (3 sig. figs.)
1 qt = 946 mL (3 sig. figs.)

Metric-Metric Conversions

It is often necessary to convert one metric unit to another. Suppose you needed to convert 4.5 cm into the equivalent length in millimeters. A conversion factor is used to do this. Conversion factor statements always have the form:

quantity being SOUGHT = quantity KNOWN × (conversion factor)

quantity KNOWN = 4.5 cm

quantity being SOUGHT = length in mm

length in mm = length in cm × (conversion factor)

The conversion factor is a fraction that relates centimeters to millimeters. The factor needed to convert centimeters to millimeters can be developed like this:

100 cm = 1 m = 1,000 mm

Since both equal 1 meter, they equal each other: 100 cm = 1,000 mm

Dividing both sides by 100 gives: 1 cm = 10 mm

Two conversion factors (fractions) can be developed from this equality: 1 cm = 10 mm

$$\frac{1\,\text{cm}}{10\,\text{mm}} \text{ and } \frac{10\,\text{mm}}{1\,\text{cm}}$$

The conversion factor chosen to change cm to mm is the one that will allow cm to be canceled (divided out) and replaced with mm. Units can be canceled in fractions just as numbers can:

$$\text{length in mm} = 4.5\,\cancel{\text{cm}}\left(\frac{10\,\text{mm}}{1\,\cancel{\text{cm}}}\right) = 45\,\text{mm}$$

Conversion factors can also be linked together to make a series of unit changes. Converting 25 kg to the equivalent mass in centigrams can be done using two conversion factors.

$$25\text{ kg} = ?\text{ cg}$$

The needed relationships between kg, g, and cg, are found in the table of metric mass units.

$$\text{mass in cg} = 25\,\cancel{\text{kg}}\left(\frac{1{,}000\,\cancel{\text{g}}}{1\,\cancel{\text{kg}}}\right)\left(\frac{100\,\text{cg}}{1\,\cancel{\text{g}}}\right) = 2{,}500{,}000\text{ cg} = 2.5\times 10^{6}\text{ cg}$$

$$25\text{ kg} = 2.5\times 10^{6}\text{ cg (2 sig. figs.)}$$

The first conversion factor changes kg to g, and the second changes g to cg. Notice that two significant figures in 25 kg dictates 2 significant figures in the final answer. Remember, the metric-to-metric conversion factors are exact and don't alter the usual rules of significant figures.

Example Problems

1. 9.35 mm = ? cm

   Answer: 0.935 cm

   Earlier it was shown that 1 cm = 10 mm. The answer is stated to 3 significant figures.

   $$\text{length in cm} = 9.35\,\cancel{\text{mm}}\left(\frac{1\,\text{cm}}{10\,\cancel{\text{mm}}}\right) = 0.935\,\text{cm}$$

2. 1.50 km = ? cm

   Answer: 1.50×10^{5} cm

   Two conversion factors are used here, the first converting km to m, the second, m to cm. The answer is stated to 3 significant figures.

   $$\text{length in cm} = 1.50\,\cancel{\text{km}}\left(\frac{1{,}000\,\cancel{\text{m}}}{1\,\cancel{\text{km}}}\right)\left(\frac{100\,\text{cm}}{1\,\cancel{\text{m}}}\right) = 150{,}000\text{ cm} = 1.50\times 10^{5}\text{ cm}$$

3. 0.0045 dg = ? mg

   Answer: 0.45 mg

   The first factor converts dg to g; the second converts g to mg. The answer is stated to 2 significant figures.

   $$\text{mass in mg} = 0.0045\,\cancel{\text{dg}}\left(\frac{0.1\,\cancel{\text{g}}}{1\,\cancel{\text{dg}}}\right)\left(\frac{1{,}000\,\text{mg}}{1\,\cancel{\text{g}}}\right) = 0.45\,\text{mg}$$

Work Problems

Make the required conversions and state the answers in the proper number of significant figures.

1. 0.025 g = _____ kg

2. 400. m = _____ cm

3. 12 μL = _____ mL

Worked Solutions

1. **2.5×10^{-5} kg**

   The answer requires 2 significant figures.

   $$\text{mass in kg} = 0.025\,\cancel{g}\left(\frac{1\,\text{kg}}{1{,}000\,\cancel{g}}\right) = 0.000025\,\text{kg} = 2.5 \times 10^{-5}\,\text{kg}$$

2. **4.00×10^{4} cm**

   The answer requires 3 significant figures.

   $$\text{length in cm} = 400.\,\cancel{m}\left(\frac{100\,\text{cm}}{1\,\cancel{m}}\right) = 40{,}000\,\text{cm} = 4.00 \times 10^{4}\,\text{cm}$$

3. **0.012 mL**

   The answer requires 2 significant figures.

   $$\text{volume in mL} = 0.012\,\cancel{\mu L}\left(\frac{1\,\cancel{L}}{1{,}000{,}000\,\cancel{\mu L}}\right)\left(\frac{1{,}000\,\text{mL}}{1\,\cancel{L}}\right) = 1.2 \times 10^{-5}\,\text{mL}$$

English-Metric Conversions—More Conversion Factors

Summarizing the English-metric conversions presented up to now:

1 in = 2.54 cm
1 m = 39.37 in
1 mi = 1.609 km

1 L = 1.06 qt
1 qt = 946 mL

1 lb = 454 g
1 kg = 2.20 lb

Conversions between the metric and English systems are done exactly the same way as converting only metric terms, but with one difference. The English-metric conversion factors, with the exception of length, are *not exact* and can affect significant figures. You will need to pay attention to this difference. Depending how pounds to grams or quarts to liters are related, there can be 3 or 4 significant figures:

1 lb = 454 g or 1 lb = 453.6 g
1 qt = 0.946 L or 1 qt = 0.9464 L

Unlike mass and volume, the inch-centimeter relationship is exact:

1 inch = 2.54 cm (exactly)

Within the English system of units, relationships between length, mass, and volume are exact:

1 mi = 5,280 ft
1 ft = 12 in
1 yd = 36 in

1 lb = 16 ounces (oz)
1 ton* (T) = 2,000 lbs

1 gallon (gal) = 4 qts

**This is the ton used in the United States and Canada, not the metric ton.*

Example Problems

Make the following conversions between the English and metric systems, stating the answers in the correct number of significant figures.

1. 6.25 yd = _____ cm

Answer: 572 cm

The answer is limited to 3 significant figures by the 3 digits in 6.25.

$$\text{length in cm} = 6.25\,\cancel{\text{yd}}\left(\frac{36\,\cancel{\text{in}}}{1\,\cancel{\text{yd}}}\right)\left(\frac{2.54\,\text{cm}}{1\,\cancel{\text{in}}}\right) = 571.5\text{ cm} = 572\text{ cm}$$

2. 3.0 gal = _____ L

Answer: 11 L

The volume in L can only have 2 significant figures.

$$\text{volume in L} = 3.0\,\cancel{\text{gal}}\left(\frac{4\,\cancel{\text{qt}}}{1\,\cancel{\text{gal}}}\right)\left(\frac{946\,\cancel{\text{mL}}}{1\,\cancel{\text{qt}}}\right)\left(\frac{1\text{ L}}{1{,}000\,\cancel{\text{mL}}}\right) = 11.352\text{ L} = 11\text{ L}$$

3. 1.25 lb = _____ mg

Answer: 5.68×10^5 mg

State the answer to 3 significant figures.

$$\text{mass in mg} = 1.25\,\cancel{\text{lb}}\left(\frac{454\,\cancel{\text{g}}}{1\,\cancel{\text{lb}}}\right)\left(\frac{1{,}000\text{ mg}}{1\,\cancel{\text{g}}}\right) = 567{,}500\text{ mg} = 5.68 \times 10^5\text{ mg}$$

Work Problems

Make the following conversions between English and metric units, stating the answers in the correct number of significant figures.

1. 2.25 ft = _____ mm
2. 20.5 mL = _____ qt
3. 750. kg = _____ oz

Worked Solutions

1. **686 mm**

$$\text{length in mm} = 2.25\,\cancel{\text{ft}}\left(\frac{12\,\cancel{\text{in}}}{1\,\cancel{\text{ft}}}\right)\left(\frac{2.54\,\cancel{\text{cm}}}{1\,\cancel{\text{in}}}\right)\left(\frac{10\,\text{mm}}{1\,\cancel{\text{cm}}}\right) = 685.8\,\text{mm} = 686\,\text{mm}$$

2. **2.17×10^{-2} qt**

$$\text{volume in qt} = 20.5\,\cancel{\text{mL}}\left(\frac{1\,\text{qt}}{946\,\cancel{\text{mL}}}\right) = 0.02167\,\text{qt} = 2.17 \times 10^{-2}\,\text{qt}$$

3. **2.64×10^{4} oz**

$$\text{mass in oz} = 750.\,\cancel{\text{kg}}\left(\frac{1{,}000\,\cancel{\text{g}}}{1\,\cancel{\text{kg}}}\right)\left(\frac{1\,\cancel{\text{lb}}}{454\,\cancel{\text{g}}}\right)\left(\frac{16\,\text{oz}}{1\,\cancel{\text{lb}}}\right) = 26{,}431.7\,\text{oz} = 2.64 \times 10^{4}\,\text{oz}$$

Density

Density equals the mass of a sample divided by its volume and is stated as mass per *unit* volume (1 mL, 1 cc, 1 qt, and so on).

$$\text{density} = \frac{\text{mass}}{\text{volume}}$$

The density of water is 1.00 g/mL. This means that 1.00 mL of water has a mass of 1.00 g. The density of gold, a solid, is 19.3 g/cc; one cubic centimeter of gold (1 cc =1 mL) has a mass of 19.3 grams. A person might say by mistake that gold is "heavier" than water, although the correct comparison would be that the density of gold is greater than that of water. Anything with a density greater than water will sink in water. If its density is less, it will float. In general, densities for liquids are expressed as g/mL, while for solids, g/cc or g/cm^3.

Example Problems

1. 5.00 mL of ethyl alcohol has a mass of 3.95 g. What is the density of ethyl alcohol?

   Answer: 0.790 g/mL

$$\text{density} = \frac{3.95\,\text{g}}{5.00\,\text{mL}} = 0.790\,\text{g/mL}$$

2. 100. lbs of sulfuric acid has a volume of 24.5 L. What is the density of sulfuric acid in pounds per quart, lb/qt?

   Answer: 3.85 lb/qt

   The first term calculates the density in lb/L and the second converts L to qt.

$$\text{density} = \left(\frac{100.\,\text{lb}}{24.5\,\cancel{\text{L}}}\right)\left(\frac{1.00\,\cancel{\text{L}}}{1.06\,\text{qt}}\right) = 3.85\,\text{lb/qt}$$

Work Problems

1. What is the density of table salt, sodium chloride, if 50.0 cc of salt has a mass of 108 g?
2. 1.00 gallon of gasoline has a mass of 5.85 pounds. What is the density of gasoline in pounds per quart?

Worked Solutions

1. **2.16 g/cc**

$$\text{density} = \frac{108\,\text{g}}{50.0\,\text{cc}} = 21.6\,\text{g/cc}$$

2. **1.46 lb/qt**

$$\text{density} = \left(\frac{5.85\,\text{lbs}}{1.00\,\cancel{\text{gal}}}\right)\left(\frac{1\,\cancel{\text{gal}}}{4\,\text{qt}}\right) = 1.46\,\text{lb/qt}$$

Densities relate the mass of a substance to its volume and for that reason can be used as conversion factors. The density of iron is 7.86 g/cc. This can be written as two conversion factors:

$$\left(\frac{7.86\,\text{g}}{1.00\,\text{cc}}\right) \text{ or inverted to be } \left(\frac{1.00\,\text{cc}}{7.86\,\text{g}}\right)$$

Volume is written to 3 significant figures, such as 1.00 cc, to match the 3 significant figures in the mass. As conversion factors, densities can be used to convert a known volume to mass or a known mass to volume.

Example Problems

1. The density of iron is 7.86 g/cc. What would be the mass of 950. cc of iron?

Answer: 7.47×10^3 g

$$\text{mass of iron} = 950.\,\cancel{\text{cc}}\left(\frac{7.86\,\text{g}}{1.00\,\cancel{\text{cc}}}\right) = 7{,}467\,\text{g} = 7.47 \times 10^3\,\text{g}$$

2. What volume, in cubic centimeters (cc), would be occupied by 855 g of iron? The density of iron is 7.86 g/cc.

Answer: 109 cc

$$\text{volume} = 855\,\cancel{\text{g}}\left(\frac{1.00\,\text{cc}}{7.86\,\cancel{\text{g}}}\right) = 108.78\,\text{cc} = 109\,\text{cc}$$

Work Problems

1. What is the mass of 775 mL of ethyl alcohol, a colorless liquid that has a density of 0.789 g/mL?
2. What volume would 800. g of vegetable oil occupy if the density of the oil is 0.933 g/mL?

Worked Solutions

1. **611 g**

$$\text{mass} = 775\,\cancel{\text{mL}}\left(\frac{0.789\,\text{g}}{1.00\,\cancel{\text{mL}}}\right) = 611.47\,\text{g} = 611\,\text{g}$$

2. **857 mL**

$$\text{volume} = 800.\,\cancel{\text{g}}\left(\frac{1.00\,\text{mL}}{0.933\,\cancel{\text{g}}}\right) = 857.45\,\text{mL} = 857\,\text{mL}$$

Measuring Temperature

Temperature is a measure of the intensity of heat energy in a sample of matter. Temperature is not heat. Heat energy is related to the motion of the particles that make up a sample. The higher the temperature, the more rapid the motion of particles.

You need to be familiar with three temperature scales, two of which are commonly used in science: the *Celsius* scale and the *Kelvin* scale, also known as the *Absolute* scale. The third is the *Fahrenheit* scale that is in everyday use in commerce in the United States but not in science. The three temperature scales are summarized in the following table.

The Temperature Scales

	Celsius Scale	*Kelvin Scale*	*Fahrenheit Scale*
Symbol of temperature	C	K	F
Unit of temperature and its symbol	degree (°)	degree, no symbol	degree (°)
Freezing point of water	0°C	273 K*	32°F
Boiling point of water	100°C	373 K*	212°F
Number of degrees between freezing and boiling points of water	100°	100	180°

**273.15 K and 373.15 K to be exact, although 273 K and 373 K will be used here.*

The number of degree units between the freezing and boiling points of water are identical in the Celsius and Kelvin scales, 100 degrees. This means the size of the Celsius degree and the Kelvin degree is the same.

$$\text{A change of } 1\,°\text{C} = \text{a change of } 1\text{ K}$$

There are 180 degrees between the freezing and boiling points on the Fahrenheit scale. Compared to the Celsius and Kelvin scales:

$$\text{A change of } 1\,°\text{C or a change of } 1\text{ K} = \text{a change of } 1.8\,°\text{F (exactly)}$$

Using the freezing point of water to compare the three temperature scales (0°C, 273 K, and 32°F) and knowing how the size of the degree compares, it is possible to develop equations to convert a temperature on one scale to the corresponding temperature on another.

A Celsius temperature can be converted to a Kelvin temperature simply by adding the number of the Celsius temperature to 273 and attaching "K" (the symbol for Kelvin temperature).

$$\text{Kelvin temperature} = (\text{temperature in °C}) + 273$$

$$K = °C + 273$$

Changing a Celsius temperature to one in Fahrenheit must take into account the two different values for the freezing point of water and the different size of the Celsius and Fahrenheit degree.

$$\text{Fahrenheit temperature} = 1.8\,(\text{Celsius temperature}) + 32$$

$$°F = 1.8\,(°C) + 32$$

The equation tells you that the numerical value of the Celsius temperature is multiplied by 1.8, and then 32 is added to the product. The symbol for degree Fahrenheit, °F, is attached to the sum.

This equation can be rearranged to ease conversion of Fahrenheit temperatures to Celsius.

$$\text{Celsius temperature} = \frac{(\text{Fahrenheit temperature} - 32)}{1.8}$$

$$°C = \frac{(°F - 32)}{1.8}$$

This equation tells you to first subtract 32 from the numerical value of the Fahrenheit temperature, then divide the difference by 1.8. The symbol for degree Celsius, °C, is added to the answer.

Example Problems

Make the following temperature conversions.

1. 125°F = ? °C

   Answer: 51.7°C Note that only the numerical value of the Fahrenheit temperature is used. 125°F is written as 125.

   $$°C = \frac{(125 - 32)}{1.8} = 51.6666°C = 51.7°C \text{ (3 sig. figs.)}$$

2. −18°C = ? K

   Answer: 255 K

   $$K = -18 + 273 = 255\,K$$

3. 45° F = ? K

   Answer: 280 K First convert Fahrenheit to Celsius; then convert Celsius to Kelvin.

   $$°C = \frac{(45 - 32)}{1.8} = 7.2°C$$

   $K = 7.2 + 273 = 280\,K$ In terms of significant figures, 7.2 is rounded to 7, and then added to 273.

Work Problems

Make the following temperature conversions.

1. 25°C = _____ °F

2. 450 K = _____ °C

3. 70°F = _____ K

Worked Solutions

1. **77°F**

$$^{\circ}F = (25 \times 1.8) + 32 = 77^{\circ}F$$

2. **177°C**

$$^{\circ}C = 450 - 273 = 177^{\circ}C$$

3. **294 K**

$$^{\circ}C = \frac{(70 - 32)}{1.8} = 21^{\circ}C \text{ then } K = 21 + 273 = 294\,K$$

Energy and the Measurement of Heat

A body possesses energy if it has the ability to move another object. So, for example, a rapidly moving atom possesses energy, since it can crash into another atom, causing it to move. The energy possessed by a body by virtue of its motion is called *kinetic energy*. A sample of TNT also has the ability to move objects. It possesses energy stored in its chemical composition—energy that has the potential to move an object. Stored energy is called *potential energy*. Both an arrow in a drawn bow and a wound clock spring possess potential energy.

There are many forms of energy: light energy, acoustic energy, electrical energy, heat energy, and nuclear energy, to name a few. Energy can change from one kind into another. The electrical energy of a battery can heat the filament of a light bulb, generating both heat and light energy. Although energy is changing form, it is important to realize that no new energy is being created nor is any energy being destroyed. This fact is summed up in the *Law of Conservation of Energy,* which states that energy can be neither created nor destroyed, although it may change form.

It is more difficult to measure energy than it is to measure length, mass, volume, or temperature. But when measured, energy is quantitatively given in the SI unit of the Joule (J) or in the calorie (cal), an older unit used for measuring heat energy. By definition:

4.184 Joules = 1 calorie (exactly)

If larger units of energy are needed, the kilojoule (kJ) and the kilocalorie (kcal) can be used.

1 kilojoule = 1,000 Joules
1 kilocalorie = 1,000 calories

The amount of heat needed to raise the temperature of 1 gram of a substance 1°C is called the *specific heat* (S.H.) of that substance. It requires 4.184 Joules (or 1 calorie) of heat to raise the temperature of 1 gram of water 1°C. It follows, then, that the specific heat of water is 4.184 J/g°C. If you know the amount of heat needed to raise the temperature of 1 g of water 1°C, you can calculate the amount of heat (symbolized, q) needed to raise the temperature of 10 g of water 10°C. The formula for doing just this is:

$$\text{amount of heat energy} = q = (\text{S.H. of water})(\text{mass of water})(\text{temperature change of water})$$

So, how many Joules of heat energy are required to raise the temperature of 10.0 g of water 10.0°C?

$$q = \left(\frac{4.184\,\text{J}}{\cancel{g}\,\cancel{^\circ C}}\right)(10.0\,\cancel{g})(10.0\,\cancel{^\circ C}) = 418\ \text{Joule}$$

Notice that the units of gram and °C cancel, leaving Joule, the unit of energy. This equation can be rewritten in a general way that applies to all substances. ΔT symbolizes a change in temperature.

$$q = (\text{S.H.})(\text{mass})(\Delta T)$$

Read Δ as "a change in." If the temperature rises from 25°C to 40°C, ΔT equals 15°C, the difference between the final temperature and the initial temperature, $\Delta T = (T_{final} - T_{initial}) = (40°C - 25°C) = 15°C$.

Example Problems

1. 125 cal = ? J; 500. J = ? cal

   Answer: 523 J; 120 cal

   The conversion factor is derived from: 4.184 J = 1 cal (exactly)

$$125\,\cancel{cal}\left(\frac{4.184\,\text{J}}{1\,\cancel{cal}}\right) = 523\,\text{J};\ 500\,\text{J}\left(\frac{1\,\text{cal}}{4.184\,\text{J}}\right) = 119.50\,\text{cal} = 120\,\text{cal}$$

2. The specific heat of water is 4.184 J/g°C. How many Joules of heat energy would be required to raise the temperature of 150. g of water from 20.0°C to 40.0°C?

   Answer: 1.26×10^4 J

   ΔT = 40.0°C – 20.0°C = 20.0°C

$$q = \left(\frac{4.184\,\text{J}}{\text{g}^\circ\text{C}}\right)(150.\,\text{g})(20.0^\circ\text{C}) = 12{,}552\,\text{J} = 1.26\times 10^4\,\text{J}$$

3. 55.1 g of aluminum absorbed 1,500 Joule of heat. Its temperature increased from 20.0°C to 50.2°C. What is the specific heat of aluminum?

Answer: 0.901 J/g°C

$\Delta T = 50.2°C - 20.0°C = 30.2°C$

Rearranging the heat equation to solve for specific heat:

$$S.H. = \frac{q}{(\text{mass})(\Delta T)}$$

$$S.H. = \frac{1,500.\ J}{(55.1 g)(30.2^\circ C)} = 0.901 \frac{J}{g^\circ C}$$

Work Problems

1. 1,750 cal = ? kJ

2. How many Joules of heat energy are required to raise the temperature of 500. g of water from 23.5°C to 75.0°C? The specific heat of water is 4.184 J/g-°C. State the answer in scientific notation.

3. The addition of 1,800. J of heat energy to a 200. g sample of copper raised the temperature of the sample 23.4°C. What is the specific heat of copper?

Worked Solutions

1. **7.322 kJ**

$$1,750.\ \cancel{cal}\left(\frac{4.184 J}{1\ \cancel{cal}}\right) = 7,322 J$$

$$7,322\ \cancel{J}\left(\frac{1 kJ}{1,000\ \cancel{J}}\right) = 7.322 kJ$$

2. **Amount of heat energy = 1.08×10^5 J**

$\Delta T = 75.0^\circ C - 23.5^\circ C = 51.5^\circ C$

$$q = \left(\frac{4.184 J}{\cancel{g}^\circ \cancel{C}}\right)(500.\ \cancel{g})(51.5^\circ \cancel{C}) = 107,738 J$$

$q = 107,738 J = 1.08 \times 10^5 J$ (3 sig. figs.)

3. **S.H. = 0.385 J/g-°C**

$$S.H. = \frac{q}{(\text{mass})(\Delta T)}$$

$$S.H. = \frac{1,800.\ J}{(200.\ g)(23.4^\circ C)} = 0.385 \frac{J}{g^\circ C}$$

Chapter Problems and Answers

Problems

1. Write the following numbers in scientific notation:

   (a) 0.0043

   (b) 2,965

   (c) 0.0000000163

   (d) 12

2. State the number of significant figures in each of the following:

   (a) 0.00410

   (b) 19.00002

   (c) 5,200.

   (d) 3,000

3. Round off each number to the requested number of significant figures, in parentheses, and express the answers in scientific notation.

   (a) 419 (2)

   (b) 0.006355 (2)

   (c) 1,047.3 (3)

   (d) 0.05055 (2)

4. Perform the following calculations expressing the answers to the allowed number of significant figures.

   (a) $2.90 \times 6.3 =$

   (b) $19.06 + 6.8 =$

   (c) $1,021 - 3.36 =$

   (d) $3.14 \div 4.280 =$

5. Perform the requested metric-metric conversions, expressing the answers in scientific notation to the allowed number of significant figures.

(a) 1.42 m = _____ μm

(b) 8.0 dm = _____ km

(c) 125 mL = _____ L

(d) 250. g = _____ mg

6. Perform the requested English-metric conversions, expressing the answers in scientific notation to the allowed number of significant figures.

(a) 100. cm = _____ in

(b) 2.5 qt = _____ mL

(c) 3.40 mi = _____ m

(d) 1.00 oz = _____ mg

7. What is the density of an alcohol-water solution if 75.0 mL of the solution has a mass of 70.3 g?

8. What is the mass of 275 mL of mercury knowing that the density of mercury is 13.6 g/mL?

9. What volume in liters will be occupied by 1.00 kg of mercury that has a density of 13.6 g/mL?

10. Perform the requested temperature conversions, expressing the answers to the allowed number of significant figures.

(a) 125°C = _____ K

(b) 65.0°C = _____ °F

(c) 500. K = _____ °F

(d) −30°F = _____ °C

11. Perform the requested energy conversions, expressing the answers to the allowed number of significant figures.

(a) 500 calories = _____ kJ

(b) 1.9×10^3 J = _____ kcal

12. What amount of heat energy is required to raise the temperature of 35.0 g of lead 40.0°C? The specific heat of lead is 0.128 J/g°C.

13. What amount of heat energy is required to raise the temperature of 1.50×10^3 g of water from 23.0°C to 70.0°C? The specific heat of water is 4.184 J/g°C.

14. What is the expected change in temperature, ΔT, if 100. J of heat energy is added to 10.0 g of gold? The specific heat of gold is 0.131 J/g°C.

15. What is the specific heat, S.H., of a metal if it requires 350.0 J of heat energy to raise 114 g of the metal 18.5°C?

Answers

1. **(a) 4.3×10^{-3} (b) 2.965×10^{3} (c) 1.63×10^{-8} (d) 1.2×10^{1}**

2. **(a) 3 significant figures (b) 7 significant figures (c) 4 significant figures (d) Only 1 digit is certain, the 3 in 3,000 (3000±1000)**

3. **(a) 4.2×10^{2} (b) 6.4×10^{-3} (c) 1.05×10^{3} (d) 5.1×10^{-2}**

4. **(a) 18, rounded to 2 sig. figs., 18.|~~27~~ = 18**

   (b) 25.9, round 19.06 to 19.1 then add to 6.8

   (c) 1,018, no digits after the decimal are allowed so, 1,021 – 3.|~~36~~ = 1,021 – 3 = 1,018

   (d) 0.734, rounded to 3 sig. figs., 0.733|~~645~~ = 0.734

5. **(a) 1.42×10^{6} µm** $1.42\,\text{m}\left(\frac{1\times 10^{6}\,\mu\text{m}}{1\,\text{m}}\right) = 1.42\times 10^{6}\,\mu\text{m}$

   (b) 8.0×10^{-4} km $8.0\,\text{dm}\left(\frac{1\,\text{m}}{10\,\text{dm}}\right)\left(\frac{1\,\text{km}}{1000\,\text{m}}\right) = 8.0\times 10^{-4}\,\text{km}$

   (c) 0.125 L $125\,\text{mL}\left(\frac{1\,\text{L}}{1{,}000\,\text{mL}}\right) = 0.125\,\text{L}$

   (d) 2.50×10^{5} mg $250.\,\text{g}\left(\frac{1{,}000\,\text{mg}}{1\,\text{g}}\right) = 2.50\times 10^{5}\,\text{mg}$

6. **(a) 3.94×10^{1} in** $100.\,\text{cm}\left(\frac{1\,\text{in}}{2.54\,\text{cm}}\right) = 3.94\times 10^{1}\,\text{in}$

   (b) 2.4×10^{3} mL $2.5\,\text{qt}\left(\frac{946\,\text{mL}}{1.00\,\text{qt}}\right) = 2{,}365\,\text{mL} = 2.4\times 10^{3}\,\text{mL}$

   (c) 5.47×10^{3} m $3.40\,\text{mi}\left(\frac{1.609\,\text{km}}{1\,\text{mi}}\right)\left(\frac{1{,}000\,\text{m}}{1\,\text{km}}\right) = 5{,}470.6\,\text{m} = 5.47\times 10^{3}\,\text{m}$

   (d) 2.84×10^{4} mg $1.00\,\text{oz}\left(\frac{1\,\text{lb}}{16\,\text{oz}}\right)\left(\frac{454\,\text{g}}{1\,\text{lb}}\right)\left(\frac{1{,}000\,\text{mg}}{1\,\text{g}}\right) = 28{,}375\,\text{mg} = 2.84\times 10^{4}\,\text{mg}$

7. **0.973 g/mL** $\text{density} = \left(\frac{70.3\,\text{g}}{75.0\,\text{mL}}\right) = 0.93733\,\text{g/mL} = 0.937\,\text{g/mL}$

8. **3.74×10^{3} g** $275\,\text{mL}\left(\frac{13.6\,\text{g}}{1.00\,\text{mL}}\right) = 3{,}740\,\text{g} = 3.74\times 10^{3}\,\text{g}$

9. **7.35×10^{-2} L** $1.00\,\text{kg}\left(\frac{1{,}000\,\text{g}}{1\,\text{kg}}\right)\left(\frac{1\,\text{mL}}{13.6\,\text{g}}\right)\left(\frac{1\,\text{L}}{1{,}000\,\text{mL}}\right) = 0.073529\,\text{L} = 7.35\times 10^{-2}\,\text{L}$

10. **(a) 398 K** $K = (125 + 273) = 398\,K$

(b) 149°F $°F = 1.8(65.0) + 32 = 149°F$

(c) 441°F $°C = (500 - 273) = 227°C;\ °F = 1.8(227) + 32 = 440.6°F = 441°F$

(d) –34°C $°C = \left(\frac{(-30 - 32)}{1.8}\right) = \left(\frac{-62}{1.8}\right) = -34.4°C = -34°C$

11. **(a) 2.09 kJ** $500.\,\cancel{cal}\left(\frac{4.184\,\cancel{J}}{1\,\cancel{cal}}\right)\left(\frac{1\,kJ}{1{,}000\,\cancel{J}}\right) = 2.092\,kJ = 2.09\,kJ$

(b) 0.45 kcal $1.9 \times 10^3\,\cancel{J}\left(\frac{1\,\cancel{cal}}{4.184\,\cancel{J}}\right)\left(\frac{1\,kcal}{1{,}000\,\cancel{cal}}\right) = 0.4541\,kcal = 0.45\,kcal$

12. **1.79×10^3 J** $q = (35.0\,\cancel{g})(0.128\,J/\cancel{g}\,°\cancel{C})(40.0°\cancel{C}) = 179.2\,J = 1.79 \times 10^3\,J$

13. **2.95×10^5 J** $\Delta T = (70.0°C - 23.0°C) = 47.0°C$

$$q = (1.50 \times 10^3\,g)\left(\frac{4.184\,J}{g\,°C}\right)(420°C)$$

$$= 297{,}972\,J$$

$$= 2.95 \times 10^5\,J$$

14. **76.3°C** $\Delta T = \frac{q}{(mass)(S.H.)} = \frac{100.\,J}{(10.0\,\cancel{g})(0.131\,\cancel{J}/\cancel{g}\,°\cancel{C})} = 76.3°C$

15. **0.166 J/g °C** $S.H. = \frac{q}{(mass)(\Delta T)} = \frac{350.0\,J}{(114\,g)(18.5°C)} = 0.16595\,J/g°C = 0.166\,J/g°C$

Supplemental Chapter Problems

Problems

1. Write the following numbers in scientific notation:

(a) 0.000195

(b) 8,407

(c) 0.0000000021

(d) 741

2. State the number of significant digits in each of the following:

(a) 0.004050

(b) 1980.1

(c) 5.2×10^{-5}

(d) 4,900

3. Round off each number to the requested number of significant figures and express the answer in scientific notation.

(a) 0.0034487 (2)

(b) 5,344,392 (4)

(c) 38,471 (1)

(d) 0.000074522 (2)

4. Perform the following calculations and round off the answers to the allowed number of significant figures. Write each answer in scientific notation.

(a) $2.38 \div 19 =$

(b) $10.005 + 3.06 =$

(c) $19.95 - 0.1234 =$

(d) $3.1416 \times 8.2 =$

5. Perform the following metric-metric conversions. State the answers in scientific notation.

(a) 0.00185 L = _____ dL

(b) 1.37×10^6 μL = _____ mL

(c) 1,548 mg = _____ kg

(d) 194 cm = _____ km

6. Perform the following English-metric conversions. State the answers in scientific notation to the allowed number of significant figures.

(a) 855 mL = _____ qt

(b) 1.00×10^2 mg = _____ oz

(c) 2.54 ft = _____ mm

(d) 70.0 in = _____ m

7. 250 ml of sulfuric acid has a mass of 453 grams. What is the density of the acid?

8. The density of a sample of antifreeze is 1.055 g/mL. What would 1,500 mL of this antifreeze weigh?

9. Consider a bar of gold that has a mass of 90.0 pounds (2.22×10^5 g). Knowing that the density of gold is 19.3 g/cc, what is the volume of this bar of gold?

10. Perform the following temperature conversions:

    (a) 525°C = _____ °F

    (b) −75°F = _____ °C

    (c) 370 K = _____ °F

    (d) 25°C = _____ K

11. How many Joules of heat energy will be needed to raise the temperature of 1.50×10^3 g of alcohol from 22.0°C to 45.0°C? The specific heat of alcohol is 2.14 J/g°C.

12. When 450.0 J of heat was added to a 50.00 g sample of magnesium, the temperature of the metal increased 8.79°C. What is the specific heat of magnesium?

13. What increase in temperature will a 225 g block of sulfur experience when exactly 1,000 J of heat energy is added to it? The specific heat of sulfur is 0.777 J/g°C.

14. Which demands the greater amount of heat energy?

    a. Heating 10.0 g of copper (S.H. = 0.38 J/g°C) from 0°C to 100°C.

    b. Heating 10.0 g of water (S.H. = 4.2 J/g°C) from 15°C to 25°C.

Answers

1. **(a) 1.95×10^{-4} (b) 8.407×10^3 (c) 2.1×10^{-9} (d) 7.41×10^2 (page 11)**

2. **(a) 4 (b) 5 (c) 2 (d) no more than 2 (4,900 ± 100) (page 13)**

3. **(a) 3.4 × 10−3 (b) 5.344 × 106 (c) 4 × 104 (d) 7.5 × 10−5 (page 15)**

4. **(a) 1.3×10^{-1} (b) 13.07 (c) 19.83 (d) 26 (page 16)**

5. **(a) 1.85 10^{-2} dL (b) 1.37 10^3 mL (c) 1.548×10^{-3} kg (d) 1.94×10^{-3} km (page 22)**

6. **(a) 9.04×10^{-1} qt (b) 3.52×10^{-3} oz (c) 7.74 10^2 mm (d) 1.78 10^0 m (page 24)**

7. **1.81 g/mL (page 26)**

8. **1,583 g (page 26)**

9. **1.15×10^4 cc (page 26)**

10. **(a) 977°F (b) −59°C (c) 207°F (d) 298 K (page 28)**

11. **7.38×10^4 J (page 30)**

12. **1.02 J/g°C (page 30)**

13. **5.72°C (page 30)**

14. **b (page 30)**

Chapter 2
The Classification of Matter, Elements, Compounds, and Mixtures

So what do we call all this stuff around us? We know specific names: This is a table, that is a loaf of bread, and water comes out of the faucet. But to a chemist, all of this—and everything else—is broadly classed as **matter,** anything that has mass and occupies space (that is, has volume). This definition is so broad that it doesn't help you understand nature very well. So, in this chapter, matter is examined in greater detail so that you can begin to organize it into categories and classes that are more easily understood. Let's start by dividing matter into its three common physical states: solid, liquid, and gas.

Solids, Liquids, and Gases

Matter can exist as solids, liquids, or gases, and it can change from one state into another with heating or cooling. If you hold a cube of ice (solid water) in your hand, it will slowly change to liquid water as it absorbs heat from your body. Put this water into a pan and heat it on a stove to bring it to a boil, and you convert the liquid water to a gas, one that is commonly known as *steam.* Solid to liquid to gas, and it is still water. Although water can exist in three different physical states, in books and tables it is common to classify the state of a substance as it exists at 25°C, a temperature just a bit higher than room temperature and normal atmospheric pressure.

A **solid** is a form of matter that has a *definite shape and volume.* The small particles that make up a solid (atoms or molecules) are strongly attracted to one another, are touching, and are held rigidly in place. Solids cannot be compressed to a smaller volume. The highly organized arrangement of the particles in solids is why many of them are crystalline with definite geometric shapes, like the crystals of table salt.

A **liquid** has a *definite volume and indefinite shape;* it takes the shape of the container that holds it. The particles that make up a liquid are touching because of their attraction for one another, but the attraction in liquids is not nearly as strong as it is in solids. For this reason, the particles are not rigidly held in place, and they can flow around one another. But since the particles are touching, liquids, like solids, are not compressible.

A **gas** has *neither a definite shape nor volume.* It will assume the shape *and* volume of the container that holds it. The particles that make up gases are very weakly attracted to one another, and for that reason are widely separated and not touching. Gases are mostly empty space, they can flow, and they can be readily compressed to smaller volumes.

Another useful way to classify matter, whether a solid, liquid, or gas, is to indicate whether it is a pure substance or a mixture. A **pure substance** (often simply referred to as a **substance**) is one that has a definite composition, such as water or copper. There are two kinds of pure substances: elements and compounds. An **element** is a substance than cannot be broken down into simpler substances by chemical means (like heating to a high temperature). The simplest particle of an element is the atom, and that will be discussed in greater detail in Chapter 3.

A **compound** is composed of two or more elements combined chemically in a definite, non-changing ratio. Unlike elements, compounds *can* be broken down into simpler substances. Water is a compound composed of two elements, hydrogen and oxygen, in a 2 to 1 ratio; 2 atoms of hydrogen combined with 1 atom of oxygen, exactly.

A **mixture** is composed of two or more substances (elements and/or compounds) mixed together, but not in any particular ratio. A good example of a mixture is a solution of sugar dissolved in water. You can start with a small amount of sugar dissolved in a glass of water, and then you can add more sugar. The ratio of sugar to water can vary widely; it's not fixed by a chemical combination between the two. Mixtures are discussed further at the end of this chapter.

A solution of sugar dissolved in water is a **homogeneous** mixture. It has the same composition throughout and it is not possible to see that two different substances are present. Solutions and pure substances (elements and compounds) are homogeneous. By contrast, a handful of sand picked up on a beach is a **heterogeneous** mixture. If you look at it carefully, you can see grains of different color sand, bits of shell, and other materials. A heterogeneous mixture has distinct **phases** that can be seen by the naked eye. A homogeneous mixture exists in a single phase. The terms used to classify matter are summarized in the following figure.

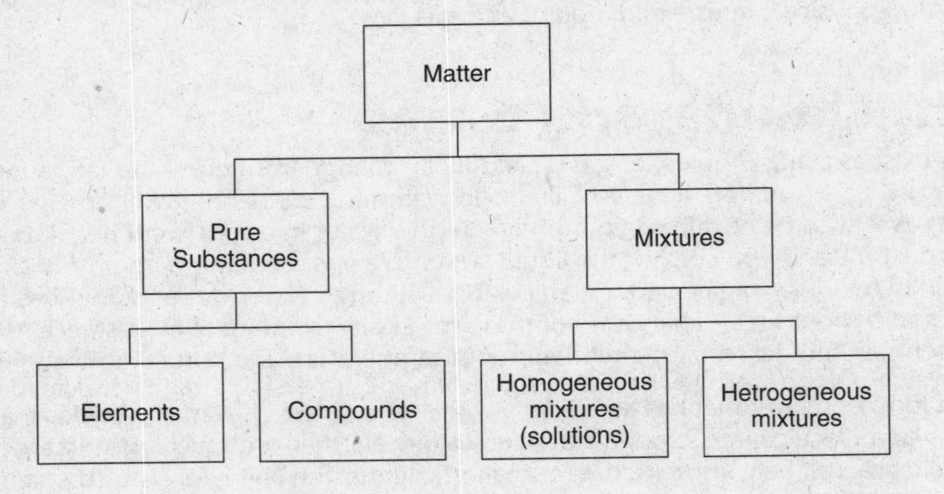

Example Problems

The following list includes several items. Which three of the following eight terms apply to each one?

Solid, liquid, gas, element, compound, mixture, homogeneous, heterogeneous

1. Air

Answer: gas, mixture (oxygen, nitrogen, water vapor, and so on), homogeneous

2. Purified water

   Answer: liquid, compound, homogeneous

3. A handful of pennies and dimes

   Answer: solid, mixture, heterogeneous

Work Problems

The following list includes several items. Which three of the following terms apply to each one?

Solid, liquid, gas, element, compound, mixture, homogeneous, heterogeneous

1. copper wire
2. oil and vinegar salad dressing
3. gold ore

Worked Solutions

1. **solid, element, homogeneous** (Copper is a pure element.)
2. **liquid, mixture, heterogeneous** (The oil and water form two phases.)
3. **solid, mixture, heterogeneous** (The small flakes of gold are visible.)

Properties and Changes

Properties are those characteristics of a substance that are responsible for its unique identity. Just as hair color, height, sex, and eye color are some of your properties, there are many properties than can be used to identify chemical substances. Chemists divide properties into two classes. **Physical properties** are those determined without changing the chemical identity of the substance, such as color, physical state, odor, hardness, freezing or boiling temperature, density, and specific heat. Observing some physical properties requires that a physical change occur. A **physical change** is one that does not change the identity of the substance, such as freezing, boiling, bending, scratching, stretching, hammering, or measuring, if it conducts electricity. **Chemical properties** are those that describe whether or not a substance can be changed into a different substance or broken down into simpler substances by a chemical change. It is a chemical property of iron that it rusts. It is a chemical property of gold that it does not. It is a chemical property of gasoline that it burns in air, and of water that it does not. A **chemical change,** also called a chemical reaction, is a process by which one or more substances are changed into different substances.

Example Problems

Identify each observation as a physical property, physical change, chemical property, or chemical change.

1. Water boils at 100°C.

   Answer: Physical property; water only changes physical state.

2. Dissolving sugar in iced tea.

   Answer: Physical change; sugar dissolves in water but does not change its identity. Sugar is still as sweet.

3. Coal is burned to heat a factory.

   Answer: Chemical change; during combustion coal is changed into other substances as heat is produced.

Work Problems

Identify each observation as a physical property, physical change, chemical property, or chemical change.

1. Silver is a good conductor of electricity.
2. A poached egg left on a silver tray will turn the tray black.
3. Carbon monoxide burns in air to form carbon dioxide.
4. An antacid can relieve heartburn by neutralizing excess stomach acid.

Worked Solutions

1. **Electrical conductivity is a physical property.**
2. **The tarnishing of silver to form a black substance is a chemical change.**
3. **It is a chemical property of carbon monoxide that it can burn in air to form a new compound.**
4. **It is a chemical property of an antacid that it can neutralize (eliminate) stomach acid.**

The Elements

There are about 115 known elements, but of that number only about 40 are involved in the formation of most compounds, and of that 40, only 10 compose nearly all of the earth's crust. Those elements that comprise over 99% of the earth's crust are listed in the following table. Each percent is percent by weight which means that 49% of the weight of the earth's crust is oxygen, and so forth.

The Top Ten Elements in the Earth's Crust			
Element	***Approximate Percent of Earth's Crust***	***Element***	***Approximate Percent of Earth's Crust***
Oxygen	49%	Sodium	2.6%
Silicon	26%	Potassium	2.4%
Aluminum	7.5%	Magnesium	2.0%
Iron	4.8%	Hydrogen	0.9%
Calcium	3.4%	Titanium	0.6%

Unique one- or two-letter symbols are used to represent each element. If a symbol is one letter, it is always capitalized, like N for nitrogen, or O for oxygen. If a symbol uses two letters, only the first letter is capitalized, as in Al for aluminum. Never capitalize both letters! The symbol Co represents the element cobalt, but CO is the compound carbon monoxide. The names and symbols of 40 elements you need to know are listed in alphabetical order in the following table.

Forty Elements You Need to Know			
Element	***Symbol***	***Element***	***Symbol***
Aluminum	Al	Lithium	Li
Argon	Ar	Magnesium	Mg
Arsenic	As	Manganese	Mn
Barium	Ba	Mercury	Hg
Boron	B	Neon	Ne
Bromine	Br	Nickel	Ni
Cadmium	Cd	Nitrogen	N
Calcium	Ca	Oxygen	O
Carbon	C	Phosphorus	P
Chlorine	Cl	Platinum	Pt
Chromium	Cr	Potassium	K
Cobalt	Co	Selenium	Se
Copper	Cu	Silicon	Si
Fluorine	F	Silver	Ag
Gold	Au	Sodium	Na
Helium	He	Sulfur	S
Hydrogen	H	Tin	Sn
Iodine	I	Titanium	Ti
Iron	Fe	Uranium	U
Lead	Pb	Zinc	Zn

Most of the symbols are clearly derived from the English name of the element, but a few come from older Latin names, such as iron (Fe, *ferrum*), potassium (K, *kalium*), and sodium (Na, *natrium*). It is important to know the names and symbols of the most common elements, because chemical symbols are used in all areas of science. Practicing with flashcards is the easiest way to memorize the symbols and names.

Example Problems

Give the symbol or the name (correctly spelled) for the elements in each set.

1. DNA is constructed with these elements: H, C, N, P, and O

   Answer: hydrogen, carbon, nitrogen, phosphorus, and oxygen

2. These metals are important nutrients: iron, calcium, sodium, and potassium

   Answer: Fe, Ca, Na, and K

3. These metals are used to make coins: Au, Ag, and Cu

   Answer: gold, silver, and copper

Work Problems

Give either the symbol or the name (correctly spelled) for these elements.

1. He
2. sulfur
3. Cl
4. bromine

Worked Solutions

1. **helium**
2. **S**
3. **chlorine**
4. **Br**

The Periodic Table of the Elements

The **periodic table** is a tabular array of the elements that lists them horizontally in order of increasing **atomic number.** Each element is represented by its symbol, and its atomic number is written above the symbol. The importance of the atomic number will be discussed in Chapter 3. In addition, the periodic table is organized so that elements with similar chemical properties are aligned in columns. This kind of organization makes the periodic table a valuable tool. If you know the chemical properties of one element, then it is reasonable to assume that the other elements in the same column will have similar properties. For this and many other reasons, the periodic table is the single most useful tool in chemistry. The modern periodic table is shown in the following figure.

The Modern Periodic Table of Elements

Main groups (1–2) | Transition Metals (3–12) | Main groups (13–18)

1 H 1.00794

Period	1 IA	2 IIA	3 IIIB	4 IVB	5 VB	6 VIB	7 VIIB	8 VIIIB	9 VIIIB	10 VIIIB	11 IB	12 IIB	13 IIIA	14 IVA	15 VA	16 VIA	17 VIIA	18 VIIIA
1	1 **H** 1.00794																	2 **He** 4.00260
2	3 **Li** 6.941	4 **Be** 9.01218											5 **B** 10.81	6 **C** 12.011	7 **N** 14.0067	8 **O** 15.9994	9 **F** 18.99840	10 **Ne** 20.1797
3	11 **Na** 22.98977	12 **Mg** 24.305											13 **Al** 26.98154	14 **Si** 28.0855	15 **P** 30.97376	16 **S** 32.066	17 **Cl** 35.453	18 **Ar** 39.948
4	19 **K** 39.0983	20 **Ca** 40.078	21 **Sc** 44.9559	22 **Ti** 47.88	23 **V** 50.9415	24 **Cr** 51.996	25 **Mn** 54.9380	26 **Fe** 55.847	27 **Co** 58.9332	28 **Ni** 58.69	29 **Cu** 63.546	30 **Zn** 65.39	31 **Ga** 69.72	32 **Ge** 72.61	33 **As** 74.9216	34 **Se** 78.96	35 **Br** 79.904	36 **Kr** 83.80
5	37 **Rb** 85.4678	38 **Sr** 87.62	39 **Y** 88.9059	40 **Zr** 91.224	41 **Nb** 92.9064	42 **Mo** 95.94	43 **Tc** (98)	44 **Ru** 101.07	45 **Rh** 102.9055	46 **Pd** 105.42	47 **Ag** 107.8682	48 **Cd** 112.41	49 **In** 114.82	50 **Sn** 118.710	51 **Sb** 121.757	52 **Te** 127.60	53 **I** 126.9045	54 **Xe** 131.29
6	55 **Cs** 132.9054	56 **Ba** 137.33	57 **La** 138.9055	72 **Hf** 178.49	73 **Ta** 180.9479	74 **W** 183.85	75 **Re** 186.207	76 **Os** 190.2	77 **Ir** 192.22	78 **Pt** 195.08	79 **Au** 196.9665	80 **Hg** 200.59	81 **Ti** 204.383	82 **Pb** 207.2	83 **Bi** 208.9804	84 **Po** (209)	85 **At** (210)	86 **Rn** (222)
7	87 **Fr** (223)	88 **Ra** 226.0254	89 **Ac** 227.0278	104 **Rf** (261)	105 **Db** (262)	106 **Sg** (263)	107 **Bh** (262)	108 **Hs** (265)	109 **Mt** (266)	110 **Uun**	111 **Uuu**	112 **Uub**		114 **Uuq**		116 **Uuh**		

58 **Ce** 140.12	59 **Pr** 140.9077	60 **Nd** 144.24	61 **Pm** (145)	62 **Sm** 150.36	63 **Eu** 151.96	64 **Gd** 157.25	65 **Tb** 158.9254	66 **Dy** 162.50	67 **Ho** 164.9304	68 **Er** 167.26	69 **Tm** 168.9342	70 **Yb** 173.04	71 **Lu** 174.967
90 **Th** 232.0381	91 **Pa** 231.0359	92 **U** 238.0289	93 **Np** 237.048	94 **Pu** (244)	95 **Am** (243)	96 **Cm** (247)	97 **Bk** (247)	98 **Cf** (251)	99 **Es** (252)	100 **Fm** (257)	101 **Md** (258)	102 **No** (259)	103 **Lr** (260)

The horizontal rows of elements are called **periods** and are numbered top to bottom, 1 to 7. The first period contains only 2 elements (H and He), the second and third contain 8, and the fourth, 18.

The vertical columns are called **groups** or **families,** and each is numbered with a Roman numeral followed by the letter A or B. Since there are 18 groups, some newer periodic tables simply number them 1 to 18, but because the Roman numerals are more often used in the United States, they are used here.

The elements in the A-groups are the **main-group** (or representative) elements; those in the middle of the table are the **transition metals.** The two rows of 14 elements at the bottom of the table are the **inner-transition metals.** They are placed below the table to keep it from becoming too wide to print on a single page.

Certain groups of elements have their own special names, because of the significant chemical similarity of the elements in them.

- **Alkali metals:** The metals in Group IA (Li, Na, K, Rb, Cs, and Fr). These soft, silvery metals are very reactive. Although hydrogen, H, is placed in group IA for good reason, its properties are so unique it is often set off by itself above the table, too.
- **Alkaline earths:** The metals in Group IIA (Mg, Ca, Sr, Ba, and Ra). These soft, silvery metals are a little less chemically reactive than the alkali metals.
- **Halogens:** The elements in Group VIIA (F, Cl, Br, and I). These nonmetals are very chemically reactive.
- **Noble gases:** The elements in Group VIIIA (He, Ne, Ar, Kr, Xe, and Rn). These colorless elements have no or very little chemical reactivity. For many years, they were thought to be totally inert.

The heavy stair-step line that starts beneath boron (B) and descends to the right separates the metals on the left from the nonmetals on the right.

Metals are solids, except mercury (Hg), which is a liquid. Metals are good conductors of electricity and heat, and many are malleable (can be hammered into thin sheets), ductile (can be drawn into wires), and most have shiny surfaces.

Nonmetals have properties that are nearly the opposite those of metals. Though most are solids (some quite brittle), bromine (Br) is a liquid, and several are gases: nitrogen (N), oxygen (O), fluorine (F), chlorine (Cl), and all the noble gases.

It should not be surprising that the elements along the stair-step, boron (B), silicon (Si), germanium (Ge), arsenic (As), antimony (Sb), and tellurium (Te), have properties between those of metals and nonmetals and are classed as semimetals or **metalloids.**

It should be noted that in groups IVA, VA, and VIA, the elements at the top are nonmetals (carbon, nitrogen, and oxygen) and those at the bottom are metals (lead, bismuth, and polonium). The stair-step cuts these groups in two, separating metals from nonmetals with metalloids in the middle.

Example Problems

Identify the group and period in which the element appears on the periodic table.

1. B

Answer: period 2, group IIIA

2. potassium

   Answer: period 4, group IA

3. Ag

   Answer: period 5, group IB

Identify the element (or elements) that meets each description.

4. A metal in group IVA

   Answer: either tin (Sn) or lead (Pb)

5. An alkali metal in the second period

   Answer: lithium (Li)

6. Group the following 8 elements in pairs that would be expected to have similar chemical properties: Na, I, Mg, Rb, P, Cl, Sr, N

   Answer: Na and Rb are both in group IA; Cl and I are in Group VIIA; Mg and Sr are in group IIA; N and P are in group VA. Main-group elements in the same group share many of the same chemical properties.

Work Problems

Identify the group and period in which each element appears on the periodic table.

1. mercury
2. Br

Identify the element(s) that meet(s) each description.

3. a metalloid in period 5
4. a main-group element that is a liquid at room temperature
5. Among As, Se, Sb, or I, which would you expect to be most like tellurium, Te?

Worked Solutions

1. **Hg, period 6, group IIB**
2. **Br, period 4, group VIIA**
3. **antimony (Sb) and tellurium (Te)**
4. **bromine (Br)**
5. **selenium, Se, the element in the same group**

Atoms and Molecules of the Elements

The smallest particle of an element is the atom, an extremely small particle. A copper atom has a diameter of 2.8×10^{-10} meter, and it would take about 3.5 million of them to form a line just 1 mm long, about the thickness of a dime. Most of the elements exist as atoms, and when an element is represented by its symbol, you can think of it (at this point) as symbolizing one atom of that element. The symbol "He" indicates 1 atom of helium, for example.

Seven elements, all of them nonmetals, do not exist as individual atoms in their pure form. Rather, for reasons you will learn later (see "The Diatomic Elements" table below), two (sometimes more) atoms join together to form a **molecule,** a particle composed of two or more atoms bonded together. These seven elements exist as diatomic molecules (*di*-meaning two, *-atomic* meaning atom), and are listed in the following table.

The Diatomic Elements

Element	*Formula of the Diatomic Molecule*	*Name of the Element as a Diatomic Molecule*
Hydrogen	H_2	dihydrogen
Nitrogen	N_2	dinitrogen
Oxygen	O_2	dioxygen
Fluorine	F_2	difluorine
Chlorine	Cl_2	dichlorine
Bromine	Br_2	dibromine
Iodine	I_2	diiodine

A problem often arises with these seven elements when it comes to their names. For example, when someone mentions "hydrogen," does he mean the atom of hydrogen, H, or the molecule of hydrogen, H_2? They are both hydrogen, but clearly they are not the same thing. To get around this problem and eliminate confusion, the name of the atom is the name of the element, and adding the prefix *di-* to the name of the element forms the name of the molecule. Hydrogen is H; dihydrogen is H_2. Bromine is Br; Br_2 is dibromine. Yet, common usage frequently contradicts this rule. Hydrogen gas, H_2, is commonly called hydrogen because the free element has to be H_2. Similarly, oxygen gas, O_2, is commonly called oxygen, but you would need to know that you are talking about the free element to translate "oxygen" as O_2 and not an atom of oxygen. The same is true for the rest of the diatomic elements. In their elemental state, *if there is no chance of confusion,* the diatomic species are often called by the name of the element and the context of use; you are to understand that it refers to the diatomic molecule. But, if there is a chance of confusion, use the *di-* prefix for the molecular species.

Some elements form still larger molecules, such as O_3 (trioxygen, ozone), P_4 (tetraphosphorus), and S_8 (octasulfur). The prefixes tri-, tetra-, and octa- mean 3, 4, and 8, respectively. Remember that the molecules of these elements exist only when they are in their pure form or in mixtures, *not* when they are part of a compound.

Compounds

The composition of compounds was the focus of much research in the early days of chemistry and, in the early 1800s, resulted in the development of two of the most important laws of nature.

The Law of Definite Composition: All samples of a pure compound, no matter where they come from, always contain the same elements in the same, nonchanging ratio(s) by mass.

Explanation: Water is water no matter where it comes from. All samples of pure water are composed of the same elements, hydrogen (H) and oxygen (O), in the same mass ratio of 1.00 g H to 7.94 g O. It never changes.

The Law of Multiple Proportions: If two elements can combine to form two or more different compounds, the mass of one element compared to a fixed mass of the other will always be in a ratio of small, whole numbers. This really requires an explanation to make sense.

Explanation: Carbon and oxygen form two compounds, carbon monoxide, CO, and carbon dioxide, CO_2. In CO, there is 1.33 g O per 1.00 g C. In CO_2, the ratio is 2.66 g O per 1.00 g C. The mass of oxygen per gram of carbon is *exactly two times* as large in CO_2 as it is in CO. The important point is "exactly two times" the oxygen per gram of carbon in CO_2 as in CO. It should not surprise you, then, that these same ratios appear in the formulas of these compounds, CO (one O per C) and CO_2 (two O per C).

Today, it is common to think of these laws in terms of "atoms" of the elements as opposed to "grams" of the elements. Instead of a ratio of masses, we talk in terms of a ratio of atoms, and the definition of compound becomes easier to understand. **Compounds** are pure substances composed of two or more elements in a definite, nonchanging ratio of atoms.

Compounds can be divided into two classes: **molecular compounds** and **ionic compounds.** They are described in the following table, along with examples of each. The chemical formulas of compounds show the elements that compose them. Formulas are discussed in the next section.

Characteristics of Molecular and Ionic Compounds

	Molecular Compounds	*Ionic Compounds*
The kinds of elements that form the compounds	Nonmetals*	Metals + nonmetals*
Smallest unit of each class of compound	Molecules	The "formula unit," that represents the simplest ratio of the elements in the compound. No molecules exist.
Description of the smallest unit	Neutral atoms bound together to form a neutral molecule	Positive and negative ions bound together in a crystalline solid.
Example compounds	Water–H_2O glucose–$C_6H_{12}O_6$	Table salt–NaCl Fluorite–CaF_2

**There are some exceptions, but you don't see them here.*

Molecular and ionic compounds are quite different in their physical properties. For example, molecular compounds can be solids, liquids, or gases at room temperature, whereas ionic compounds are always solids. Molecular compounds that are solids usually melt at temperatures much lower than ionic solids. Solid water, ice, melts at 0°C; table salt, NaCl, melts at 801°C. Ionic compounds are composed of **ions,** particles formed from atoms that bear an electrical charge. Ions that bear a positive charge are classed as cations, and those that bear a negative charge are anions. Both cations and anions are described further in Chapter 3. In an ionic compound, ions of opposite charge, Na^+ and Cl^-, for example, are strongly attracted to one another and arrange themselves in highly ordered, three-dimensional arrays that lead to the formation of crystalline solids, like NaCl crystals. In the crystal, the negative charges from the anions are exactly balanced by the positive charges from the cations, so the crystal, as a whole, is electrically neutral, that is, it has no charge at all. Ionic and molecular compounds, their formulas, and names are discussed further in Chapter 4.

Example Problems

1. Examine the elements in each compound and predict which would exist as molecules: $CaCl_2$, NO_2, BaO, CO, SCl_2, Na_2SO_4.

   Answer: NO_2, CO, and SCl_2 are composed of only nonmetals and are predicted to exist as molecules (molecular compounds).

2. Examine the elements that compose these compounds and predict which are ionic compounds: K_2O, PCl_3, $BaSO_4$, NiF_2, CS_2, CH_2O.

   Answer: K_2O, $BaSO_4$, and NiF_2 are composed of both metals *and* nonmetals and are expected to be ionic compounds.

3. What is wrong with the phrase, "molecules of NaCl"?

   Answer: NaCl is an ionic compound which exists as a three-dimensional array of ions. Each positive ion is associated with several negative ions, and vice versa. There are no specific, individual molecules in ionic compounds.

4. Glucose is a molecular compound containing carbon, oxygen, and hydrogen. In a 15.0 g sample of glucose, there are 6.0 g of carbon, 1.0 g of hydrogen, and 8.0 g of oxygen. How many grams of glucose would contain only 2.0 g of carbon?

   Answer: 5.0 g of glucose. 2.0 g of carbon is exactly one-third the amount of carbon in 15.0 grams of glucose, so 2.0 g of carbon should be in exactly one-third of 15.0 g of glucose, which is 5.0 g. This demonstrates the Law of Definite Composition.

Work Problems

1. Classify each as either an ionic or a molecular compound: CO_2, MgF_2, $SeBr_2$, Al_2O_3.

2. Do the following composition values support the Law of Multiple Proportions? Explain your answer.

   CuCl 1.00 g Cu to 0.558 g Cl

   $CuCl_2$ 1.00 g Cu to 1.12 g Cl

3. Which of the following are elements? Which are molecules but not compounds? Which are both compounds and molecules? Which are compounds but not molecules?

   NO_2, P_4, FeO, Na, N_2O_3, F, Cl_2, O_3, C_2H_4, S_8, CaF_2

Worked Solutions

1. **CO_2; molecular compound (only nonmetallic elements)**

   MgF_2; ionic compound (metal + nonmetal)

   $SeBr_2$; molecular compound

   Al_2O_3; ionic compound

2. **Yes.** There is exactly twice the mass of chlorine per gram of copper in $CuCl_2$ as in CuCl.

3. **Elements: P_4, Na, F, Cl_2, O_3, and S_8**

   Molecule, not compound: P_4, Cl_2, O_3, and S_8

   Both molecule and compound: NO_2, N_2O_3, and C_2H_4

   Compound, not molecule: CaF_3 and FeO

Chemical Formulas

Compounds are represented by chemical formulas. A **chemical formula** shows the elements that make up the compound and the numbers of atoms of each element in the smallest unit of that compound, be it a molecule or a formula unit. Water is a molecular compound composed of hydrogen and oxygen in a nonchanging ratio of two atoms of hydrogen to one atom of oxygen. The chemical formula of water, H_2O, shows this fact. The subscript "2" after H indicates two atoms of hydrogen, and the symbol O, without a subscript, indicates one atom of oxygen in one molecule of water. The formula N_2O_4 represents a molecule made up of 2 atoms of nitrogen and 4 atoms of oxygen. MgF_2 is the formula of an ionic compound that is made up of ions from one magnesium atom and 2 fluorine atoms per formula unit (no molecules here). The **formula unit** of an ionic compound shows the simplest ratio of the elements that make it up. Contrast that to the **molecular formula,** which shows not only the ratio of elements but also all the atoms in one molecule of the compound. For example, both NO_2 and N_2O_4 represent two different molecular compounds. Both show a 1 to 2 ratio of nitrogen to oxygen, but NO_2 is made up of three atoms, N_2O_4, six atoms. Table salt, NaCl, is an ionic compound and would never be represented by a formula like Na_2Cl_2, even though it shows the 1 to 1 ratio of Na to Cl. It can show only the *simplest* ratio of the two elements, 1 Na to 1 Cl.

Although a molecular formula tells a great deal about a compound, it doesn't tell you how the atoms are joined together or how they are arranged in three-dimensional space. But that will come later (see Chapter 10).

Pronouncing a formula in conversation is not too difficult. The formula of water, H_2O, is pronounced H-two-O and that for glucose, $C_6H_{12}O_6$, is C-six-H-twelve-O-six.

Example Problems

1. Arrange the following compounds in order of increasing number of atoms per molecule: C_2H_5OH, $C_7H_6O_3$, SF_6, C_6H_6.

   Answer: SF_6: 7 atoms; C_2H_5OH: 9 atoms; C_6H_6: 12 atoms; $C_7H_6O_3$: 16 atoms

2. Which represent molecules, which formula units: CaS, $NaNO_3$, SO_3, $CuCl_2$?

   Answer: CaS, $NaNO_3$, $CuCl_2$. All are ionic compounds; therefore, each formula represents a formula unit. SO_3 is a molecular compound; therefore, its formula represents a molecule.

Work Problems

1. Which formulas represent molecules, which formula units: $BaBr_2$, NH_3, BF_3, Na_2O, C_2H_2?

2. What is wrong with this statement: Calcium chloride is an ionic compound with a formula of Ca_2Cl_4?

Worked Solutions

1. **NH_3, BF_3, and C_2H_2 are molecular compounds, and their formulas represent the composition of the molecules.** $BaBr_2$ and Na_2O are ionic compounds, and their formulas represent one formula unit of the compound.

2. **The formula of an ionic compound shows only the *simplest* ratio of the elements that make it up, so it should be $CaCl_2$, not Ca_2Cl_4.** The following figure shows the breakdown of pure substances.

Structural Units of Elements and Compounds

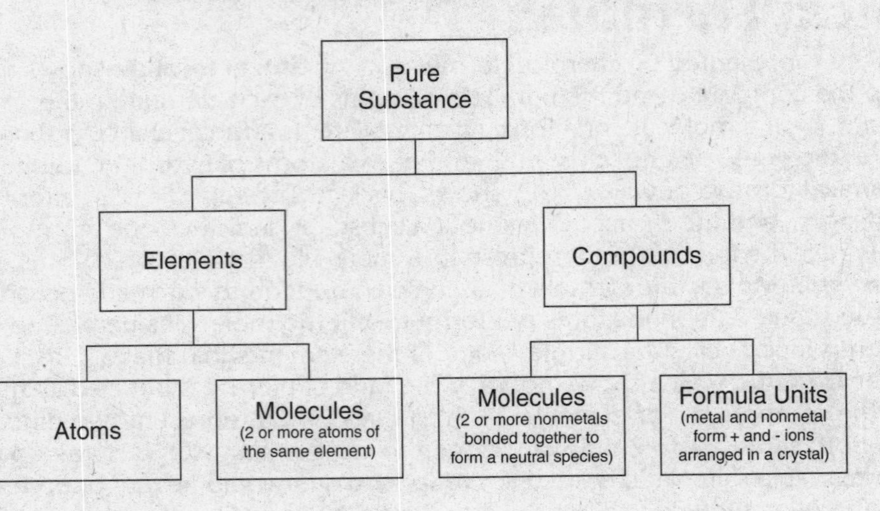

Naming Molecular Compounds

Several molecular compounds have common names that you use every day, such as water, ammonia, and alcohol. But common names can cause confusion. Does "ammonia" mean ammonia the pure gaseous compound or a solution of ammonia in water, as it is sold in stores? And, there are several compounds known as "alcohol." Clearly, a more accurate system of naming compounds is necessary. So here, and in Chapter 4, you can learn how compounds are named following the international rules of chemical nomenclature. The goal of nomenclature is to develop a name that gives all the necessary information to allow someone else to construct the formula of the compound, and vice versa.

Since this is the book's first look at nomenclature, we limit ourselves to **molecular binary compounds,** those made up of only two elements, both nonmetals. Other binary compounds are named in Chapter 4, after you learn more about ions. Let's start by naming four molecular binary compounds to get an idea how names are formed, and then follow this with some rules.

CS_2	carbon disulfide
N_2O_5	dinitrogen pentaoxide
PCl_3	phosphorus trichloride
H_2O	dihydrogen oxide

Rule 1: The names of all binary compounds end in "-ide." The end of the name of the second element is dropped, and "-ide" is attached to the stem that remains, as shown in the following table.

Element Name Changes in Binary Compounds

Element	***Stem***	***Element's Name in the Compound***
Fluorine	fluor-	fluor*ide*
Chlorine	chlor-	chlor*ide*
Bromine	brom-	brom*ide*
Iodine	iod-	iod*ide*
Oxygen	ox-	ox*ide*
Sulfur	sulf-	sulf*ide*
Nitrogen	nitr-	nitr*ide*
Phosphorus	phosph-	phosph*ide*
Boron	bor-	bor*ide*
Carbon	carb-	carb*ide*
Hydrogen	hydr-	hydr*ide*

Rule 2: The elements are named in the order they appear in the formula, left to right. The names of the two elements are separated with a space.

Rule 3: The number of atoms of each element is indicated using numerical prefixes:

1 = mono-	6 = hexa-
2 = di-	7 = hepta-
3 = tri-	8 = octa-
4 = tetra-	9 = nona-
5 = penta-	10 = deca-

The mono- prefix is usually not used unless it is important to distinguish between compounds of the same two elements.

CO	carbon monoxide*	(a very toxic gas)
CO_2	carbon dioxide	(the bubbles in a soft drink)

**the "o" in mono- is dropped, because oxide begins with an "o."*

Let's look at more names of binary molecular compounds to see how these rules are applied:

N_2O	dinitrogen oxide
ClF_3	chlorine trifluoride
B_2O_3	diboron trioxide
P_2S_3	diphosphorus trisulfide
CH_4	carbon tetrahydride (common names: methane, natural gas)

Doing the reverse, writing a formula from the name of a compound, requires knowing the meaning of the numerical prefixes and the stems of the elements. Examine these examples to follow the logic from name to formula:

nitrogen trifluoride	1-N, 3-F	NF_3
diphosphorus hexaoxide	2-P, 6-O	P_2O_6
tricarbon octahydride	3-C, 8-H	C_3H_8 (common name: propane)

Example Problems

Name these molecular binary compounds:

1. PBr_3

   Answer: phosphorus tribromide

2. N_2O_4

   Answer: dinitrogen tetraoxide

3. BrF_3

   Answer: bromine trifluoride

Write the formula of these molecular binary compounds:

4. sulfur hexafluoride

   Answer: SF_6

5. nitrogen monoxide

   Answer: NO

6. tetraphosphorus decaoxide

   Answer: P_4O_{10}

Work Problems

Name these molecular binary compounds:

1. CCl_4
2. N_2O_3

Write the formula of these molecular binary compounds:

3. iodine heptafluoride
4. hexacarbon hexahydride (common name: benzene)

Worked Solutions

1. **carbon tetrachloride**

2. **dinitrogen trioxide**

3. IF_7

4. C_6H_6

Mixtures

In nature, it is rare for an element or a compound to occur in its pure state. Some do, like gold, but most come to us in mixtures of several substances. A **mixture** is composed of two or more substances that, unlike a compound, can be mixed in varying ratios to one another. Mixtures can be homogeneous (solutions) or heterogeneous. The foods we eat are mixtures, as is the soil, the ocean, the atmosphere, and most everything else around us.

Mixtures can be separated by *physical means* like evaporation, distillation, filtering, or dissolving one component into a liquid that does not dissolve the others. Sometimes, it is possible to separate heterogeneous mixtures by mechanical means using tweezers, comb, or magnet. In the end, the separated components obtained from a mixture are still the same substances they were before they were separated. A **physical separation** does not change the identity of the substances separated. In contrast, a **chemical separation** changes the identity of the substance being separated. For example, separating a compound into its component elements requires its destruction, because chemically combined elements require *chemical means* (chemical reactions), to separate them. Separating the elements destroys the compound that was made up of those elements.

A summary comparing elements, compounds, and mixtures is shown in the following table.

A Comparison of Elements, Compounds, and Mixtures

	Elements	*Compounds*	*Mixtures*
Composition	A pure substance; the simplest building blocks of compounds.	A pure substance composed of two or more elements chemically combined in a definite ratio of atoms.	Composed of two or more pure substances physically mixed in no particular ratio of substances.
Physical state at 25°C	Homogeneous; solid, liquid, or gas	Homogeneous: Solid, liquid, or gas	Homogeneous or heterogeneous; solid, liquid, or gas
Separation into simpler substances	No, not possible	Yes, by chemical means	Yes, by physical or mechanical means
Nature and result of the separation		The compound is destroyed as its component elements are separated.	Elements and/or compounds that made up the mixture retain their identity.

Chapter Problems and Answers

Problems

Questions 1 through 6 concern the three physical states of matter; solid, liquid, and gas. Which one or two states are described by each characteristic?

1. Particles are touching one another.
2. Particles can flow.
3. Particles are very difficult to compress to a smaller volume.
4. Particles are widely separated.
5. Particles are rigidly held in place.
6. Particles are easily compressed into a smaller volume.

Questions 7 through 12 list common items and you are to assign three of the following eight terms to each one: solid, liquid, or gas; element, compound, or mixture; homogeneous or heterogeneous.

7. orange juice
8. nitrogen gas
9. a slice of salami or pepperoni
10. cadmium yellow oil paint
11. windshield washer fluid
12. lacquer thinner, that is pure ethyl acetate

Questions 13 through 18 ask you to identify each example as a physical property, chemical property, physical change, or chemical change.

13. Diamond is extremely hard.
14. Diamond burns in oxygen to form carbon dioxide.
15. Copper does not react with hydrochloric acid.
16. Copper conducts electricity and heat very well.
17. Ammonia gas can be turned into a liquid at –33°C.
18. Sodium metal must be stored under oil to prevent reaction with moisture in the air.

Write the correct symbol for the elements in each set.

19. carbon, oxygen, hydrogen, helium

20. aluminum, iron, copper, tin

21. fluorine, chlorine, bromine, iodine

22. nickel, calcium, lead, zinc

23. magnesium, barium, neon, chromium

Write the name, spelled correctly, for the elements in each set.

24. P, K, Se

25. Si, Ar, Li

26. Bi, B, Sr

Questions 27 through 31 concern the periodic table of the elements.

27. In what order are the elements arranged in the periodic table?

28. What is the term used for those elements aligned in columns on the periodic table?

29. What is the term used for those elements aligned in rows on the periodic table?

30. The elements in the A groups are termed the _______ elements.

31. Is the element in the 4th period and group VB a metal, nonmetal, or a metalloid?

32. Would an element that is a solid, is a good conductor of electricity, and is able to be hammered into thin sheets likely be classed as a metal, a nonmetal, or a metalloid?

33. Which of the following elements exist as diatomic molecules in their pure states: boron, oxygen, carbon, chlorine, silicon, sulfur, neon, nitrogen?

34. Correctly name the following species: O, O_2, I_2, and F.

35. Which of the following compounds would be expected to exist as molecules: NaBr, NO_2, C_2H_6, NiO, BaF_2, $C_{12}H_{22}O_{11}$, PF_3?

36. Which of the following compounds would be expected to exist as ionic compounds: CCl_4, BrF_5, SO_2, $MgCl_2$, CaO?

Correctly name the following binary, molecular compounds.

37. P_2O_3, B_2H_6, NF_3

38. N_2O_5, SeO_3, CF_4

Write the correct formula for the following binary, molecular compounds.

39. nitrogen oxide, diboron trioxide, dinitrogen oxide

40. iodine heptafluoride, diphosphorus trisulfide, selenium dioxide

Answers

1. **solids and liquids**
2. **liquids and gases**
3. **solids and liquids**
4. **gases**
5. **solids**
6. **gases**
7. **liquid, mixture, heterogeneous (if it has orange pulp)**
8. **gas, element, homogeneous**
9. **solid, mixture, heterogeneous (different meats and spices are visible)**
10. **liquid, mixture, homogeneous**
11. **liquid, mixture, homogeneous**
12. **liquid, compound, homogeneous**
13. **physical property**
14. **chemical change**
15. **chemical property**
16. **physical property**
17. **physical property**
18. **chemical property**
19. **C, O, H, He**
20. **Al, Fe, Cu, Sn**
21. **F, Cl, Br, I**
22. **Ni, Ca, Pb, Zn**
23. **Mg, Ba, Ne, Cr**

24. **phosphorus, potassium, selenium**

25. **silicon, argon, lithium**

26. **bismuth, boron, strontium**

27. **The elements are arranged in order of increasing atomic number.**

28. **groups or families**

29. **periods**

30. **main-group elements (also, representative elements)**

31. **The element is arsenic, As; that is a metalloid.**

32. **metal**

33. **oxygen, O_2, chlorine, Cl_2, and nitrogen, N_2**

34. **oxygen, dioxygen, diiodine, fluorine**

35. **NO_2, C_2H_6, $C_{12}H_{22}O_{11}$, and PF_3 are composed of only nonmetallic elements.**

36. **$MgCl_2$ and CaO are composed of metals and nonmetals; they are ionic.**

37. **diphosphorus trioxide, diboron hexahydride, nitrogen trifluoride**

38. **dinitrogen pentaoxide, selenium trioxide, carbon tetrafluoride**

39. **NO, B_2O_3, N_2O**

40. **IF_7, P_2S_3, SeO_2**

Supplemental Chapter Problems

Problems

Questions 1 through 4 concern the three physical states of matter; solid, liquid, and gas. Which one or two states are described by each characteristic?

1. The particles that compose it are very strongly attracted to one another.

2. These substances can be poured from one container into another.

3. Very difficult to compress to a smaller volume.

Questions 4 and 5 list common items and you are to assign three of the following eight terms to each one: solid, liquid, or gas; element, compound, or mixture; homogeneous or heterogeneous.

4. a 2-pound silver paperweight

5. the atmosphere on a clear day on Mt. Everest

Identify each example as a physical property, chemical property, physical change, or chemical change:

6. A ruby is red.

7. A fresh aluminum surface slowly forms an oxide coating when exposed to air.

8. Ammonia is very soluble in cold water.

Write the correct symbol for the elements in each set:

9. potassium, silver, lithium, argon

10. calcium, platinum, bromine, mercury

Write the name, correctly spelled, for the elements in each set:

11. Ni, Ne, Al

12. Mg, Fe, Co

Questions 13 and 14 concern the periodic table of the elements.

13. Is the element in the third period and group IIA a metal, nonmetal, or a metalloid?

14. Which one of these elements does not belong with the others: Ge, S, C, Sn, Pb?

15. Those characteristics of elements and compounds that allow them to be identified are termed ________.

16. Elements in the same A-group have similar ________.

17. What is the difference between a species called bromine and one called dibromine?

18. Which of the following compounds would be expected to exist as molecules: K_2O, CO_2, CuO, N_2O_3, P_2O_5, SF_6

Correctly name the following binary, molecular compounds.

19. $BrCl_3$, C_4H_{10}, C_2N_2

Write the correct formula for the following binary, molecular compounds.

20. boron nitride, carbon monoxide, carbon tetrabromide

21. sulfur dichloride, silicon dioxide

Answers

Questions 1 through 3, states of matter. (page 39)

1. **Solids have the strongest forces of attraction between the particles that compose them.**
2. **liquids and gases**
3. **solids and liquids**

Questions 4 and 5, descriptions of matter. (page 40)

4. **solid, element, homogeneous**
5. **gas, mixture, homogeneous**

Questions 6 through 8, properties of matter. (page 41)

6. **physical property**
7. **chemical change**
8. **physical property**

Questions 9 through 12, symbols and names of elements. (page 43)

9. **K, Ag, Li, and Ar**
10. **Ca, Pt, Br, and Hg**
11. **nickel, neon, and aluminum**
12. **magnesium, iron, and cobalt**

Questions 13 through 16, periodic table. (page 45)

13. **metal**
14. **S**
15. **properties**
16. **chemical properties**

Questions 17 and 18, diatomic elements and ionic and molecular compounds. (page 48)

17. **Br and Br_2, atom and molecule**
18. **CO_2, N_2O_3, P_2O_5, and SF_6**

Questions 19 through 21, names and formulas of binary molecular compounds. (page 52)

19. **bromine trichloride, tetracarbon decahydride, dicarbon dinitride**

20. **BN, CO, CBr_4**

21. **SCl_2, SiO_2**

Chapter 3

Atoms I—The Basics

The atom is the smallest particle of an element. Every atom of an element has the same chemical properties, so it would be reasonable to think all the atoms of that element must be exactly the same. John Dalton (1766–1844), a British schoolmaster and scientist, proposed this idea in 1808 along with several others that accelerated the development of chemistry.

Dalton's Theory of Atoms

Dalton's ideas are summarized below, each followed by a brief explanation or an update.

1. **The smallest particle of any element is the atom, which is indestructible.**

   This statement is true as far as chemistry is concerned, but today we know that atoms can be broken down in high-energy physics experiments and nuclear reactions—but that's not chemistry. It still holds for chemistry.

2. **All atoms of a given element are identical.**

   The second statement needs modification, too, since today we know about subatomic particles, the particles that make up atoms. Looking only at the element, carbon, we find there are three different kinds of carbon atoms in nature, each with a slightly different composition and weight. Yet, all carbon atoms have the same *chemical properties*. How this can happen will be explained in this chapter.

3. **Chemical compounds result from the combination of two or more atoms of different elements in simple whole number ratios: 1 to 1 (AB), 1 to 2 (AB_2), 2 to 3 (A_2B_3), and so on.**

   This idea is better known today as the Law of Definite Composition, described in Chapter 2.

4. **Atoms of the same elements can combine in two or more different ratios to form two or more different compounds.**

   This statement embodies the core of the Law of Multiple Proportions, which was also described in Chapter 2.

The basic ideas of Dalton's theory have stood the test of time, and are foundation principles of chemistry. Without mentioning the name of John Dalton, these ideas were applied many times in the previous chapter.

Electrons, Protons, and Neutrons

Although an atom is an unimaginably small particle, still smaller particles make up atoms: the **subatomic particles.** Three fundamental subatomic particles have been characterized during the last 130 years: the electron, the proton, and the neutron. The electron and proton bear opposite electrical charge: negative and positive, respectively. The neutron bears no charge, as its name (from the same root as "neutral") implies.

There are only two kinds of electrical charge: negative (−) and positive (+). Opposite charges attract one another, and like charges repel. In addition, the closer the charges are together, the greater the force of attraction or repulsion. If the distance between a + and − charge is cut in half, the force of attraction increases four-fold. If the distance between the charges doubles, the force of attraction is then only one-fourth of what it was. This is referred to as **inverse-square behavior.** If the separation of a + and a − charge increases three-fold, the force of attraction changes to $\left(\frac{1}{3}\right)^2$, or $\frac{1}{9}$ of what it was.

- The **electron,** e^-, bears a negative electrical charge and has a mass of 9.1094×10^{-28} g. The size of the negative charge is exceedingly small and is simply stated as 1− or −1, a single unit of negative charge.
- The **proton,** p^+, bears a positive electrical charge and has a mass about 1,840 times greater than that of the electron, 1.6726×10^{-24} g. The positive charge on the proton is exactly the same size as the negative charge on the electron, but opposite in sign, and is stated as 1+ or +1, a single unit of positive charge.
- The **neutron,** n, has no electrical charge and a mass of 1.6749×10^{-24} g, slightly greater than that of the proton.

Just as the sizes of the charge on the electron and proton have been reduced to a convenient 1− and 1+, the extremely small masses of the three subatomic particles are often stated in **atomic mass units (amu)** to bring the mass of the proton and neutron numerically close to 1. One amu = 1.6606×10^{-24} g. The mass in both grams and amu for the three subatomic particles is given in the following table.

The Electron, Proton, and Neutron

Particle	*Mass in Grams*	*Mass in amu*	*Charge*
Electron, e^-	9.1094×10^{-28} g	0.00055 amu	1−
Proton, p^+	1.6726×10^{-24} g	1.00728 amu	1+
Neutron, n	1.6749×10^{-24} g	1.00866 amu	0

The Nuclear Atom

How are the subatomic particles arranged to form an atom? This puzzled scientists for many years until a British physicist, Ernest Rutherford, and his colleagues discovered the answer in 1911. Their experiments led them to conclude that nearly all the mass of an atom is packed into a tiny, positive-charged body called the **nucleus,** centered in a very much larger cloud of electrons. The atom was seen to be mostly empty space!

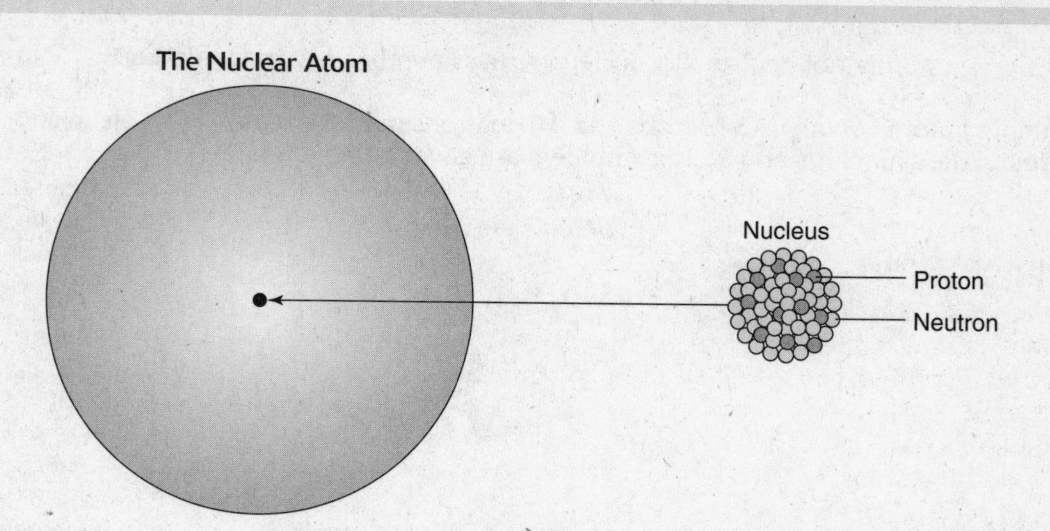

Protons and neutrons, the two massive subatomic particles, make up the nucleus and are responsible for nearly all the mass of the atom. The size of the enveloping cloud of electrons is from 10,000 to 20,000 times the size of the nucleus, and the size of this electron cloud represents the size of the atom. Although the electron cloud of an iron atom, Fe, is immensely larger than the nucleus, the mass of the nucleus is about 4,000 times greater than the mass of all the electrons surrounding it.

The Atomic Number and Mass Number

Chapter 2 shows you that elements are arranged in the periodic table in order of increasing atomic number. The **atomic number** of an element equals the number of protons in the nucleus of an atom of that element. *The atomic number identifies an element*. The atomic number appears above the symbol for each element in the periodic table. Any atom with only 1 proton in its nucleus is a hydrogen atom, the element with an atomic number of 1. The atomic number of carbon, for example, is 6. Every atom of carbon has 6 protons in its nucleus. Likewise, any atom that has 6 protons in its nucleus is a carbon atom. If an atom has 7 protons in its nucleus, it's not carbon; it's the next element, nitrogen. For a neutral atom, the atomic number also equals the number of electrons about the nucleus. An electrically neutral atom must have the identical number of positive and negative charges.

atomic number = number of p^+ in the nucleus, and the number of e^- about the nucleus in the neutral atom

The **mass number** of an atom (notice "of an atom" not "of an element") equals the sum of the number of protons *plus* the number of neutrons in the nucleus. The mass number equals the total number of heavy (massive) particles in the nucleus.

mass number = number of p^+ + number of neutrons in the nucleus

Symbolically, the atomic number and mass number of an atom are displayed on the lower-left and upper-left side of the symbol of the element, respectively.

$$\text{mass number} \rightarrow 27 \leftarrow \text{(equals number of } p^+ + n)$$
$$\text{Al}$$
$$\text{atomic number} \rightarrow 13 \leftarrow \text{(equals number of } p^+)$$

Showing both the atomic number and the symbol for aluminum may seem redundant since both indicate the same element, but the presence of the atomic number does serve a useful purpose. Subtracting the atomic number from the mass number above it equals the number of neutrons in the nucleus.

number of neutrons in nucleus = mass number − atomic number

The aluminum atom contains 13 protons and 14 neutrons in its nucleus and 13 electrons outside the nucleus. Mass numbers and atomic numbers are always whole numbers.

The Aluminum Atom

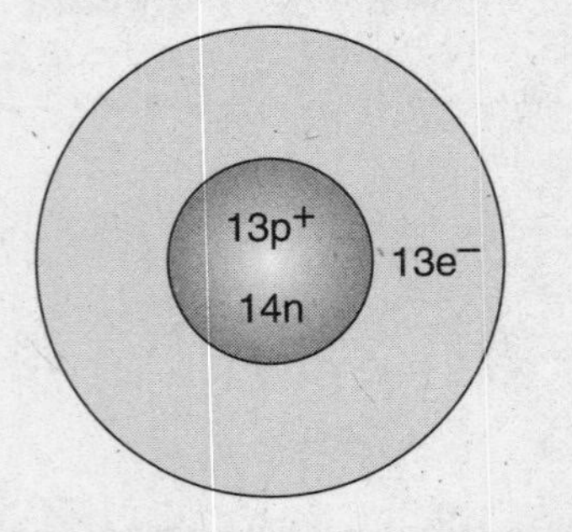

It is important to realize that the chemical properties of an element are determined by its atomic number, not its mass number. This is why the atomic numbers are shown on the periodic table and mass numbers are not. Chemical reactions involve the electrons outside the nucleus; the nucleus itself is not directly involved. Because the atomic number also equals the number of electrons an atom possesses, it becomes the more important number of the two in terms of the chemical properties of an element. Yet, even if an atom loses or gains an electron in a reaction to form an ion (an electrically charged species), the identity of that ion is still determined by the number of protons in the nucleus, which would be the atomic number of the parent element.

Example Problems

1. Describe the composition of a fluorine atom: $^{19}_{9}F$

   Answer: The atomic number, 9, indicates 9 protons in the nucleus and 9 electrons about it. The difference between the mass number, 19, and the atomic number shows there are 10 neutrons in the nucleus with the protons.

2. An atom of uranium, U, has 146 neutrons. Write the correct symbol for this atom with its atomic number and mass number. Consult the periodic table for the atomic number.

   Answer: The atomic number of uranium from the periodic table is 92. The mass number equals 92 + 146 = 238. The symbol is $^{238}_{92}U$.

3. Complete the table for each neutral atom.

Symbol of the Element	*Atomic Number*	*Mass Number*	*Number of Electrons*	*Number of Protons*	*Number of Neutrons*
	11	23			
P					16
				53	74

Answer:

Symbol of the Element	*Atomic Number*	*Mass Number*	*Number of Electrons*	*Number of Protons*	*Number of Neutrons*
Na	11	23	*11*	*11*	*12*
P	*15*	*31*	*15*	*15*	16
I	*53*	*127*	*53*	53	74

Work Problems

1. Describe the composition of an atom of iron: $^{53}_{26}Fe$.
2. An atom of bromine, Br, has 46 neutrons in its nucleus. Write the symbol for this atom of bromine showing its composition.
3. Complete the table for each neutral atom.

Symbol of the Element	*Atomic Number*	*Mass Number*	*Number of Electrons*	*Number of Protons*	*Number of Neutrons*
S		32			
	19				20
			18		22

Worked Solutions

1. **26 p^+, 26 e^-, and 27 n (53 – 26 = 27).** The neutrons and protons are in the nucleus and the electrons are about it.
2. **The atomic number for bromine from the periodic table is 35, and the mass number equals 81, the sum of the protons plus neutrons; $^{81}_{35}Br$.**
3.

Symbol of the Element	*Atomic Number*	*Mass Number*	*Number of Electrons*	*Number of Protons*	*Number of Neutrons*
S	*16*	32	*16*	*16*	*16*
K	19	*39*	*19*	*19*	20
Ar	*18*	*40*	18	*18*	22

Isotopes of the Elements

Dalton reasoned that every atom of an element was identical because they had the same chemical properties. But atoms of an element can differ from one another in a way that has minimal effect on their chemical properties—they can vary in the number of neutrons in the nucleus forming closely related species called isotopes.

Isotopes are atoms with the same atomic number (atoms of the same element) that have different mass numbers. They have the same number of protons in the nucleus, but the number of neutrons will not be the same. Adding neutrons to an atom will make it heavier, but that won't change the chemistry of the atom.

Most elements exist in nature in two or more isotopic forms. Chlorine, Cl, is a typical example. All samples of chlorine obtained from nature are composed of the same nonchanging mixture of two isotopes, chlorine-35 (Cl-35) and chlorine-37 (Cl-37). The 35 and 37 are the mass numbers of the two isotopes. Of course, both isotopes have the same atomic number, 17.

$^{35}_{17}Cl$ chlorine-35 or Cl-35

$^{37}_{17}Cl$ chlorine-37 or Cl-37

The two isotopes of chlorine differ *only* in the number of neutrons in the nucleus.

	Cl-35		Cl-37
	$17\ e^-$		$17\ e^-$
	$17\ p^+$		$17\ p^+$
	18 n		20 n

In nature, 75.771% of all chlorine atoms, by weight, are Cl-35, and 24.229% are the heavier isotope, Cl-37. The percentage distribution of the isotopes of an element is called the **percent natural abundance** or just **natural abundance** of the isotopes. All samples of chlorine, no matter the source, have the same percent natural abundance of the two isotopes. It is a nonchanging mixture of isotopes, but remember that even though they are isotopes of one another, they are still atoms of the same element and have the same chemical properties. For comparison, the isotope composition of two other elements, carbon and magnesium, are given in the following table.

The Isotopes of Carbon and Magnesium

Element	*Isotopes*	*%-Natural Abundance*
Carbon	C-12	99.89%
	C-13	1.11%
	C-14	Trace
Magnesium	Mg-24	78.99%
	Mg-25	10.00%
	Mg-26	11.01%

Some elements, like sodium, aluminum, and fluorine, exist in nature in only *one* isotopic form. All sodium atoms are identical: same atomic number, same mass number. For other elements, carbon is a good example, one or more isotopes may be radioactive as a result of unstable nuclei. Carbon-14 is a radioactive isotope of carbon. Although carbon-14 is an infinitesimally small fraction of carbon atoms, its presence in all living things has led to its use in the radiocarbon dating of ancient artifacts.

Example Problems

1. Which of these six isotopic symbols represent isotopes?

   $^{19}_{9}X$ $^{19}_{10}X$ $^{20}_{9}X$ $^{21}_{11}X$ $^{19}_{8}X$

   Answer: $^{19}_{9}X$ and $^{20}_{9}X$, same atomic number, different mass numbers.

2. There are three isotopes of hydrogen. This is the only element for which the isotopes have their own unique names.

   protium $^{1}_{1}H$ deuterium $^{2}_{1}H$ tritium $^{3}_{1}H$

   What subatomic particles are in the nucleus of each isotope?

   Answer: protium, 1 p^+; deuterium, 1 p^+, 1 n; tritium, 1p^+, 2 n (protium is the only atom of *any* element that does not have at least one neutron in its nucleus)

3. How should the isotopes of mercury be arranged from left to right, in order of increasing mass of the atoms?

   $^{202}_{80}Hg$, $^{199}_{80}Hg$, mercury-196, $^{204}_{80}Hg$, $^{198}_{80}Hg$, Hg-200

   Answer: The isotopes should be arranged in order of increasing mass number, the number of heavy particles in the atoms.

   mercury-196, $^{198}_{80}Hg$, $^{199}_{80}Hg$, Hg-200, $^{202}_{80}Hg$, $^{204}_{80}Hg$

Work Problems

Correct the following statements:

1. Mass numbers can have fractional values, atomic numbers cannot.
2. What is the isotopic symbol for a neutral atom, X, which has 23 electrons and 26 neutrons?
3. If an element appears in nature in three isotopic forms, one-third of all the atoms of that element will be of each isotope.

Worked Solutions

1. **Neither mass numbers nor atomic numbers ever have fractional values.**

2. **$^{49}_{23}X$, the mass number is the sum of the protons (same as the number of electrons in a neutral atom) and neutrons.**

3. **The fraction of atoms of an element that is a particular isotope can range from near 0% to 100%.** For many elements, one isotope usually dominates and the other isotopes are present in smaller amounts.

The Atomic Mass (Atomic Weight) of the Elements

Individual atoms are so infinitesimally small it is impossible to weigh them on a balance. A single atom of oxygen-16 has a mass of 2.67×10^{-23} g. (If you could weigh one atom a second, it would take over 10^{13} centuries before you would reach 1.00 g of oxygen.) In the early 1800s, it was known that atoms of oxygen were heavier than atoms of hydrogen, that atoms of iron were heavier than atoms of oxygen, and so forth. John Dalton proposed a method of describing the weights of elements on a relative basis by comparing the weight of all the known elements to the weight of one element that was chosen as a standard. Oxygen became that standard, and the weights of other elements, as compared to the weight of oxygen, were called **atomic weights.** Oxygen, the benchmark, was *assigned* an atomic weight of 16.00. Since experiments indicated that carbon atoms were only $\frac{3}{4}$ as heavy as oxygen atoms, the atomic weight of carbon was then $\frac{3}{4}$ of 16.00, or 12.0. Sulfur atoms were about twice as heavy as oxygen atoms, so the atomic weight of sulfur was twice 16.00, or 32.0. Hydrogen was given an atomic weight of 1.00. Dalton's ingenious scheme of relative weights resulted in the determination of many, quite accurate, atomic weights, but in time, it became clear that a more refined system was needed, one that could deal with individual isotopes of an element. Yet, Dalton's idea of "relative weights of elements" has remained and is at the core of the modern atomic mass scale.

Dalton's atomic weight concept has been markedly improved with the development of the *mass spectrometer,* a sophisticated instrument that can accurately determine the mass of individual atoms. It is now more convenient to express the mass of atoms in terms of the atomic mass unit, amu.

$$1 \text{ amu} = 1.6606 \times 10^{-24} \text{ g}$$

The standard used in the modern scale of atomic mass is the carbon-12 isotope, which is assigned an atomic mass of *exactly* 12.000 amu. Oxygen is no longer used.

$$^{12}_{6}C = 12.000 \text{ amu (exactly)}$$

An atom with a mass exactly twice as large as that of the carbon-12 atom would be assigned an atomic mass exactly twice as large as that of carbon-12, 24.000 amu.

In nature, elements come to us as nonchanging mixtures of isotopes, and each isotope has its own atomic mass. In order to have an atomic mass that is accurate for an element and accounts for the natural mixture of its isotopes, it is necessary to determine what is called a "weighted-average" atomic mass. This sounds complicated, but it is the best way to get exact atomic masses of the elements.

To determine the weighted-average atomic mass of chlorine, which has two isotopes, Cl-35 and Cl-37, the natural abundance and the atomic mass of each isotope must be known. Both are given in the following table.

The Isotopes of Chlorine		
Isotope	***Natural Abundance***	***Atomic Mass***
Cl-35	75.771%	34.969 amu
Cl-37	24.229%	36.966 amu

The weighted-average atomic mass takes into account the fact that about 76% of the chlorine atoms are of the lighter isotope, Cl-35, and about 24%, the heavier. Here is how the atomic mass of chlorine is calculated, taking into account the fact that it is composed of two isotopes:

1. **Convert the percent natural abundance of each isotope into a decimal number by dividing each by 100%.**

   75.771% becomes 0.75771 and 24.229% becomes 0.24229

2. **Multiply the atomic mass of each isotope by the decimal equivalent of its natural abundance.**

   Cl-35 0.75771×34.969 amu = 26.496 amu

   Cl-37 0.24229×36.966 amu = 8.956 amu

3. **The atomic mass of chlorine is the sum of the contributions from the two isotopes.**

   atomic weight of Cl = 26.496 amu + 8.956 amu = 35.453 amu

   This is the "weighted-average" atomic mass of chlorine that appears on the periodic table.

   The modern definition of atomic mass (called atomic weight by John Dalton) can now be stated:

   The **atomic mass** of an element is the sum of the atomic masses of its isotopes, in amu, accounting for their natural abundance, relative to the atomic mass of carbon-12.

It is important *not* to confuse the mass number of an element with its atomic mass. Mass numbers are whole numbers that equal the number of protons plus neutrons in an atom. Atomic masses compare the relative masses of atoms. Atomic masses will always be decimal numbers.

The atomic mass is an extremely important property of an element and is listed along with its symbol on all periodic tables. On the other hand, the unit of atomic mass, the amu, is usually dropped in periodic tables. The unit is "understood" to be there.

17	←	atomic number
Cl	←	symbol
35.453	←	atomic mass

The smallest atomic mass is that of hydrogen, 1.00794 amu, and the largest values exceed 250 amu. Many atomic masses are expressed to several places after the decimal; others, for several reasons, may appear as whole numbers in parentheses, as is done with the highly radioactive elements that come after uranium on the periodic table. The true power and usefulness of the atomic mass is shown in Chapters 5 and 6.

Example Problems

These problems have both answers and solutions given.

1. Consulting the periodic table, what are the atomic masses of lithium, nitrogen, and silicon? Approximately how many times heavier are silicon atoms than lithium atoms or nitrogen atoms?

   Answer: The atomic masses are: Li = 6.941 amu, N = 14.00674 amu, and Si = 28.0855 amu, roughly a 7:14:28 ratio. Since the atomic mass of silicon is approximately four times that of lithium, silicon atoms are approximately four times heavier than lithium atoms. Using the same reasoning, silicon atoms are approximately twice as heavy as nitrogen atoms.

2. Knowing that 1 amu = 1.6606×10^{-24} g, what is the mass on one atom of $^{12}_{6}C$?

   Answer: By definition, the atomic mass of carbon-12 is exactly 12.000 amu. Its mass in grams is:

$$\text{mass} = (12.000\,\cancel{\text{amu}}) \times \left(\frac{1.6606 \times 10^{-24}\text{ g}}{1\,\cancel{\text{amu}}}\right) = 1.9927 \times 10^{-23}\text{ g}$$

3. Consider a hypothetical element, X, that has three isotopes. Calculate the weighted-average atomic mass of the element from the following data.

Isotope	**Abundance**	**Atomic Mass**
X-85	10.00%	85.32 amu
X-87	70.00%	87.51 amu
X-88	20.00%	88.10 amu

   Answer:

X-85	0.1000×85.32 amu =	8.532 amu
X-87	0.7000×87.51 amu =	61.257 amu
X-88	0.2000×88.10 amu =	17.620 amu
		87.409 amu

   The weighted-average atomic mass of X = 87.41 amu (to 4 sig. figs.)

4. An element occurs in nature in two isotopic forms. From the data given below, calculate the atomic mass of the element and compare your value with those on the periodic table to determine the identity of the element.

Atom	**Abundance**	**Atomic Mass**
isotope 1	19.78%	10.0128 amu
isotope 2	80.22%	11.0093 amu

   Answer: isotope 1: 0.1978×10.0128 amu = 1.9806 amu

   isotope 2: 0.8022×11.0093 amu = 8.8317 amu

   10.8123 amu

To 4 significant figures: 10.81 amu

Checking the periodic table shows the element to be boron, B.

Work Problems

1. Consult the periodic table to answer the following:

   (a) What element has atoms that are approximately twice as heavy as those of neon?

   (b) The atoms of what element are only about one-third as heavy as those of titanium?

2. Knowing that 1 amu = 1.6606×10^{-24} g, what is the atomic mass, in amu, of an isotope of copper that has a mass of 1.045×10^{-22} g?

3. There are two naturally occurring isotopes of lithium, $^{6}_{3}Li$ and $^{7}_{3}Li$. Calculate the atomic mass of lithium as it would appear on the periodic table.

Isotope	**Abundance**	**Atomic Mass**
Li-6	7.49%	6.0151 amu
Li-7	92.51%	7.0160 amu

Worked Solutions

1. **(a) The atomic mass of neon, Ne, is approximately 20 amu, so the element with an atomic mass approximately twice as large, 40 amu, is calcium.**

   (b) The atomic mass of titanium, Ti, is 47.88 amu. One-third of this is approximately 16. The element is oxygen, atomic mass 15.9994 amu.

2. $$\textbf{atomic mass} = \left(1.045 \times 10^{-22}\,\cancel{g}\right) \times \left(\frac{1\,\text{amu}}{1.6606 \times 10^{-24}\,\cancel{g}}\right) = 62.93\,\text{amu}$$

3. **For Li-6 0.0749×6.0151 amu = 0.45053 amu**

   For Li-7 0.9251×7.0160 amu = 6.49050 amu

   Atomic mass = 0.45053 amu + 6.49050 amu = 6.94103 amu = 6.94 amu (to 3 sig. figs.)

Ions from Atoms

Atoms can gain or lose 1 or more electrons in chemical reactions to form ions, species bearing an electrical charge. When this happens, the composition of the nucleus is not affected. The identity of the element does not change, because the number of protons in the nucleus does not change. Ions that bear a positive charge arise when an atom *loses* 1 or more electrons. Consider the element sodium, atomic number 11. A sodium atom, Na, commonly loses 1 electron in chemical reactions to become a positive ion, Na^+, a sodium ion.

$$\underset{\substack{(11\,e^-,\ 11\,p^+)\\ \text{(neutral)}}}{Na} \rightarrow \underset{\substack{(10\,e^-,\ 11\,p^+)\\ \text{(+ ion)}}}{Na^+} + e^-$$

The neutral sodium atom has 11 protons in its nucleus and 11 electrons about it. Neutrons are neutral and don't affect the charge and, for that reason, aren't considered in this discussion. Losing 1 electron creates a charge imbalance; 11 protons in the nucleus with only 10 electrons about it. The ion has an excess of 1 positive charge, thus it is an ion with a single positive charge and is written as Na^+. But note that it is still the same *element* it was before losing the electron. The atomic number has not changed; there are still 11 protons in the nucleus, and that means it is an ion of sodium.

Sodium is in group IA on the left edge of the periodic table. All the metals in group IA lose 1 electron in chemical reactions, just like sodium, to form ions with a single positive charge, 1+. Each 1+ ion has 1 less electron *about* its nucleus than it has protons *in* its nucleus. The names of the ions are simply the names of the metal followed by "ion", like lithium ion, Li^+, or cesium ion, Cs^+.

The Ions of the Metals of Group IA

The Element			***The Ion***			
Lithium	Li	→	Li^+	+	e^-	Lithium ion
Sodium	Na	→	Na^+	+	e^-	Sodium ion
Potassium	K	→	K^+	+	e^-	Potassium ion
Rubidium	Rb	→	Rb^+	+	e^-	Rubidium ion
Cesium	Cs	→	Cs^+	+	e^-	Cesium ion

The metals in group IIA lose 2 electrons in chemical reactions. Calcium, a group IIA metal, loses 2 electrons and forms an ion with a 2+ charge.

$$Ca \rightarrow Ca^{2+} + 2\,e^-$$

$$\begin{array}{cc} (20\ e^-,\ 20\ p^+) & (18\ e^-,\ 20\ p^+) \\ \text{(neutral)} & \text{(2+ ion)} \end{array}$$

All the elements in group IIA behave the same way, losing two electrons in chemical reactions to form ions with a 2+ charge. Each ion has two more protons in the nucleus than electrons outside the nucleus. The names of the ions are the names of the metal followed by "ion" as in barium ion, Ba^{2+}.

The Ions of the Metals of Group IIA

The Element			***The Ion***			
Beryllium	Be	→	Be^{2+}	+	$2e^-$	Beryllium ion
Magnesium	Mg	→	Mg^{2+}	+	$2e^-$	Magnesium ion
Calcium	Ca	→	Ca^{2+}	+	$2e^-$	Calcium ion
Strontium	Sr	→	Sr^{2+}	+	$2e^-$	Strontium ion
Barium	Ba	→	Ba^{2+}	+	$2e^-$	Barium ion

Ions that carry a positive charge are collectively called **cations.** Metals tend to lose electrons in chemical reactions to become cations.

Ions that bear a negative charge arise when atoms *gain* 1 or more electrons. The element fluorine, F, readily gains 1 electron in chemical reactions with metals to form the fluor*ide* ion, F^-. The fluoride ion has 10 negative electrons around a nucleus with only 9 positive protons, an excess of a single negative charge.

$$F + e^- \rightarrow F^-$$

$$(9\ e^-, 9\ p^+) \quad (10\ e^-, 9\ p^+)$$

$$\text{(neutral)} \quad (1-\ \text{ion})$$

Fluorine is in group VIIA on the periodic table. All the elements in group VIIA (the **halogens**) form ions with a single negative charge when they react with metals. The names of the ions end in *-ide,* as in fluoride, F^-, chloride, Cl^-, bromide, Br^-, and iodide, I^-.

The Ions of the Nonmetals of Group VIIA

The Element	*The Ion*					
Fluorine	F	+	e^-	→	F^-	Fluoride ion
Chlorine	Cl	+	e^-	→	Cl^-	Chloride ion
Bromine	Br	+	e^-	→	Br^-	Bromide ion
Iodine	I	+	e^-	→	I^-	Iodide ion

Ions that carry a negative charge are collectively called **anions.** Nonmetals tend to gain electrons in reactions with metals to form anions. Remember, whether electrons are lost or gained by an atom, the name of the resulting ion is still associated with the name of the element from which it came. Again, only the electrons are involved, not the nucleus.

When an atom loses or gains electrons, it forms a **monatomic ion,** an ion from a single atom. (All the ions described in this section are monatomic ions.) The chemical properties of an ion are not at all like those of the atom from which it came; the electrical charge becomes the dominant property. A sodium atom, Na, and a sodium ion, Na^+, are chemically very different. Sodium atoms could never exist in contact with water, for example, because they react vigorously with water, but sodium ions exist in water without a problem.

Example Problems

1. Give the number of protons and electrons in each of the following:

Ion	*Number of Protons*	*Number of Electrons*
Al^{3+}		
O^{2-}		
P^{3-}		
Fe^{2+}		
Fe^{3+}		
Ag^+		

Answer: The number of protons equals the atomic number of the element (found on the periodic table), and the number of electrons is less than the atomic number for cations and greater than the atomic number for anions.

Ion	Number of Protons	Number of Electrons
Al^{3+}	13	10
O^{2-}	8	10
P^{3-}	15	18
Fe^{2+}	26	24
Fe^{3+}	26	23
Ag^{+}	47	46

2. When determining the charge on an ion, only protons and electrons are considered, not the neutrons in the nucleus. Why?

   Answer: The neutron has no charge and cannot affect the charge or lack of charge of a species. It's the comparison of positive protons and negative electrons that determines the charge on an ion.

3. Which pair of species would be expected to have similar chemical properties?

   (a) ${}^{6}_{3}Li$ and ${}^{7}_{3}Li^{+}$ (b) ${}^{16}_{8}O$ and ${}^{18}_{8}O$ (c) ${}^{37}_{17}Cl$ and ${}^{35}_{17}Cl^{-}$

   Answer: (b) ${}^{16}_{8}O$ and ${}^{18}_{8}O$ are both neutral isotopes of oxygen and have the same chemical properties. The other pairs are also isotopes of one another, but in each case, one is neutral and the other an ion. Monatomic ions are not similar chemically to neutral atoms of the same element.

Work Problems

1. Give the number of protons and electrons in each of the following:

Ion	Number of Protons	Number of Electrons
Se^{2-}		
Ni^{2+}		
Bi^{3+}		
N^{3-}		
Mn^{2+}		
H^{+}		

2. Why do the number of protons in the nucleus remain constant when the atom becomes an ion?

3. Which pair of species should most resemble each other chemically?

(a) $^{79}_{35}Br$ and $^{81}_{35}Br^-$ (b) $^{52}_{24}Cr$ and $^{52}_{24}Cr^{3+}$ (c) $^{28}_{14}Si$ and $^{30}_{14}Si$

Worked Solutions

1.

Ion	*Number of Protons*	*Number of Electrons*
Se^{2-}	34	36
Ni^{2+}	28	26
Bi^{3+}	83	80
N^{3-}	7	10
Mn^{2+}	25	23
H^+	1	0

Note, the hydrogen ion, H^+, is just the nucleus of the hydrogen atom.

2. **Altering the composition of a nucleus requires much more energy than is involved in chemical reactions, and so the nucleus remains unchanged.** We don't have nuclear reactions in chemistry. Much less energy is involved in the gain or loss of electrons.

3. **(c) $^{28}_{14}Si$ and $^{30}_{14}Si$ are both neutral isotopes of silicon and have the same chemical properties.** The other choices are neutral atoms and ions of the same elements, which do not share the same chemical properties.

Chapter Problems and Answers

Problems

1. What is an atom?

2. Describe the three subatomic particles, including

   (a) Its name and symbol

   (b) The location in an atom

   (c) The electrical charge of the particle

   (d) The mass of the particle compared to the others

3. In terms of subatomic particles, all atoms of a given element have one thing in common. What is it?

4. In the early 1800s, John Dalton thought that all atoms of an element were identical in all ways. In what way was his idea correct? In what way was it incorrect?

5. In what way do isotopes of an element differ from one another? In what way are they similar?

6. Which of the following are isotopes of one another?

$^{176}_{72}X$ $^{180}_{71}X$ $^{177}_{72}X$ $^{180}_{70}X$ $^{176}_{72}X$ $^{176}_{70}X$

7. Write the correct isotopic symbol for a neutral atom that contains 39 electrons around a nucleus containing 44 neutrons.

8. How many neutrons are in the nucleus of an atom of copper-63?

9. What is the mass number of a tin atom, Sn, that has 68 neutrons?

10. How many protons are in the nucleus of the C-12 isotope?

11. Complete the table for each neutral atom. Consult the periodic table as needed.

Symbol of the Element	*Atomic Number*	*Mass Number*	*Number of Electrons*	*Number of Protons*	*Number of Neutrons*
Ba					82
	8				10
			33		42
Ca		40			

12. Atomic masses of the elements are called "relative masses." What does this mean and to what isotope are the atomic masses relative?

13. Which two go together? $^{27}_{13}Al$, aluminum-13, Al-27, aluminum-14

14. What is meant by the term "weighted-average atomic mass?"

15. Consider a hypothetical element, M, which has three isotopes. Calculate the weighted-average atomic mass of the element from the following data.

Isotope	**Abundance**	**Atomic Mass**
M-92	18.00%	91.98 amu
M-94	55.00%	93.82 amu
M-95	27.00%	94.79 amu

16. Consider another hypothetical element, D, that has three isotopes. Calculate the weighted-average atomic mass of the element from the following data.

Isotope	**Abundance**	**Atomic Mass**
D-179	7.000%	178.91 amu
D-177	42.00%	176.84 amu
D-176	51.00%	175.92 amu

17. An element exists in nature as two isotopes, which are described below. Calculate the weighted-average atomic mass of the element then consult the periodic table and identify the element.

Atom	Abundance	Atomic Mass
Isotope 1	72.15%	84.912 amu
Isotope 2	27.85%	86.909 amu

18. One isotope of a nonmetal has a mass number of 79, has 45 neutrons in its nucleus, and has a charge of 2–. Write the complete isotopic symbol of the ion.

19. Give the number of protons and electrons in each of the following:

Ion	*Number of Protons*	*Number of Electrons*
In^{3+}		
V^{2+}		
I^{-}		
K^{+}		
Ti^{2+}		
Br^{-}		

20. Six species are described in terms of the subatomic particles that make them up.

(a) Which are neutral species?

(b) Which are cations?

(c) Which are anions?

(d) Write the complete symbol for each species, A through F.

Species	*Number of Electrons*	*Number of Protons*	*Number of Neutrons*
A	18	15	16
B	18	18	22
C	10	13	14
D	78	80	122
E	2	3	4
F	76	78	117

Answers

1. **An atom is the smallest particle of an element.**

2. The **electron,** e^-, is located outside the nucleus and bears a single negative charge. It is much lighter than the proton or neutron.

   The **proton,** p^+, is located in the nucleus and bears a single positive charge. It is over 1,800 times heavier than the electron and a little lighter then the neutron.

   The **neutron,** n, is located in the nucleus and bears no electrical charge. Its mass is just a little greater than that of the proton.

3. **All atoms of the same element have the same atomic number, the same number of protons in the nucleus.**

4. **The atoms of an element are similar in their chemical properties but they may differ from one another in the number of neutrons in the nucleus; that is, the element may exist as a constant mixture of isotopes.** Dalton did not know about isotopes.

5. **Isotopes have the same atomic number but different mass number.** They differ only in the number of neutrons in the nucleus.

6. **Only $^{176}_{72}X$ and $^{177}_{72}X$ are isotopes.** Two of the species are identical and not isotopes.

7. **$^{83}_{39}Y$; in a neutral atom the number of electrons equals the number of protons, which is the atomic number, 39.** The mass number is the number of protons plus the number of neutrons, 83.

8. **34 neutrons; mass number = 63, atomic number = 29; (63 – 29) = 34.**

9. **Mass number = 118; atomic number of Sn = 50, add to that the number of neutrons.**

10. **6 neutrons; mass number = 12, atomic number = 6; (12 – 6) = 6 neutrons.**

11.

Element Symbol	*Atomic Number*	*Mass Number*	*Number of Electrons*	*Number of Protons*	*Number of Neutrons*
Ba	*56*	*138*	*56*	*56*	82
O	8	*18*	*8*	*8*	10
As	*33*	*75*	33	*33*	42
Ca	*20*	40	*20*	*20*	*20*

12. **A relative weight is one compared to the weight of an adopted standard to which a weight is assigned.** The standard for atomic mass is the carbon-12 isotope, which is assigned a mass of exactly 12.000 amu. If an atom had a mass only half as large as the carbon-12 atom, its atomic mass would be half of 12.000 amu, or 6.000 amu.

13. **$^{27}_{13}Al$ and Al-27 both represent the same isotope.** The others indicate the number of protons or neutrons after the name of the element.

14. **A weighed-average atomic mass is one that is determined from the atomic mass of all the isotopes that make up that element according to their natural abundance.** The contribution of each isotope to the final atomic mass of the element is directly related to the percent that isotope contributes to the total, as well as the specific atomic mass of that isotope.

15. **M-92 0.1800 × 91.98 amu = 16.556 amu**

M-94 0.5500 × 93.82 amu = 51.601 amu

M-95 0.2700 × 94.79 amu = 25.593 amu

93.750 amu

The weighted-average atomic mass of M = 93.75 amu (to 4 sig. figs.).

16. **D-179 0.07000 × 178.91 amu = 12.524 amu**

D-177 0.4200 × 176.84 amu = 74.273 amu

D-176 0.5100 × 175.92 amu = 89.719 amu

176.516 amu

The weighted-average atomic mass of D = 176.5 amu (to 4 sig. figs.).

17. **isotope 1 0.7215 × 84.912 amu = 61.264 amu**

isotope 2 0.2785 × 86.909 amu = 24.204 amu

85.468 amu

The weighted-average atomic mass equals 85.47 amu (to 4 sig. figs.).

The element is rubidium, Rb.

18. $^{79}_{34}Se^{2-}$**; subtracting the number of neutrons from the mass number gives the atomic number of the element, selenium.**

19.

Ion	*Number of Protons*	*Number of Electrons*
In^{3+}	*49*	*46*
V^{2+}	*23*	*21*
I^-	*53*	*54*
K^+	*19*	*18*
Ti^{2+}	*22*	*20*
Br^-	*35*	*36*

20. **Six species are described in terms of the subatomic particles that make them up.**

(a) B is the only neutral atom.

(b) C, D, E, and F are all cations.

(c) A is the only anion.

(d) A is $^{31}_{15}P^{3-}$; B is $^{40}_{18}Ar$; C is $^{27}_{13}Al^{3+}$; D is $^{202}_{80}Hg^{2+}$; E is $^{7}_{3}Li^{+}$; F is $^{195}_{78}Pt^{2+}$.

Supplemental Problems

Problems

1. What subatomic particle gives the identity to an atom?

2. Fill the spaces in the following table:

Symbol	*Protons*	*Neutrons*	*Electrons*	*Charge*
$^{108}_{47}Ag^{+}$				
	20	20		2+
	5	6	5	
	79	117	76	
		16	18	3–

3. Which symbol provides more information about the atom? ^{195}Pt or $_{78}Pt$

4. What do isotopes have in common?

5. Show three ways you can indicate an isotope of tin that contains 69 neutrons.

6. Why is the atomic mass of carbon not exactly 12.000 amu on the periodic table?

7. Silicon has three naturally occurring isotopes listed below. Calculate the weighted-average atomic mass of the element to 4 significant figures.

Isotope	**Abundance**	**Atomic Mass**
Si-28	92.18%	27.997 amu
Si-29	4.71%	28.977 amu
Si-30	3.12%	29.974 amu

8. What do Ca and Ca^{2+} have in common? What do they not have in common?

9. In what way does the number of neutrons in a nucleus affect the charge on an ion?

10. Many elements exist in nature as a mixture of its isotopes, yet this mixture has a unique property that other mixtures, like one of black and orange jellybeans, do not share. Thinking in terms of the definition of a mixture, what is that property?

Answers

1. **The number of protons in the nucleus.** (page 65)

2. **The completed table.** (page 65)

Symbol	*Protons*	*Neutrons*	*Electrons*	*Charge*
$^{108}_{47}Ag^{+}$	*47*	*61*	*46*	*1+*
$^{40}_{20}Ca^{2+}$	20	20	*18*	2+
$^{11}_{5}B$	5	6	5	*0*
$^{196}_{79}Au^{3+}$	79	117	76	*3+*
$^{31}_{15}P^{3-}$	*15*	16	18	3–

3. **^{195}Pt shows both the mass number and the symbol of the element, which allows you to get the atomic number from the periodic table.** There is no way of knowing the mass number for the second symbol. (page 65)

4. **They have the same atomic number**. (page 65)

5. **The mass number is 119 (69 + 50); Sn-119, tin-119, and $^{119}_{50}Sn$.** (page 65)

6. **The atomic mass values on the periodic table are the weighted-average values obtained from the mixture of isotopes of the element.** Carbon-12 is not the only isotope of carbon; there are small fractions of two heavier isotopes. (page 68)

7. **28.11 amu.** (page 70)

8. **They are both calcium, same atomic number; they do not have similar chemical properties, because one is the neutral atom and the other a 2+ charged ion.** (page 74)

9. **The number of neutrons has no effect on the charge of an ion.** (page 74)

10. **Whereas mixtures of black and orange jellybeans can vary widely in terms of percent black and percent orange, mixtures of isotopes as found in nature do not change; the percent composition of isotopes is constant over time.** (page 68)

Chapter 4

Formulas and Names of Ionic Compounds, Acids, and Bases

Although many compounds are molecular (as presented in Chapter 2), a very large number are best described as ionic, those composed of ions. In this chapter, the focus is on the formulas and names of ionic compounds. Nearly all ionic compounds are composed of both metals and nonmetals. Remember that metals have a great tendency to lose electrons in chemical reactions to form positive ions (cations), while nonmetals, in their reaction with metals, have a great tendency to gain electrons to form negative ions (anions). These ions of opposite charge combine to form ionic compounds.

Formulas of Binary Ionic Compounds

A **binary ionic compound** is one composed of ions from only two elements, one a metal, the other a nonmetal. NaCl, CaF_2, and $FeCl_3$ are all binary ionic compounds. Before starting to develop formulas for ionic compounds, it is necessary to know the monatomic ions (ions formed from a single atom) of the common metals and nonmetals.

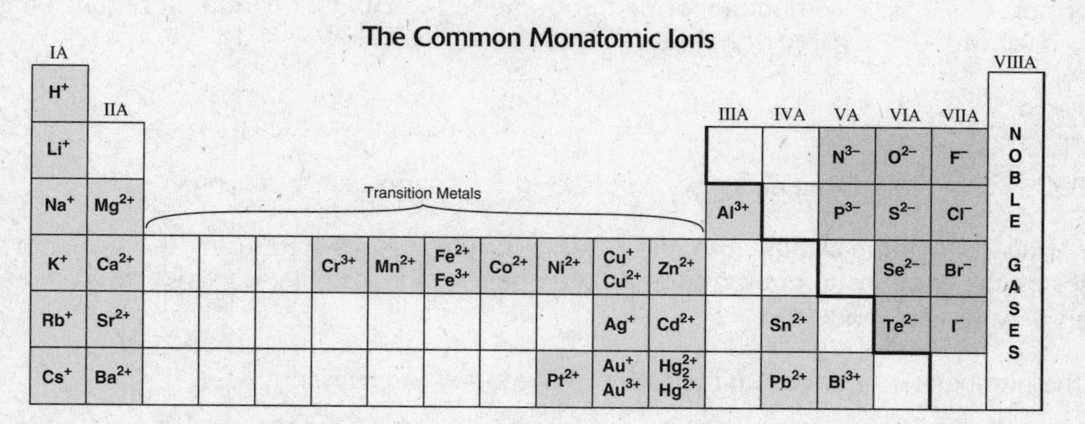

Displaying the ions in their locations on the periodic table can help you remember the charge on the ions. First, look only at the main-group elements, those in the A-groups. All the cations in group IA have a charge of 1+, a single positive charge. Those in group IIA have a 2+ charge. The only frequently seen metal ion from group IIIA is the aluminum ion with a 3+ charge. The two metals at the bottom of group IVA, tin and lead, form 2+ ions. If you have noticed the trend, you might expect tin and lead, because they are in group IVA, to form ions with a 4+ charge, but it is exceedingly difficult to form ions with a 4+ charge; those with a 2+ charge form readily.

From Group VA through VIIA, the elements that are most important are the nonmetals that form negative ions (anions). The elements in group VIIA, the halogens, form ions with a 1– charge; those in group VIA, form ions with a 2– charge, and nitrogen and phosphorus in group VA form ions with a 3– charge.

A Summary of the Ions of the Main-Group Elements

	Metals				*Nonmetals*		
Group	IA	IIA	IIIA	IVA	VA	VIA	VIIA
Charge	1+	2+	3+	2+	3–	2–	1–

The transition metals in the center of the periodic table form cations with charges from 1+ to 3+, but not in any particular easy-to-remember way. Several form ions with two different charges, like iron, Fe^{2+} and Fe^{3+}, copper, Cu^{+} and Cu^{2+}, and mercury, Hg_2^{2+} and Hg^{2+}. (Note that Hg_2^{2+} is a diatomic cation, two Hg^{+} ions bonded together acting as a single unit, Hg^{+}—Hg^{+}.)

There are three rules to follow when writing formulas of binary ionic compounds:

1. The symbol of the cation always comes before the symbol of the anion in the formula.

   There is no space between the ions.
2. The sum of the *positive charges* on the cations *must equal* the sum of the *negative charges* on the anions so that the formula of the compound is electrically neutral.
3. Whole numbers, written as subscripts, indicate the number of each ion in the formula unit. The *smallest* set of numbers must be used.

Table salt is a familiar ionic compound formed from sodium and chlorine. Sodium forms an ion with a single positive charge, Na^{+}, and chlorine, an ion with a single negative charge, Cl^{-}. A neutral formula for the compound requires one sodium ion, Na^{+}, for each chloride ion, Cl^{-}, and so the formula is written, NaCl. The formula tells us that in a crystal of NaCl, there is 1 Na^{+} for every Cl^{-}, and vice versa.

The formula of the ionic compound formed from the Fe^{3+} ion and Cl^{-} is $FeCl_3$. It requires three Cl^{-} ions to balance the 3+ charge of a single Fe^{3+} ion, as shown below:

Fe^{3+}	= 3+	
Cl^{-} Cl^{-} Cl^{-}	= 3–	
$FeCl_3$	0	A net charge of zero shows the charges are balanced.

Another way of figuring out the formulas of ionic compounds involves using the size of the charges on the two ions to get the subscript numbers for the formula. Consider the compound formed between Fe^{3+} and O^{2-}:

The number of + charges on the cation, 3, becomes the subscript

Start with Fe^{3+} O^{2-} and add arrows Fe^{3+} O^{2-}

of the anion, O_3, and the number of – charges on the anion, 2, becomes the subscript of the cation Fe_2.

Fe_2O_3 is the correct formula for the compound.

Fe^{3+} Fe^{3+}	= 6+	
O^{2-} O^{2-} O^{2-}	= 6 –	
Fe_2O_3	0	A net charge of zero shows the charges are balanced.

Example Problems

Use the two ions to write a correct formula for an ionic compound.

1. Mg^{2+} and O^{2-}

   Answer: MgO

   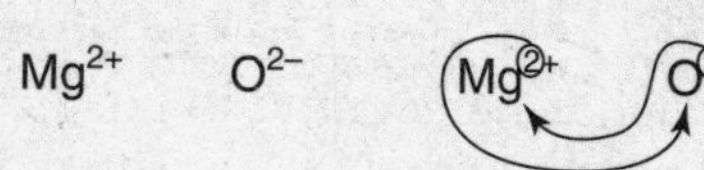

   Mg_2O_2 = MgO (smallest set of subscript numbers)

2. Na^{+} and N^{3-}

   Answer: Na_3N

   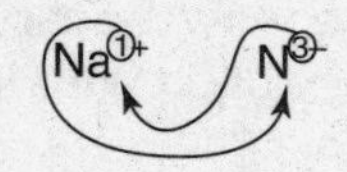

   Na_3N_1 = Na_3N (the subscript 1 is usually omitted)

3. Hg_2^{2+} and Cl^{-}

   Answer: Hg_2Cl_2

   Hg_2^{2+} Cl^{1-}

   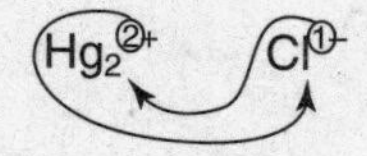

   Hg_2^{2+} is a diatomic cation. Its 2+ charge is balanced by the charge of two chloride ions. One Hg_2^{2+} ion and two Cl^{-} ions give the smallest set of subscripts numbers. Hg_2^{2+} is the only common diatomic cation.

Work Problems

Use the two ions to write a correct formula for an ionic compound.

1. Cd^{2+} and Br^{-}
2. Cu^{2+} and P^{3-}
3. Ag^{+} and S^{2-}

Worked Solutions

1. **$CdBr_2$; it requires 2 Br^- to balance the charge of 1 Cd^{2+} ion.**

2. **Cu_3P_2**

   Cu^{2+} P^{3-} Cu^{2+} P^{3-}

   3 Cu^{2+} ions balance 2 P^{3-} ions (6+, 6−).

3. **Ag_2S; it requires 2 Ag^+ to balance the charge of 1 S^{2-} ion.**

Naming Binary Ionic Compounds

Over the centuries, many ionic compounds have come to be known by common names, such as lime (CaO), salt (NaCl), galena (PbS), and alumina (Al_2O_3). The names are fine as long as everyone knows what they stand for, but they do not say anything about the chemical makeup of the compound. In chemistry, the name of a compound must convey enough information to allow a person to write the correct formula of the compound from that name. Today, proper names of compounds are derived using rules set forth by international committees, most notably the International Union of Pure and Applied Chemistry (IUPAC). Names derived using these rules are called **systematic names,** and they are understood the world over. Developing systematic names for binary ionic compounds is not done in the same way described in Chapter 2 for binary molecular compounds. The Greek numerical prefixes are not used with ionic compounds.

There are two classes of metals when it comes to the ions they form.

- **Single-cation metals** are those that form only a single ion. These are the main-group metals, and a few others. Sodium, Na, is a typical example. It forms only one cation, Na^+, in chemical reactions. It never forms a cation with any other charge.
- **Multiple-cation metals** are those that form ions of two or more charges. Most of the transition metals are of this type. Iron, a typical transition metal, can form ions with both a 2+ and 3+ charge, Fe^{2+} and Fe^{3+}.

Each category of metals requires a different scheme for naming their compounds. You start by learning to name ionic compounds involving metals that only form only a single cation.

Naming Compounds Involving Single-Cation Metals

Naming binary ionic compounds in which the metal forms only a single cation is not difficult. The name of the compound is the name of the cation followed by the name of the anion. Doing the reverse—that is, writing the formula of the compound from its name—can be a bit more challenging. In order to write a correct formula, you must know the charge on both the cation and the anion, and this isn't revealed in the name. It is assumed that the person reading the name would know about the charges on the cations and anions. So you can see why it is important to know both the names and charge of the main-group ions.

The names of the cations of the main-group elements, no matter what charge the ion bears, is the name of the metal followed by the word "ion":

Li^+ lithium ion Ca^{2+} calcium ion Al^{3+} aluminum ion

On the other hand, when you read "aluminum ion" or "aluminum" in a name, you have to *know* that the charge on that aluminum ion is 3+; for "calcium ion" it's 2+; and so forth. A review of the charges of the single-cation metals is in order. Although zinc, cadmium, and silver are in the B-groups and not considered main-group elements, they are metals that form a single cation, and are included here also.

- **Group IA metals:** Ions are always 1+ (Li^+, Na^+, K^+, Rb^+, Cs^+)
- **Group IIA metals:** Ions are always 2+ (Be^{2+}, Mg^{2+}, Ca^{2+}, Sr^{2+}, Ba^{2+})
- **Aluminum:** Always 3+ (Al^{3+})
- **Zinc and cadmium:** Always 2+ (Zn^{2+}, Cd^{2+})
- **Silver:** Always 1+ (Ag^+)

The names of the anions are derived from the name of the element by adding *-ide* to the stem of the element's name. A list of the monatomic anions and their names are given in the following table.

The Monatomic Anions

Anion	*Stem*	*Ion Name*
N^{3-}	nitr-	Nitride ion
P^{3-}	phos-	Phosphide ion
O^{2-}	ox-	Oxide ion
S^{2-}	sulf-	Sulfide ion
Se^{2-}	selen-	Selenide ion
Te^{2-}	tellur-	Telluride ion
F^-	fluor-	Fluoride ion
Cl^-	chlor-	Chloride ion
Br^-	brom-	Bromide ion
I^-	iod-	Iodide ion

The rules for naming binary ionic compounds with single-cation metals are as follows:

1. The name of the cation comes first, followed by the name of the anion separated by a space; this is the same order in which they appear in the formula.
2. The name of the cation is the name of the metal.
3. The name of the anion ends in *-ide,* as shown in the preceding table.

KBr – potassium bromide; Al_2O_3 – aluminum oxide; Na_2S – sodium sulfide

It is not necessary to capitalize the names of compounds unless they are at the beginning of a sentence. They may be capitalized in tables and lists of compounds.

Example Problems

Correctly name these ionic compounds:

1. LiCl and $CaCl_2$

   Answer: lithium chloride and calcium chloride

2. Mg_3N_2 and Na_3N

   Answer: magnesium nitride and sodium nitride

3. AlP and AgCl

   Answer: aluminum phosphide and silver chloride

Work Problems

Correctly name these ionic compounds:

1. KBr and ZnO
2. Rb_2O and SrI_2
3. Ba_3P_2 and CdS

Worked Solutions

1. **potassium bromide and zinc oxide**
2. **rubidium oxide and strontium iodide**
3. **barium phosphide and cadmium sulfide**

Naming Compounds Involving Multiple-Cation Metals

Most of the transition metals are able to form two or more ions, and so there needs to be a way to indicate which ion is present in the name of the compound. The systematic name of the cation is developed by following the name of the metal with its charge *written as a Roman numeral in parentheses*. This is the **Stock system** for naming ions. Yet an older, nonsystematic way of naming these ions still hangs on, the **Classical system,** which uses the *-ous* and *-ic* ending on the name of the metal to indicate the lower and higher charge, respectively. Its use is discouraged, although it is still seen. The ions of several transition metals with their Stock and Classical names appear in the following table.

Several Multiple-Cation Metals and Their Names		
Cation	***Stock Name***	***Classical Name***
Fe^{2+}	Iron(II) ion	Ferrous ion
Fe^{3+}	Iron(III) ion	Ferric ion

Cation	*Stock Name*	*Classical Name*
Cu^{+}	Copper(I) ion	Cuprous ion
Cu^{2+}	Copper(II) ion	Cupric ion
Hg_2^{2+}	Mercury(I) ion	Mercurous ion
Hg^{2+}	Mercury(II) ion	Mercuric ion
Cr^{2+}	Chromium(II) ion	Chromous ion
Cr^{3+}	Chromium(III) ion	Chromic ion
Co^{2+}	Cobalt(II) ion	Cobaltous ion
Co^{3+}	Cobalt(III) ion	Cobaltic ion
Ni^{2+}	Nickel(II) ion*	None
Sn^{2+}	Tin(II) ion*	Stannous ion
Pb^{2+}	Lead(II) ion*	Plumbous ion

**You will not see other ions of these metals, but Roman numerals will be used.*

Three ionic compounds are named below using both the Stock and Classical systems for naming cations.

Cation	*Stock Name*	*Classical Name*
$CuCl$	copper(I) chloride	cuprous chloride
Co_2S_3	cobalt(III) sulfide	cobaltic sulfide
$CrBr_3$	chromium(III) bromide	chromic bromide

Clearly, the Stock names give more information, because they tell you the charge on the metal ion. The Classical method is useful only if you know beforehand what the *-ous* and *-ic* endings mean for each metal. For clarity, the Stock names are emphasized here.

Example Problems

Use the Stock method to name these ionic compounds involving multiple-cation metals.

1. FeO and Fe_2O_3

   Answer: iron(II) oxide and iron(III) oxide; knowing that the oxide ion bears a 2– charge, O^{2-}, the charge of the iron ion must be 2+ in FeO to give a neutral formula. In Fe_2O_3, 3 oxide ions total to 6 negative charges that must be balanced by 2 iron ions bearing a 3+ charge.

2. $CrCl_3$ and Hg_2Cl_2

   Answer: chromium(III) chloride and mercury(I) chloride; the chloride ion, Cl^{-}, bears a single negative charge, and 3 of them total 3 negative charges. A neutral formula then requires that the chromium ion is Cr^{3+}. In the second compound, the subscript 2 on Hg indicates the mercury(I) ion, Hg_2^{2+}. This is reinforced by the need for 2 chloride ions to total 2 negative charges.

3. CuO and NiS

Answer: copper(II) oxide and nickel(II) sulfide; both the oxide and sulfide ions bear a 2– charge, O^{2-} and S^{2-}. Both copper and nickel must be 2+ to form a neutral formula.

Work Problems

Use the Stock method to name these ionic compounds involving multiple-cation metals.

1. $CoCl_2$ and Ni_3N_2
2. FeI_2 and $HgBr_2$
3. Co_2O_3 and MnO

Worked Solutions

The names are given with the set of ions needed to give a neutral formula.

1. **cobalt(II) chloride; Co^{2+} and 2 Cl^-**

   nickel(II) nitride; 3 Ni^{2+} and 2 N^{3-}

2. **iron(II) iodide; Fe^{2+} and 2 I^-**

   mercury(II) bromide; Hg^{2+} and 2 Br^-

3. **cobalt(III) oxide; 2 Co^{3+} and 3 O^{2-}**

   manganese(II) oxide; Mn^{2+} and O^{2-}

The Polyatomic Ions

Up to this point, only ions derived from single atoms have been used since the focus has been on binary ionic compounds. But there are many ions, almost all of them anions, that contain two or more atoms, and they are the polyatomic ions. **Polyatomic ions** contain two or more atoms chemically bound into a single unit that bears an electrical charge. The most common polyatomic ions are listed in the following table.

The Polyatomic Ions

Single Charged Ions	*Double Charged Ions*	*Triple Charged Ions*
Ammonium ion, NH_4^+	Sulfate ion, – SO_4^{2-}	Phosphate ion, PO_4^{3-}
Hydroxide ion, OH^-	Sulfite ion, SO_3^{2-}	
Cyanide ion, CN^-	Carbonate ion, CO_3^{2-}	
Nitrate ion, NO_3^-	Oxalate ion, $C_2O_4^{2-}$	
Nitrite ion, NO_2^-	Chromate ion, CrO_4^{2-}	

Single Charged Ions	*Double Charged Ions*	*Triple Charged Ions*
Hydrogencarbonate or bicarbonate ion, HCO_3^-	Dichromate ion, $Cr_2O_7^{2-}$	
Acetate ion, $C_2H_3O_2^-$	Hydrogenphosphate ion, HPO_4^{2-}	
Permanganate ion, MnO_4^-		
Hypochlorite ion, ClO^-		
Chlorite ion, ClO_2^-		
Chlorate ion, ClO_3^-		
Perchlorate ion, ClO_4^-		
Hydrogensulfate ion, HSO_4^-		
Dihydrogenphosphate ion, $H_2PO_4^-$		

There is one polyatomic ion that has a positive charge: the ammonium ion, NH_4^+. All the others bear negative charges. Most of the anions are **oxyions,** which combine one or more oxygen atoms with another element as seen in the carbonate ion, CO_3^{2-}, and the nitrate ion, NO_3^-. Though over 20 polyatomic ions are listed, and should be learned, there are relationships between ions than can help remember them. The phosphate ion is PO_4^{3-}, but adding a hydrogen ion, H^+, to it forms the hydrogenphosphate ion HPO_4^{2-}; adding one more H^+ gives the dihydrogenphosphate ion, $H_2PO_4^-$. Note that the negative charge on the ion decreases by one each time a hydrogen ion is added. The same relationship is seen for the carbonate and sulfate ions, too.

PO_4^{3-} phosphate ion HPO_4^{2-} hydrogenphosphate ion $H_2PO_4^-$ dihydrogenphosphate ion
CO_3^{2-} carbonate ion HCO_3^- hydrogencarbonate ion (also called the bicarbonate ion)
SO_4^{2-} sulfate ion HSO_4^- hydrogensulfate ion

The number of oxygen atoms in an oxyion of the same element has an affect on the name of the ion. The most common oxyion of a given element ends in *-ate*. One less oxygen than the most common ion ends in *-ite*.

NO_3^- nitr*ate* SO_4^{2-} sulf*ate* ClO_3^- chlor*ate* (the most common oxyions of N, S, and Cl)
NO_2^- nitr*ite* SO_3^{2-} sulf*ite* ClO_2^- chlor*ite* (one less oxygen than most common ion)

In the case of the chlorate ion, ClO_3^-, the oxyion with one *more* oxygen, ClO_4^-, adds a *per-* prefix to chlorate, becoming *per*chlorate ion. The oxyion with one less oxygen than the chlorite, ClO_2^-, is the *hypo*chlorite ion, ClO^-. Chlorine is one of the few elements that can form four different oxyions.

Example Problems

Give either the formula, with charge, or the name of these polyatomic ions.

1. NO_3^-, NO_2^-, SO_4^{2-}, and SO_3^{2-}

   Answer: nitrate ion, nitrite ion, sulfate ion, and sulfite ion

2. Hydroxide ion, ammonium ion, and carbonate ion

   Answer: OH^-, NH_4^+, and CO_3^{2-}

3. $C_2H_3O_2^-$, $C_2O_4^{2-}$, and CrO_4^{2-}

Answer: Acetate ion, oxalate ion, and chromate ion

Work Problems

Give either the formula, with charge, or the name of these polyatomic ions.

1. Hypochlorite ion, chlorite ion, chlorate ion, and perchlorate ion
2. PO_4^{3-}, HPO_4^{2-}, and $H_2PO_4^-$
3. MnO_4^-, CN^-, and $Cr_2O_7^{2-}$

Worked Solutions

1. **ClO^-, ClO_2^-, ClO_3^-, and ClO_4^-**
2. **Phosphate ion, hydrogenphosphate ion, and dihydrogenphosphate ion**
3. **Permanganate ion, cyanide ion, and dichromate ion**

Compounds that contain three elements are classed as **ternary compounds.** Most of the ionic compounds that contain a polyatomic ion, such as Na_2SO_4, are ternary ionic compounds.

Formulas of Ionic Compounds with Polyatomic Ions

Writing formulas for ionic compounds that include polyatomic ions is done in the same way formulas are written for binary compounds. The polyatomic ions are treated as single units that bear a charge. The goal is to get the right ratio of cation to anion so that an overall electrically neutral combination is obtained. The same rules used for binary compounds apply here, with one modification:

1. The symbol of the cation always comes before the symbol of the anion in the formula.
2. The sum of the *positive charges* on the cations must equal the sum of the *negative charges* on the anions so the formula of the compound is electrically neutral.
3. If two or more of a particular polyatomic ion are needed in a formula, the ion is enclosed in parentheses, followed by a subscript number to indicate the number of that ion in the formula unit. The *smallest* set of numbers must be used.

The formula of the ionic compound incorporating the Fe^{3+} ion and the nitrate ion, NO_3^-, would be:

$$\begin{array}{ll} Fe^{3+} & = 3+ \\ NO_3^-\ NO_3^-\ NO_3^- & = 3- \\ \hline Fe(NO_3)_3 & \quad 0 \end{array}$$

A net charge of zero shows the charges are balanced.

The nitrate ion is enclosed in parentheses. The subscript 3 outside parentheses indicates 3 NO_3^- ions.

The formula of the ionic compound containing the ammonium ion, NH_4^+, and the sulfate ion, SO_4^{2-}, is:

NH_4^+ NH_4^+	= 2+	
SO_4^{2-}	= 2–	
$(NH_4)_2SO_4$	0	A net charge of zero shows the charges are balanced.

Parentheses are required for the two ammonium ions, but they are not needed for the single sulfate ion.

Example Problems

Use the two ions to construct a correct formula for a ternary ionic compound.

1. Na^+ and HCO_3^-

   Answer: $NaHCO_3$; a single hydrogen carbonate ion balances the sodium ion and does not require parentheses.

2. NH_4^+ and PO_4^{3-}

   Answer: $(NH_4)_3PO_4$; three ammonium ions are required to balance one phosphate ion.

3. K^+ and CrO_4^{2-}

   Answer: K_2CrO_4; two potassium ions, K^+, balance the 2– charge of one chromate ion.

Work Problems

Use the two ions to construct a correct formula for a ternary ionic compound.

1. Ca^{2+} and $C_2H_3O_2^-$
2. Al^{3+} and SO_4^{2-}
3. Cu^{2+} and CN^-
4. Ca^{2+} and $H_2PO_4^-$

Worked Solutions

1. **$Ca(C_2H_3O_2)_2$; 2 acetate ions, $C_2H_3O_2^-$, balance the 2+ charge of one calcium ion.**
2. **$Al_2(SO_4)_3$; 2 Al^{3+} = 6+ and 3 SO_4^{2-} = 6–**
3. **$Cu(CN)_2$; 2 cyanide ions, CN^-, balance the 2+ charge of one copper(II) ion.**
4. **$Ca(H_2PO_4)_2$; 2 dihydrogenphosphate ions, $H_2PO_4^-$, balance Ca^{2+}.**

Naming Ionic Compounds with Polyatomic Ions

Ionic compounds that contain polyatomic ions are named in the same way used to name binary ionic compounds. Attention must be paid as to whether the cation is a metal that forms a single cation or multiple cations, and the name of the polyatomic ion is that given in the table. Note how these six ionic compounds are named:

$NaHCO_3$, sodium hydrogencarbonate
$(NH_4)_3PO_4$, ammonium phosphate
$Fe_2(SO_4)_3$, iron(III) sulfate
$FeSO_4$, iron(II) sulfate
$Ba(OH)_2$, barium hydroxide
CaC_2O_4, calcium oxalate

Analysis of four of these formulas shows how their names are obtained:

- **$NaHCO_3$:** The reason it is important to know the formulas of the polyatomic ions is so that you can recognize the unique pattern of letters and subscript numbers for each ion quickly. When you know the characteristic letter pattern of the hydrogencarbonate ion, H-C-O-3, the formula, $NaHCO_3$, can be quickly separated into the two ions, $Na^+ | HCO_3^-$, and the name of the compound follows, sodium hydrogencarbonate.
- **$(NH_4)_3PO_4$:** Parentheses, if present, isolate polyatomic ions for quick identification. In $(NH_4)_3PO_4$, there are two polyatomic ions: the ammonium ion, NH_4^+, and the phosphate ion, PO_4^{3-}. The parentheses isolate the ammonium ion, leaving the P-O-4 pattern, which is recognized as the phosphate ion.

  It is important to remember that numerical prefixes are *not* used in the names of ionic compounds. It would be incorrect to name $(NH_4)_3PO_4$ *tri*ammonium phosphate. The only time you would see a numerical prefix is if it is part of the name of a polyatomic ion, as in the *di*hydrogenphosphate ion.
- **$Fe_2(SO_4)_3$:** This compound contains a multiple-cation transition metal, and the charge on the cation must be determined to write a correct name. There are three sulfate ions in parentheses, giving a total negative charge of 6– (3 × 2– = 6–). The two iron ions then must be iron(III) ions, Fe^{3+} (2 × 3+ = 6+), to balance the negative charge of the sulfate ions. The systematic name of the compound is iron(III) sulfate.
- **$FeSO_4$:** In the same vein as the first compound, recognizing the S-O-4 pattern of the sulfate ion in $FeSO_4$ shows you where to separate the formula, $Fe | SO_4$, and how to analyze the charges on the ions; one SO_4^{2-} (2–) requiring a positive charge of 2+ on the single iron ion, making it the iron(II) ion, Fe^{2+}. The compound is iron (II) sulfate.

You may remember that the names of binary ionic compounds (two elements) always ended in *-ide*. The *-ide* ending tells you it is a binary compound. But there are three exceptions to this rule:

- Compounds containing the cyanide ion, CN^-
- Hydroxide ion, OH^-
- Ammonium ion, NH_4^+

The names of the first two ions end in *-ide*, and the ammonium ion can be paired up with a monatomic anion with an *-ide* ending to form a compound that has more than two elements. The examples that follow all contain three elements but have names with the binary *-ide* ending.

- NaCN, sodium cyan*ide*
- KOH, potassium hydrox*ide*
- NH_4Cl, ammonium chlor*ide*

Example Problems

Correctly name the following ionic compounds that contain polyatomic ions.

1. $BaCrO_4$

   Answer: barium chromate; Ba | CrO_4 → Ba^{2+} and CrO_4^{2-}, the chromate ion.

2. $Ni(ClO_4)_2$

   Answer: nickel(II) perchlorate; ClO_4^- is the perchlorate ion, and having two of them requires nickel to be Ni^{2+}, the nickel(II) ion.

3. NH_4NO_3

   Answer: ammonium nitrate; NH_4 | NO_3 → the ammonium and nitrate ions, NH_4^+, NO_3^-.

4. $Ca(ClO)_2$

   Answer: calcium hypochlorite; → Ca^{2+} and ClO^- is the hypochlorite ion.

5. Li_2CO_3

   Answer: lithium carbonate; → Lc^+ and CO_3^{2-} is the carbonate ion.

Work Problems

Correctly name the following ionic compounds that contain polyatomic ions.

1. $NaC_2H_3O_2$
2. $(NH_4)_2CO_3$
3. $Al(NO_3)_3$
4. NaOH
5. KCN

Worked Solutions

1. **sodium acetate; Na | $C_2H_3O_2$ → Na^+ and $C_2H_3O_2^-$, the acetate ion**
2. **ammonium carbonate; → NH_4^+, ammonium ion, and CO_3^{2-}, carbonate ion**
3. **aluminum nitrate; → Al^{3+} and NO^{3-}, the nitrate ion**
4. **sodium hydroxide; Na | OH → Na^+ and OH^-, the hydroxide ion**
5. **potassium cyanide; K | CN → K^+ and CN^-, the cyanide ion**

The Common Acids and Bases

Two of the most important classes of chemical compounds are the acids and the bases. **Acids** are compounds that produce a hydrogen ion, H^+, when dissolved in water. The proper way to symbolize the hydrogen ion in water is $H^+(aq)$. The (aq) stands for **aqueous,** meaning dissolved in water. Acids have a sour taste like that of vinegar or lemon juice. **Bases** are compounds that produce hydroxide ion, $OH^-(aq)$, when dissolved in water. Bases have a bitter taste that you experience if you get soap in your mouth.

Acids in water → $H^+(aq)$
Bases in water → $OH^-(aq)$

Additional material dealing with acids and bases appears in Chapter 14. Only the names of the common acids and bases are considered here.

The Binary Acids

A distinction must be made between the name of a binary acid (the solution in water) and the pure compound itself. For example, pure HCl is a gas at room temperature and is named hydrogen chloride. But when dissolved in water, it forms the acid, HCl(aq), which ionizes to form $H^+(aq)$ and $Cl^-(aq)$.

$$HCl(aq) \rightarrow H^+(aq) + Cl^-(aq)$$

The solution, HCl(aq), is called hydrochloric acid. Remember that the (aq) means dissolved in water.

The binary acids are named by adding the prefix *hydro-* to the stem of the nonmetal followed by the *-ic* ending then adding the word "acid." Hydrogen chloride, HCl, becomes *hydro-* chlor *-ic* acid, HCl(aq).

The names of the most common binary acids appear in the following table.

The Common Binary Acids

The Pure Compound	*The Acid*	*Name of the Binary Acid*
Hydrogen fluoride	HF(aq)	Hydrofluoric acid
Hydrogen chloride	HCl(aq)	Hydrochloric acid
Hydrogen bromide	HBr(aq)	Hydrobromic acid
Hydrogen iodide	HI(aq)	Hydroiodic acid

The Oxyacids

The **oxyacids** are compounds made up of hydrogen, oxygen, and another nonmetal, like S, C, N, Cl, or P. They become acids when they are dissolved in water. The formula of an oxyacid looks like a hydrogen compound of a polyatomic ion. The most widely used oxyacid is sulfuric acid, $H_2SO_4(aq)$, a compound that looks like it is the combination of two hydrogen ions, H^+, and a sulfate ion, SO_4^{2-}. As pure compounds out of water, the oxyacids are not considered to be ionic compounds, but when dissolved in water, they will ionize and produce hydrogen ions. Nitric acid is a good example:

$$HNO_3(aq) \rightarrow H^+(aq) + NO_3^-(aq)$$

If the oxyion of the acid has a name ending in *-ate*, like the nitrate ion, NO_3^-, the name of the acid of that oxyion replaces *-ate* with *-ic*, followed by the word "acid." Nitr*ate*, NO_3^-, becomes nitr*ic* acid, $HNO_3(aq)$. In the same way, acet*ate* ion, $C_2H_3O_2^-$, becomes acet*ic* acid, $HC_2H_3O_2(aq)$.

If the oxyion of the acid has a name ending in *-ite*, like the nitrite ion, NO_2^-, the name of the acid of that oxyion replaces *-ite* with *-ous* followed by the word "acid." Nitr*ite* ion, NO_2^-, becomes nitr*ous* acid, $HNO_2(aq)$. Sulf*ite* ion, SO_3^{2-}, when in the acid, becomes $H_2SO_3(aq)$, sulfur*ous* acid.

For oxyacids, if the name of the oxyion ends in *-ate*, the name of the acid ends in *-ic*. If the name of the oxyion ends in *-ite*, the name of the acid ends in *-ous*. A list of several oxyacids is given in the following table.

The Common Oxyacids

Formula	*Name of the Oxyacid*
$HC_2H_3O_2(aq)$*	Acetic acid
$HNO_3(aq)$	Nitric acid
$HNO_2(aq)$	Nitrous acid
$H_2SO_4(aq)$	Sulfuric acid
$H_2SO_3(aq)$	Sulfurous acid
$H_2CO_3(aq)$	Carbonic acid
$H_3PO_4(aq)$	Phosphoric acid
$H_3BO_3(aq)$	Boric acid
$H_2C_2O_4(aq)$	Oxalic acid

**The formula of acetic acid separates one hydrogen from the other three since only one hydrogen ionizes in water forming $H^+(aq)$ and $C_2H_3O_2^-(aq)$.*

The Bases

Bases are compounds that produce hydroxide ions, OH^-, when dissolved in water. Most bases are ionic compounds containing the hydroxide ion combined with a Group IA or IIA metal ion. Ammonia, NH_3, which is a gas at room temperature, is also a base because, when dissolved in water, it reacts weakly with water to produce hydroxide ion. As a base, it is called aqueous ammonia and symbolized $NH_3(aq)$, though many chemists still refer to it simply as "ammonia."

Several common bases are listed in the following table.

Several Common Bases

Formula	*Name of the Base*
$NaOH(aq)$*	Sodium hydroxide
$KOH(aq)$*	Potassium hydroxide
$Ca(OH)_2(aq)$*	Calcium hydroxide
$Ba(OH)_2(aq)$*	Barium hydroxide
$NH_3(aq)$	Aqueous ammonia

**NaOH, KOH, $Ca(OH)_2$, and $Ba(OH)_2$ have the same name as the pure solid in solution compound.*

Hydrates

Hydrates are ionic compounds that contain a definite number of water molecules associated with the formula unit of one compound, as in $FeCl_3{\cdot}6\ H_2O$. When water evaporates from a solution of copper(II) sulfate, beautiful blue crystals remain behind that have the formula $CuSO_4{\cdot}5\ H_2O$. Five molecules of water are associated with one formula unit of copper(II) sulfate, and they are separated from the formula of $CuSO_4$, with a dot (·). This hydrate is named copper(II) sulfate pentahydrate. Numerical prefixes are used to indicate the number of water molecules in one formula unit. Three other hydrates are listed and named as follows:

$NiCl_2{\cdot}6\ H_2O$	nickel(II) chloride hexahydrate
$MgSO_4{\cdot}7\ H_2O$	magnesium sulfate heptahydrate
$CaSO_4{\cdot}2\ H_2O$	calcium sulfate dihydrate

Hydrates often form as solutions of ionic compounds evaporate to dryness. All hydrates are solids. The water in a hydrate can be driven off with heating, leaving the water-free compound behind.

Example Problems

1. Write the formula and name of the oxyacid based on the oxyion, CO_3^{2-}.

   Answer: Two H^+ added to CO_3^{2-} gives H_2CO_3, carbonic acid, $H_2CO_3(aq)$.

2. Write the formulas of the following compounds:

   (a) sulfuric acid (c) chlorous acid

   (b) hydrobromic acid (d) iron(III) nitrate hexahydrate

   Answer: (a) $H_2SO_4(aq)$ (b) HBr(aq) (c) $HClO_2(aq)$ (d) $Fe(NO_3)_3{\cdot}6\ H_2O$

Work Problems

1. Give the formula of the oxyion on which these acids are based:

   (a) $HNO_3(aq)$ (c) $H_3PO_4(aq)$

   (b) $HC_2H_3O_2(aq)$ (d) $H_2C_2O_4(aq)$

2. Name the following compounds:

   (a) $LiCl{\cdot}2\ H_2O$ (c) $HNO_2(aq)$

   (b) NaOH or NaOH(aq) (d) $NH_3(aq)$

Worked Solutions

1. **(a) NO_3^-; removing the hydrogen ion leaves the nitrate ion, $H^+ | NO_3^-$.**

   (b) $C_2H_3O_2^-$; removing the one hydrogen ion leaves the acetate ion, $H^+ | C_2H_3O_2^-$.

(c) PO_4^{3-}; removing the 3 hydrogen ions leaves the phosphate ion, $3H^+ | PO_4^{3-}$.

(d) $C_2O_4^{2-}$; removing the 2 hydrogen ions leaves the oxalate ion, $2H^+ | C_2O_4^{2-}$.

2. **(a) lithium chloride dihydrate** **(c) nitrous acid**

   (b) sodium hydroxide **(d) aqueous ammonia**

Chapter Problems and Answers

Problems

1. Each of the following elements forms a monatomic ion with a single charge. What is that charge?

   (a) Na (b) F (c) Cd (d) S (e) Mg (f) K (g) P (h) Al

2. For the main-group elements, those in the A-groups, elements in the same group form monatomic ions with the same charge. What is that charge?

   (a) The ions of the elements in Group IA have a _____ charge.

   (b) The ions of the elements in Group IIA have a _____ charge.

   (c) The ions of the elements in Group VIIA have a _____ charge.

   (d) The ions of the elements in Group VIA have a _____ charge.

3. Make a correct formula for the ionic compound composed of the following pairs of ions:

   (a) Ca^{2+} and P^{3-} (f) Cd^{2+} and I^-

   (b) K^+ and S^{2-} (g) Al^{3+} and Se^{2-}

   (c) Cr^{3+} and O^{2-} (h) Co^{2+} and Cl^-

   (d) Cu^{2+} and N^{3-} (i) Ni^{2+} and F^-

   (e) Hg_2^{2+} and Br^- (j) Hg^{2+} and S^{2-}

4. Give the systematic name for each of the following ionic compounds:

   (a) Al_2O_3 (f) CdSe

   (b) K_2O (g) SnF_2

   (c) CuS (h) $ZnBr_2$

   (d) HgI_2 (i) $MnCl_2$

   (e) $MgBr_2$ (j) $PbCl_2$

5. Write the correct formula for the following ionic compounds:

(a) sodium bromide
(b) iron(III) sulfide
(c) lead(II) oxide
(d) aluminum fluoride
(e) calcium phosphide
(f) copper(II) selenide
(g) tin(II) sulfide
(h) barium nitride
(i) cobalt(III) iodide
(j) manganese(II) oxide

6. Name the following polyatomic ions:

(a) OH^-
(b) NO_3^-
(c) SO_4^{2-}
(d) PO_4^{3-}
(e) CO_3^{2-}
(f) NH_4^+
(g) $H_2PO_4^-$
(h) HCO_3^-
(i) $C_2O_4^{2-}$
(j) CN^-

7. Give the correct formula with charge for each of the following polyatomic ions:

(a) sulfite ion
(b) chlorate ion
(c) permanganate ion
(d) chromate ion
(e) acetate ion
(f) hydrogenphosphate ion

8. Use the ions in each pair to make a correct formula for an ionic compound:

(a) Na^+ and HCO_3^-
(b) Ca^{2+} and ClO^-
(c) Fe^{3+} and SO_4^{2-}
(d) K^+ and HPO_4^{2-}
(e) Ba^{2+} and OH^-
(f) Cr^{3+} and $C_2O_4^{2-}$
(g) Pb^{2+} and NO_3^-
(h) Na^+ and HSO_4^-
(i) Co^{3+} and $C_2H_3O_2^-$
(j) K^+ and CrO_4^{2-}

9. Correctly name these ionic compounds involving polyatomic ions:

(a) Na_2SO_4
(b) $Fe(NO_3)_3$
(c) $NaNO_2$
(d) K_2CrO_4
(e) $CaCO_3$
(h) $Pb(C_2H_3O_2)_2$
(i) Li_2CO_3
(j) $(NH_4)_2SO_4$
(k) $Na_2Cr_2O_7$
(l) $Ca_3(PO_4)_2$

(f) $NaHCO_3$ (m) $NiSO_4$

(g) $Al(OH)_3$ (n) $Al_2(SO_4)_3$

10. Write the correct formula for the following ionic compounds:

(a) cobalt(III) phosphate

(b) barium hydrogencarbonate

(c) copper(II) sulfate

(d) lead(II) hydroxide

(e) iron(III) chromate

(f) calcium acetate

(g) chromium(III) chlorate

(h) sodium carbonate

(i) strontium nitrate

(j) potassium permanganate

11. What ion do all acids produce when dissolved in water?

12. What ion do all bases produce when dissolved in water?

13. Name the following acids and bases:

(a) $HCl(aq)$

(b) $H_2SO_4(aq)$

(c) $HC_2H_3O_2(aq)$

(d) $H_2C_2O_4(aq)$

(e) $NaOH(aq)$

(f) $KOH(aq)$

(g) $Ca(OH)_2(aq)$

(h) $NH_3(aq)$

(i) $H_2CO_3(aq)$

(j) $H_3PO_4(aq)$

Answers

1. **(a) 1+; all Group IA cations are 1+.**

(b) 1−; all Group VIIA anions are 1−.

(c) 2+; cadmium is always 2+.

(d) 2−; all Group VIA monatomic anions are 2−.

(e) 2+; all Group IIA cations are 2+.

(f) 1+; a Group IA cation.

(g) 3−; the Group VA anions are 3−.

(h) 3+; aluminum, Group IIIA, is always 3+.

2. **(a) 1+; (b) 2+; (c) 1−; (d) 2−**

3. **(a)** Ca_3P_2

(b) K_2S

(c) Cr_2O_3

(d) Cu_3N_2

(e) Hg_2Br_2

(f) CdI_2

(g) Al_2Se_3

(h) $CoCl_2$

(i) NiF_2

(j) HgS

4. (a) aluminum oxide
(b) potassium oxide
(c) copper(II) sulfide
(d) mercury(II) iodide
(e) magnesium bromide
(f) cadmium selenide
(g) tin(II) fluoride
(h) zinc bromide
(i) manganese(II) chloride
(j) lead(II) chloride

5. (a) $NaBr$
(b) Fe_2S_3
(c) PbO
(d) AlF_3
(e) Ca_3P_2
(f) $CuSe$
(g) SnS
(h) Ba_3N_2
(i) CoI_3
(j) MnO

6. (a) hydroxide ion
(b) nitrate ion
(c) sulfate ion
(d) phosphate ion
(e) carbonate ion
(f) ammonium ion
(g) dihydrogenphosphate ion
(h) hydrogencarbonate ion
(i) oxalate ion
(j) cyanide ion

7. (a) SO_3^{2-}
(b) ClO_3^-
(c) MnO_4^-
(d) CrO_4^{2-}
(e) $C_2H_3O_2^-$
(f) HPO_4^{2-}

8. (a) $NaHCO_3$
(b) $Ca(ClO)_2$
(c) $Fe_2(SO_4)_3$
(d) K_2HPO_4
(e) $Ba(OH)_2$
(f) $Cr_2(C_2O_4)_3$
(g) $Pb(NO_3)_2$
(h) $NaHSO_4$
(i) $Co(C_2H_3O_2)_3$
(j) K_2CrO_4

9. (a) sodium sulfate
(b) iron(III) nitrate
(c) sodium nitrite
(d) potassium chromate
(e) calcium carbonate
(h) lead(II) acetate
(i) lithium carbonate
(j) ammonium sulfate
(k) sodium dichromate
(l) calcium phosphate

(f) sodium hydrogencarbonate **(m) nickel(II) sulfate**

(g) aluminum hydroxide **(n) aluminum sulfate**

10. **(a) $CoPO_4$** **(f) $Ca(C_2H_3O_2)_2$**

(b) $Ba(HCO_3)_2$ **(g) $Cr(ClO_3)_3$**

(c) $CuSO_4$ **(h) Na_2CO_3**

(d) $Pb(OH)_2$ **(i) $Sr(NO_3)_2$**

(e) $Fe_2(CrO_4)_3$ **(j) $KMnO_4$**

11. **Acids produce H^+ in water, $H^+(aq)$.**

12. **Bases produce OH^- in water, $OH^-(aq)$.**

13. **(a) hydrochloric acid** **(f) potassium hydroxide**

(b) sulfuric acid **(g) calcium hydroxide**

(c) acetic acid **(h) aqueous ammonia**

(d) oxalic acid **(i) carbonic acid**

(e) sodium hydroxide **(j) phosphoric acid**

Supplemental Chapter Problems

Problems

1. Use the pairs of ions to make a correct formula for an ionic compound:

(a) K^+ and NO_2^- (k) Mn^{2+} and ClO_4^-

(b) Ag^+ and CN^- (l) Sn^{2+} and OH^-

(c) Al^{3+} and ClO_4^- (m) Mg^{2+} and NO_3^-

(d) Ca^{2+} and MnO_4^- (n) Cu^+ and Cl^-

(e) Co^{3+} and $C_2H_3O_2^-$ (o) Fe^{3+} and S^{2-}

(f) Ba^{2+} and PO_4^{3-} (p) Na^+ and HPO_4^{2-}

(g) NH_4^+ and CO_3^{2-} (q) Li^+ and SO_3^{2-}

(h) Na^+ and HPO_4^{2-} (r) Cu^{2+} and $C_2H_3O_2^-$

(i) Fe^{2+} and HCO_3^- (s) NH_4^+ and ClO_3^-

(j) Cs^+ and SO_4^{2-}

2. Give the correct name for each of the following ionic compounds:

(a) NaI
(b) CdO
(c) Fe_2O_3
(d) Cu_2O
(e) $Al_2(SO_4)_3$
(f) CoF_2
(g) FeS
(h) NH_4HSO_4
(i) Rb_2HPO_4
(j) $CaSO_3$
(k) $Cr(CN)_3$
(l) $BaSO_4$
(m) $Ca(ClO)_2$
(n) $K_2Cr_2O_7$
(o) $Sn(NO_2)_2$
(p) $Cu(C_2H_3O_2)_2$
(q) KCl
(r) $SrCO_3$
(s) $(NH_4)_2SO_4$
(t) $Pb(NO_3)_2$

3. Write the correct formulas for the ionic compounds:

(a) ammonium chloride
(b) iron(III) sulfate
(c) barium chlorate
(d) tin(II) bromide
(e) manganese(II) carbonate
(f) silver nitrate
(g) potassium hydrogensulfate
(h) copper(II) oxide
(i) lithium hydrogencarbonate
(j) sodium oxalate
(k) copper(I) cyanide
(l) aluminum acetate
(m) sodium nitrite
(n) lead(II) chromate
(o) ammonium dihydrogenphosphate
(p) strontium fluoride
(q) chromium(III) sulfide
(r) calcium hydroxide
(s) barium dichromate
(t) sodium acetate

4. Write the correct name for those compounds that are incorrectly named:

(a) Cr_2O_3; chromium(II) oxide
(b) Na_2SO_4; sodium sulfate
(c) $NaClO_3$; sodium chlorate
(d) $Fe(NO_3)_2$; iron(III) nitrate
(e) $CdSO_3$; calcium sulfate
(f) CuS; copper(I) sulfide

5. Name for the following acids, bases, and hydrates:

(a) $HCl(aq)$

(b) $NaOH(aq)$

(c) $CrCl_2 \cdot 6\ H_2O$

(d) $H_2SO_4(aq)$

(e) $H_2C_2O_4(aq)$

(f) $Sr(NO_3)_2 \cdot 4\ H_2O$

Answers

1. **(a) KNO_2**

(b) $AgCN$

(c) $Al(ClO_4)_3$

(d) $Ca(MnO_4)_2$

(e) $Co(C_2H_3O_2)_3$

(f) $Ba_3(PO_4)_2$

(g) $(NH_4)_2CO_3$

(h) Na_2HPO_4

(i) $Fe(HCO_3)_2$

(j) Cs_2SO_4

(k) $Mn(ClO_4)_2$

(l) $Sn(OH)_2$

(m) $Mg(NO_3)_2$

(n) $CuCl$

(o) Fe_3S_2

(p) Na_2HPO_4

(q) Li_2SO_3

(r) $Cu(C_2H_3O_2)_2$

(s) NH_4ClO_3

(page 85)

2. **(a) sodium iodide**

(b) cadmium oxide

(c) iron(III) oxide

(d) copper(I) oxide

(e) aluminum sulfate

(f) cobalt(II) fluoride

(g) iron(II) sulfide

(h) ammonium hydrogensulfate

(i) rubidium hydrogenphosphate

(j) calcium sulfite

(k) chromium(III) cyanide

(l) barium sulfate

(m) calcium hypochlorite

(n) potassium dichromate

(o) tin(II) nitrite

(p) copper(II) acetate

(q) potassium chloride

(r) strontium carbonate

(s) ammonium sulfate

(t) lead(II) nitrate

(page 88)

3. (a) NH_4Cl
(b) Fe_2S_3
(c) $Ba(ClO_3)_2$
(d) $SnBr_2$
(e) $MnCO_3$
(f) $AgNO_3$
(g) $KHSO_4$
(h) CuO
(i) $LiHCO_3$
(j) $Na_2C_2O_4$
(k) $CuCN$
(l) $Al(C_2H_3O_2)_3$
(m) $NaNO_2$
(n) $PbCrO_4$
(o) $NH_4H_2PO_4$
(p) SrF_2
(q) Cr_2S_3
(r) $Ca(OH)_2$
(s) $BaCr_2O_7$
(t) $NaC_2H_3O_2$

(page 94)

4. (a) The correct name is chromium(III) oxide.
(b) The name is correct.
(c) The name is correct.
(d) The correct name is iron(II) nitrate.
(e) The correct name is cadmium sulfite.
(f) The correct name is copper(II) sulfide.

(page 96)

5. (a) hydrochloric acid
(b) sodium hydroxide
(c) chromium(II) chloride hexahydrate
(d) sulfuric acid
(e) oxalic acid
(f) strontium nitrate tetrahydrate

(page 98)

Chapter 5

The Mole—Elements and Compounds

If there is a single area of chemistry that makes people uncomfortable more than any other, it is chemical calculations. But it is absolutely possible to make these calculations straightforward, understandable, and workable. All calculations in this chapter use conversion factors that change one quantity into another. Conversion factors require a keen awareness of the units on numbers so that unwanted units are canceled and replaced by those desired. Using units, along with the rules of significant figures, are conventions followed throughout the chapter.

Because it is convenient to have quick access to the atomic masses of the elements, the atomic mass of the 40 most often used elements appears in the following table. Of course, the atomic mass for any element can be found on any periodic table. The unit of atomic mass, molecular mass, and formula mass is the **atomic mass unit, amu.** In some texts, this unit is omitted when solving problems, while in other texts it is used consistently. Because it is important to get accustomed to using units, the amu is used in this text.

The Atomic Masses of 40 Elements (In amu)

Element	*Symbol*	*Atomic Mass*	*Element*	*Symbol*	*Atomic Mass*
Aluminum	Al	26.98	Magnesium	Mg	24.31
Argon	Ar	39.95	Manganese	Mn	59.94
Arsenic	As	74.92	Mercury	Hg	200.6
Barium	Ba	137.3	Neon	Ne	20.18
Boron	B	10.81	Nickel	Ni	58.69
Bromine	Br	79.90	Nitrogen	N	14.01
Cadmium	Cd	112.4	Oxygen	O	16.00
Calcium	Ca	40.08	Phosphorus	P	30.97
Carbon	C	12.01	Platinum	Pt	195.1
Chlorine	Cl	35.45	Potassium	K	39.10
Chromium	Cr	52.00	Selenium	Se	78.96
Cobalt	Co	58.93	Silicon	Si	28.09

(continued)

The Atomic Masses of 40 Elements (In amu) *(continued)*

Element	*Symbol*	*Atomic Mass*	*Element*	*Symbol*	*Atomic Mass*
Copper	Cu	63.55	Silver	Ag	107.9
Fluorine	F	19.00	Sodium	Na	22.99
Gold	Au	197.0	Strontium	Sr	87.62
Hydrogen	H	1.008	Sulfur	S	32.07
Iodine	I	126.9	Tin	Sn	118.7
Iron	Fe	55.85	Titanium	Ti	47.88
Lead	Pb	207.2	Uranium	U	238.0
Lithium	Li	6.941	Zinc	Zn	65.39

Nearly all calculations in chemistry are based on the **mole** and after you understand what this term means, you'll be well on your way to success in solving many kinds of chemical calculations.

The Mole and Avogadro's Number

In the arithmetic of chemistry, a lot of calculations come down to counting atoms and molecules. The international unit for the amount of substance isn't the kilogram (which is used for mass), it's the **mole.** Chemists need to think in terms of numbers of particles, and this is what the mole allows. The mole is a very large number: 6.022×10^{23}. It is the number of carbon-12 atoms in exactly 12.00 grams of that isotope. It is such an important number in chemistry that it has its own special name, **Avogadro's number,** named after Amadeo Avogadro (1776–1856), one of the most famous early scientists.

$$\text{Avogadro's number} = 1 \text{ mole} = 6.022 \times 10^{23}$$

Think of the mole as a number of objects:

$$1 \text{ mole C-12 atoms} = 6.022 \times 10^{23} \text{ C-12 atoms}$$
$$1 \text{ mole of water molecules} = 6.022 \times 10^{23} \text{ water molecules}$$
$$1 \text{ mole donuts} = 6.022 \times 10^{23} \text{ donuts}$$
$$1 \text{ mole dollars} = 6.022 \times 10^{23} \text{ dollars}$$

Just to give you an idea of how big 1 mole is, if you could spend $1,000,000 every minute (Bill Gates, move over), it would take you 1,145 *billion* years to spend just 1 mole of dollars, and this is almost 250 times longer than the age of the earth, which geologists estimate to be right around 4.6 billion years. One mole is a very, very large number, but it has to be large if it is to be used to count things as infinitesimally small as atoms and molecules.

Molar Mass

The term **molar mass** is used all the time in chemical calculations, so you need to know what it means. First, it is a mass, so it has units of mass, commonly the gram. Second, it concerns the mole (Avogadro's number). Whether you're dealing with elements or compounds, the molar mass of a species is the mass in grams of one mole (6.022×10^{23}) of that species: one mole of atoms, one mole of molecules, or one mole of formula units.

Here's a good example. What is the molar mass of a penny? A single American penny has a mass of about 3 grams. The molar mass of the American penny would be the mass of Avogadro's number of pennies, in grams. Multiplying Avogadro's number by the mass of one penny shows this to be about 2×10^{24} grams. One molar mass of pennies is about 2×10^{24} grams of pennies.

But we won't be dealing with pennies here. We start with atoms of the elements, and then move onto molecules and formula units of compounds. As the discussion progresses, the definition of important terms is repeated to help you stay on course.

Elements and the Mole—Molar Mass

By definition, the atomic mass of the carbon-12 atom is *exactly* 12.00 amu. One mole of carbon-12 atoms has a mass of *exactly* 12.00 g, and that 12.00 g mass contains *exactly* 6.022×10^{23} carbon-12 atoms. This statement sets the benchmark for all chemical calculations involving the mole. One mole of *any* element is an amount of that element equal to its atomic mass in grams (its molar mass), and that mass contains 6.022×10^{23} atoms of that element. Using atomic masses, you can apply these relationships to the elements hydrogen and nitrogen.

The atomic mass of hydrogen is 1.008 amu, so 1.008 grams of hydrogen is 1 mole of hydrogen, and that 1.008 g mass contains 6.022×10^{23} atoms of hydrogen.

$$1 \text{ mole H} = 1.008 \text{ g H} = 6.022 \times 10^{23} \text{ H atoms}$$

Likewise, the atomic mass of nitrogen, N, is 14.01 amu. 14.01 grams of nitrogen is 1 mole of nitrogen, and 14.01 grams of nitrogen contain 6.022×10^{23} atoms of nitrogen.

$$1 \text{ mole N} = 14.01 \text{ g N} = 6.022 \times 10^{23} \text{ N atoms}$$

You may wonder why it takes 14.01 g of nitrogen to have the same number of atoms as there are in 1.008 g of hydrogen. It's because nitrogen atoms are about 14 times heavier than hydrogen atoms. Remember that atomic masses are relative masses. If atoms of element X are 50 times heavier than those of element Z, the atomic mass of X will be 50 times larger than that of Z.

The molar mass of an element is the mass in grams of 1 mole of that element. One molar mass of hydrogen is 1.008 g of hydrogen. One molar mass of nitrogen is 14.01 g of nitrogen. The term molar mass also applies to compounds, as will be seen later in this chapter.

Two very important facts to remember:

- One mole of an element is 6.022×10^{23} atoms of that element.
- One mole of an element is an amount of that element equal to its molar mass (an amount equal to its atomic mass in grams).

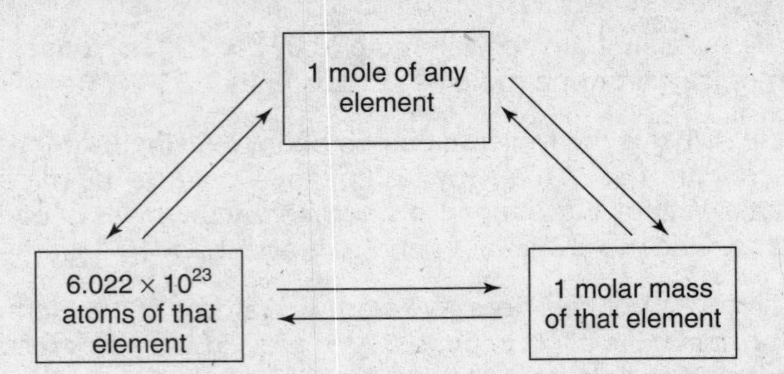

The three most common problems that involve the mole, molar mass, and numbers of atoms, as they apply to elements, are discussed in the three following sections. As each problem is put forth, keep in mind that the methods used to solve them will also be applied to compounds later in the chapter. Each problem is discussed in greater detail here to ensure that you understand the principles.

Elements: Determining Molar Mass

Because we are concerned with elements, let's use iron, Fe, as our example. From the table of atomic masses, we find that the atomic mass of iron is 55.85 amu. In terms of iron:

1 mole Fe = molar mass Fe = the atomic mass of Fe in grams

It follows, then, that because the atomic mass of Fe is 55.85 amu,

1 mole Fe = 55.85 g Fe = the molar mass Fe

Remember, 1 mole of *any* element is an amount equal to its atomic mass in grams, its molar mass.

Example Problems

Fill in the blanks for the indicated element:

1. 1 mole Na = _______ g Na = _______ atoms Na

   Answer: 1 mole Na = 22.99 g Na = 6.022×10^{23} atoms Na

2. 118.7 g Sn = _______ mole Sn = _______ atoms Sn

   Answer: 118.7 g Sn = 1.000 mole Sn = 6.022×10^{23} atoms Sn

3. 6.022×10^{23} atoms O = _______ g O = _______ mole O

   Answer: 6.022×10^{23} atoms O = 16.00 g O = 1.000 mole O

Work Problems

Fill in the blanks for the indicated element:

1. 1 mole Cu = _______ g Cu = _______ atoms Cu

2. 10.81 g B = ________ mole B = ________ atoms B

3. 6.022×10^{23} atoms S = ________ g S = ________ mole S

Worked Solutions

1. **1 mole Cu = 63.55 g Cu = 6.022×10^{23} atoms Cu**

2. **10.81 g B = 1.000 mole B = 6.022×10^{23} atoms B**

3. **6.022×10^{23} atoms S = 32.07 g S = 1.000 mole S**

Elements: Mass-to-Mole and Mole-to-Mass Conversions

These can also be called gram-to-mole and mole-to-gram conversions. They are done using conversion factors derived from the molar mass of the element. Again, using iron as our element:

$$\text{molar mass Fe} = 55.85 \text{ g Fe}$$

therefore,

$$1 \text{ mole Fe} = 55.85 \text{ g Fe}$$

Two conversion factors can be written from this equality, one the inverse of the other:

$$\left(\frac{1 \text{ mole Fe}}{55.85 \text{ g Fe}}\right) \text{ and } \left(\frac{55.85 \text{ g Fe}}{1 \text{ mole Fe}}\right)$$

Both factors are used in the following problems, which involve converting a known mass of an element to moles:

Example 1: 10.0 g iron = _____ mole iron

As described in Chapter 1, conversions are set up in terms of sought and known values.

the SOUGHT value = (the KNOWN value) × (conversion factor)
mole Fe = (10.0 g Fe) × (mass-to-mole conversion factor)

The units of the conversion factor must cancel "g Fe" and replace it with "mole Fe:"

$$\text{mole Fe} = (10.0 \text{ g}\cancel{\text{Fe}}) \times \left(\frac{1 \text{ mole Fe}}{55.85 \text{ g}\cancel{\text{Fe}}}\right) = 0.179 \text{ mole Fe}$$

Answer: 10.0 g iron = 0.179 mole iron

Notice how the units guided which conversion factor was used. The upper unit of the conversion factor becomes the unit of the answer. The lower unit cancels out.

Example 2: 0.350 mole iron = _____ g iron

the SOUGHT value = (the KNOWN value) × (conversion factor)
mass of Fe = (0.350 mole Fe) × (mole-to-mass conversion factor)

The conversion factor must cancel "mole Fe" and replace it with "g Fe," so the conversion factor with "g Fe" on top and "mole Fe" on the bottom is used:

$$\text{mass of Fe} = (0.350\,\cancel{\text{mole Fe}}) \times \left(\frac{55.85\text{ g Fe}}{1\,\cancel{\text{mole Fe}}}\right) = 19.5\text{ g Fe}$$

Answer: 0.350 mole iron = 19.5 g iron

When doing mass-to-mole or mole-to-mass conversions, you will always use conversion factors that relate 1 mole of the species to its molar mass. These kinds of conversions are done over and over in chemistry, and you need to be able to do them with ease.

Elements: Moles and Counting Atoms

If you want to count atoms, Avogadro's number must be part of the solution. Keep in mind that 1 mole of any element represents 6.022×10^{23} atoms of that element. This fact provides the necessary equality to develop two conversion factors. In terms of iron, Fe:

$$1\text{ mole of Fe} = 6.022 \times 10^{23}\text{ atoms of Fe}$$

The conversion factors are: $\left(\frac{1\text{ mole Fe}}{6.022\times 10^{23}\text{ atoms}}\right)$ and $\left(\frac{6.022\times 10^{23}\text{ atoms}}{1\text{ mole Fe}}\right)$.

Example 1: 25.0 g iron = _____ atoms of iron

Because the conversion factors are in terms of "mole Fe," the mass of iron (gram) must first be converted to mole of iron:

$$\text{mole Fe} = (25.0\,\text{g}\,\cancel{\text{Fe}}) \times \left(\frac{1\text{ mole Fe}}{55.85\,\text{g}\,\cancel{\text{Fe}}}\right) = 0.448\text{ mole Fe}$$

Then, the mole of iron is converted to atoms of iron using the conversion factor that cancels "mole Fe" and retains "atoms of Fe."

$$\text{atoms Fe} = (0.448\,\cancel{\text{mole Fe}}) \times \left(\frac{6.022\times 10^{23}\text{ atoms of Fe}}{1\,\cancel{\text{mole Fe}}}\right) = 2.70\times 10^{23}\text{ atoms of Fe}$$

Answer: 25.0 g iron = 2.70×10^{23} atoms of iron

Remember that because this conversion factor was in terms of mole of iron, the amount of iron also had to be changed to mole to be unit-consistent with the conversion factor.

But there is a faster way to solve this problem, eliminating the need for the conversion from grams Fe to mole Fe. A conversion factor in terms of "g Fe" and "atoms Fe" can be assembled from the following equality that allows a one-step solution to the problem:

$$1\text{ mole Fe} = 55.85\text{ g Fe} = 6.022 \times 10^{23}\text{ atoms Fe}$$

Using the fact that 55.85 g Fe represents 6.022×10^{23} atoms of Fe, the conversion factor is written

$$\left(\frac{6.022\times 10^{23}\text{ atoms Fe}}{55.85\text{ g Fe}}\right)$$

and the problem is solved in one step.

$$\text{atoms Fe} = (25.0\ \cancel{\text{g Fe}}) \times \left(\frac{6.022 \times 10^{23}\ \text{atoms Fe}}{55.85\ \cancel{\text{g Fe}}}\right) = 2.70 \times 10^{23}\ \text{atoms Fe}$$

There is a lesson here: Choose or develop the most efficient conversion factor to solve the problem. If you're converting grams to atoms, find the equality that relates grams and atoms for the conversion factor. If it's mole to atoms, find the connection between mole and atoms. Use the units "g Fe," "mole Fe," and "atoms Fe" to guide the correct use of the conversion factor.

Example 2: 1,000 atoms of iron = _____ gram iron

$$1\ \text{mole Fe} = 55.85\ \text{g Fe} = 6.022 \times 10^{23}\ \text{atoms Fe}$$
$$55.85\ \text{g Fe} = 6.022 \times 10^{23}\ \text{atoms Fe}$$

The conversion factor is derived so that "atoms Fe" cancel and are replaced with "g Fe."

$$\text{mass of } 1{,}000.\ \text{atoms Fe} = (1{,}000.\ \cancel{\text{atoms Fe}}) \times \left(\frac{55.85\ \text{g Fe}}{6.022 \times 10^{23}\ \cancel{\text{atoms Fe}}}\right) = 9.274 \times 10^{-20}\ \text{g Fe}$$

Answer: 1,000 atoms of Fe = 9.274×10^{-20} g Fe

Example Problems

These problems have both answers and solutions given. The required atomic masses are taken from the table of atomic masses.

1. 18.0 g of carbon = _____ mole carbon

   Answer: 1.50 mole C

   The atomic mass of C is 12.01 amu; therefore, 1 mole C = 12.01 g C. This equality provides the needed gram-to-mole conversion factor.

   $$\text{mole C} = (18.0\ \cancel{\text{g C}}) \times \left(\frac{1\ \text{mole C}}{12.01\ \cancel{\text{g C}}}\right) = 1.50\ \text{mole C}$$

2. 1.35 mole Mg = _____ g Mg

   Answer: 32.8 g Mg

   The atomic mass of Mg is 24.31 amu; therefore, 1 mole Mg = 24.31 g Mg

   $$\text{mass Mg} = (1.35\ \cancel{\text{mole Mg}}) \times \left(\frac{24.31\ \text{g Mg}}{1\ \cancel{\text{mole Mg}}}\right) = 32.8\ \text{g Mg}$$

3. 8.50 g Cu = _____ atoms of Cu

   Answer: 8.05×10^{22} atoms

   The atomic mass of Cu is 63.55 amu; therefore, 63.55 g Cu = 1 mole Cu = 6.022×10^{23} atoms Cu.

   The conversion factor relating "g Cu" to "atoms Cu" allows a one-step solution:

   $$\text{atoms Cu} = (8.50\ \cancel{\text{g Cu}}) \times \left(\frac{6.022 \times 10^{23}\ \text{atoms Cu}}{63.55\ \cancel{\text{g Cu}}}\right) = 8.05 \times 10^{22}\ \text{atoms Cu}$$

Work Problems

Use these problems for additional practice.

1. 45.5 g silver = _____ mole silver

2. 0.700 mole Ne = _____ g Ne

3. 21.5 g B = _____ atoms B

Worked Solutions

1. **0.422 mole silver (Ag); 1 mole Ag = 107.9 g Ag**

$$\text{mole Ag} = (45.5\,\text{g Ag}) \times \left(\frac{1\,\text{mole Ag}}{107.9\,\text{g Ag}}\right) = 0.422\,\text{mole Ag}$$

2. **14.1 g Ne; 1 mole Ne = 20.18 g Ne**

$$\text{mass Ne} = (0.700\text{ mole Ne}) \times \left(\frac{20.18\,\text{g Ne}}{1\,\text{mole Ne}}\right) = 14.1\,\text{g Ne}$$

3. **1.20×10^{24} atoms B; 1 mole B = 10.81 g B = 6.022×10^{23} atoms B**

$$\text{atoms B} = (21.5\,\text{g B}) \times \left(\frac{6.022 \times 10^{23}\,\text{atoms B}}{10.81\,\text{g B}}\right) = 1.20 \times 10^{24}\,\text{atoms B}$$

Diatomic Elements and the Mole

Seven elements in their pure states exist not as individual atoms but as diatomic (two-atom) molecules:

$$H_2,\ N_2,\ O_2,\ F_2,\ Cl_2,\ Br_2,\ \text{and } I_2$$

For these elements, there are times when it is more appropriate to speak in terms of molecules of the element as opposed to atoms, so instead of the atomic mass, you need the molecular mass of the molecule. The **molecular mass** of a molecule is the sum of the atomic masses of all the atoms in the molecule. Dihydrogen, H_2, is composed of two hydrogen atoms, so its molecular mass is two times the atomic mass of hydrogen: 2×1.008 amu, 2.016 amu. The molar mass of H_2, then, is 2.016 g.

$$1\text{ mole } H_2 = 2.016\text{ g } H_2 = 6.022 \times 10^{23}\text{ molecules of } H_2$$

The molecular mass of O_2 is two times the atomic mass of oxygen: 2×16.00 amu = 32.00 amu. The molar mass of dioxygen is 32.00 g.

$$1\text{ mole } O_2 = 32.00\text{ g } O_2 = 6.022 \times 10^{23}\text{ molecules of } O_2$$

Notice that one mole of molecules is still Avogadro's number of molecules, the number of molecules in one molar mass of the species. One mole of anything is *always* Avogadro's number of "things."

All the problems worked in the previous section with atoms are worked the same way when you are using molecules, but with one difference; molecular masses must be used instead of atomic masses.

Example 1: What is the mass of 1 mole of Br_2?

The molecular mass of Br_2 is 2 times the atomic mass of Br: 2×79.90 amu = 159.8 amu

1 mole Br_2 = the molar mass of Br_2 = 159.8 g

Answer: 1 mole Br_2 = 159.8 g Br_2

Example 2: How many molecules of Br_2 are in 0.045 mole of dibromine?

1 mole of Br_2 = 159.8 g Br_2 = 6.022×10^{23} molecules of Br_2

$$\text{molecules } Br_2 = (0.045 \text{ mole } Br_2) \times \left(\frac{6.022 \times 10^{23} \text{ molecules}}{1 \text{ mole } Br_2}\right) = 2.7 \times 10^{22} \text{ molecules}$$

Answer: 0.045 mole Br_2 = 2.7×10^{22} molecules Br_2

Example Problems

These problems have both answers and solutions given. The required atomic masses are taken from the table of atomic masses.

1. 25.0 g of N_2 = _____ mole N_2

   Answer: 0.892 mole N_2; 1 mole N_2 = 28.02 g N_2 (2 times the atomic mass of N).

   $$\text{mole } N_2 = (25.0 \text{ g } N_2) \times \left(\frac{1 \text{ mole } N_2}{28.02 \text{ g } N_2}\right) = 0.892 \text{ mole } N_2$$

2. 1.00 g I_2 = _____ molecules I_2

   Answer: 2.37×10^{21} molecules I_2

   1 mole I_2 = 253.8 g I_2 = 6.022×10^{23} molecules I_2

   $$\text{molecules } I_2 = (1.00 \text{ g } I_2) \times \left(\frac{6.022 \times 10^{23} \text{ molecules } I_2}{253.8 \text{ g } I_2}\right) = 2.37 \times 10^{21} \text{ molecules } I_2$$

Work Problems

Use these problems for additional practice.

1. 8.25 g dichlorine = _____ mole Cl_2
2. 1.25 g H_2 = _____ molecules dihydrogen

Worked Solutions

1. **0.116 mole Cl_2;** 1 mole Cl_2 = 70.90 g Cl_2

   $$\text{mole } Cl_2 = (8.25 \text{ g } Cl_2) \times \left(\frac{1 \text{ mole } Cl_2}{70.90 \text{ g } Cl_2}\right) = 0.116 \text{ mole } Cl_2$$

2. 3.73×10^{23} molecules of dihydrogen

$$1 \text{ mole } H_2 = 2.016 \text{ g } H_2 = 6.022 \times 10^{23} \text{ molecules } H_2$$

$$\text{molecules } H_2 = (1.25 \text{ g } H_2) \times \left(\frac{6.022 \times 10^{23} \text{ molecules } H_2}{2.016 \text{ g } H_2}\right) = 3.73 \times 10^{23} \text{ molecules } H_2$$

Compounds and the Mole

Whether you're dealing with elements or compounds, one fundamental relationship never changes:

1 mole of X = 1 molar mass of X = Avogadro's number of X

1 mole of a compound = 1 molar mass of the compound = 6.022×10^{23} units of the compound

Note that the unit is the molecule for molecular compounds and the formula unit for ionic compounds.

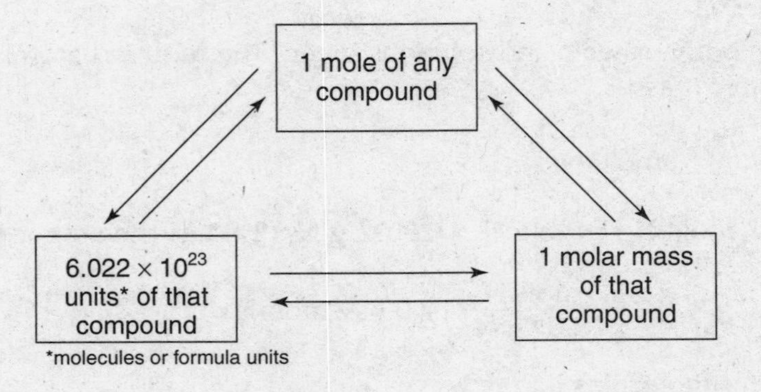

Determining the Molar Mass of Compounds

Whether you are dealing with elements or compounds, the **molar mass** of a species is the mass in grams of 1 mole (6.022×10^{23}) of that species: 1 mole of atoms, 1 mole of molecules, or 1 mole of formula units. With compounds, you're dealing with molecules and formula units, so it is necessary to calculate the molecular or formula mass of each compound to get its molar mass. It's easy to get confused by the language:

- Molecular mass equals the sum of the atomic masses of all the atoms in the molecule. The unit of molecular mass is the amu.
- Molar mass is an amount of the compound equal to its molecular mass in grams. The unit of molar mass is the gram.

What is the molar mass of sulfur trioxide, SO_3, a molecular compound? By definition, the molar mass of SO_3 equals its molecular mass in grams. So you first need to calculate the molecular mass of SO_3.

The **molecular mass** (or **formula mass**) of a compound is the sum of the atomic masses of all the atoms in the molecule (or formula). The molecular mass of SO_3, with 4 atoms in its molecule, is:

$$1 \text{ S} = 1 \times 32.07 \text{ amu} = 32.07 \text{ amu}$$

$$3 \text{ O} = 3 \times 16.00 \text{ amu} = 48.00 \text{ amu}$$

$$\text{The molecular mass of } SO_3 = 80.07 \text{ amu}$$

The molar mass of SO_3 is its molecular mass in grams, 80.07g, so it follows that:

$$1 \text{ mole } SO_3 = 80.07\text{g } SO_3 = 6.022 \times 10^{23} \text{ molecules of } SO_3$$

Ionic compounds are composed of ions of opposite charge packed together in a highly organized way to form crystals. Molecules do not exist in ionic compounds. The formulas of ionic compounds give the simplest ratio of the ions in the crystal. Calcium chloride, $CaCl_2$, is an ionic compound and its formula shows a 1:2 ratio of Ca^{2+} to Cl^- ions.

The formula, $CaCl_2$, represents one **formula unit** of the compound and represents the simplest ratio of its ions. One mole of an ionic compound contains Avogadro's number of "formula units" of the compound, so instead of molecular mass, the term **formula mass** is used. The formula mass of $CaCl_2$ is the sum of the atomic masses of all the atoms in its formula. The atomic masses of the calcium and chloride ions are the same as those of the neutral atoms. The formula mass of $CaCl_2$, with 3 atoms in its formula, is as follows:

$$\begin{aligned} 1\,Ca = 1 \times 40.08\,\text{amu} &= 48.08\,\text{amu} \\ 2\,Cl = 2 \times 35.45\,\text{amu} &= 70.90\,\text{amu} \\ \text{The formula mass of } CaCl_2 &= \overline{110.98\,\text{amu}} \end{aligned}$$

Note that although calcium and chlorine exist here as ions, the formula mass is the sum of the atomic masses. For this reason, calcium and chlorine are shown as atoms when adding the atomic masses of the ions. The formula mass in grams is the molar mass of $CaCl_2$, 110.98 g.

$$1 \text{ mole } CaCl_2 = 110.98 \text{ g } CaCl_2 = 6.022 \times 10^{23}\ CaCl_2 \text{ formula units}$$

Example 1: Find the molar mass of $(NH_4)_2SO_4$. Counting all the atoms in the formulas of some ionic compounds requires care. This is because many ionic compounds contain multiple polyatomic ions. The formula of ammonium sulfate, $(NH_4)_2SO_4$, is made up of three polyatomic ions, two ammonium ions, NH_4^+, and one sulfate ion, SO_4^{2-}. There is a total of 15 atoms in the formula unit, and each atom contributes to the formula mass of the compound. Two NH_4^+ ions = 2 N atoms and 8 H atoms; one SO_4^{2-} ion = 1 S atom and 4 O atoms. The sum of the atomic masses can be symbolized:

$$\text{Formula} = 2\,N + 8\,H + S + 4\,O$$

The formula mass of $(NH_4)_2SO_4$, with 15 atoms, is:

$$\begin{aligned} 2\,N = 2 \times 14.01\,\text{amu} &= 28.02\,\text{amu} \\ 8\,H = 8 \times 1.008\,\text{amu} &= 8.064\,\text{amu} \\ 1\,S = 1 \times 32.07\,\text{amu} &= 32.07\,\text{amu} \\ 4\,O = 4 \times 16.00\,\text{amu} &= 64.00\,\text{amu} \\ \text{formula mass of } (NH_4)_2SO_4 &= \overline{132.15\,\text{amu}} \end{aligned}$$

Answer: The molar mass of $(NH_4)_2SO_4$ is 132.15 g, and:

$$1 \text{ mole } (NH_4)_2SO_4 = 132.15 \text{ g } (NH_4)_2SO_4 = 6.022 \times 10^{23} \text{ formula units of } (NH_4)_2SO_4$$

Example 2: Find the molar mass of $Ca_3(PO_4)_2$. Determining the formula mass of calcium phosphate, $Ca_3(PO_4)_2$, which has 3 calcium ions, Ca^{2+}, and 2 phosphate ions, PO_4^{3-}, starts by counting the atoms of each element in the formula—3 Ca, 2 P, and 8 O—and ends by adding the atomic masses of the 13 atoms.

$$\text{The formula of } Ca_3(PO_4)_2 = 3\ Ca + 2\ P + 8\ O$$
$$(3\ Ca \times 40.08\ amu) + (2\ P \times 30.97\ amu) + (8\ O \times 16.00\ amu) = 310.18\ amu$$
$$\text{The formula mass of } Ca_3(PO_4)_2 = 310.18\ amu$$
$$\text{The molar mass of } Ca_3(PO_4)_2 = 310.18\ g$$
$$1\ \text{mole of } Ca_3(PO_4)_2 = 310.18\ g \text{ of } Ca_3(PO_4)_2 = 6.022 \times 10^{23} \text{ formula units}$$

Example 3: Find the molar mass of $MgSO_4{\cdot}7\ H_2O$. Hydrates are also ionic compounds that contain a definite amount of water in the crystalline solid. The hydrate of magnesium sulfate, $MgSO_4{\cdot}7\ H_2O$, has 1 Mg^{2+} ion, 1 SO_4^{2-} ion, and 7 molecules of water in the formula unit. Seven moles of water are present for each 1 mole of $MgSO_4$. There are 27 atoms in the formula unit: 1 Mg, 1 S, 11 O, and 14 H. The formula mass of $MgSO_4{\cdot}7\ H_2O$ is:

$$\text{The formula of } MgSO_4{\cdot}7\ H_2O = 1\ Mg + 1\ S + 11\ O + 14\ H$$
$$(1\ Mg \times 24.31\ amu) + (1\ S \times 32.07\ amu) + (11\ O \times 16.00\ amu) + (14\ H \times 1.008\ amu) = 246.49\ amu$$
$$\text{The molar mass of } MgSO_4{\cdot}7\ H_2O = 246.49\ g$$
$$1\ \text{mole of } MgSO_4{\cdot}7\ H_2O = 246.49\ g \text{ of } MgSO_4{\cdot}7\ H_2O = 6.022 \times 10^{23} \text{ formula units}$$

Example Problems

These problems have both answers and solutions given.

1. Calculate the molecular mass and molar mass (the mass of 1 mole) of each of the following molecular compounds. Round off the values to one place after the decimal.

   (a) H_2O

   Answer: molecular mass = 18.0 amu; molar mass = 18.0 g H_2O

   $H_2O = 2\ H + O = (2 \times 1.008\ amu) + 16.00\ amu = 18.016\ amu = 18.0\ amu$

   (b) CCl_4

   Answer: molecular mass = 153.8 amu; molar mass = 153.8 g CCl_4

   $CCl_4 = C + 4\ Cl = 12.01\ amu + (4 \times 35.45\ amu) = 153.81\ amu = 153.8\ amu$

   (c) $C_6H_{12}O_6$

   Answer: molecular mass = 180.1 amu; molar mass = 180.1 g $C_6H_{12}O_6$

   $C_6H_{12}O_6 = 6\ C + 12\ H + 6\ O = (6 \times 12.01\ amu) + (12 \times 1.008\ amu) + (6 \times 16.00\ amu) =$ 180.068 amu = 180.1 amu

2. Calculate the formula mass and molar mass of each of the following ionic compounds. Round off the values to one place after the decimal.

   (a) NaCl

Answer: formula mass = 58.4 amu; molar mass = 58.4 g NaCl

NaCl = Na + Cl = 22.99 amu + 35.45 amu = 58.44 amu = 58.4 amu

(b) $FeSO_4$

Answer: formula mass = 151.9 amu; molar mass = 151.9 g $FeSO_4$

$FeSO_4$ = Fe + S + 4 O = 55.85 amu + 32.07 amu + (4 × 16.00 amu) = 151.92 amu = 151.9 amu

(c) $K_3PO_4 \cdot 12H_2O$

Answer: formula mass = 428.5 amu; molar mass = 428.5 g $K_3PO_4 \cdot 12H_2O$

$K_3PO_4 \cdot 12H_2O$ = 3 K + P + 16 O + 24 H = (3 × 39.10 amu) + 30.97 amu + (16 × 16.00 amu) + (24 × 1.008 amu) = 428.462 amu = 428.5 amu

(d) $Ni(HCO_3)_2$

Answer: formula mass = 180.7 amu; molar mass = 180.7 g $Ni(HCO_3)_2$

$Ni(HCO_3)_2$ = Ni + 2 H + 2 C + 6 O = 58.69 amu + (2 × 1.008 amu) + (2 × 12.01 amu) + (6 × 16.00 amu) = 180.716 amu = 180.7 amu

Work Problems

Use these problems for additional practice.

1. Calculate the molecular mass and molar mass of each of the following molecular compounds. State all values to one place after the decimal:

   (a) SO_2; (b) P_4O_{10}; (c) N_2O_5; (d) C_8H_{18}

2. Calculate the formula mass and molar mass of each of the following ionic compounds. State all values to one place after the decimal:

   (a) LiCl; (b) Na_2CO_3; (c) $NiCl_2 \cdot 6H_2O$; (d) $Ba(OH)_2$

Worked Solutions

1. **Add the atomic masses of all the atoms in the molecule to get the molecular mass.**

   (a) molecular mass = 64.1 amu; molar mass = 64.1 g SO_2

   SO_2 = S + 2 O = 32.07 amu = (2 × 16.00 amu) = 64.07 amu = 64.1 amu

   (b) molecular mass = 283.9 amu; molar mass = 284.9 g P_4O_{10}

   P_4O_{10} = 4 P + 10 O = (4 × 30.97 amu) + (10 × 16.00 amu) = 283.88 amu = 283.9 amu

(c) molecular mass = 108.0 amu; molar mass = 108.0 g N_2O_5

N_2O_5 = 2 N + 5 O = (2 × 14.01 amu) + (5 × 16.00 amu) = 108.02 amu = 108.0 amu

(d) molecular mass = 114.2 amu; molar mass = 114.2 g C_8H_{18}

C_8H_{18} = 8 C + 18 H = (8 × 12.01 amu) + (18 × 1.008 amu) = 114.224 amu = 114.2 amu

2. **The formula mass is the sum of the atomic masses of all the atoms in the formula. The molar mass is the formula mass in grams.**

(a) formula mass = 42.4 amu; molar mass = 42.2 g LiCl

LiCl = Li + Cl = 6.941 amu + 35.45 amu = 42.391 amu = 42.4 amu

(b) formula mass =106.0 amu; molar mass = 106.0 g Na_2CO_3

Na_2CO_3 = 2 Na + C + 3 O = (2 × 22.99 amu) + 12.01 amu + (3 × 16.00) = 105.99 amu = 106.0 amu

(c) formula mass = 237.7 amu; molar mass = 237.7 g $NiCl_2{\cdot}6H_2O$

$NiCl_2{\cdot}6H_2O$ = Ni + 2 Cl + 12 H + 6 O = 58.69 amu + (2 × 35.45 amu) + (12 × 1.008 amu) + (6 × 16.00 amu) = 237.686 amu = 237.7 amu

(d) formula mass = 171.3 amu; molar mass = 171.3 g $Ba(OH)_2$

$Ba(OH)_2$ = Ba + 2 O + 2 H = 137.3 amu + (2 × 16.00 amu) + 2 × 1.008 amu) = 171.316 amu = 171.3 amu

Determining molecular and formula masses of compounds is a skill you will use over and over as you master chemical calculations. Be certain you can do it with ease before moving on.

Compounds: Mass-to-Mole and Mole-to-Mass Conversions

Just as with the elements, these are gram-to-mole and mole-to-gram conversions, and they are done in exactly the same way. The key is getting the appropriate conversion factor. Carbon dioxide, CO_2, will be used as a typical compound in the following problems.

Example 1: 75.0 g of CO_2 = _____ mole CO_2

First, calculate the molecular mass of CO_2. In all mass-to-mole or mole-to-mass conversions, the molecular mass of the compound must be known.

molecular mass CO_2 = C + 2 O = 12.01 amu + (2 × 16.00 amu) = 44.01 amu

The molar mass of CO_2 is 44.01 g, and the necessary conversion factor is derived from:

1 mole CO_2 = 44.01 g CO_2

The conversion factor must cancel "g CO_2" and replace it with "mole CO_2."

$$\text{mole } CO_2 = (75.0 \text{ g } CO_2) \times \left(\frac{1 \text{ mole } CO_2}{44.01 \text{ g } CO_2}\right) = 1.70 \text{ mole } CO_2$$

Answer: 75.0 g of CO_2 = 1.70 mole CO_2

Example 2: 0.0550 mole CO_2 = _____ gram CO_2

The conversion factor for this conversion is the inverse of that used in the preceding example, so that "mole CO_2" cancels and is replaced by "g CO_2."

$$\text{mass } CO_2 = 0.0550 \text{ mole } CO_2 \times \left(\frac{44.01 \text{ g } CO_2}{1 \text{ mole } CO_2}\right) = 2.42 \text{ g } CO_2$$

Answer: 0.0550 mole CO_2 = 2.42 g CO_2

Compounds: Moles and Counting Molecules

The key fact is that 1 mole of any compound = 6.022×10^{23} units of that compound.

Example 1: 5.00 g CO_2 = _____ molecules CO_2

To convert "g CO_2" to "molecules CO_2," you need a conversion factor that relates these terms. Knowing that the molar mass of CO_2 is 44.0 g:

$$1 \text{ mole } CO_2 = 44.0 \text{ g } CO_2 = 6.022 \times 10^{23} \text{ molecules } CO_2$$

The equality, 44.0 g CO_2 = 6.022×10^{23} molecules CO_2, will provide the conversion factor.

$$\text{molecules } CO_2 = (5.00 \text{ g } CO_2) \times \left(\frac{6.022 \times 10^{23} \text{ molecules}}{44.0 \text{ g } CO_2}\right) = 6.84 \times 10^{22} \text{ molecules}$$

Answer: 5.00 g CO_2 = 6.84×10^{22} molecules of CO_2

Example 2: 7.75×10^{10} molecules of CO_2 = _____ g CO_2

The conversion factor used in the previous problem can be used here, but it must be inverted to cancel "molecules" and keep "gram CO_2."

$$\text{mass } CO_2 = (7.75 \times 10^{10} \text{ molecules}) \times \left(\frac{44.0 \text{ g } CO_2}{6.022 \times 10^{23} \text{ molecules}}\right) = 5.66 \times 10^{-12} \text{ g } CO_2$$

Answer: 7.75×10^{10} molecules of CO_2 = 5.66×10^{-12} g CO_2

As stated earlier, ionic compounds are not made up of molecules; rather, the smallest unit of an ionic compound is the formula unit that is given by the formula of the compound. Barium fluoride, BaF_2, is an ionic compound and its formula unit, BaF_2, is composed of one barium ion, Ba^{2+}, and two fluoride ions, F^-. The molar mass of BaF_2 is 175.3 g.

$$1 \text{ mole } BaF_2 = 175.3 \text{ g } BaF_2 = 6.022 \times 10^{23} \text{ formula units of } BaF_2$$

Example 3: 10.5 g BaF_2 = _____ formula units BaF_2

The relationship between the molar mass of BaF_2 and Avogadro's number of formula units provides the conversion factor needed to solve the problem.

$$\text{formula units } BaF_2 = (10.5 \text{ g } \cancel{BaF_2}) \times \left(\frac{6.022 \times 10^{23} \text{ formula units}}{175.3 \text{ g } \cancel{BaF_2}}\right) = 3.61 \times 10^{22} \text{ formula units}$$

Answer: 10.5 g BaF_2 = 3.61×10^{22} formula units of BaF_2

Counting Atoms in Molecules

In 1 molecule of glucose, $C_6H_{12}O_6$, there are 6 atoms of carbon, 12 atoms of hydrogen, and 6 atoms of oxygen. The subscript numbers tell the number of atoms of each element in 1 molecule. The number of carbon atoms in 2.25×10^4 glucose molecules is 6 times the number of glucose molecules:

$$\text{number of C atoms} = (2.25 \times 10^4 \cancel{\text{molecules}}) \times \left(\frac{6 \text{ C atoms}}{1 C_6H_{12}O_6 \cancel{\text{ molecule}}}\right) = 1.35 \times 10^5 \text{ C atoms}$$

Because there are twice as many hydrogen atoms as carbon atoms in 1 molecule, the number of hydrogen atoms in 2.25×10^4 glucose molecules would be $2(1.35 \times 10^5$ atoms) or 2.70×10^5 H atoms.

Example 1: How many fluorine atoms are in 62.5 g of SF_4? The molar mass of SF_4 is 108.1 g.

Because there are 4 fluorine atoms in 1 molecule of SF_4, the number of fluorine atoms will be 4 times the number of molecules of SF_4 in 62.5 g of the compound.

$$\text{molecules of } SF_4 = 62.5 \text{ g } \cancel{SF_4} \times \left(\frac{6.022 \times 10^{23} \text{ molecules } SF_4}{108.1 \text{ g } \cancel{SF_4}}\right) = 3.48 \times 10^{23} \text{ molecules}$$

$$\text{number of F atoms} = (3.48 \times 10^{23} \cancel{\text{molecules}}) \times \left(\frac{4 \text{ F atoms}}{1 \cancel{\text{ molecule}}}\right) = 1.39 \times 10^{24} \text{ F atoms}$$

Answer: 1.39×10^{24} F atoms are in 62.5 g SF_4

Example 2: What mass of acetic acid, $HC_2H_3O_2$, contains 1.00×10^6 atoms of hydrogen? The molar mass of acetic acid is 60.0 g.

The formula, $HC_2H_3O_2$, shows that 1 molecule contains 4 atoms of H. The molar mass of $HC_2H_3O_2$ is 60.0 grams so, by definition, 60.0 g of $HC_2H_3O_2$ contains Avogadro's number of molecules. These two facts give us the necessary conversion factors.

$$\text{molecules } HC_2H_3O_2 = (1.00 \times 10^6 \cancel{\text{ H atoms}}) \times \left(\frac{1 \text{ molecule } HC_2H_3O_2}{4 \cancel{\text{ H atoms}}}\right) = 2.50 \times 10^5 \text{ molecules}$$

$$\text{mass } HC_2H_3O_2 = (2.50 \times 10^5 \cancel{\text{ molecules}}) \times \left(\frac{60.0 \text{ g } HC_2H_3O_2}{6.022 \times 10^{23} \cancel{\text{ molecules}}}\right) = 2.49 \times 10^{-17} \text{ g } HC_2H_3O_2$$

Answer: 2.49×10^{-17} g $HC_2H_3O_2$ contain 1.00×10^6 atoms of H

Example Problems

These problems have both answers and solutions given.

1. Which quantity represents the largest number of moles?

   (a) 355 g of $FeCl_3$; molar mass $FeCl_3$ = 295.6 g

   (b) 78.0 g $HC_2H_3O_2$; molar mass $HC_2H_3O_2$ = 60.0 g

   (c) 1.25 moles CS_2

   (d) 6.022×10^{23} formula units of NaCl

   Answer: (b) 1.30 moles $HC_2H_3O_2$. Compare the number of moles of each compound.

   (a) $\text{mole FeCl}_3 = (355\ \text{g FeCl}_3) \times \left(\frac{1\ \text{mole}}{295.6\ \text{g FeCl}_3}\right) = 1.20\ \text{moles FeCl}_3$

   (b $\text{mole HC}_2\text{H}_3\text{O}_2 = (78.0\ \text{g HC}_2\text{H}_3\text{O}_2) \times \left(\frac{1\ \text{mole}}{60.0\ \text{g HC}_2\text{H}_3\text{O}_2}\right) = 1.30\ \text{moles HC}_2\text{H}_3\text{O}_2$

   (c) 1.25 moles CS_2

   (d) 6.022×10^{23} formula units of NaCl = 1.00 mole NaCl

2. 255 g PCl_5 = _____ mole PCl_5; molar mass PCl_5 = 208.2 g

   Answer: 1.22 moles PCl_5

   $\text{mole PCl}_5 = (255\ \text{g PCl}_5) \times \left(\frac{1\ \text{mole}}{208.2\ \text{g PCl}_5}\right) = 1.22\ \text{moles PCl}_5$

3. 0.650 mole $Ni(NO_3)_2$ = _____ g $Ni(NO_3)_2$; molar mass $Ni(NO_3)_2$ = 182.7 g

   Answer: 119 g $Ni(NO_3)_2$

   $\text{mass Ni(NO}_3)_2 = (0.650\ \text{mole Ni(NO}_3)_2) \times \left(\frac{182.7\ \text{g Ni(NO}_3)_2}{1\ \text{mole Ni(NO}_3)_2}\right) = 118.8\ \text{g} = 119\ \text{g Ni(NO}_3)_2$

4. 3.50×10^{22} molecules Cl_2 = _____ g Cl_2; molar mass Cl_2 = 70.9 g

   Answer: 4.12 g Cl_2

   Knowing that one molar mass contains Avogadro's number of molecules, the mass is:

   $\text{mass Cl}_2 = (3.50 \times 10^{22}\ \text{molecules}) \times \left(\frac{70.9\ \text{g Cl}_2}{6.022 \times 10^{23}\ \text{molecules}}\right) = 4.12\ \text{g Cl}_2$

5. 85.0 g BCl_3 = _____ mole BCl_3 = _____ molecules BCl_3; molar mass BCl_3 = 117.2 g

   Answer: 0.725 mole, 4.32×10^{23} molecules

   $\text{mole BCl}_3 = (85.0\ \text{g BCl}_3) \times \left(\frac{1\ \text{mole BCl}_3}{117.2\ \text{g BCl}_3}\right) = 0.725\ \text{mole BCl}_3$

   $\text{molecules BCl}_3 = (0.725\ \text{mole BCl}_3) \times \left(\frac{6.022 \times 10^{23}\ \text{molecules}}{1\ \text{mole BCl}_3}\right) = 4.37 \times 10^{23}\ \text{molecules}$

6. The formula of ascorbic acid, vitamin C, is $C_6H_8O_6$. How many hydrogen atoms are in 0.125 mole of the compound?

Answer: 6.02×10^{23} H atoms

$$\text{number of molecules} = (0.125\,\text{mole}) \times \left(\frac{6.022\times 10^{23}\ \text{molecules}}{1\ \text{mole}}\right) = 7.53\times 10^{22}\ \text{molecules}$$

One molecule of acorbic acid contains 8 hydrogen atoms.

$$\text{number of H atoms} = (7.53\times 10^{22}\ \text{molecules}) \times \left(\frac{8\ \text{H atoms}}{1\ \text{molecule}}\right) = 6.02\times 10^{23}\ \text{H atoms}$$

Work Problems

Use these problems for additional practice.

1. Which quantity represents the largest number of moles?

   (a) 65.0 g of CrI_3; molar mass CrI_3 = 432.7 g

   (b) 24.8 g N_2O_5; molar mass N_2O_5 = 108.0 g

   (c) 0.140 mole $Ba(OH)_2$

   (d) 5.12×10^{22} molecules N_2

2. 7.35 g CuO = _____ mole CuO; molar mass CuO = 79.6 g

3. 0.800 mole $C_6H_{12}O_6$ = _____ g $C_6H_{12}O_6$; molar mass $C_6H_{12}O_6$ = 180.2 g

4. 9.35×10^{24} molecules of H_2O = _____ g H_2O; molar mass H_2O = 18.0 g

5. 42.6 g CO = _____ mole CO = _____ molecules CO; molar mass CO = 28.0 g

6. How many atoms of all elements are in 0.500 mole of aspirin, $C_9H_8O_4$?

Worked Solutions

1. **24.8 g N_2O_5 = 0.227 mole N_2O_5**

   (a) $\text{mole}\ CrI_3 = (65.0\ \text{g}\ CrI_3) \times \left(\frac{1\ \text{mole}}{432.7\ \text{g}\ CrI_3}\right) = 0.150\ \text{mole}\ CrI_3$

   (b) $\text{mole}\ N_2O_5 = (24.8\ \text{g}\ N_2O_5) \times \left(\frac{1\ \text{mole}}{108.0\ \text{g}\ N_2O_5}\right) = 0.230\ \text{mole}\ N_2O_5$

   (c) 0.140 mole $Ba(OH)_2$

   (d) $\text{mole}\ N_2 = (5.12\times 10^{22}\ \text{molecules}) \times \left(\frac{1\ \text{mole}\ N_2}{6.022\times 10^{23}\ \text{molecules}}\right) = 0.0850\ \text{mole}\ N_2$

2. **0.0923 mole CuO**

$$\text{mole CuO} = (7.35\ \text{g}\ \cancel{\text{CuO}}) \times \left(\frac{1\ \text{mole}}{79.6\ \text{g}\ \cancel{\text{CuO}}}\right) = 0.0923\ \text{mole CuO}$$

3. **144 g $C_6H_{12}O_6$**

$$\text{mass}\ C_6H_{12}O_6 = (0.800\ \cancel{\text{mole}\ C_6H_{12}O_6}) \times \left(\frac{180.2\ \text{g}\ C_6H_{12}O_6}{1\ \cancel{\text{mole}\ C_6H_{12}O_6}}\right) = 144\ \text{g}\ C_6H_{12}O_6$$

4. **279 g H_2O**

$$\text{mass}\ H_2O = (9.35 \times 10^{24}\ \cancel{\text{molecules}}) \times \left(\frac{18.0\ \text{g}\ H_2O}{6.022 \times 10^{23}\ \cancel{\text{molecules}}}\right) = 279\ \text{g}\ H_2O$$

5. **1.52 moles, 9.16 × 10^{23} molecules**

$$\text{mole CO} = (42.6\ \text{g}\ \cancel{\text{CO}}) \times \left(\frac{1\ \text{mole CO}}{28.0\ \text{g}\ \cancel{\text{CO}}}\right) = 1.52\ \text{moles CO}$$

$$\text{molecules CO} = (1.52\ \cancel{\text{mole CO}}) \times \left(\frac{6.022 \times 10^{23}\ \text{molecules}}{1\ \cancel{\text{mole CO}}}\right) = 9.16 \times 10^{23}\ \text{molecules}$$

6. **0.500 mole contains 6.32 × 10^{24} atoms**

The conversion of mole to molecule and molecules to atoms can be done in a single calculation:

$$\text{number of atoms} = (0.500\ \cancel{\text{mole}}) \times \left(\frac{6.022 \times 10^{23}\ \cancel{\text{molecules}}}{1\ \cancel{\text{mole}}}\right)\left(\frac{21\ \text{atoms}}{1\ \cancel{\text{molecule}}}\right) = 6.32 \times 10^{24}\ \text{atoms}$$

Chapter Problems and Answers

Consult the table of atomic masses or a periodic table for the necessary atomic masses.

Problems

1. Determine the molar mass of the following elements.

   (a) I, I_2 (c) Cd, Cu

   (b) Mn, Mg (d) O_2, N_2

2. Determine the molecular mass and the molar mass of these molecular compounds.

   (a) B_2H_6 (c) SF_6

   (b) $HC_3H_5O_2$ (d) $C_6H_4Br_2$

3. Determine the formula mass and molar mass of these ionic compounds.

   (a) $BaBr_2$ (c) $(NH_4)_2CO_3$

   (b) $Fe(NO_3)_3$ (d) $CuSO_4 \cdot 5\ H_2O$

For Questions 4 through 11, perform the requested mole-to-mass or mass-to-mole conversions.

4. 125 g Cu = _____ mole Cu

5. 1.65 moles Cl_2 = _____ g Cl_2

6. 4.55 g of N_2O_5 = _____ mole N_2O_5

7. 0.850 mole $Cu(NO_3)_2$ = _____ g $Cu(NO_3)_2$

8. 21.6 g NaCl = _____ mole NaCl

9. 10.0 moles $Sn(OH)_2$ = _____ g $Sn(OH)_2$

10. 9.55 g $CuSO_4 \cdot 5\ H_2O$ = _____ mole $CuSO_4 \cdot 5\ H_2O$

11. 0.5552 mole Al_2O_3 = _____ g Al_2O_3

For Questions 12 through 21, perform the requested conversions involving Avogadro's number.

12. 1.45 g of O_2 = _____ molecules of O_2 = _____ atoms of oxygen

13. 7.75×10^{25} molecules of H_2O = _____ g H_2O

14. 1.00 mole $Fe_2(SO_4)_3$ = _____ mole Fe^{3+} ions = _____ O atoms

15. 25.0 g H_2O = _____ molecules of H_2O = _____ atoms of hydrogen

16. 0.355 mole $C_6H_{12}O_6$ = _____ molecules of $C_6H_{12}O_6$

17. 2.35×10^{-3} g $CaCl_2$ = _____ Ca^{2+} ions and _____ Cl^- ions

18. 3.82×10^{-2} mole SO_3 = _____ SO_3 molecules = _____ oxygen atoms

19. 8.10×10^{25} atoms of H are in _____ mole of C_2H_6

20. 24 atoms of O are in _____ molecules of $C_2H_6O_2$

21. 8.00 moles of $Fe_2(SO_4)_3$ = _____ atoms of oxygen

Questions 22 through 25 require a solid understanding of the mole

22. 250 molecules of CO_2 contain _____ mole C

23. 750 atoms of phosphorus are in _____ g P_2O_5

24. 1.05×10^{-7} g $Fe(NO_3)_2$ contain _____ mole O

25. 500 O atoms are in _____ g $Ca_3(PO_4)_2$

Answers

1. **The molar mass is the atomic mass in grams.** The molar mass of a diatomic species is the molecular mass in grams.

   (a) $I = 126.9$ g, $I_2 = 253.8$ g

   (b) $Mn = 59.94$ g, $Mg = 24.31$ g

   (c) $Cd = 112.4$ g, $Cu = 63.55$ g

   (d) $O_2 = 32.00$ g, $N_2 = 28.02$ g

2. **The molecular mass is the sum of the atomic masses of all the atoms in the molecule.**

   (a) For B_2H_6, the molecular mass equals 2 B + 6 H = 2(10.81 amu) + 6(1.008 amu) = 27.67 amu and the molar mass is 27.67 g B_2H_6.

   (b) $HC_3H_5O_2$; molecular mass = 6 H + 3 C + 2 O = 74.08 amu; molar mass = 74.08 g

   (c) SF_6; molecular mass = S + 6 F = 146.07 amu; molar mass = 146.07 g

   (d) $C_6H_4Br_2$; molecular mass = 6 C + 4 H + 2 Br = 235.89 amu; molar mass = 235.89 g

3. **The formula mass of an ionic compound is the sum of the atomic masses of all the atoms in the formula.** Although present as ions, the atomic mass of the ion and atom are the same.

   (a) For $BaBr_2$, the formula mass equals Ba + 2 Br = 137.3 amu + 2(79.90 amu) = 297.10 amu and the molar mass is 297.10 g $BaBr_2$.

   (b) $Fe(NO_3)_3$; formula mass = Fe + 3 N + 9 O = 241.88 amu; molar mass = 241.88 g

   (c) $(NH_4)_2CO_3$; formula mass = 2 N + 8 H + C + 3 O = 96.09 amu; molar mass = 96.09 g

   (d) $CuSO_4 \cdot 5\ H_2O$; formula mass = Cu + S + 9 O + 10 H = 249.70 amu; molar mass = 249.70 g

4. **125 g Cu = 1.97 moles Cu**

$$\text{mole Cu} = 125\ \cancel{g} \times \left(\frac{1\ \text{mole}}{63.55\ \cancel{g\ Cu}}\right) = 1.97\ \text{moles}$$

5. **1.65 moles Cl_2 = 117 g Cl_2**

$$\text{mass } Cl_2 = 1.65\ \cancel{\text{moles } Cl_2} \times \left(\frac{70.90\ g}{1\ \cancel{\text{mole } Cl_2}}\right) = 117\ g$$

6. **4.55 g of N_2O_5 = 0.0421 mole N_2O_5**

$$\text{mole } N_2O_5 = 4.55\ \cancel{g\ N_2O_5} \times \left(\frac{1\ \text{mole}}{108.02\ \cancel{g\ N_2O_5}}\right) = 0.0421\ \text{mole}$$

7. **0.850 mole $Cu(NO_3)_2$ = 159 g $Cu(NO_3)_2$**

$$\text{mass } Cu(NO_3)_2 = 0.850\ \cancel{\text{mole } Cu(NO_3)_2} \times \left(\frac{187.6\ g}{1\ \cancel{\text{mole } Cu(NO_3)_2}}\right) = 159\ g$$

8. **21.6 g NaCl = 0.370 mole NaCl**

$$\text{mole NaCl} = 21.6\ \text{g NaCl} \times \left(\frac{1\ \text{mole}}{58.45\ \text{g NaCl}}\right) = 0.370\ \text{mole}$$

9. **10.0 moles $Sn(OH)_2$ = 1.53×10^3 g $Sn(OH)_2$**

$$\text{mass Sn(OH)}_2 = 10.0\ \text{moles Sn(OH)}_2 \times \left|\frac{152.7\ \text{g}}{1\ \text{mole Sn(OH)}_2}\right| = 1.53 \times 10^3\ \text{g}$$

10. **9.55 g $CuSO_4 \cdot 5H_2O$ = 3.82×10^{-2} mole $CuSO_4 \cdot 5H_2O$**

$$\text{mole CuSO}_4 \cdot 5\text{H}_2\text{O} = (9.55\ \text{g CuSO}_4 \cdot 5\text{H}_2\text{O}) \times \left|\frac{1\ \text{mole CuSO}_4 \cdot 5\text{H}_2\text{O}}{249.70\ \text{g CuSO}_4 \cdot 5\text{H}_2\text{O}}\right| = 3.82 \times 10^{-2}$$

11. **0.552 mole Al_2O_3 = 56.3 g Al_2O_3**

$$\text{mass Al}_2\text{O}_3 = 0.552\ \text{mole Al}_2\text{O}_3 \times \left(\frac{101.96\ \text{g}}{1\ \text{mole Al}_2\text{O}_3}\right) = 56.3\ \text{g}$$

12. **1.45 g of O_2 = 2.73×10^{22} molecules of O_2 = 5.46×10^{22} atoms of oxygen**

1 mole O_2 = 32.00 g O_2 = 6.022×10^{23} molecules O_2

$$\text{molecules O}_2 = 1.45\ \text{g O}_2 \times \left(\frac{6.022 \times 10^{23}\ \text{molecules}}{32.00\ \text{g O}_2}\right) = 2.73 \times 10^{23}\ \text{molecules}$$

number of oxygen atoms = 2 × (number of molecules of O_2) = 5.46×10^{22} atoms O

13. **7.75×10^{25} molecules of H_2O = 2.32×10^3 g H_2O**

1 mole H_2O= 18.00 g H_2O = 6.022×10^{23} molecules H_2O

$$\text{mass H}_2\text{O} = (7.75 \times 10^{25}\ \text{molecules H}_2\text{O}) \times \left(\frac{18.0\ \text{g H}_2\text{O}}{6.022 \times 10^{23}\ \text{molecules}}\right) = 2.32 \times 10^3\ \text{g}$$

14. **1.00 mole $Fe_2(SO_4)_3$ = 2 mole Fe^{3+} ions = 7.23×10^{24} O atoms**

1 mole $Fe_2(SO_4)_3$ contains 2 mole Fe, 3 mole S, and 12 mole O

12 mole O atoms = 12(6.022×10^{23} atoms) = 7.23×10^{24} O atoms

15. **25.0 g H_2O = 8.36×10^{23} molecules of H_2O =1.67×10^{24} atoms of hydrogen**

1 mole H_2O= 18.00 g H_2O = 6.022×10^{23} molecules H_2O

$$\text{molecules H}_2\text{O} = 25.0\ \text{g H}_2\text{O} \times \left(\frac{6.022 \times 10^{23}\ \text{molecules}}{18.0\ \text{g H}_2\text{O}}\right) = 8.36 \times 10^{23}\ \text{molecules}$$

$$\text{atoms of H} = 8.36 \times 10^{23}\ \text{molecules} \times \left(\frac{2\ \text{H atom}}{\text{molecule}}\right) = 1.67 \times 10^{24}\ \text{atoms}$$

16. **0.355 mole $C_6H_{12}O_6$ = 2.14×10^{23} molecules of $C_6H_{12}O_6$**

$$\text{molecules C}_6\text{H}_{12}\text{O}_6 = 0.355\ \text{mole C}_6\text{H}_{12}\text{O}_6 \times \left(\frac{6.022 \times 10^{23}\ \text{molecules}}{1\ \text{mole C}_6\text{H}_{12}\text{O}_6}\right) = 2.14 \times 10^{23}\ \text{molecules}$$

17. 2.35×10^{-3} g $CaCl_2$ = 1.28×10^{19} Ca^{2+} ions and 2.56×10^{19} Cl^- ions

1 mole $CaCl_2$ = 110.98 g $CaCl_2$ = 6.022×10^{23} formula units $CaCl_2$

$$\text{for.units } CaCl_2 = 2.35 \times 10^{-3}\,\cancel{g} \times \left(\frac{6.022 \times 10^{23}\text{ for.units}}{110.98\,\cancel{g}}\right) = 1.28 \times 10^{19}\text{ for.units}$$

$$\text{number } Ca^{2+}\text{ ions} = 1.28 \times 10^{19}\,\cancel{\text{for.units}} \times \left(\frac{1\,Ca^{2+}}{1\,\cancel{\text{for.unit}}}\right) = 1.28 \times 10^{19}\,Ca^{2+}\text{ ions}$$

number of Cl^- ions is twice the number of Ca^{2+} ions = 2.56×10^{19} Cl^- ions

18. 3.82×10^{-2} mole SO_3 = 2.30×10^{22} SO_3 molecules = 6.90×10^{22} oxygen atoms

1 mole SO_3 = 80.07 g SO_3 = 6.022×10^{22} molecules SO_3

$$\text{numbers } SO_3\text{ molecules} = \left(3.82 \times 10^{-2}\,\cancel{\text{mole}}\right) \times \left(\frac{6.022 \times 10^{22}\text{ molecules}}{1\,\cancel{\text{mole}}}\right) = 2.30 \times 10^{22}\text{ molecules}$$

$$\text{number O atoms} = \left(2.30 \times 10^{22}\,\cancel{\text{molecules}}\right) \times \left(\frac{3\text{ O atoms}}{1\,\cancel{\text{molecule}}}\right) = 6.90 \times 10^{22}\text{ O atoms}$$

19. 8.10×10^{25} atoms of H are in 22.4 moles of C_2H_6

1 mole C_2H_6 = 6 mole H atoms = $6(6.022 \times 10^{23})$ H atoms = 3.61×10^{24} H atoms

$$\text{mole } C_2H_6 = 8.10 \times 10^{25}\,\cancel{\text{H atoms}} \times \left(\frac{1\text{ mole } C_2H_6}{3.61 \times 10^{24}\,\cancel{\text{H atoms}}}\right) = 22.4\text{ moles}$$

20. 24 atoms of O are in 12 molecules of $C_2H_6O_2$

1 molecule $C_2H_6O_2$ contains 2 atoms of oxygen

$$\text{molecules of } C_2H_6O_2 = 24\,\cancel{\text{atoms O}} \times \left(\frac{1\text{ molecule}}{2\,\cancel{\text{atoms O}}}\right) = 12\text{ molecules}$$

21. 8.00 mole of $Fe_2(SO_4)_3$ = 5.78×10^{25} = atoms of oxygen

1 mole $Fe_2(SO_4)_3$ contains 12 moles O = $12(6.022 \times 10^{23})$ O atoms =7.23×10^{24} O atoms

$$\text{atoms of O} = 8\,\cancel{\text{moles } Fe_2(SO_4)_3} \times \left(\frac{7.23 \times 10^{24}\text{ O atoms}}{1\,\cancel{\text{mole } Fe_2(SO_4)_3}}\right) = 5.78 \times 10^{25}\text{ O atoms}$$

22. 4.15×10^{-22} mole C is in 250 molecules of CO_2

1 mole CO_2 = 6.022×10^{23} molecules CO_2

First, calculate the number of mole of CO_2 represented by 250 molecules of CO_2.

$$\text{mole } CO_2 = 250.\,\cancel{\text{molecules } CO_2} \times \left(\frac{1\text{ mole } CO_2}{6.022 \times 10^{23}\,\cancel{\text{molecules } CO_2}}\right) = 4.15 \times 10^{-22}\text{ mole } CO_2$$

The formula of CO_2 shows that 1 mole CO_2 contains 1 mole C; therefore, 4.15×10^{-22} mole CO_2 contains the identical number of mole of C, 4.15×10^{-22} mole C.

23. **8.84×10^{-20} g P_2O_5 contains 750 atoms of P**

First, calculate the number of molecules of P_2O_5 containing 750 atoms of P.

$$750 \text{ atoms P} \times \left(\frac{1 \text{ molecule } P_2O_5}{2 \text{ atoms P}}\right) = 375 \text{ molecules } P_2O_5$$

Then, calculate the mass of 375 molecules of P_2O_5. The molecular mass of P_2O_5 is 141.9 g.

$$\text{mass } P_2O_5 = 375 \text{ molecules} \times \left(\frac{141.9 \text{ g } P_2O_5}{6.022 \times 10^{23} \text{ molecules}}\right) = 8.84 \times 10^{-20} \text{ g } P_2O_5$$

24. **3.50×10^{-9} mole O is in 1.05×10^{-7} g $Fe(NO_3)_2$**

The formula shows that 1 mole $Fe(NO_3)_2$ contains 6 moles O.

First, calculate the number of mole $Fe(NO_3)_2$. The molar mass of $Fe(NO_3)_2$ is 179.9 g.

$$\text{mole } Fe(NO_3)_2 = 1.05 \times 10^{-7} \text{ g} \times \left(\frac{1 \text{ mole } Fe(NO_3)_2}{179.9 \text{ g}}\right) = 5.84 \times 10^{-10} \text{ mole } Fe(NO_3)_2$$

The number of mole of O is six times the number of mole of $Fe(NO_3)_2$.

$$\text{mole O} = 5.84 \times 10^{-10} \text{ mole } Fe(NO_3)_2 \times \left(\frac{6 \text{ mole O}}{1 \text{ mole } Fe(NO_3)_2}\right) = 3.50 \times 10^{-9} \text{ mole O}$$

25. **3.22×10^{-20} g $Ca_3(PO_4)_2$ contains 500 atoms of O**

One mole of $Ca_3(PO_4)_2$ contains 8 moles O which is $8 \times (6.022 \times 10^{23})$ oxygen atoms.

The molar mass of $Ca_3(PO_4)_2 = 310.18$ g.

Therefore, 310.18 g $Ca_3(PO_4)_2$ contains 4.818×10^{24} O atoms. This gives the conversion factor.

$$\text{mass } Ca_3(PO_4)_2 = 500 \text{ O atoms} \times \left(\frac{310.18 \text{ g } Ca_3(PO_4)_2}{4.818 \times 10^{24} \text{ O atoms}}\right) = 3.22 \times 10^{-20} \text{ g } Ca_3(PO_4)_2$$

Supplemental Chapter Problems

Problems

1. Determine the molar mass of the following elements.

   (a) B, Hg (b) Br, Br_2

2. Determine the molecular mass and the molar mass of these molecular compounds.

   (a) $C_4H_8Cl_2$ (b) $H_4P_2O_4$ (c) $SiCl_4$

3. Determine the formula mass and molar mass of these ionic compounds.

   (a) AlI_3 (b) $Ca(HCO_3)_2$ (c) $(NH_4)_3AsO_4$

For Questions 4 through 9, perform the requested mole-to-mass or mass-to-mole conversion.

4. 12.0 g K = _____ mole K

5. 0.450 mole H_2SO_4 = _____ g H_2SO_4

6. 125 g of $Ca(OH)_2$ = _____ mole $Ca(OH)_2$

7. 2.25 mole $Ni(NO_3)_2 \cdot 6\ H_2O$ = _____ g $Ni(NO_3)_2 \cdot 6\ H_2O$

8. 25.0 g Al_2O_3 = _____ mole Al_2O_3

9. 0.350 mole $C_2H_6O_2$ = _____ g $C_2H_6O_2$

For Questions 10 through 16, perform the requested conversions involving Avogadro's number.

10. 6.95 g of I_2 = _____ molecules of I_2 = _____ atoms of iodine

11. 4.35×10^{24} molecules of NO_2 = _____ mole NO_2 = _____ g NO_2

12. 0.750 mole H_3PO_4 contains _____ mole P and _____ g oxygen

13. 1.50×10^{12} atoms of Pb = _____ mole Pb = _____ g Pb

14. 7.35 g CH_3F contains _____ C atoms and _____ g H

15. 8.45×10^{-2} mole $Ba(OH)_2$ contains _____ OH^- ions and _____ Ba^{2+} ions

16. 0.800 mole C_4H_9OH contains _____ g C and _____ mole H atoms

Questions 17 through 20 require a solid understanding of the mole

17. 1,500 molecules of N_2O_5 contain_____ mole O

18. 350 atoms of H are in _____ g C_2H_5Cl

19. 2.45 g $Ba(OH)_2$ contains _____ mole H

20. 2.85 g $C_{12}H_{22}O_{11}$ contains _____ atoms

Answers

1. **(a) 10.81 g B, 200.6 g Hg** **(b) 79.90 g Br, 159.8 g Br_2** (page 112)

2. **(a) 127.00 amu, 127.00g** **(b) 129.97 amu, 129.97 g** **(c) 169.89 amu, 169.89 g** (page 118)

3. **(a) 407.7 amu, 407.7 g** **(b) 162.12 amu, 162.12 g** **(c) 193.05 amu, 193.05 g** (page 119)

4. **0.307 mole** (page 122)

5. **44.1 g** (page 122)
6. **1.69 moles** (page 122)
7. **654 g** (page 122)
8. **0.245 mole** (page 122)
9. **21.7 g** (page 122)
10. **1.65×10^{22} molecules, 3.30×10^{22} atoms** (page 110)
11. **7.22 moles, 332 g** (page 110)
12. **0.750 mole, 48.0 g** (page 110)
13. **2.49×10^{-12} mole, 5.16×10^{-10} g** (page 110)
14. **1.30×10^{23} atoms, 0.653 g** (page 110)
15. **1.02×10^{23} OH^-, 5.09×10^{22} Ba^{2+}** (page 110)
16. **38.4 g, 8.00 moles** (page 110)
17. **1.245×10^{-20} moles O**
18. **7.50×10^{-21} g C_2H_5Cl**
19. **2.86×10^{-2} moles H**
20. **2.26×10^{23} atoms**

Chapter 6

Percent Composition, Empirical and Molecular Formulas

The composition of compounds is given by their formulas. The formula of carbon dioxide, CO_2, tells us it is composed of carbon and oxygen, and that each molecule is composed of one atom of carbon and two atoms of oxygen. In this chapter, you will learn another way to describe the composition of compounds, in terms of the mass percent of each element in the compound. Suppose you are given one kilogram of each of two copper compounds, one that is about 66% copper by mass (CuS) and the other that is about 80% copper by mass (CuO). Which kilogram sample contains the greater mass of copper? It's the one with the greater percent copper, CuO. Knowing the mass-percent copper in each compound quickly provides the answer.

If you know the formula of a compound, you can calculate its percent composition. Just the reverse can be done too. If you know the percent composition of a compound, you can calculate a formula for the compound. A formula calculated from percent composition data is called an **empirical formula** (one calculated from experimental data). The formulas of ionic compounds are always empirical formulas. The formulas of molecular compounds may be the same as their empirical formulas or they may be some whole-number multiple of it. You will learn how to do composition-from-formula and formula-from-composition calculations in this chapter.

Calculating the Percent Composition of Compounds

The **percent composition** of a compound is the percent by mass of each element in the compound. Every sample of sulfur trioxide, SO_3, is 60.0% sulfur and 40.0% oxygen, by mass. This means that a 100.0-gram sample of sulfur trioxide contains 60.0 g of sulfur and 40.0 g of oxygen. Sodium chloride, NaCl, is 39.3% Na and 60.7% Cl, by mass. One hundred grams of NaCl contain 39.3 g of sodium and 60.7 g of chlorine.

The percent composition of a compound can be determined from its formula and the molar masses of the elements that make it up. Let's do a careful analysis of the formula of N_2O_5 to see how it is possible to calculate its percent composition is calculated.

The formula shows that in one molecule of N_2O_5 there are 2 atoms of N and 5 atoms of O:

$$1 \text{ molecule } N_2O_5 = N_{2\text{ atoms}}O_{5\text{ atoms}}$$

In 1 mole of N_2O_5 (6.022×10^{23} molecules of N_2O_5), there are 2 moles of N ($2 \times 6.022 \times 10^{23}$ atoms of N) and 5 moles of O ($5 \times 6.022 \times 10^{23}$ atoms of O):

$$1 \text{ mole } N_2O_5 = N_{2 \text{ moles}}O_{5 \text{ moles}}$$

Notice that the subscript numbers in the formula are equal to the number of moles of each element in *one* mole of the compound. This is a key fact that allows the calculation of percent composition from formulas.

Using the molar mass of nitrogen (14.01 g) and oxygen (16.00 g), the mass of each element in 1 mole of N_2O_5 is:

$$\begin{array}{ll} 2 \text{ moles N} = 2 \times 14.01 \text{ g N} = & 28.02 \text{ g N} \\ 5 \text{ moles O} = 5 \times 16.00 \text{ g O} = & 80.00 \text{ g O} \\ \hline 1 \text{ mole } N_2O_5 & = 108.02 \text{ g } N_2O_5 \end{array}$$

The calculation of the molar mass of N_2O_5 provides all the information needed to calculate the percent composition of N_2O_5. The calculation shows that 28.02 g of nitrogen and 80.00 g of oxygen are in 108.02 g of N_2O_5.

$$\% N = \frac{28.02 \cancel{g} \text{ N}}{108.02 \cancel{g} \text{ } N_2O_5} \times 100\% = 25.94\% \text{N}$$

$$\% O = \frac{80.00 \cancel{g} \text{ O}}{108.02 \cancel{g} \text{ } N_2O_5} \times 100\% = 74.06\% \text{O}$$

To summarize the calculation of percent composition from the formula of a compound:

1. Calculate the molar mass of the compound from its formula. This also provides the mass of each element in 1 mole of the compound.
2. Divide the mass of each element by the molar mass of the compound and multiply by 100%.

$$\% \text{ of an element} = \left(\frac{\text{mass of the element in 1 mole of the compound}}{\text{molar mass of the compound}} \right) \times 100\%$$

Percent numbers are always in terms of parts per hundred, so in the case of N_2O_5, the percent values can be interpreted to mean that 25.94 g of nitrogen are in 100.00 g of N_2O_5 and that 74.06 g of oxygen are in 100.00 g of N_2O_5. Two other useful ideas concerning percent composition are:

- The sum of the percents of all the elements in a compound equals 100%. If there are two elements in a compound, A and B, when you calculate the percent A, then the percent B equals (100% – %A). If there are three elements, calculate the percent of two and subtract those values from 100% for the third.
- When you know the percent of a compound that is a particular element, you can calculate the amount of that element in any size sample of that compound. To do this, change the percent of the element to its decimal equivalent (divide by 100%) and multiply it by the sample size. Suppose you need to know the mass of copper in a 525 g sample of $CuCO_3$. It is known that 51.4% of $CuCO_3$ is copper. Converting the percent copper to its decimal equivalent, 0.514, and multiplying by the sample mass shows that 270 g of copper are in 525 g of $CuCO_3$.

$$\text{mass of Cu} = 525\text{ g} \times 0.541 = 270.\text{ g Cu}$$

Or, in terms of percent:

$$\text{mass of element in a sample} = \text{mass of compound} \times \left(\frac{\%\text{ element in the compound}}{100\%}\right)$$

$$\text{mass of Cu} = 525\text{ g} \times \left(\frac{51.4\%}{100\%}\right) = 270.\text{ g Cu}$$

Let's apply both ideas in the following problem: (a) Calculate the percent composition of carbon dioxide, CO_2; (b) Determine the mass of carbon in 2,200 g of carbon dioxide. From the periodic table, the molar masses are C = 12.01 g and O = 16.00 g.

(a) One mole of CO_2 contains 1 mole of C and 2 moles of O, ($C_{1\text{ mole}}O_{2\text{ moles}}$):

$$\begin{array}{ll} 1\text{ mole C} = 1 \times 12.01\text{ g C} & = 12.01\text{ g C} \\ 2\text{ moles O} = 2 \times 16.00\text{ g O} & = 32.00\text{ g O} \\ \hline 1\text{ mole } CO_2 & = 44.01\text{ g } CO_2 \end{array}$$

12.01 g of carbon are in 44.01 g of CO_2. The percent carbon in CO_2 is:

$$\%C = \left(\frac{12.01\cancel{g}\text{ C}}{44.01\cancel{g}\text{ CO}_2}\right) \times 100\% = 27.29\%\text{C}$$

By difference, the percent oxygen is:

$$\%\text{O} = (100.00\% - \%\text{C}) = (100.00\% - 27.29\%) = 72.71\%\text{O}$$

(b) To calculate the mass of carbon in 2,200. g of CO_2, change the percent carbon to its decimal equivalent and multiply it by the mass of CO_2:

$$\text{mass of C} = 2200.\text{ g} \times 0.2729 = 600.4\text{ g C}$$

There are 600.4 g of carbon in 2,200 g of CO_2. The remainder is oxygen.

Calculating the percent composition of ionic compounds that contain polyatomic ions requires a careful counting of all the atoms of each element. For example, in 1 mole of ammonium phosphate, $(NH_4)_3PO_4$, there are 3 moles of ammonium ions, NH_4^+, and 1 mole of phosphate ion, PO_4^{3-}. In terms of the elements:

$$1\text{ mole } (NH_4)_3PO_4 = N_{3\text{ moles}}\, H_{12\text{ moles}}\, P_{1\text{ mole}}\, O_{4\text{ moles}}$$

The contribution of each element is calculated and summed to obtain the molar mass of $(NH_4)_3PO_4$. The molar masses of the elements are N = 14.01 g, H = 1.008 g, P = 30.97 g, and O = 16.00 g.

$$\begin{array}{ll} 3\text{ moles N} = 3 \times 14.01\text{ g N} & = 42.03\text{ N} \\ 12\text{ moles H} = 12 \times 1.008\text{ g H} & = 12.10\text{ g H} \\ 1\text{ mole P} = 1 \times 30.97\text{ g P} & = 30.97\text{ g P} \\ 4\text{ moles O} = 4 \times 16.00\text{ g O} & = 64.00\text{ g O} \\ \hline 1\text{ mole } (NH_4)_3PO_4 & = 149.10\text{ g } (NH_4)_3PO_4 \end{array}$$

The mass percent of nitrogen in $(NH_4)_3PO_4$ is:

$$\% N = \left(\frac{42.03 \cancel{g} N}{149.10 \cancel{g} (NH_4)_3 PO_4} \right) \times 100\% = 28.19\% N$$

The mass percents of the other three elements are H = 8.12%, P = 20.77%, and O = 42.92%. You should calculate each of these percentages to ensure they are correct.

Example Problems

1. Calculate the percent composition of aluminum oxide, Al_2O_3.

   From the periodic table, the molar masses of the elements are Al = 26.98 g and O = 16.00 g.

   Answer: Al = 52.92%; O = 47.08%

   $$\begin{array}{ll} 2 \text{ moles Al} = 2 \times 26.98 \text{ g Al} = & 53.96 \text{ g Al} \\ 3 \text{ moles O} = 3 \times 16.00 \text{ g O} = & 48.00 \text{ g O} \\ \hline 1 \text{ mole } Al_2O_3 & = 101.96 \text{ g } Al_2O_3 \end{array}$$

   $$\% Al = \frac{53.96 \cancel{g} Al}{101.96 \cancel{g} Al_2O_3} \times 100\% = 52.92\% Al$$

   $$\% O = \frac{48.00 \cancel{g} O}{101.96 \cancel{g} Al_2O_3} \times 100\% = 47.08\% O$$

2. (a) Calculate the percent aluminum in aluminum sulfate, $Al_2(SO_4)_3$.

   (b) How many grams of aluminum are in 500 g of $Al_2(SO_4)_3$?

   The molar masses of the elements are Al = 26.98 g, S = 32.07 g, and O = 16.00 g.

   Answer: Al = 15.77%; 78.85 g Al

   (a) $$\begin{array}{ll} 2 \text{ moles Al} = 2 \times 26.98 \text{ g Al} = & 53.96 \text{ g Al} \\ 3 \text{ moles S} = 3 \times 32.07 \text{ g S} = & 96.21 \text{ g S} \\ 12 \text{ moles O} = 12 \times 16.00 \text{ g O} = & 192.00 \text{ g O} \\ \hline 1 \text{ mole } Al_2(SO_4)_3 & = 342.17 \text{ g } Al_2(SO_4)_3 \end{array}$$

   $$\% Al = \frac{53.96 \cancel{g} Al}{342.17 \cancel{g} Al_2(SO_4)_3} \times 100\% = 15.77\% Al$$

   (b) The mass of Al in 500.0 g of $Al_2(SO_4)_3$ = (500.0 g × 0.1577) = 78.85 g Al.

Work Problems

1. What is the percent composition of diphosphorus pentaoxide, P_2O_5?

   The molar masses of the elements are P = 30.97 g and O = 16.00 g.

2. (a) What is the percent iron in Fe_2O_3?

   (b) How many grams of iron are in 5,000 g of Fe_2O_3?

   The molar masses of the elements are Fe = 55.85 g and O = 16.00 g.

Worked Solutions

1. **P = 43.64%; O = 56.36%**

$$\begin{array}{ll} \text{2 moles P} = 2 \times 30.97\text{ g P} = & 61.96\text{ g P} \\ \text{5 moles O} = 5 \times 16.00\text{ g O} = & 80.00\text{ g O} \\ \hline \text{1 mole } P^2O^5 & = 141.94\text{ g } P^2O^5 \end{array}$$

$$\%\,P = \frac{61.94\,\cancel{g}\,P}{141.94\,\cancel{g}\,P_2O_5} \times 100\% = 43.64\%\,P$$

and by difference: $\%\,O = (100.00\% - 43.64\%) = 56.36\%\,O$

2. **Fe = 69.94%; 3,497 g Fe**

(a)
$$\begin{array}{ll} \text{2 moles Fe} = 2 \times 55.85\text{ g Fe} = 111.70\text{ g Fe} & \\ \text{3 moles O} = 3 \times 16.00\text{ g O} = \;\; 48.00\text{ g O} & \\ \hline \text{1 mole } Fe_2O_3 & = 159.70\text{ g } Fe_2O_3 \end{array}$$

$$\%\,Fe = \frac{111.70\,\cancel{g}\,Fe}{159.70\,\cancel{g}\,Fe_2O_3} \times 100\% = 69.94\%\,Fe$$

(b) The mass of iron in 5,000 g of Fe_2O_3 = (5,000 g × 0.6994) = 3,497 g Fe.

Percent Water in a Hydrate

Hydrates were introduced and discussed in Chapter 4. They are ionic compounds that contain a definite number of moles of water in 1 mole of a compound. The formula of copper(II) sulfate pentahydrate, $CuSO_4{\cdot}5H_2O$, shows that 5 moles of water are associated with 1 formula unit of copper(II) sulfate, $CuSO_4$. Said another way, there are five moles of water in 1 mole of $CuSO_4{\cdot}5H_2O$. The 5 moles of water are separated from the formula of copper(II) sulfate with a dot (·). Some other hydrates are $NiCl_2{\cdot}6H_2O$, $MgSO_4{\cdot}7H_2O$, and $CaCl_2{\cdot}2H_2O$. The percent water in a hydrate is calculated in a manner similar to that used to calculate the percent of an element except that the mass of water in 1 mole of the hydrate is used instead of the mass of a single element.

$$\%\,H_2O\text{ in a hydrate} = \left(\frac{\text{mass of } H_2O \text{ in the hydrate}}{\text{molar mass of the hydrate}}\right) \times 100\%$$

Let's calculate the percent water in calcium chloride dihydrate, $CaCl_2 \cdot 2H_2O$. One mole of $CaCl_2 \cdot 2H_2O$ contains 1 mole of Ca, 2 moles of Cl, and 2 moles of H_2O. The molar mass of H_2O is 18.02 g. The molar masses of the elements are Ca = 40.08 g and Cl = 35.46 g.

$$\begin{array}{ll} 1 \text{ mole Ca} = 1 \times 40.08 \text{ g Ca} & = 40.08 \text{ g Ca} \\ 2 \text{ moles Cl} = 2 \times 35.45 \text{ g Cl} & = 70.90 \text{ g Cl} \\ 2 \text{ moles } H_2O = 2 \times 18.02 \text{ g } H_2O = & 36.04 \text{ g } H_2O \\ \hline 1 \text{ mole } CaCl_2 \cdot 2H_2O & = 147.02 \text{ g } CaCl_2 \cdot 2H_2O \end{array}$$

The calculation of the molar mass of the hydrate shows that 36.04 g of water are in 147.02 g of the compound. The percent water in $CaCl_2 \cdot 2H_2O$ is:

$$\% H_2O = \left(\frac{36.04 \not{g} \, H_2O}{147.02 \not{g} \, CaCl_2 \cdot 2H_2O} \right) \times 100\% = 24.51\% H_2O$$

What mass of water is in 475 g of $CaCl_2 \cdot 2H_2O$, knowing that 24.51% of the hydrate is water? Multiplying 475 g by the decimal equivalent of 24.51% shows that 116 g of water are in 475 g of $CaCl_2 \cdot 2H_2O$.

$$\text{Mass of water} = 475 \text{ g} \times 0.2451 = 116 \text{ g } H_2O$$

Because hydrates contain molecules of water along with the positive and negative ions, it is useful to describe the composition of hydrates only in terms of the water they contain. Of course, as with any compound, the percent composition in terms of each element can be calculated just as was done in the previous section with CO_2 and $(NH_4)_3PO_4$.

Example Problems

1. What is the percent water in $NiCl_2 \cdot 2H_2O$?

   The molar masses are Ni = 58.69 g, Cl = 35.45 g, and H_2O = 18.02 g.

   Answer: 21.76% H_2O

$$\begin{array}{ll} 1 \text{ mole Ni} = 1 \times 58.69 \text{ g Ni} & = 58.69 \text{ g Ni} \\ 2 \text{ moles Cl} = 2 \times 35.45 \text{ g Cl} & = 70.90 \text{ g Cl} \\ 2 \text{ moles } H_2O = 2 \times 18.02 \text{ g } H_2O = & 36.04 \text{ g } H_2O \\ \hline 1 \text{ mole } NiCl_2 \cdot 2H_2O & = 165.63 \text{ g } NiCl_2 \cdot 2H_2O \end{array}$$

$$\% H_2O = \left(\frac{36.04 \not{g} \, H_2O}{165.63 \not{g} \, NiCl_2 \cdot 2H_2O} \right) \times 100\% = 21.76\% H_2O$$

2. What mass of water is in 1.500×10^3 g of $NiCl_2 \cdot 2H_2O$?

   Answer: 326.4 g H_2O. The mass of water = $(1.500 \times 10^3 \text{ g}) \times 0.2176 = 326.4$ g H_2O.

Work Problems

1. What is the percent water in $CuSO_4 \cdot 5H_2O$?

   The molar masses are Cu = 63.55 g, S = 32.07 g, O = 16.00 g, and H_2O = 18.02 g.

2. What mass of water is in 125 g of $CuSO_4 \cdot 5H_2O$?

Worked Solutions

1. The percent water in $CuSO_4 \cdot 5H_2O$ is 36.08%.

$$\begin{aligned}
&1 \text{ mole Cu} = 1 \times 63.55 \text{ g Cu} &&= 63.55 \text{ g Cu}\\
&1 \text{ mole S} = 1 \times 32.07 \text{ g S} &&= 32.07 \text{ g S}\\
&4 \text{ moles O} = 4 \times 16.00 \text{ g O} &&= 64.00 \text{ g O}\\
&5 \text{ moles } H_2O = 5 \times 18.02 \text{ g O} &&= 90.10 \text{ g } H_2O\\
\hline
&1 \text{ mole } CuSO_4 \cdot 5H_2O &&= 249.72 \text{ g } CuSO_4 \cdot 5H_2O
\end{aligned}$$

$$\% H_2O = \left(\frac{90.10 \cancel{g}\, H_2O}{249.72 \cancel{g}\, CuSO_4 \cdot 5H_2O}\right) \times 100\% = 36.08\% H_2O$$

2. 45.1 g H_2O. The mass of water = 125 g × 0.3608 = 45.1 g H_2O.

Empirical and Molecular Formulas

There are two broad classes of formulas for compounds: empirical formulas and molecular formulas. The **empirical formula** shows the *simplest* ratio of elements in a compound and uses the smallest possible set of subscript numbers. Empirical formulas are also called **simple formulas.** The formulas of *all* ionic compounds are empirical formulas. Since ionic compounds do not exist as molecules, their formulas are not molecular formulas. The formulas calculated from percent composition data are empirical formulas. (We'll get to these calculations later.)

Molecular formulas are the formulas of molecules, and they show all the atoms of each element in the molecule. A molecular formula need not have the smallest set of subscript numbers; rather, it can be either the same as or a whole-number multiple of its empirical formula. For example, the molecular formula of glucose is $C_6H_{12}O_6$. The empirical formula of glucose is CH_2O, which shows the *simplest* ratio of the three elements. You can see that the molecular formula is six times the empirical formula: $C_6H_{12}O_6 = 6 \times (CH_2O)$. Sometimes the molecular and the empirical formulas of a molecular compound are the same, as they are for H_2O, CO_2, and BCl_3.

As a general statement:

$$\text{molecular formula} = \mathbf{n} \times (\text{empirical formula})$$

where **n** is a whole number from 1 to as large as necessary. Several molecular compounds are listed in the following table in terms of both their empirical and molecular formulas.

Comparing Empirical and Molecular Formulas			
Compound	***Empirical Formula***	***n***	***Molecular Formula***
Carbon dioxide	CO_2	1	CO_2
Oxalic acid	HCO_2	2	$H_2C_2O_4$
Dinitrogen tetraoxide	NO_2	2	N_2O_4
Benzene	CH	6	C_6H_6
Glucose	CH_2O	6	$C_6H_{12}O_6$

When an empirical formula is known for a compound, one additional piece of information is needed to determine whether it is also the molecular formula: the molecular mass or molar mass of the compound. For example, the empirical formula of a molecular compound of boron is BH_3. The sum of the atomic masses of the four atoms in BH_3 is 13.83 amu (B + 3H = 10.81 amu + 3(1.008 amu) = 13.83 amu). In a separate measurement, the molecular mass of the boron compound is found to be 27.66 amu. To summarize:

$$\text{empirical formula: } BH_3$$
$$\text{empirical formula mass} = 13.83 \text{ amu}$$
$$\text{measured molecular mass} = 27.66 \text{ amu}$$

The same whole number ratio, **n,** that holds for molecular and empirical formulas also holds for the molecular and empirical formula masses.

$$n = \frac{\text{molecular formula}}{\text{empirical formula}} = \frac{\text{molecular mass}}{\text{empirical formula mass}} = \frac{27.66\,\text{amu}}{13.83\,\text{amu}} = 2$$

Knowing that **n** equals 2, the molecular formula is then two times the empirical formula:

$$\text{molecular formula} = 2 \times (\text{empirical formula})$$
$$\text{molecular formula} = 2(BH_3) = B_2H_6$$

The only time you need to be concerned with the conversion of an empirical formula to a molecular formula is with molecular compounds. Remember, molecular compounds are composed of nonmetallic elements; ionic compounds are composed of both metallic and nonmetallic elements.

Example Problems

1. A compound has been isolated and analyzed. The empirical formula of the compound is CH_2. Its molecular mass is 42.08 amu. What is the molecular formula of the compound?

   Answer: C_3H_6

   The empirical formula mass of CH_2 = C + 2H = 12.01 amu + 2(1.008 amu) = 14.03 amu.

   $$n = \frac{\text{molecular mass}}{\text{empirical formula mass}} = \frac{42.08\,\text{amu}}{14.03\,\text{amu}} = 2.999 = 3$$

   molecular formula = $3(CH_2) = C_3H_6$

2. The empirical formula of colorless gas is $COCl_2$. Its molecular mass is 98.9 amu. What is the molecular formula of the compound?

Answer: $COCl_2$

The empirical formula mass of $COCl_2$ = C + O + 2Cl =

12.01 amu + (16.00 amu) + 2(35.45 amu) = 98.91 amu.

The empirical formula mass is the same as the molecular mass of the compound, n = 1. This means that the empirical formula and molecular formula are the same.

molecular formula = $COCl_2$

Work Problems

1. A compound used to whiten teeth has the empirical formula of HO. The molecular mass of this compound is 34.02 amu. What is its molecular formula?

2. Xylene, a solvent used in industry, has the empirical formula of C_4H_5. The molecular mass of xylene is 106.2 amu. What is the molecular formula of xylene?

Worked Solutions

1. **The compound is H_2O_2, hydrogen peroxide.**

The empirical formula mass of HO = (H + O) = 1.008 amu + 16.00 amu = 17.008 amu.

$$n = \frac{\text{molecular mass}}{\text{empirical formula mass}} = \frac{34.02\,\text{amu}}{17.008\,\text{amu}} = 2.000 = 2$$

molecular formula = 2(HO) = H_2O_2

2. **The molecular formula of xylene is C_8H_{10}.**

The empirical formula mass of C_4H_5 = 4C + 5H = 4(12.01 amu) + 5(1.008 amu) = 53.08 amu.

$$n = \frac{\text{molecular mass}}{\text{empirical formula mass}} = \frac{106.2\,\text{amu}}{53.08\,\text{amu}} = 2.001 = 2$$

molecular formula = $2(C_4H_5)$ = C_8H_{10}

Calculating Empirical Formulas

Empirical formulas can be calculated from the percent composition of a compound. You have already learned that the subscript numbers in formulas can be read in terms of the number of moles of each element. In 1 mole of N_2O_5, there are 2 moles of N and 5 moles of O. Read the formula as $N_{2\,\text{moles}}O_{5\,\text{moles}}$. The reason this is mentioned again is because any data that allows the number of moles of each element in a compound to be known allows the calculation of the empirical formula of that compound. Percent composition data does this. Here's the step-by-step way it is done:

1. Convert the percent numbers to grams for each element.
2. Convert the mass of each element to moles of that element.
3. The number of moles for each element become their subscript in the empirical formula.
4. Convert the subscript numbers to whole numbers.

Let's follow these steps and determine the empirical formula of an ionic compound of chromium and chlorine, which has a percent composition of Cr = 42.31% and Cl = 57.69%.

1. Convert the percent numbers to grams of each element.

   Cr = 42.31% becomes Cr = 42.31 g

   Cl = 57.69% becomes Cl = 57.69 g

2. Convert the mass of each element to moles of that element. (The molar masses of the elements are Cr = 52.00 g and Cl = 35.45 g.)

   $$\text{mole Cr} = (42.31\,\text{g Cr}) \times \left(\frac{1\,\text{mole Cr}}{52.00\,\text{g Cr}}\right) = 0.8137\,\text{mole Cr}$$

   $$\text{mole Cl} = (57.69\,\text{g Cl}) \times \left(\frac{1\,\text{mole Cl}}{35.45\,\text{g Cl}}\right) = 1.627\,\text{moles Cl}$$

3. The number of moles of each element become the subscripts in the empirical formula.

   $Cr_{0.8137\,\text{mole}}Cl_{1.627\,\text{mole}}$

4. Convert the subscript numbers to whole numbers.

   Since formulas are written using whole numbers, not decimals, a little arithmetic is necessary to convert decimals to whole numbers. Divide *both* numbers of moles by the *smaller* number. This converts the smaller number to one, so you have at least one whole number. This may also convert the subscript number of the other element to a whole number. If so, you have the empirical formula.

   $$Cr_{\frac{0.814\,\text{mole}}{0.837\,\text{mole}}}Cl_{\frac{1.627\,\text{mole}}{0.837\,\text{mole}}} = Cr_1Cl_2 = CrCl_2$$

   The compound is chromium (II) chloride, $CrCl_2$.

Here's another example. Determine the empirical formula of an oxide of iron from its percent composition: Fe = 69.94%; O = 30.06%.

1. Change the percent numbers to mass: Fe = 69.94 g and O = 30.06 g.
2. Convert the mass of each element to moles of that element:

   $$\text{mole Fe} = (69.94\,\text{g Fe}) \times \left(\frac{1\,\text{mole Fe}}{55.85\,\text{g Fe}}\right) = 1.252\,\text{moles Fe}$$

   $$\text{mole O} = (30.06\,\text{g O}) \times \left(\frac{1\,\text{mole O}}{16.00\,\text{g O}}\right) = 1.879\,\text{moles O}$$

3. Using these mole values, the empirical formula can be written as:

   $Fe_{1.252\,\text{mole}}O_{1.879\,\text{mole}}$

4. Dividing each mole quantity by the smaller value gives:

$$Fe_{\frac{1.252\ mole}{1.252\ mole}}O_{\frac{1.879\ mole}{1.252\ mole}} = Fe_1O_{1.5}$$

This time all the subscripts did not turn out to be whole numbers, but one more bit of arithmetic solves the problem: Multiply both subscripts by 2.

$$Fe_{1\times 2}O_{1.5\times 2} = Fe_2O_3$$

And the empirical formula of the compound is obtained: Fe_2O_3, iron (III) oxide.

When you have determined the number of moles of each element in the empirical formula, the rest of the problem is the arithmetic needed to make the subscripts whole numbers. Divide all subscripts by the smaller or smallest one, and if they are not whole numbers, begin multiplying them *all* by 2 or 3 or 4, and so on, until they all become whole numbers.

Empirical formulas can also be calculated from the *mass* of each element in a sample of a compound. Analysis of a sample of a compound shows that it contains 1.179 g Na and 0.821 g S. You could calculate the percent of each element from these numbers, but that's not necessary. The mass of each element is converted directly to moles for the empirical formula. From the table of atomic masses, we find the molar masses are 22.99 g for Na and 32.07 g for S.

$$\text{mole Na} = (1.179\ \cancel{\text{g Na}}) \times \left(\frac{1\ \text{mole Na}}{22.99\ \cancel{\text{g Na}}}\right) = 0.05128\ \text{mole Na}$$

$$\text{mole S} = (0.821\ \cancel{\text{g S}}) \times \left(\frac{1\ \text{mole S}}{32.07\ \cancel{\text{g S}}}\right) = 0.0256\ \text{mole S}$$

The empirical formula is $Na_{0.0528\ mole}S_{0.0256\ mole} = Na_{\frac{0.0528\ mole}{0.0256\ mole}}S_{\frac{0.0256\ mole}{0.0256\ mole}} = Na_2S_1 = Na_2S$.

It is important to realize that only empirical formulas are calculated from percent composition data or mass data. That's fine for ionic compounds because their formulas are always empirical formulas. But the complete formulas of molecular compounds, the molecular formulas, can only be obtained from the empirical formulas if the molecular mass or molar mass of the compound is also known.

The following table lists three different molecular compounds that have the same empirical formula, CH. Notice that the percent composition of each of the three compounds is the same, 92.26% C and 7.74% H. The molecular formulas are different but, because the empirical formulas are the same, the percent compositions are the same.

Different Molecular Formulas—Same Percent Composition

Empirical Formula	***n***	***Molecular Formula***	***Molecular Mass***	***Percent Composition***
CH	2	C_2H_2	26.04 amu	92.26% C; 7.74% H
CH	4	C_4H_4	52.08 amu	92.26% C; 7.74% H
CH	6	C_6H_6	78.12 amu	92.26% C; 7.74% H

Calculating Molecular Formulas

Let's put all the ideas presented in the previous sections together and calculate the molecular formula of borazine. The percent composition of borazine is 40.29% boron, 52.21% nitrogen, and 7.50% hydrogen.

The molecular mass of borazine is 80.50 amu. The atomic masses of the elements are B = 10.81amu, N = 14.01 amu, and H = 1.008 amu.

1. Derive the empirical formula of borazine. Determine the number of moles of each element.

$$\text{mole B} = (40.29\ g\cancel{B}) \times \left(\frac{1\ \text{mole B}}{10.81\ g\cancel{B}}\right) = 3.727\ \text{moles B}$$

$$\text{mole N} = (52.21\ g\cancel{N}) \times \left(\frac{1\ \text{mole N}}{14.01\ g\cancel{N}}\right) = 3.727\ \text{moles N}$$

$$\text{mole H} = (7.50\ g\cancel{H}) \times \left(\frac{1\ \text{mole H}}{1.008\ g\cancel{H}}\right) = 7.44\ \text{moles H}$$

These mole values are used to write an empirical formula.

$B_{3.727\ mole}H_{3.727\ mole}H_{7.44\ mole}$

2. Dividing each decimal number by the smaller number, 3.727 moles, converts each to a whole number. The empirical formula is:

$$B_{\frac{3.727\ mole}{3.727\ mole}}N_{\frac{3.727\ mole}{3.727\ mole}}H_{\frac{7.44\ mole}{3.727\ mole}} = B_1N_1H_2 = BNH_2$$

The empirical formula mass = B + N + 2H = 10.81 amu + 14.01 amu + 2(1.008 amu) = 26.84 amu.

3. Dividing the molecular mass of borazine, 80.50 amu, by the empirical formula mass shows that the formula of the molecule is three times the empirical formula.

$$n = \frac{\text{molecular mass}}{\text{empirical formula mass}} = \frac{80.50\ \text{amu}}{26.84\ \text{amu}} = 2.999 = 3$$

The molecular formula of borazine is $3(BNH_2) = B_3N_3H_6$.

Example Problems

Obtain the necessary atomic masses from the table of atomic masses in Chapter 5 or a periodic table.

1. Calculate the empirical formula of a compound that is 63.6% nitrogen and 36.4% oxygen.

Answer: N_2O

Change the percent values to mass, and calculate the number of moles of each element.

$$\text{mole N} = (63.6\ g\cancel{N}) \times \left(\frac{1\ \text{mole N}}{14.01\ g\cancel{N}}\right) = 4.54\ \text{moles N}$$

$$\text{mole O} = (36.4\ g\cancel{O}) \times \left(\frac{1\ \text{mole O}}{16.00\ g\cancel{O}}\right) = 2.28\ \text{moles O}$$

Write the empirical formula of the compound using these mole values and convert them to small, whole numbers.

$$N_{4.54\ \text{mole}}O_{2.28\ \text{mole}} = N_{\frac{4.54\ \text{mole}}{2.28\ \text{mole}}}O_{\frac{2.28\ \text{mole}}{2.28\ \text{mole}}} = N_2O_1 = N_2O$$

2. A 1.500 g sample of a compound contains 0.467 g sulfur and 1.033 g chlorine. What is the empirical formula of the compound?

Answer: SCl_2

Determining the number of moles of each element:

$$\text{mole S} = 0.467\ \text{g S} \times \left(\frac{1\ \text{mole S}}{32.07\ \text{g S}}\right) = 0.0146\ \text{mole S}$$

$$\text{mole Cl} = 1.033\ \text{g Cl} \times \left(\frac{1\ \text{mole Cl}}{35.45\ \text{g Cl}}\right) = 0.0291\ \text{mole Cl}$$

The empirical formula is $S_{0.0146\ \text{mole}}Cl_{0.0291\ \text{mole}} = S_{\frac{0.0146\ \text{mole}}{0.0146\ \text{mole}}}Cl_{\frac{0.0291\ \text{mole}}{0.0146\ \text{mole}}} = S_1Cl_2 = SCl_2$.

3. The empirical formula of a compound is $PNCl_2$, and its molecular mass is 347.6 amu. What is the molecular formula of the compound?

Answer: $P_3N_3Cl_6$

First, calculate the mass of the empirical formula, $PNCl_2$.

$$P + N + 2\ Cl = 30.97\ \text{amu} + 14.01\ \text{amu} + 2(35.45\ \text{amu}) = 115.9\ \text{amu}$$

Dividing the molecular mass of the compound by the empirical formula mass shows that the formula of the molecule is three times the empirical formula.

$$n = \frac{\text{molecular mass}}{\text{empirical formula mass}} = \frac{347.6\ \text{amu}}{115.9\ \text{amu}} = 2.99 = 3$$

The molecular formula is $3(PNCl_2) = P_3N_3Cl_6$.

4. Calculate the molecular formula of a compound that is 40.68% C, 5.12% H, and 54.20% O. The molecular mass of the compound is 118.1 amu.

Answer: $C_4H_6O_4$

Convert each percent value to mass, and determine the number of moles of each element:

$$\text{mole C} = 40.68\ \text{g C} \times \left(\frac{1\ \text{mole C}}{12.01\ \text{g C}}\right) = 3.387\ \text{moles C}$$

$$\text{mole H} = 5.12\ \text{g H} \times \left(\frac{1\ \text{mole H}}{1.008\ \text{g H}}\right) = 5.079\ \text{moles H}$$

$$\text{mole O} = 54.20\ \text{g O} \times \left(\frac{1\ \text{mole O}}{16.00\ \text{g O}}\right) = 3.387\ \text{moles O}$$

Using the number of moles of each element, the empirical formula is:

$$C_{3.387\text{ mole}}H_{5.079\text{ mole}}O_{3.387\text{ mole}} = C_{\frac{3.387\text{ mole}}{3.387\text{ mole}}}H_{\frac{5.079\text{ mole}}{3.387\text{ mole}}}O_{\frac{3.387\text{ mole}}{3.387\text{ mole}}}$$

Multiplying each subscript by two gives whole number subscripts in the empirical formula.

$$C_{1\times 2}H_{2\times 1.5}O_{1\times 2} = C_2H_3O_2$$

Adding the atomic masses of all the atoms in the empirical formula gives the empirical formula mass.

$$2\text{ C} + 3\text{ H} + 2\text{ O} = 2(12.01\text{ amu}) + 3(1.008\text{ amu}) = 2(16.00\text{ amu}) = 59.04\text{ amu}$$

Dividing the molecular mass of the compound by the empirical formula mass shows that the formula of the molecule is two times the empirical formula.

$$n = \frac{118.1\text{amu}}{59.04\text{ amu}} = 2$$

The formula of the molecule is $2(C_2H_3O_2) = C_4H_6O_4$.

Work Problems

Obtain the necessary atomic masses from the table of atomic masses in Chapter 5 or from a periodic table.

1. Calculate the empirical formula of a compound that is 43.64% phosphorus and 56.36% oxygen.
2. A sample of a brown-black compound is composed of 2.477 g manganese and 1.323 g oxygen. What is the empirical formula of the compound?
3. The empirical formula of an organic compound is $C_3H_5O_2$. The molecular mass of the compound is 146.1 amu. What is the molecular formula of the compound?
4. The percent composition of moth balls is C = 49.0%, H = 2.80%, and Cl = 48.2%. The molecular mass of the compound is 147 amu. What is its molecular formula?

Worked Solutions

1. **The empirical formula is P_2O_5.**

$$\text{mole P} = 43.64\text{ g}\cancel{P} \times \left(\frac{1\text{ mole P}}{30.97\text{ g}\cancel{P}}\right) = 1.409\text{ moles P}$$

$$\text{mole O} = 56.36\text{ g}\cancel{O} \times \left(\frac{1\text{ mole O}}{16.00\text{ g}\cancel{O}}\right) = 3.523\text{ moles O}$$

The number of moles of each element is used to develop the empirical formula.

$$P_{1.409\text{ mole}}O_{3.523\text{ mole}} = P_{\frac{1.409\text{ mole}}{1.409\text{ mole}}}O_{\frac{3.523\text{ mole}}{1.409\text{ mole}}} = P_1O_{2.5}$$

Multiplying both subscripts by two converts them to whole numbers. The empirical formula is P_2O_5.

2. **The empirical formula is MnO_2.**

Determining the number of moles of each element from the mass of each:

$$\text{mole Mn} = 2.477\,\text{g Mn} \times \left(\frac{1\,\text{mole Mn}}{59.94\,\text{g Mn}}\right) = 0.04132\,\text{mole Mn}$$

$$\text{mole O} = 1.323\,\text{g O} \times \left(\frac{1\,\text{mole O}}{16.00\,\text{g O}}\right) = 0.08269\,\text{mole O}$$

The number of moles of each element is used to develop the empirical formula.

$$Mn_{0.04132\text{ mole}}O_{0.08269\text{ mole}} = Mn_{\frac{0.04132\text{ mole}}{0.04132\text{ mole}}}O_{\frac{0.08269\text{ mole}}{0.04132\text{ mole}}} = Mn_1O_2 = MnO_2$$

3. **The molecular formula is $C_6H_{10}O_4$.**

The empirical formula mass of $C_3H_5O_2$ is:

3 C + 5 H + 2 O = 3(12.01 amu) + 5(1.008 amu) + 2(16.00 amu) = 73.1 amu

Dividing the molecular mass of the compound by the empirical formula mass shows that the formula of the molecule is two times the empirical formula.

$$n = \frac{146.0\,\text{amu}}{73.1\,\text{amu}} = 2$$

The molecular formula is $2(C_3H_5O_2) = C_6H_{10}O_4$.

4. **The molecular formula is $C_6H_4Cl_2$.**

Determining the number of moles of each element from the percent values:

$$\text{mole C} = 49.0\,\text{g C} \times \left(\frac{1\,\text{mole C}}{12.01\,\text{g C}}\right) = 4.08\,\text{moles C}$$

$$\text{mole H} = 2.80\,\text{g H} \times \left(\frac{1\,\text{mole H}}{1.008\,\text{g H}}\right) = 2.78\,\text{moles H}$$

$$\text{mole Cl} = 48.2\,\text{g Cl} \times \left(\frac{1\,\text{mole Cl}}{35.45\,\text{g Cl}}\right) = 1.36\,\text{moles Cl}$$

The number of moles of each element is used to develop the empirical formula.

$$C_{4.08\text{ mole}}H_{2.78\text{ mole}}Cl_{1.36\text{ mole}} = C_{\frac{4.08\text{ mole}}{1.36\text{ mole}}}H_{\frac{2.78\text{ mole}}{1.36\text{ mole}}}Cl_{\frac{1.36\text{ mole}}{1.36\text{ mole}}} = C_3H_2Cl_1$$

The empirical formula mass of C_3H_2Cl is:

3 C + 2 H + Cl = 3(12.01 amu) + 2(1.008amu) + 35.45 amu = 73.5 amu

Dividing the molecular mass of the compound by the empirical formula mass shows that the formula of the molecule is two times the empirical formula.

$$n = \frac{147.0\,\text{amu}}{73.2\,\text{amu}} = 2$$

The molecular formula is $2(C_3H_2Cl) = C_6H_4Cl_2$.

Chapter Problems and Answers

Problems

For Questions 1 through 5, determine the percent composition by mass of each compound.

Obtain the necessary atomic masses from the table of atomic masses in Chapter 5 or a periodic table.

1. H_2O %H = _____; %O = _____
2. BCl_3 %B = _____; %Cl = _____
3. $Ni(ClO_4)_2$ %Ni = _____; %Cl = _____; %O = _____
4. Na_2S %Na = _____; %S = _____
5. $Ca(OH)_2$ %Ca = _____; %O = _____; %H = _____

For Questions 6 and 7, calculate the percent water in the hydrate.

6. $MgSO_4{\cdot}7H_2O$ $\%H_2O$ = _____
7. $CoCl_2{\cdot}6H_2O$ $\%H_2O$ = _____

For Questions 8 through 12, determine the empirical formula for each compound.

8. a compound that is 47.27% Cu and 52.73% Cl
9. a compound that is 36.84% N and 63.16% O
10. a compound that contains 1.52 g N and 3.47 g O in an analyzed sample
11. a compound that contains 6.00 g C, 1.51 g H, and 4.00 g O in an analyzed sample
12. a compound that is 25.94% N and 74.06% O

For Questions 13 through 17, calculate the molecular formula for each compound.

13. The molar mass of caffeine is 194.2 g. Its empirical formula is $C_4H_5N_2O$. What is its molecular formula?
14. Analysis of a compound shows its composition to be 40.00% C, 6.72% H, and 53.29% O, and its molar mass is 180 g. What is the molecular formula of the compound?
15. A molecular compound is 55.77% C, 11.70% H, and 32.53% N. Its molar mass is 172.3 g. What is the molecular formula of the compound?
16. Write the empirical formula for each of the following: C_2H_6, C_6H_5F, $C_6H_4F_2$, B_2H_6, and $C_{12}H_{26}$.
17. Both urea, $(NH_2)_2CO$, and ammonium nitrate, NH_4NO_3, are used as sources of nitrogen in fertilizer. Which has the higher percent nitrogen?

Answers

1. **H = 11.2%; O = 88.8%**

The molar masses are H_2O = 18.02 g, H = 1.008 g, and O = 16.00 g.

In 1 mole of H_2O (18.02 g H_2O), there are 2 moles of H (2.016 g H) and 1 mole of O (16.00 g O).

$$\%\text{H} = \left(\frac{2.016\,\not{g}\,\text{H}}{18.02\,\not{g}\,\text{H}_2\text{O}}\right) \times 100\% = 11.19\%\text{H} = 11.2\%\,\text{H}$$

%O = (100.00% – %H) = (100.00% – 11.2%) = 88.8% O

2. **B = 9.23%; Cl = 90.77%**

The molar masses are BCl_3 = 117.16 g, B = 10.81 g, and Cl = 35.45 g.

In 1 mole of BCl_3 (117.2 g BCl_3), there is 1 mole of B (10.81 g B) and 3 moles of Cl (106.35 g Cl).

$$\%\text{B} = \left(\frac{10.81\,\not{g}\,\text{B}}{117.16\,\not{g}\,\text{BCl}_3}\right) \times 100\% = 9.23\%\,\text{B}$$

%Cl = (100.00% – %B) = (100.00% – 9.23%) = 90.77% Cl

3. **Ni = 22.78%; Cl = 27.52%; O = 49.70%**

The molar masses are $Ni(ClO_4)_2$ = 257.59 g, Ni = 58.69 g, Cl = 35.45 g, and O = 16.00 g.

In 1 mole of $Ni(ClO_4)_2$ (257.59 g $Ni(ClO_4)_2$), there is 1 mole of Ni (58.69 g Ni), 2 moles of Cl (70.90 g Cl), and 8 moles of O (128.0 g O).

$$\%\,\text{Ni} = \left(\frac{58.69\,\not{g}\,\text{Ni}}{257.59\,\not{g}\,\text{Ni}(\text{ClO}_4)_2}\right) \times 100\% = 22.78\%\,\text{Ni}$$

$$\%\,\text{Cl} = \left(\frac{70.90\,\not{g}\,\text{Cl}}{257.59\,\not{g}\,\text{Ni}(\text{ClO}_4)_2}\right) \times 100\% = 27.52\%\,\text{Cl}$$

%O = (100.00% – %Ni – %Cl) = (100.00% – 22.78% – 27.52%) = 49.70% O

4. **Na = 58.91%; S = 41.09%**

The molar masses are Na_2S = 78.05 g, Na = 22.99 g, and S = 32.07 g.

In 1 mole of Na_2S (78.05 g Na_2S), there are 2 moles of Na (45.98 g Na) and 1 mole of S (32.07 g S).

$$\%\,\text{Na} = \left(\frac{45.98\,\not{g}\,\text{Na}}{78.05\,\not{g}\,\text{Na}_2\text{S}}\right) \times 100\% = 58.91\%\,\text{Na}$$

%S = 100.00 – 58.91% = 41.09% S

5. Ca = 54.09%; O = 43.18%; H = 2.73%

The molar masses are $Ca(OH)_2$ = 74.10 g, Ca = 40.08 g, O = 16.00 g, and H = 1.008 g.

In 1 mole of $Ca(OH)_2$ (74.10 g $Ca(OH)_2$), there is 1 mole of Ca (40.08 g Ca), 2 moles of O (32.00 g O), and 2 moles of H (2.016 g H).

$$\% Ca = \left(\frac{40.08 \cancel{g}\ Ca}{74.10 \cancel{g}\ Ca(OH)_2}\right) \times 100\% = 54.09\% Ca$$

$$\% O = \left(\frac{32.00 \cancel{g}\ O}{74.10 \cancel{g}\ Ca(OH)_2}\right) \times 100\% = 43.18\% O$$

%H = (100.00% – %Ca – %O) = (100.00% – 54.09% – 43.18%) = 2.73% H

6. The percent water in $MgSO_4 \cdot 7H_2O$ = 51.15%.

The molar masses are $MgSO_4 \cdot 7H_2O$ = 246.52 g and H_2O = 18.02 g.

In 1 mole of $MgSO_4 \cdot 7H_2O$ (246.52 g $MgSO_4 \cdot 7H_2O$), there are 7 moles of H_2O (7 × 18.02 g = 126.1 g H_2O).

$$\% H_2O = \left(\frac{126.1 \cancel{g}\ H_2O}{246.52 \cancel{g}\ MgSO_4 \cdot 7H_2O}\right) \times 100\% = 51.15\% H_2O$$

7. The percent H_2O in $CoCl_2 \cdot 6H_2O$ = 45.43%.

The molar masses are $CoCl_2 \cdot 6H_2O$ = 237.95 g and H_2O = 18.02 g.

In 1 mole of $CoCl_2 \cdot 6H_2O$ (237.95 g $CoCl_2 \cdot 6H_2O$), there are 6 moles of H_2O (6 × 18.02 g = 108.1 g H_2O).

$$\% H_2O = \left(\frac{108.1 \cancel{g}\ H_2O}{237.95 \cancel{g}\ CoCl_2 \cdot 6H_2O}\right) \times 100\% = 45.43\% H_2O$$

8. The empirical formula is $CuCl_2$.

The molar masses are Cu = 63.55 g and Cl = 35.45 g.

Converting percent to mass: 47.27 g Cu and 52.73 g Cl are in 100.00 g of the compound.

47.27 g Cu = 0.744 mole of Cu (You should check these conversions to ensure each is correct.)

52.73 g Cl = 1.487 moles of Cl

$$Cu_{0.7435\ mole}Cl_{1.487\ mole} = Cu_{\frac{0.7435\ mole}{0.7435\ mole}}Cl_{\frac{1.487\ mole}{0.7435\ mole}} = Cu_1Cl_2 = CuCl_2$$

9. The empirical formula is N_2O_3.

The molar masses are N = 14.01 g and O = 16.00 g.

Converting percent to mass: 36.84 g N and 63.16 g O are in 100.00 g of the compound.

36.84 g N = 2.63 moles of N

63.16 g O = 3.95 moles of O

$N_{2.630\ mole}O_{3.95\ mole} = N_{\frac{2.630\ mole}{2.630\ mole}}O_{\frac{3.95\ mole}{2.630\ mole}} = N_1O_{1.5}$

To obtain whole-number subscripts, multiply each by 2.

$N_{1\times 2}O_{1.5\times 2} = N_2O_3$

10. **The empirical formula is NO_2.**

The molar masses are N = 14.01 g and O = 16.00 g.

There are 1.52 g N and 3.47 g O in the sample of compound.

1.52 g N = 0.108 mole of N

3.47 g O = 0.217 mole of O

$N_{0.108\ mole}O_{0.217\ mole} = N_{\frac{0.108\ mole}{0.108\ mole}}O_{\frac{0.217\ mole}{0.108\ mole}} = N_1O_2 = NO_2$

11. **The empirical formula is C_2H_6O.**

The molar masses are C = 12.01 g, H = 1.008 g, and O = 16.00 g.

There are 6.00 g C, 1.51 g H, and 4.00 g O in the sample.

6.00 g C = 0.500 mole of C

1.51 g H = 1.50 moles of H

4.00 g O = 0.250 mole of O

$C_{0.500\ mole}H_{1.50\ mole}O_{0.250\ mole} = C_{\frac{0.500\ mole}{0.250\ mole}}H_{\frac{1.50\ mole}{0.250\ mole}}O_{\frac{0.250\ mole}{0.250\ mole}} = C_2H_6O_1 = C_2H_6O$

12. **The empirical formula is N_2O_5.**

The molar masses are N = 14.01 g and O = 16.00 g.

Converting percent to mass: 25.94 g N and 74.06 g O are in 100.00 g of the compound.

25.94 g N = 1.85 moles of N

74.06 g O = 4.63 moles of O

$N_{1.852\ mole}O_{4.629\ mole} = N_{\frac{1.852\ mole}{1.852\ mole}}O_{\frac{4.629\ mole}{1.852\ mole}} = N_1O_{2.5}$

(Multiply each subscript by 2.) $N_{1\times 2}O_{2.5\times 2} = N_2O_5$

13. **The molecular formula is $C_8H_{10}N_4O_2$.**

The molar mass of the empirical formula $C_4H_5N_2O$ is 97.1 g. That of the molecular formula is 194.2 g. These allow the calculation of **n** in the following equation: molecular formula = **n** (empirical formula).

$$n = \frac{\text{molar mass of molecular formula}}{\text{molar mass of empirical formula}} = \frac{194.2\,\not{g}}{97.1\,\not{g}} = 2$$

The molecular formula is twice the empirical formula: $2(C_4H_5N_2O) = C_8H_{10}N_4O_2$.

14. **The molecular formula is $C_6H_{12}O_6$.**

The molar masses are C = 12.01 g, H = 1.008 g, and O = 16.00 g.

Converting the percent values to mass: 40.00 g C, 6.72 g H, and 53.29 g O

40.00 g C = 3.331 moles of C

6.72 g H = 6.67 moles of H

53.29 g O = 3.331 moles of O

The empirical formula is:

$$C_{3.331\text{ mole}}H_{6.67\text{ mole}}O_{3.331\text{ mole}} = C_{\frac{3.331\text{ mole}}{3.331\text{ mole}}}H_{\frac{6.67\text{ mole}}{3.331\text{ mole}}}O_{\frac{3.331\text{ mole}}{3.331\text{ mole}}} = C_1H_2O_1 = CH_2O$$

The molar mass of the empirical formula CH_2O is 30.0 g.

$$n = \frac{\text{molar mass of molecular formula}}{\text{molar mass of empirical formula}} = \frac{180\,\not{g}}{30.0\,\not{g}} = 6$$

The molecular formula is six times the empirical formula: $6(CH_2O) = C_6H_{12}O_6$.

15. **The molecular formula is $C_8H_{20}N_4$.**

The molar masses are C = 12.01 g, H = 1.008 g, and N = 14.01 g.

Converting the percent values to mass: 55.77 g C, 11.70 g H, and 32.53 g N

55.77 g C = 4.644 moles of C

11.70 g H = 11.61 moles of H

32.53 g N = 2.322 moles of N

The empirical formula is:

$$C_{4.644\text{ mole}}H_{11.61\text{ mole}}N_{2.322\text{ mole}} = C_{\frac{4.644\text{ mole}}{2.322\text{ mole}}}H_{\frac{11.61\text{ mole}}{2.322\text{ mole}}}N_{\frac{2.322\text{ mole}}{2.322\text{ mole}}} = C_2H_5N_1 = C_2H_5N$$

The molar mass of the empirical formula C_2H_5N is 43.07 g.

$$n = \frac{\text{molar mass of molecular formula}}{\text{molar mass of empirical formula}} = \frac{172.3\,\not{g}}{43.07\,\not{g}} = 4$$

The molecular formula is four times the empirical formula: $4(C_2H_5N) = C_8H_{20}N_4$.

16. **molecular formula** **empirical formula**

molecular formula	empirical formula
C_2H_6	CH_3
C_6H_5F	C_6H_5F
$C_6H_4F_2$	C_3H_2F
B_2H_6	BH_3
$C_{12}H_{26}$	C_6H_{13}

17. **Urea has the greater nitrogen content, 46.65%, compared to ammonium nitrate, 35.00%.**

Molar mass of urea: $(NH_2)_2CO$ = 2 N + 4 H + C + O =

(2 × 14.01 g N) + (4 × 1.008 g H) + 12.01 g C + 16.00 g O = 60.06 g

1 mole of $(NH_2)_2CO$ (60.06 g $(NH_2)_2CO$) contains 2 moles of N (28.02 g N).

$$\%N = \left(\frac{28.02\,\cancel{g}\,N}{60.06\,\cancel{g}\,(NH_2)_2CO}\right) \times 100\% = 46.65\%N$$

Molar mass of ammonium nitrate: NH_4NO_3 = 2 N + 4 H + 3 O =

(2 × 14.01 g N) + (4 × 1.008 g H) + (3 × 16.00 g O) = 80.05 g

1 mole of NH_4NO_3 (80.05 g NH_4NO_3) contains 2 moles of N (28.02 g N).

$$\%N = \left(\frac{28.02\,\cancel{g}\,N}{80.05\,\cancel{g}\,NH_4NO_3}\right) \times 100\% = 35.00\%N$$

Supplemental Chapter Problems

Problems

For Questions 1 through 4, determine the percent composition by mass of each compound.

1. $ZnCl_2$ %Zn = _____; %Cl = _____
2. $CHCl_3$ %C = _____; %H = _____; %Cl = _____
3. $Al_2(SO_4)_3$ %Al = _____; %S = _____; %O = _____
4. $C_2H_3NO_5$ %C = _____; %H = _____; %N = _____; %O = _____

For Questions 5 and 6, calculate the percent water in the hydrate.

5. $Na_2CO_3 \cdot 10H_2O$ $\%H_2O$ = _____
6. $Ba(OH)_2 \cdot 8H_2O$ $\%H_2O$ = _____

For Questions 7 through 9, determine the empirical formula for each compound.

7. a compound that is 52.66% Ca, 12.30% Si, and 35.04% O

8. a compound that is 18.4% C, 21.5% N, and 60.1% K

9. a compound that contains 3.47 g N, 1.00 g H, 3.98 g S, and 7.94 g O in an analyzed sample

For Questions 10 through 12, calculate the molecular formula for each compound.

10. The molecular mass of a compound used to prepare silicone rubber is 296.6 amu. The empirical formula of the compound is C_2H_6SiO. What is its molecular formula?

11. Analysis of a compound shows its composition to be 50.7% C, 9.9% H, and 39.4% N. The molar mass of the compound is 142 g. What is the molecular formula of the compound?

12. Two compounds are analyzed, and both have the same percent composition: 85.62% C and 14.38% H. One of the compounds has a molar mass of 28.03 g, the other 56.06 g. What are the molecular formulas for the two compounds?

Answers

1. **Zn = 47.98% and Cl = 52.02%** (page 135)

2. **C = 10.06%, H = 0.84%, and Cl = 89.09%** (page 135)

3. **Al = 15.77%, S = 28.12%, and O = 56.11%** (page 135)

4. **C = 19.84%, H = 2.50%, N = 11.57%, and O = 66.09%** (page 135)

5. **62.97% H_2O** (page 139)

6. **45.70% H_2O** (page 139)

7. **Ca_3SiO_5** (page 143)

8. **KCN** (page 143)

9. **$N_2H_8SO_4$ = $(NH_4)_2SO_4$** (page 143)

10. **$C_8H_{24}Si_4O_4$** (page 146)

11. **$C_6H_{14}N_4$** (page 146)

12. **C_2H_4 and C_4H_8** (page 146)

Chapter 7
Chemical Reactions and Chemical Equations

At the core of chemistry are chemical reactions, those processes that change one substance into another. Chemists have developed a shorthand way of stating chemical changes using chemical equations. Knowing how to write these equations correctly and interpreting them correctly is a necessary skill every student of chemistry must acquire. Let's begin by learning how to write chemical equations.

Writing Chemical Equations

A **chemical equation** is an abbreviated way of representing a chemical reaction on paper. It shows the species present at the beginning of the reaction (the **reactants**) and those that are formed in the reaction (the **products**). The reactants and products are separated by an arrow, which indicates the direction of the chemical change.

$$\text{reactants} \rightarrow \text{products}$$

Chemical equations can be given as **word equations,** which use the names of the reactants and products, or they can be written as **formula equations,** which use symbols and formulas of the species involved. The reaction of calcium oxide (agricultural lime) with water to form calcium hydroxide is written here as a word equation.

$$\text{calcium oxide} + \text{water} \rightarrow \text{calcium hydroxide}$$

This word equation shows that calcium oxide and water are reacting to form calcium hydroxide. The reactants are on the left; the product is on the right. Though word equations can be useful, they can only provide the minimum amount of information about a reaction. Chemists prefer formula equations because they make it possible to keep track of the atoms of each element as reactants become products. A properly written and balanced formula equation can be used quantitatively, allowing the calculation of amounts of reactants and products. (This is the topic of Chapter 8.) The formula equation for the reaction of calcium oxide with water is:

$$CaO + H_2O \rightarrow Ca(OH)_2$$

The formula equation goes beyond the word equation in that it gives the composition of each species and shows how the atoms have rearranged in the reaction. Here's how you could "read" this equation in words:

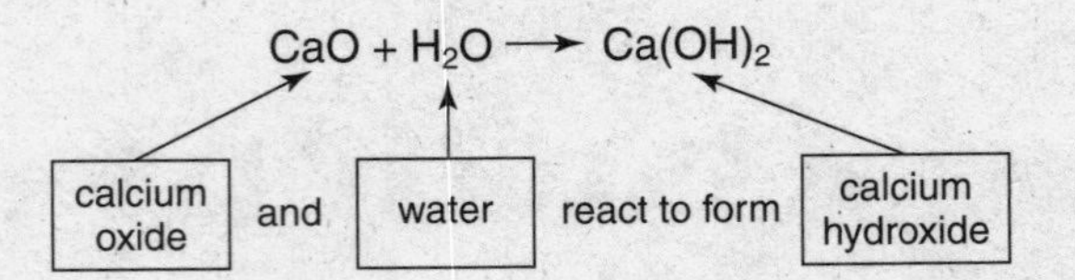

The formula equation can be modified further to show the physical state of each species. If a species is a solid, its formula is followed by *(s)*; if a liquid, by *(l)*; and if a gas, by *(g)*. If a species is dissolved in water, the formula is followed with *(aq)*, indicating it is in an *aq*ueous solution. Going back to our equation, if calcium oxide (a solid), is added to a beaker of water (a liquid), to form a solution of calcium hydroxide, the equation can be modified using the *(s)*, *(l)*, and *(aq)* attachments to show this.

$$CaO(s) + H_2O(l) \rightarrow Ca(OH)_2(aq)$$

Now we know even more about the reaction. If you need to include *solid*, *liquid*, and *in water* in your reading of this equation, it can be read as follows: "Solid calcium oxide and liquid water react to form a solution of calcium hydroxide in water." That's a mouthful, but it's accurate.

A summary of the symbols used in writing equations is given in the following table.

Symbols Used in Chemical Equations

Symbol	*Meaning*	*Use*
+	and	used between formulas of reactants or products
→	reacts to form, or yields	used between reactants and products
(s)	solid	used after a formula or symbol to indicate a solid
(l)	liquid	used after a formula or symbol to indicate a liquid
(g)	gas	used after a formula or symbol to indicate a gas
(aq)	aqueous	used after a formula or symbol to indicate it is dissolved in water
Δ	heat	placed above or below the arrow symbol (→) to indicate heating is required

When writing chemical equations:

1. The reactants are written on the left and separated from the products on the right by an arrow (→), which shows the direction of the reaction. The arrow is read as "reacts to form" or "yields."
2. If there are two or more reactants or two or more products, they are separated by a plus sign (+). The plus sign can be read as "and."
3. The physical state of each species may be indicated using *(s)* for solid, *(l)* for liquid, *(g)* for gas, and *(aq)* if dissolved in water. The symbol is written immediately after the formula.
4. If a reaction requires heating to take place, a Greek delta (Δ) is placed above or below the arrow separating reactants and products.

   $CaCO_3(s) \xrightarrow{\Delta} CaO(s) + CO_2(g)$

5. If special conditions are required for a reaction to take place, such as the presence of a catalyst, the conditions or the catalyst can be written above or below the arrow. Platinum metal catalyzes (speeds up) the decomposition of nitric oxide to its constituent elements in catalytic converters.

   $2\ NO(g) \xrightarrow{Pt} N_2(g) + O_2(g)$

6. The equation must be balanced. Whole numbers are placed before formulas and symbols to ensure that there are the same numbers of atoms of each element on both sides of the equation. These numbers are the **coefficients** of a balanced equation. The 2 in front of $NO(g)$ in the preceding equation balances the nitrogen and oxygen atoms: two of each on either side of the arrow. Balancing equations is the topic of the next section.

Example Problems

1. Convert the following word equation into a formula equation. The reaction takes place in water.

   hydrochloric acid + sodium hydroxide → sodium chloride + water

   Answer: $HCl(aq) + NaOH(aq) \rightarrow NaCl(aq) + H_2O(l)$ Note: If water is part of a reaction that takes place in water, it is written as $H_2O(l)$ in the equation.

2. Write a description of each of the following reactions, naming the reactants and products, indicating their physical states and describing any special reaction conditions. Consult Chapter 4 if you need help naming chemical compounds.

   (a) $H_2(g) + O_2(g) \rightarrow 2\ H_2O(l)$

   (b) $2\ KClO_3(s) \xrightarrow[MnO_2]{\Delta} 2\ KCl(s) + 3\ O_2(g)$

   Answer:

   (a) Hydrogen (or dihydrogen) gas and oxygen (or dioxygen) gas react to form liquid water.

   (b) Solid potassium chlorate is heated in the presence of a catalyst, MnO_2, to form solid potassium chloride and oxygen gas. Note: The presence of those elements that exist as diatomic molecules, H_2, O_2, N_2, Cl_2, and so on, when not in a compound, can be indicated with the name of the element. If they are not in a compound, it is understood they exist as diatomic molecules, as described in Chapter 2.

Work Problems

1. Convert the following word equation into a formula equation. Carbon dioxide is a gas, the other compounds are solids.

   Sodium carbonate $\xrightarrow{\Delta}$ sodium oxide + carbon dioxide

2. Write a description of each of the following reactions by naming the reactants and products, indicating their physical states and describing any special reaction conditions.

   (a) $2\ AgNO_3(aq) + Cu(s) \rightarrow Cu(NO_3)_2(aq) + 2\ Ag(s)$

   (b) $NH_4Cl(s) \xrightarrow{\Delta} NH_3(g) + HCl(g)$

Worked Solutions

1. $Na_2CO_3(s) \xrightarrow{\Delta} Na_2O(s) + CO_2(g)$

2. **(a) Solid copper is added to a solution of silver nitrate to form solid silver and a solution of copper(II) nitrate.**

   (b) Solid ammonium chloride is heated to form ammonia gas and hydrogen chloride gas.

Balancing Chemical Equations

A **balanced equation** has the same number of atoms of each element on both sides of the arrow. Because atoms are neither destroyed nor created in chemical reactions, every atom must be accounted for. If there are 10 carbon atoms in the reactants, there must be 10 carbon atoms in the products. Because every atom is accounted for, a balanced equation always obeys the Law of Conservation of Mass. It is important that you know how to balance chemical equations. Most equations are not difficult to balance, but you have to pay attention to the details of counting atoms.

Equations are balanced by placing whole-number coefficients in front of one or more species in the equation, using the smallest set of numbers that does the job. The following equation describing the reaction of aluminum with oxygen is correctly balanced. Inspecting the equation shows the same number of aluminum and oxygen atoms on both sides of the equation.

$$4\ Al(s) + 3\ O_2(g) \rightarrow 2\ Al_2O_3(s)$$
$$4 \text{ Al atoms} \rightarrow (2 \times 2) = 4 \text{ Al atoms}$$
$$(3 \times 2) = 6 \text{ O atoms} \rightarrow (2 \times 3) = 6 \text{ O atoms}$$
$$\text{a total of 10 atoms} \rightarrow \text{a total of 10 atoms}$$

The coefficients indicate the number of each species (atoms of Al, molecules of O_2, or formula units of Al_2O_3) in the balanced equation. To count atoms of a particular element, multiply the coefficient in front of the formula by the subscript of that element in the formula. Thus, 3 O_2 indicates 3×2 or 6 oxygen atoms, and 4 Al is 4 atoms of aluminum. Likewise, 2 Al_2O_3 indicates 2×2 or 4 atoms of aluminum *and* 2×3 or 6 atoms of oxygen. Remember, equations are balanced using coefficients. *Never change a subscript number to balance an element* because that changes the substance. You can change coefficients but never subscript numbers.

Choosing the right coefficients is done by trial and error, but if you carefully check the numbers of atoms each time a coefficient is changed, you can reduce the trial and eliminate the error. The equation describing the combustion of methane, the principal component of natural gas, follows. Counting the atoms of each element on both sides of the equation shows it is not a balanced equation.

$$CH_4(g) + O_2(g) \rightarrow CO_2(g) + H_2O(l)$$
$$1 \text{ C atom} \rightarrow 1 \text{ C atom (balanced)}$$
$$4 \text{ H atoms} \rightarrow 2 \text{ H atoms (not balanced)}$$
$$2 \text{ O atoms} \rightarrow 3 \text{ O atoms (not balanced)}$$

On the left side, we start with 1 molecule of CH_4, 1 C atom, and 4 H atoms. On the right side of the equation, there is 1 C atom in 1 CO_2 molecule (C is balanced), but only 2 H atoms in 1 H_2O molecule. Hydrogen can be balanced by placing a 2 in front of H_2O. This doubles both H and O, (2 H_2O = 4 H and 2 O). Let's check the results.

$$CH_4(g) + O_2(g) \rightarrow CO_2(g) + 2\ H_2O(l)$$
1 C atom → 1 C atom (balanced)
4 H atoms → 4 H atoms (balanced)
2 O atoms → 4 O atoms (not balanced)

Carbon and hydrogen are now balanced. Oxygen needs to be balanced. There are 4 O atoms on the right. On the left, there are 2 oxygen atoms in the O_2 molecule. Placing a 2 in front of O_2 doubles the number of oxygen atoms on the left side, balancing the 4 O atoms on the right. Let's add the coefficient to O_2 and check the balance again.

$$CH_4(g) + 2\ O_2(g) \rightarrow CO_2(g) + 2\ H_2O(l)$$
1 C atom → 1 C atom (balanced)
4 H atoms → 4 H atoms (balanced)
4 O atoms → 4 O atoms (balanced)

There are an identical number of atoms of each element on both sides of the arrow: The equation is balanced!

Let's try another, this time showing how a fraction (½) can be used to get to the final equation.

$$H_2O_2(aq) \rightarrow H_2O(l) + O_2(g)$$
2 H atoms → 2 H atoms (balanced)
2 O atoms → 3 O atoms (not balanced)

Again, let's start with 1 H_2O_2 on the left side of the equation. Two H atoms on the left side are balanced by 2 H atoms on the right side. Oxygen is not balanced; there is 1 more oxygen on the right side than on the left. Because 2 oxygen atoms are packaged together in O_2, place a coefficient of ½ in front of O_2 so that only 1 O atom (½ O_2 = 1 O) is added to the right side, bringing oxygen into balance.

$$H_2O_2(aq) \rightarrow H_2O(l) + 1/2\ O_2(g)$$
2 H atoms → 2 H atoms (balanced)
2 O atoms → 2 O atoms (balanced)

Though the equation is balanced, fractional coefficients are not normally used in equations. To remove the 1/2, multiply *all* coefficients by 2. This doubles all species and changes ½ to 1, $(2 \times 1/2 = 1)$.

$$2 \times [H_2O_2(aq) \rightarrow H_2O(l) + 1/2\ O_2(g)]$$
$$2\ H_2O_2(aq) \rightarrow 2\ H_2O(l) + O_2(g)$$
4 H atoms → 4 H atoms (balanced)
4 O atoms → 4 O atoms (balanced)

A fractional coefficient is a device to aid the balancing process, but fractions like ½, 3/2, 5/2, and so on can only be used for species that have an even number of atoms of each element. Most commonly, fractions are only used with the diatomic elements: H_2, O_2, N_2, Cl_2, and so on.

In many chemical reactions, polyatomic ions, like NO_3^- or SO_4^{2-}, pass from reactants to products unchanged, which allows them to be treated as single units like atoms. In the following reaction, the nitrate ion remains unchanged as reactants are converted to products. It is treated as a single unit.

$$AgNO_3(aq) + Cu(s) \rightarrow Cu(NO_3)_2(aq) + Ag(s)$$

1 Ag atom → 1 Ag atom (balanced)

1 NO_3^- ion → 2 NO_3^- ions (not balanced)

1 Cu atom → 1 Cu atom (balanced)

Clearly, the nitrate ion is not balanced. Placing a coefficient of 2 before $AgNO_3$ brings nitrate into balance but doubles Ag, putting it out of balance. Let's count the atoms:

$$2\ AgNO_3(aq) + Cu(s) \rightarrow Cu(NO_3)_2(aq) + Ag(s)$$

2 Ag atoms → 1 Ag atom (not balanced)

2 NO_3^- ions → 2 NO_3^- ions (balanced)

1 Cu atom → 1 Cu atom (balanced)

A coefficient of 2 in front of Ag on the right side balances Ag and gives a final, balanced equation.

$$2\ AgNO_3(aq) + Cu(s) \rightarrow Cu(NO_3)_2(aq) + 2\ Ag(s)$$

2 Ag atoms → 2 Ag atoms (balanced)

2 NO_3^- ions → 2 NO_3^- ions (balanced)

1 Cu atom → 1 Cu atom (balanced)

The scheme used to balance the next two equations is given in a sequence of steps. You should follow the progress by doing an atom count after each step.

$$Mg(s) + Fe_2(SO_4)_3(aq) \rightarrow MgSO_4(aq) + Fe(s)$$

Because Fe and Mg appear as monatomic elements, start with the one that is out of balance, Fe.

1. Balance Fe: Place a 2 before Fe on the right side of the equation.
2. Balance SO_4^{2-}: Place a 3 before $MgSO_4$ on the right side of the equation.
3. Balance Mg: Place a 3 in front of Mg on left side of the equation, and the equation is balanced.

The balanced equation is $3\ Mg(s) + Fe_2(SO_4)_3(aq) \rightarrow 3\ MgSO_4(aq) + 2\ Fe(s)$.

$$C_4H_{10}(g) + O_2(g) \rightarrow CO_2(g) + H_2O(l)$$

Since there are two species containing oxygen on the right side, it is easier to begin with C.

1. Balance C: Place a 4 in front of CO_2 on the right side of the equation.
2. Balance H: Place a 5 in front of H_2O on the right side of the equation.
3. Balance O: At this point there are 13 O atoms on the right side of the equation and 2 on the left. Place a coefficient of $1\frac{3}{2}$ in front of O_2 to get 13 atoms of O ($1\frac{3}{2} \times O_2 = 13$ O).
4. Multiply all coefficients by 2 to remove the fractional coefficient.

The final, balanced equation is $2\ C_4H_{10}(g) + 13\ O_2(g) \rightarrow 8\ CO_2(g) + 10\ H_2O(l)$.

There is no rule that says we all have to balance equations the same way. After all, it is a trial-and-error method. Often you have to backtrack and try a different idea or change a coefficient that was fine a minute ago but no longer works. The important thing is to end with a balanced equation, not which path you take to get there.

Example Problems

1. Two equations are given to a student to balance. The unbalanced equations and the student's balanced equations are given. Comment on the student's work.

   (a) Unbalanced equation: $Mg(s) + O_2(g) \rightarrow MgO(s)$

   Student's balanced equation: $Mg(s) + O_2(g) \rightarrow MgO_2(s)$

   (b) Unbalanced equation: $K(s) + H_2O(l) \rightarrow KOH(aq) + H_2(g)$

   Student's balanced equation: $K(s) + H_2O(l) \rightarrow KOH(aq) + H(g)$

   Answer: In both cases, the student altered the formula of a species to balance the equation. In (a) MgO became MgO_2, and in (b) H_2 became H. These are completely different substances. Equations must be balanced using coefficients. When correctly balanced, the equations are:

   $2\ Mg(s) + O_2(g) \rightarrow 2\ MgO(s)$

   $2\ K(s) + 2\ H_2O(l) \rightarrow 2\ KOH(aq) + H_2(g)$

2. Correctly balance the following equations:

   (a) $H_2(g) + N_2(g) \rightarrow NH_3(g)$

   Answer: $3\ H_2(g) + N_2(g) \rightarrow 2\ NH_3(g)$

   First balance N: Place a coefficient of 2 in front of NH_3 on the right side of the equation.

   Then, balance H: Place a coefficient of 3 in front of H_2 on the left side. It's balanced!

   (b) $N_2O_5(g) \rightarrow NO_2(g) + O_2(g)$

   Answer: $2\ N_2O_5(g) \rightarrow 4\ NO_2(g) + O_2(g)$

   First balance N: Place a coefficient of 2 in front of NO_2 on the right side.

   Then, balance O: Place a coefficient of ½ in front of O_2 on the right side.

   Multiply through side by 2 to remove the fractional coefficient, and you're done.

3. Correctly balance the following equations involving polyatomic ions:

   (a) $CrCl_3(aq) + AgNO_3(aq) \rightarrow Cr(NO_3)_3(aq) + AgCl(s)$

   Answer: $CrCl_3(aq) + 3\ AgNO_3(aq) \rightarrow Cr(NO_3)_3(aq) + 3\ AgCl(s)$

   You could start with either Cl or the nitrate ion. Let's pick Cl.

   Balance Cl: Place a coefficient of 3 in front of AgCl on the right side.

   Then, balance Ag: Place a coefficient of 3 in front of $AgNO_3$ on the left side. That's it!

   (b) $Ba(NO_3)_2(aq) + Na_3PO_4(aq) \rightarrow NaNO_3(aq) + Ba_3(PO_4)_2(s)$

Answer: 3 $Ba(NO_3)_2$*(aq)* + 2 Na_3PO_4*(aq)* → 6 $NaNO_3$*(aq)* + $Ba_3(PO_4)_2$*(s)*

Start by balancing the barium and phosphate ions since they both appear in a single product. Place a 3 in front of $Ba(NO_3)_2$*(aq)* and a 2 in front of Na_3PO_4*(aq)*.

With 6 nitrate ions and 6 sodium ions on the left, placing a 6 in front of $NaNO_3$*(aq)* on the right side balances the equation.

Work Problems

1. Correctly balance the following equations:

   (a) $NaHCO_3$*(s)* → Na_2CO_3*(s)* + CO_2*(g)* + H_2O*(l)*

   (b) FeS*(s)* + O_2*(g)* → Fe_2O_3*(s)* + SO_2*(g)*

2. Correctly balance the following equations involving polyatomic ions:

   (a) Al*(s)* + H_2SO_4*(aq)* → $Al_2(SO_4)_3$*(aq)* + H_2*(g)*

   (b) $Pb(NO_3)_2$*(aq)* + NH_4Cl*(aq)* → NH_4NO_3*(aq)* + $PbCl_2$*(s)*

Worked Solutions

1. **(a) 2 $NaHCO_3$*(s)* → Na_2CO_3*(s)* + CO_2*(g)* + H_2O*(l)***

   Balance Na: Place a coefficient of 2 in front of $NaHCO_3$ on the left side of the equation.

   (b) 4 FeS*(s)* + 7 O_2*(g)* → 2 Fe_2O_3*(s)* + 4 SO_2*(g)*

   Balance Fe: Place a coefficient of 2 in front of FeS on the left side.

   Balance S: Place a coefficient of 2 in front of SO_2 on the right side.

   Balance O: Place a coefficient of 7/2 in front of O_2 on the left side.

   Multiply through by 2 to remove the fractional coefficient, and it's done.

2. **(a) 2 Al*(s)* + 3 H_2SO_4*(aq)* → $Al_2(SO_4)_3$*(aq)* + 3 H_2*(g)***

   Balance Al: Place a coefficient of 2 in front of Al on the left side.

   Balance SO_4^{2-}: Place a coefficient of 3 in front of H_2SO_4 on the left side.

   Balance H: Place a coefficient of 3 in front of H_2 on the right side.

   (b) $Pb(NO_3)_2$*(aq)* + 2 NH_4Cl*(aq)* → 2 NH_4NO_3*(aq)* + $PbCl_2$*(s)*

   Balance NO_3^-: Place a coefficient of 2 in front of NH_4NO_3 on the right side.

   Balance NH_4^+: Place a coefficient of 2 in front of NH_4Cl on the left side.

Types of Chemical Reactions

You have seen many different kinds of chemical reactions in the equations you have studied up to now, and you might think the number of possible reactions is endless. There are a lot of reactions, but instead of trying to learn each one, things can be made easier by separating them into groups of similar reactions. In this way reactions that might first appear to be unrelated reactions are seen as variations of a common theme. Though there are several ways to gather reactions into groups, most can be placed in one of four classes of chemical reactions:

- Synthesis (combination) reactions
- Decomposition reactions
- Single-replacement reactions
- Double-replacement reactions

Let's look at the nature of each class and some reactions that belong to each.

Synthesis (Combination) Reactions

Synthesis reactions are those in which two or more substances combine to form a third. Using letters of the alphabet, synthesis reactions can be symbolized as:

$$A + B \rightarrow C$$

in which A and B could be either elements or compounds, and C is the new, synthesized compound. There are three variations of synthesis reactions, but the overriding factor is that in each a new compound (one new compound) is formed. Synthesis reactions may or may not take place in solution. Many do not.

1. Elements combine to form a new compound:

   $N_2(g) + 3\ F_2(g) \rightarrow 2\ NF_3(g)$

   $4\ Al(s) + 3\ O_2(g) \rightarrow 2\ Al_2O_3(s)$

2. An element and a compound combine to form a new compound:

   $O_2(g) + 2\ CO(g) \rightarrow 2\ CO_2(g)$

   $Cl_2(aq) + 2\ FeCl_2(aq) \rightarrow 2\ FeCl_3(aq)$

3. Compounds combine to form a new compound:

   $NH_3(g) + HCl(g) \rightarrow NH_4Cl(s)$

   $H_2O(l) + SO_3(g) \rightarrow H_2SO_4(aq)$

How can you tell whether an equation is describing a synthesis reaction? Look for two or more reactants and a single product. The following is a summary of the three kinds of synthesis reactions:

Synthesis —
1. Elements combine to form a compound.
2. An element and a compound combine.
3. Two compounds combine.

Decomposition Reactions

Decomposition reactions are those in which a compound is broken down into simpler substances, either smaller compounds, elements, or both. Many compounds can be decomposed by heating them to high temperatures, often to hundreds of degrees Celsius. Some compounds are decomposed by light energy. Decomposition reactions can by symbolized as:

$$AB \rightarrow A + B$$

in which AB is a compound and A and B are elements or simpler compounds (compounds with fewer atoms than AB). There are three variations of the decomposition reaction, but the common theme in each is the decomposition of a compound. Decomposition reactions rarely take place in solution.

1. Decomposition of a compound into two or more elements:

   $2\ HgO(s) \xrightarrow{\Delta} 2\ Hg(l) + O_2(g)$

   $2\ AgBr(s) \xrightarrow{light} 2\ Ag(s) + Br_2(l)$

2. Decomposition of a compound into elements and compounds:

   $2\ KClO_3(s) \xrightarrow{\Delta} 2\ KCl(s) + 3\ O_2(g)$

   $2\ N_2O_5(g) \xrightarrow{\Delta} 2\ NO_2(g) + O_2(g)$

3. Decomposition of a compound into two or more compounds:

   $CaCO_3(s) \xrightarrow{\Delta} CaO(s) + CO_2(g)$

   $2\ NaHCO_3(g) \xrightarrow{\Delta} Na_2CO_3(s) + CO_2(g) + H_2O(l)$

How can you tell whether an equation is describing a decomposition reaction? There is one reactant and two or more products. An additional clue could be the Δ above or below the arrow.

Decomposition (compound) —
1. Two or more elements are formed.
2. Elements and compounds form.
3. Two or more compounds form.

Single-Replacement Reactions

Single-replacement reactions are those in which a free element replaces a different element in a compound, forming a new compound and freeing the replaced element. Single-replacement reactions can be symbolized as:

$$A + BC \rightarrow AC + B$$

in which A and B are two elements and BC and AC are two compounds. Most single-replacement reactions occur in solution at room temperature. Those that are not in solution usually take place at elevated temperatures. There are three categories of single-replacement reactions. In the first, a metal replaces a metal. In the second, a metal replaces hydrogen, a nonmetal. And in the third, a nonmetal replaces a nonmetal. Notice in the first equation how Cu (a free element) takes the place of silver, Ag, in $AgNO_3$, freeing Ag as a new compound; $Cu(NO_3)_2$ is formed. In each of the following equations, look for one element replacing another in a compound.

1. A metal replaces another metal in a compound:

   $Cu(s) + 2\ AgNO_3(aq) \rightarrow Cu(NO_3)_2(aq) + 2\ Ag(s)$

   $Fe(s) + CuSO_4(aq) \rightarrow FeSO_4(aq) + Cu(s)$

2. A metal replaces hydrogen in an acid or water to produce $H_2(g)$:

   $Zn(s) + 2\ HCl(aq) \rightarrow ZnCl_2(aq) + H_2(g)$

   $Ca(s) + 2\ H_2O(l) \rightarrow Ca(OH)_2(aq) + H_2(g)$

   $2\ Al(s) + 3\ H_2SO_4(aq) \rightarrow Al_2(SO_4)_3(aq) + 3\ H_2(g)$

   Note that only the most chemically active elements (such as Li, Na, K, Ba, Ca) can react with cold water to produce $H_2(g)$. Metals like Mg, Zn, Fe, Sn, and others can liberate $H_2(g)$ from acids. They do not react with cold water. Platinum, silver, and gold do not produce H_2 when combined with acids.

3. A halogen (the nonmetals in Group VIIA) replaces another halogen in a compound:

   $Cl_2(aq) + 2\ NaI(aq) \rightarrow 2\ NaCl(aq) + I_2(aq)$

   $Cl_2(aq) + CaBr_2(aq) \rightarrow CaCl_2(aq) + Br_2(aq)$

   Note that the capability of one halogen to replace another from a compound depends on the chemical reactivity of the halogens. F_2 is more active than Cl_2, which is more active than Br_2, which is more active then I_2. This means Br_2 can only replace iodine in a compound. Cl_2 can replace both bromine and iodine, and F_2 can replace chlorine, bromine, and iodine. These reactions take place in aqueous solution.

Here is how you can tell whether an equation is describing a single-replacement reaction. There are two reactants (one element, one compound) and two products (one element, one compound). The two elements are different, and the replacement of one element by the other is evident in the two compounds. The following is a summary of the three types of single-replacement reactions:

Single-Replacement	1. A metal replaces another metal.
	2. A metal replaces hydrogen to form $H_2(g)$.
	3. A nonmetal replaces another nonmetal.

Double-Replacement Reactions

Double-replacement reactions are those in which two compounds exchange ions or atoms to form two new compounds. These reactions are also called **exchange reactions** since the equations show the exchange of "partners" when reactants and products are compared. The partner exchange is easily seen in the equation that symbolizes double-replacement reactions.

$$AB + CD \rightarrow AD + BC$$

In the equation, AB and CD are the original compounds and AD and BC are the new compounds. Some important double-replacement processes do not occur in solution, but *most* do, and those that do either (a) produce a precipitate (an insoluble solid), (b) produce a gas, or (c) produce water. These are the three categories of double-replacement reactions.

1. Double-replacement reactions that form a precipitate.

   In these reactions, both reactants and one of the products are soluble, but the other product is not and settles out of solution as a solid, *(s)*. Insoluble solids produced in reactions are called **precipitates.** In the following two equations, silver chloride, $AgCl(s)$, and iron(III) hydroxide, $Fe(OH)_3(s)$, are precipitates. Underlining a formula, $\underline{AgCl}$, or drawing a down-pointing arrow after the formula, $AgCl\downarrow$, can also indicate a precipitate. Study these equations to see how the cations, (Ag^+ and Na^+) or (Fe^{3+} and Na^+), exchange places. One of the new combinations in each equation is insoluble and labeled as a solid, *(s)*.

   $AgNO_3(aq) + NaCl(aq) \rightarrow NaNO_3(aq) + AgCl(s)$

   $FeCl_3(aq) + 3\ NaOH(aq) \rightarrow 3\ NaCl(aq) + Fe(OH)_3(s)$

2. Double-replacement reactions in which a gas is evolved.

   In the next four reactions, either $CO_2(g)$, $H_2S(g)$, or $HCl(g)$ is produced. Each is a gas, *(g)*. Drawing an upward-pointing arrow after a formula, $H_2S\uparrow$, can also be used to indicate a gaseous product.

   $K_2CO_3(s) + H_2SO_4(aq) \rightarrow K_2SO_4(aq) + H_2O(l) + CO_2(g)$

   $NaHCO_3(aq) + HCl(aq) \rightarrow NaCl(aq) + H_2O(l) + CO_2(g)$

   $Na_2S(aq) + 2\ HCl(aq) \rightarrow 2\ NaCl(aq) + H_2S(g)$

   $2\ NaCl(s) + H_2SO_4(l) \rightarrow Na_2SO_4(s) + 2\ HCl(g)$

   The first two equations, which show the reaction of potassium carbonate, K_2CO_3, and sodium hydrogencarbonate, $NaHCO_3$, with acid, do not look like double-replacement reactions at first glance, though they are. The products, $H_2O(l)$ and $CO_2(g)$, are the result of the rapid decomposition of carbonic acid, $H_2CO_3(aq)$, which forms initially. Rewriting the first equation with carbonic acid as the product clearly shows the reaction is double-replacement. The ions have exchanged places.

   $K_2CO_3(s) + H_2SO_4(aq) \rightarrow K_2SO_4(aq) + H_2CO_3(aq)$

   But $H_2CO_3(aq)$ is very unstable and quickly decomposes to $H_2O(l)$ and $CO_2(g)$.

   $H_2CO_3(aq) \rightarrow H_2O(l) + CO_2(g)$

   Because carbonic acid is so unstable, the equations are written showing the observed products, $H_2O(l)$ and $CO_2(g)$, instead of $H_2CO_3(aq)$. As was mentioned earlier in this chapter, if a reaction takes place in water, the water in equations is written as $H_2O(l)$ as opposed to $H_2O(aq)$. Water that is consumed or produced is ultimately part of the liquid water in which the reaction occurs.

3. Double-replacement reactions that form water as a product.

   Most all reactions of this type are those between acids and bases. (See Chapter 4 for a list of the common acids and bases.) These reactions are called **neutralization reactions** because the acid and base neutralize the properties of the other. Water is formed in neutralization reactions.

   $HNO_3(aq) + NaOH(aq) \rightarrow NaNO_3(aq) + H_2O(l)$

   $H_2SO_4(aq) + 2\ KOH(aq) \rightarrow K_2SO_4(aq) + 2\ H_2O(l)$

   The oxides of the Group IA and IIA metals (the basic oxides, such as Na_2O, K_2O, and CaO) also react with acids to form water. These reactions are also neutralization processes that form water.

   $K_2O(s) + 2\ HCl(aq) \rightarrow 2\ KCl(aq) + H_2O(l)$

   $BaO(s) + 2\ HNO_3(aq) \rightarrow Ba(NO_3)_2(aq) + H_2O(l)$

The ionic compounds produced in neutralization reactions, such as $NaNO_3$ and KCl, are collectively referred to as **salts.** A salt is the ionic product of a neutralization reaction. Normally when you hear the word *salt* you think of table salt, sodium chloride, NaCl, but NaCl is one of hundreds of compounds classed as salts. Many ionic compounds are referred to simply as salts. It's not a very descriptive term, but it is widely used.

How you can tell whether an equation is describing a double-replacement reaction? There are two reactants (both compounds) and two (or three) products (all compounds). If one of the reactants is a carbonate or hydrogencarbonate salt, there are three products. In addition, one of the products in double-replacement reactions is a precipitate, a gas or water. The following is a summary of the three types of double-replacement reactions:

Double-Replacement (two compounds)

1. Reactions in solution that form a precipitate.
2. Reactions that form a gas.
3. Reactions that form water.

Oxidation-Reduction Reactions

An **oxidation-reduction reaction** is one in which electrons are transferred from one reactant to another. They are often called **redox reactions** for short. **Oxidation** is the *loss* of one or more electrons by a species. The species losing electrons is oxidized. **Reduction** is the *gain* of one or more electrons by a species, and that species is reduced. Oxidation and reduction always occur simultaneously. The single-displacement reaction of copper metal with silver nitrate solution is both a single replacement reaction and an oxidation-reduction reaction.

$$Cu(s) + 2\ AgNO_3(aq) \rightarrow Cu(NO_3)_2(aq) + 2\ Ag(s)$$

It might be easier to see this as oxidation reduction when the equation is rewritten to show only the two metals. (The nitrate ions do not change in the reaction.)

$$Cu(s) + 2\ Ag^+(aq) \rightarrow Cu^{2+}(aq) + 2\ Ag(s)$$

Comparing the left and right sides of the equation, it is evident that each copper atom loses two electrons as the reaction proceeds. Copper is oxidized.

$$Cu(s) \rightarrow Cu^{2+}(aq) + 2\ e^-$$

The electrons lost by copper are gained by the silver ions. Each silver ion gains one electron to become an atom of silver. Silver ion is reduced. Two silver ions are needed to take the two electrons lost by each atom of copper.

$$2\ Ag^+(aq) + 2\ e^- \rightarrow 2\ Ag(s)$$

These two equations showing the loss and gain of electrons are called **half-reactions;** each describes half of the electron-transfer process. Adding the two half-reactions together gives the overall reaction. The two electrons lost and gained subtract out and do not appear in the overall equation, but you can see from the half-reactions that the core of the reaction is the transfer of electrons from copper metal to silver ions.

$$\begin{array}{ll} Cu(s) \rightarrow Cu^{2+}(aq) + 2\,e^- & \text{(oxidation)} \\ 2\,Ag^+(aq) + 2\,e^- \rightarrow 2\,Ag(s) & \text{(reduction)} \\ \hline Cu(s) + 2\,Ag^+(aq) \rightarrow Cu^{2+}(aq) + 2\,Ag(s) & \end{array}$$

Neither oxidation nor reduction occurs without the other. If a species loses an electron, there *must* be a species to accept it and vice versa. The species being oxidized is causing the reduction of the other, so it is called the reducing agent. The species being reduced is causing the oxidation of the other, so it is called the oxidizing agent. The language of redox chemistry can be confusing, so let's organize the terms.

The **reducing agent** loses electrons, is oxidized, and causes the other reactant to be reduced by giving it electrons. In the previous equation, Cu*(s)* is the reducing agent.

The **oxidizing agent** gains electrons, is reduced, and causes the other reactant to be oxidized by taking electrons from it. In the previous equation, Ag^+*(aq)* is the oxidizing agent.

It is not difficult identifying the reducing agent and the oxidizing agent in reactions involving atoms of elements and monatomic ions of metals and nonmetals.

- If the charge on an individual atom or monatomic ion *increases* in the reaction, that is, becomes more positive, that species is being oxidized and is the reducing agent. In the following, Zn*(s)*, I^-*(aq)*, and Fe^{2+} are being oxidized. Follow the change in charge.

  $Zn(s) \rightarrow Zn^{2+}(aq)$ $2\,I^-(aq) \rightarrow I_2(s)$ $Fe^{2+}(aq) \rightarrow Fe^{3+}(aq)$

- If the charge on an individual atom or monatomic ion *decreases* in the reaction, that is, becomes less positive (or negative), that species is being reduced and is the oxidizing agent. In the following, Pb^{2+}*(aq)*, Cl_2*(g)*, and Cu^{2+} are being reduced.

  $Pb^{2+}(aq) \rightarrow Pb(s)$ $Cl_2(g) \rightarrow 2\,Cl^-(aq)$ $Cu^{2+}(aq) \rightarrow Cu^+(aq)$

Problem: Identify the reducing agent and the oxidizing agent in the following equations:

(a) $Fe(s) + I_2(aq) \rightarrow Fe^{2+}(aq) + 2\,I^-(aq)$

(b) $2\,Br^-(aq) + Cl_2(aq) \rightarrow Br_2(aq) + 2\,Cl^-(aq)$

Answer:

(a) Fe is losing two electrons to become Fe^{2+}. It is being oxidized and is called the reducing agent. I_2 is gaining two electrons to form 2 I^- ions. It is being reduced and is therefore the oxidizing agent.

(b) The bromide ions, Br^-, are losing electrons and molecular chlorine is gaining them to form chloride ions, Cl^-. The bromide ion is oxidized and is the reducing agent that causes the reduction of the chloride ion, which then is classed as the oxidizing agent.

The chemical processes of corrosion, batteries, and cellular respiration are oxidation-reduction processes. Though these processes might be more complex, they all come down to the same thing: the transfer of electrons from one species (the reducing agent) to another (the oxidizing agent). Many synthesis and single-replacement reactions are redox processes, even though they are not always described that way.

Example Problems

1. Classify the following as either a synthesis reaction or a decomposition reaction.

   (a) $2\ Al(s) + 3\ Br_2(l) \rightarrow 2\ AlBr_3(s)$

   (b) $2\ Cl_2O_5(g) \xrightarrow{\Delta} 2\ Cl_2(g) + 5\ O_2(g)$

   (c) $2\ Zn(NO_3)_2(s) \xrightarrow{\Delta} 2\ ZnO(s) + 4\ NO_2(g) + O_2(g)$

   (d) $MgSO_4{\cdot}7H_2O(s) \xrightarrow{\Delta} MgSO_4(s) + 7\ H_2O(g)$

   (e) $CaO(s) + SO_2(g) \rightarrow CaSO_3(s)$

   Answers:

   (a) Synthesis; two elements combine to form a compound.

   (b) Decomposition; a compound is decomposed to its elements.

   (c) Decomposition; a compound is decomposed to simpler compounds and an element.

   (d) Decomposition; a compound is decomposed to simpler compounds.

   (e) Synthesis; two compounds combine to form a new compound.

2. Classify the following as either a single-replacement reaction or a double-replacement reaction.

   (a) $BaCl_2(aq) + H_2SO_4(aq) \rightarrow 2\ HCl(aq) + BaSO_4(s)$

   (b) $2\ K(s) + 2\ H_2O(l) \rightarrow 2\ KOH(aq) + H_2(g)$

   (c) $2\ KCl(s) + H_2SO_4(l) \rightarrow K_2SO_4(s) + 2\ HCl(g)$

   (d) $Al_2O_3(s) + 3\ H_2SO_4(aq) \rightarrow Al_2(SO_4)_3(aq) + 3\ H_2O(l)$

   (e) $F_2(g) + 2\ NaCl(aq) \rightarrow 2\ NaF(aq) + Cl_2(g)$

   Answers:

   (a) Double-replacement reaction; two compounds exchange ions to form two different compounds, and one is a precipitate.

   (b) Single-replacement reaction; an element and a compound react to form a new compound and $H_2(g)$.

   (c) Double-replacement reaction; two compounds exchange ions to form two different compounds, and one is a gas.

(d) Double-replacement reaction; two compounds react to form two new compounds, and one is water.

(e) Single-replacement reaction; one halogen replaces another in a compound to form a new compound and another halogen.

3. Define oxidation and reduction. Which species is oxidized and which is reduced in the following equation?

$Zn(s) + 2\ H^{+}(aq) \rightarrow Zn^{2+}(aq) + H_2(g)$

Answer: Oxidation is the loss of electron(s) by a species, and reduction is the gain of electron(s) by a species. Zinc metal, $Zn(s)$, is being oxidized, and hydrogen ion, $H^{+}(aq)$, is being reduced.

Work Problems

Classify each of the following reactions as a synthesis, decomposition, single-replacement reaction, or double-replacement reaction.

1. $2\ SO_2(g) + O_2(g) \rightarrow 2\ SO_3(g)$
2. $Mg(s) + 2\ HNO_3(aq) \rightarrow Mg(NO_3)_2(aq) + H_2(g)$
3. $CaC_2O_4 \cdot 2H_2O(s) \xrightarrow{\Delta} CaC_2O_4(s) + 2\ H_2O(g)$
4. $C(s) + Fe_2O_3(s) \rightarrow 3\ CO_2(g) + 2\ Fe(l)$
5. $Pb(NO_3)_2(aq) + 2\ NaCl(aq) \rightarrow 2\ NaNO_3(aq) + PbCl_2(s)$
6. $FeS(s) + H_2SO_4(l) \rightarrow FeSO_4(s) + H_2S(g)$
7. $Zn(s) + Pb(NO_3)_2(aq) \rightarrow Zn(NO_3)_2(aq) + Pb(s)$
8. $2\ HCl(aq) + Ca(OH)_2(aq) \rightarrow CaCl_2(aq) + 2\ H_2O(l)$
9. $BaO(s) + 2\ HC_2H_3O_2(aq) \rightarrow Ba(C_2H_3O_2)_2(aq) + H_2O(l)$

Worked Solutions

1. **Synthesis reaction; a compound and an element combine to form a new compound.**
2. **Single-replacement reaction; magnesium replaces hydrogen in nitric acid, and hydrogen gas is formed.**
3. **Decomposition reaction; a compound is heated and decomposes into two simpler compounds.**
4. **Single-replacement reaction; carbon replaces iron in Fe_2O_3 forming CO_2 and freeing iron metal.**
5. **Double-replacement reaction; two compounds exchange ions. A precipitate is formed.**

6. **Double-replacement reaction; iron replaces hydrogen in sulfuric acid forming iron(II) sulfate and $H_2S(g)$.**

7. **Single-replacement reaction; zinc replaces Pb in $Pb(NO_3)_2$, forming $Zn(NO_3)_2$ and lead metal.**

8. **Double-replacement reaction; this is a neutralization reaction. Water is formed.**

9. **Double-replacement reaction; a basic metal oxide reacts with acetic acid, and water is formed.**

Predicting Reactants and Products in Replacement Equations

If you are told the reactants and products of a reaction, it is not too difficult to write a balanced chemical equation for the process. But what if you are given only the reactants or only the products, or some combination of the two. Could you write the balanced equation? It sounds like a difficult thing to do, but if you know the type of reaction and some or most of the reactants and products, it's likely you can complete the equation. You might want to consult the tables of ions and formula writing in Chapter 4 if you need to brush up on these topics. The focus here is on single- and double-replacement equations.

Completing Single-Replacement Equations

Four incomplete equations for single-replacement reactions follow. Each equation has one or more blanks and needs to be completed with the correct species and balanced.

1. The incomplete equation is $Al(s) + NiCl_2(aq) \rightarrow$ _____ + _____.

   The first category of single-replacement reactions is one metal replacing another in a compound. Aluminum must replace nickel in $NiCl_2$ to produce nickel metal, $Ni(s)$, and aluminum chloride ($Al^{3+} + Cl^- = AlCl_3$). Place the formula for aluminum chloride and the symbol for nickel metal in the blanks, and balance. The completed equation is:

   $2\ Al(s) + 3\ NiCl_2(aq) \rightarrow 2\ AlCl_3(aq) + 3\ Ni(s)$

   You might not know the physical state of some compounds, whether they are solids, liquids, or gases or in solution, but the more important task here is determining the correct formulas for the missing species. Reviewing the three categories of single-replacement reactions shows that most of them occur in solution.

2. The incomplete equation is $Mg(s) + HNO_3(aq) \rightarrow$ _____ + _____.

   Acids react with many metals to produce hydrogen gas, $H_2(g)$, the second category of single-replacement reactions. Magnesium replaces hydrogen in nitric acid, $HNO_3(aq)$, forming magnesium nitrate and hydrogen gas. The completed equation is:

   $Mg(s) + 2\ HNO_3(aq) \rightarrow Mg(NO_3)_2(aq) + H_2(g)$

3. This equation has a missing reactant *and* a missing product. This reaction represents the third category of single-replacement reactions.

   _____ + $KI(aq) \rightarrow KCl(aq) +$ _____

   In this single-replacement reaction, iodine, in KI, is replaced by chlorine, so the missing reactant must be $Cl_2(aq)$ and the missing product has to be elemental iodine, $I_2(aq)$. The completed equation is:

   $Cl_2(aq) + 2\ KI(aq) \rightarrow 2\ KCl(aq) + I_2(aq)$

4. The incomplete equation is $Al(s) + Fe_2O_3(s) \rightarrow$ ____ + $Fe(s)$.

   Aluminum is replacing iron in Fe_2O_3. The missing product has to be the oxide of aluminum, Al_2O_3. The completed equation is:

   $2\ Al(s) + Fe_2O_3(s) \rightarrow Al_2O_3(s) + 2\ Fe(s)$

Completing Double-Replacement Equations

Completing equations of double-replacement reactions is done in a similar way. There are always two reactants, which exchange partners to give two (or three) products, one being a precipitate, a gas or water. Neutralization reactions produce a salt and water. You might first want to examine the formulas of compounds and separate them into ions. This might make it easier to see the exchange. Here are several incomplete equations for double-replacement reactions.

1. The incomplete equation is $NaOH(aq) + H_2SO_4(aq) \rightarrow$ ____ + ____.

   Sodium hydroxide is a base reacting with sulfuric acid. This is a neutralization reaction, and one product is water. The other is the salt formed as sodium replaces hydrogen in the acid, $Na_2SO_4(aq)$. The completed equation is:

   $2\ NaOH(aq) + H_2SO_4(aq) \rightarrow Na_2SO_4(aq) + 2\ H_2O(l)$

2. The incomplete equation is ____ + ____ $\rightarrow KBr(aq) + BaSO_4(s)$.

   This equation forms a precipitate of barium sulfate, $BaSO_4(s)$. Because this is a double-replacement reaction, potassium, K^+, must have been associated with the sulfate ion, SO_4^{2-}, in one reactant and barium, Ba^{2+}, must have been associated with the bromide ion, Br^-, in the other. All species except the precipitate are in solution. The completed equation is:

   $K_2SO_4(aq) + BaBr_2(aq) \rightarrow 2\ KBr(aq) + BaSO_4(s)$

3. The incomplete equation is $(NH_4)_2SO_4(aq) + Ba(NO_3)_2(aq) \rightarrow$ ____ + ____.

   Exchanging ions predicts the products will be NH_4NO_3 and $BaSO_4$. In equation 2, barium sulfate was shown to be insoluble in water, so it is a precipitate. Knowing that, the completed equation is:

   $(NH_4)_2SO_4(aq) + Ba(NO_3)_2(aq) \rightarrow 2\ NH_4NO_3(aq) + BaSO_4(s)$

4. The incomplete equation is $Na_2CO_3(s) + HCl(aq) \rightarrow$ ____ + ____ + ____.

   Exchanging the ions predicts the products will be NaCl and carbonic acid, $H_2CO_3(aq)$, which breaks down to form carbon dioxide gas and water. The completed equation is:

   $Na_2CO_3(s) + 2\ HCl(aq) \rightarrow 2\ NaCl(aq) + H_2O(l) + CO_2(g)$

5. The incomplete equation is $Ca(OH)_2(aq)$ + ____ $\rightarrow$ ____ + $H_2O(l)$.

   First note that water is a product. That suggests neutralization, especially when one reactant is a base, $Ca(OH)_2(aq)$. Though the base is known, there are no hints in the equation as to the identity of the acid, so we are free to choose one. If we pick hydrochloric acid, $HCl(aq)$, the completed equation is:

   $Ca(OH)_2(aq) + 2\ HCl(aq) \rightarrow CaCl_2(aq) + 2\ H_2O(l)$

   The most important thing about this problem was realizing right away that it was a neutralization reaction and that the missing reactant had to be an acid, one of our own choosing. If acetic acid had been chosen, $HC_2H_3O_2(aq)$, the completed equation would be:

   $Ca(OH)_2(aq) + 2\ HC_2H_3O_2(aq) \rightarrow Ca(C_2H_3O_2)_2(aq) + 2\ H_2O(l)$

Example Problems

1. Complete and balance the following single-replacement equations. Compare these equations with those of the same type to determine the missing species and physical states.

   (a) $Ca(s)$ + _____ $\rightarrow Ca(OH)_2(aq) + H_2(g)$

   (b) $Mg(s) + AgNO_3(aq) \rightarrow$ _____ + $Ag(s)$

   (c) $Cl_2(aq) + MgBr_2(aq) \rightarrow MgCl_2(aq)$ + _____

   (d) _____ + _____ $\rightarrow Zn(C_2H_3O_2)_2(aq) + H_2(g)$

Answers:

(a) $Ca(s) + 2\ H_2O(l) \rightarrow Ca(OH)_2(aq) + H_2(g)$

Hydrogen has been displaced from the missing reactant by calcium and forming $Ca(OH)_2$. The missing reactant must be water, H_2O.

(b) $Mg(s) + 2\ AgNO_3(aq) \rightarrow Mg(NO_3)_2(aq) + 2\ Ag(s)$

Magnesium replaces silver in $AgNO_3$, so the formula of the missing product combines Mg^{2+} and NO_3^-, magnesium nitrate.

(c) $Cl_2(aq) + MgBr_2(aq) \rightarrow MgCl_2(aq) + Br_2(aq)$

Chlorine, $Cl_2(aq)$ displaces bromine from $MgBr_2$, so the missing species is free bromine, $Br_2(aq)$.

(d) $Zn(s) + 2\ HC_2H_3O_2(aq) \rightarrow Zn(C_2H_3O_2)_2(aq) + H_2(g)$

Hydrogen is being replaced by zinc in acetic acid, forming $H_2(g)$. The acetate ion in $Zn(C_2H_3O_2)_2(aq)$ indicates the acid is acetic acid.

2. Complete and balance the following double-replacement equations. Compare these equations with those of the same type to determine the missing species and physical states:

   (a) $Na_2S(aq) + CuSO_4(aq) \rightarrow$ _____ + $CuS(s)$

   (b) _____ + $H_2C_2O_4(aq) \rightarrow HCl(aq) + CaC_2O_4(s)$

   (c) $Mg(OH)_2(s) + HNO_3(aq) \rightarrow$ _____ + _____

   (d) $BaCO_3(s)$ + _____ $\rightarrow BaCl_2(aq)$ + _____ + _____

Answers:

(a) $Na_2S(aq) + CuSO_4(aq) \rightarrow Na_2SO_4(aq) + CuS(s)$

If the exchange of partners in this reaction forms CuS, the other product has to be Na_2SO_4.

(b) $CaCl_2(aq) + H_2C_2O_4(aq) \rightarrow 2\ HCl(aq) + CaC_2O_4(s)$

The calcium and chloride ions on the product side must have come from $CaCl_2$, the best reactant to supply those ions and fit the form of a double-replacement reaction.

(c) $Mg(OH)_2(s) + 2\ HNO_3(aq) \rightarrow Mg(NO_3)_2(aq) + H_2O(l)$

This is an acid-base neutralization, forming magnesium nitrate and water.

(d) $BaCO_3(s) + 2\ HCl(aq) \rightarrow BaCl_2(aq) + H_2O(l) + CO_2(g)$

A carbonate treated with acid forms a salt, water, and carbon dioxide gas. The chloride ion in the salt, $BaCl_2$, indicates that the acid is hydrochloric acid.

Work Problems

Use these problems for additional practice.

1. Complete each of the following single-replacement equations by adding the correct species in the blank spaces, then balance the equation:

   (a) $AlI_3(aq)$ + ____ $\rightarrow AlBr_3(aq)$ + ____

   (b) $Zn(s) + NiCl_2(aq) \rightarrow$ ____ + $Ni(s)$

   (c) ____ + $H_2O(l) \rightarrow KOH(aq)$ + ____

2. Complete each of the following double-replacement equations by adding the correct species in the blank spaces, then balance the equation:

   (a) $Li_2CO_3(s) + H_2SO_4(aq) \rightarrow$ ____ + ____ + ____

   (b) $CaCl_2(aq) + H_3PO_4(aq) \rightarrow$ ____ + $HCl(aq)$

   (c) $NaOH(aq) + H_2C_2O_4(aq) \rightarrow$ ____ + $H_2O(l)$

Worked Solutions

1. **(a)** $2\ AlI_3(aq) + 3\ Br_2(aq) \rightarrow 2\ AlBr_3(aq) + 3\ I_2(aq)$

   (b) $Zn(s) + NiCl_2(aq) \rightarrow ZnCl_2(aq) + Ni(s)$

   (c) $2\ K(s) + 2\ H_2O(l) \rightarrow 2\ KOH(aq) + H_2(g)$

2. **(a)** $Li_2CO_3(s) + H_2SO_4(aq) \rightarrow Li_2SO_4(aq) + H_2O(l) + CO_2(g)$

   (b) $3\ CaCl_2(aq) + 2\ H_3PO_4(aq) \rightarrow Ca_3(PO_4)_2(s) + 6\ HCl(aq)$

   (c) $2\ NaOH(aq) + H_2C_2O_4(aq) \rightarrow Na_2C_2O_4(aq) + 2\ H_2O(l)$

Balanced Equations and the Mole

The coefficients in a balanced equation can be interpreted as moles of the respective substance. Consider the reaction between methane gas, CH_4, and chlorine gas. The balanced equation is:

$$CH_4(g) + 4\ Cl_2(g) \rightarrow CCl_4(l) + 4\ HCl(g)$$

In terms of moles of each substance:

$$1 \text{ mole } CH_4(g) + 4 \text{ moles } Cl_2(g) \rightarrow 1 \text{ mole } CCl_4(l) + 4 \text{ moles } HCl(g)$$

In a balanced equation, the coefficients equal the number of moles of each substance. Interpreting a balanced equation in terms of moles allows it to be used quantitatively, making it possible to determine the amount of product that forms for a given amount of reactant, or to determine how many grams of one reactant are needed to completely react with a known amount of another. More could be said about this here, but Chapter 8 is devoted to calculations of this type. For now, remember that a balanced equation is a mole statement!

Chemical Reactions and Heat

Most every chemical reaction is accompanied by the absorption or evolution of heat. Those reactions that absorb heat as they take place are called **endothermic reactions.** If the source of heat is removed from an endothermic reaction, it stops. Many decomposition reactions are endothermic processes, and for each individual decomposition reaction the amount of heat needed to decompose 1 mole of that specific compound is always the same. For example, it requires 178 kJ (kilojoules) of heat energy to decompose 1 mole of $CaCO_3$ to CaO and CO_2. The heat quantity is added to the reactant side of the equation to indicate it is absorbed (consumed) as 1 mole of calcium carbonate decomposes.

$$178 \text{ kJ} + CaCO_3(s) \rightarrow CaO(s) + CO_2(g)$$

If 2 moles of $CaCO_3$ decompose, 2×178 kJ of heat, 356 kJ, is required. The amount of heat is directly related to the amount of $CaCO_3$ decomposed. If 0.10 mole of $CaCO_3$ is decomposed, 0.10×178 kJ of heat, 17.8 kJ, is required.

The balanced equation for the decomposition of mercury(II) oxide is:

$$2\ HgO(s) \rightarrow 2\ Hg(l) + O_2(g)$$

It requires 90.7 kJ of heat energy to decompose 1 mole of $HgO(s)$. Since the balanced equation is written showing the decomposition of 2 moles of $HgO(s)$, the amount of heat energy that must be added to the reactant side to make it quantitatively correct is 181.4 kJ (2×90.7 kJ).

$$181.4 \text{ kJ} + 2\ HgO(s) \rightarrow 2\ Hg(l) + O_2(g)$$

The fact that heat is needed to carry out a reaction is usually indicated with a delta sign, Δ, above or below the arrow, but if an equation shows the number of kilojoules of heat consumed, the delta sign isn't needed.

Reactions that produce heat as they take place are called **exothermic reactions.** Neutralization reactions are exothermic processes. The reaction of NaOH(*aq*) with HCl(*aq*) produces 56.2 kJ of heat energy for each mole of HCl consumed (or for each mole of NaOH consumed).

$$NaOH(aq) + HCl(aq) \rightarrow NaCl(aq) + H_2O(l) + 56.2\ kJ$$

As the neutralization reaction takes place, the heat produced warms the solution. The reaction of calcium oxide, CaO(*s*), with water is also an exothermic reaction.

$$CaO(s) + H_2O(l) \rightarrow Ca(OH)_2(aq) + 81.4\ kJ$$

Each mole of CaO(*s*) consumed produces 81.4 kJ of heat energy. If 2 moles react with water, the amount of heat produced is twice 81.4 kJ, or 162.8 kJ.

Another way to indicate the heat associated with a process uses the symbol ΔH, which stands for the *heat of reaction*. Chemists call ΔH an enthalpy change, which is equal to the amount of heat consumed or produced in a process. If you come across the term **enthalpy,** think of it as heat, and think of ΔH as a symbol for an amount of heat energy. The equation for the endothermic decomposition of $CaCO_3(s)$ follows, this time using ΔH to indicate the heat consumed.

$$CaCO_3(s) \rightarrow CaO(s) + CO_2(g)\ \Delta H = +178\ kJ$$

The positive sign (+) before 178 kJ indicates that heat is *added to* the reaction. All endothermic reactions have *positive* (heat added) values for ΔH. The reaction of CaO(*s*) with water has already been shown to be exothermic. The sign of ΔH for an exothermic reaction is negative (–).

$$CaO(s) + H_2O(l) \rightarrow Ca(OH)_2(aq)\ \Delta H = -81.4\ kJ$$

The negative sign (–) written before 81.4 kJ indicates that heat is *evolved by* the reaction as it proceeds. All exothermic reactions have *negative* (heat evolved) values for ΔH.

In photosynthesis, the energy of the sun is absorbed as $CO_2(g)$ and $H_2O(l)$ are converted to glucose, $C_6H_{12}O_6(s)$ and $O_2(g)$. The sign of ΔH, which accompanies the balanced equation, indicates that photosynthesis is an endothermic process.

$$6\ CO_2(g) + 6\ H_2O(l) \rightarrow C_6H_{12}O_6(s) + 6\ O_2(g)\ \Delta H = +2{,}801\ kJ$$

In metabolism, glucose reacts with the oxygen we breathe into our bodies and is converted back to $CO_2(g)$ and $H_2O(l)$. The equation for this is exactly the reverse of the preceding equation, and what was an endothermic process is now an exothermic one. The sign of ΔH is negative indicating an exothermic process.

$$C_6H_{12}O_6(s) + 6\ O_2(g) \rightarrow 6\ CO_2(g) + 6\ H_2O(l)\ \Delta H = -2{,}801\ kJ$$

Reversing the reaction changes it from one that absorbs energy to one that produces energy, but notice, the quantity of energy, 2,801 kJ, does not change, only the sign of ΔH changes. One mole of glucose is associated with 2,801 kJ of energy in this reaction, be it written as an exothermic ($\Delta H = -2{,}801$ kJ) or an endothermic ($\Delta H = +2{,}801$ kJ) process.

Summarizing the heat involved in chemical reactions:

1. The amount of heat energy accompanying a particular reaction or process is the same each time the reaction is carried out the same way.

2. The amount of heat energy is *directly* related to the amount of reactant. Burning 2 grams of coal produces twice the heat energy as burning 1 gram.
3. The heat of reaction, ΔH, is + for endothermic processes and − for exothermic processes.
4. If an equation describing an exothermic reaction is reversed, it becomes an endothermic reaction and vice versa. The amount of heat energy does not change, only the sign of ΔH changes.

Petroleum-based fuels are used to produce heat energy. As they burn, they react with oxygen, much as glucose does in our bodies. The burning of a fuel is called combustion. **Combustion reactions** are exothermic and consume oxygen as the elements in the fuel are converted to their oxides. The combustion of methane gas, $CH_4(g)$, the primary component of natural gas, produces 890 kJ of heat energy for each mole of methane consumed. The carbon in methane combines with oxygen to become CO_2 and the hydrogen, H_2O.

$$CH_4(g) + 2\ O_2(g) \rightarrow CO_2(g) + 2\ H_2O(l)\ \Delta H = -890\ kJ$$

Combustion reactions occur rapidly and frequently with visible flames. The combustion of acetylene gas, $C_2H_2(g)$, in an acetylene torch produces a great deal of heat per mole of acetylene, which makes it an excellent fuel for reaching the high temperatures needed to cut sheets of iron.

$$2\ C_2H_2(g) + 5\ O_2(g) \rightarrow 4\ CO_2(g) + 2\ H_2O(g)\ \Delta H = -2512\ kJ$$

Example Problems

1. Which are exothermic? Which are endothermic?

   (a) $6.01\ kJ + H_2O(s) \rightarrow H_2O(l)$

   (b) $2\ H_2(g) + O_2(g) \rightarrow 2\ H_2O(l)\ \Delta H = -572\ kJ$

   (c) $NaHCO_3(aq) + HCl(aq) \rightarrow NaCl(aq) + H_2O(l) + CO_2(g)\ \Delta H = +11.8\ kJ$

   (d) $2\ Na(s) + 2\ H_2O(l) \rightarrow 2\ NaOH(aq) + H_2(g) + 367\ kJ$

   Answers:

   (a) Heat is absorbed as solid water melts to liquid water. It is an endothermic process.

   (b) The negative sign of ΔH indicates the synthesis of water is exothermic.

   (c) The positive sign of ΔH indicates heat is absorbed. The reaction is endothermic.

   (d) Heat is produced when sodium metal reacts with water. It is an exothermic process.

2. The balanced equation describing the decomposition of $PbO_2(s)$ is:

   $277\ kJ + PbO_2(s) \rightarrow Pb(s) + O_2(g)$

   How much heat is required to decompose 5 moles of $PbO_2(s)$?

Answer: 1,385 kJ. If 1 mole of $PbO_2(s)$ requires 277 kJ, 5 moles require 5×277 kJ of heat energy, 1,385 kJ.

3. Write the balanced equation for the combustion of isobutane, $C_4H_{10}(g)$, the fuel in a butane cigarette lighter.

Answer: $2\ C_4H_{10}(g) + 13\ O_2(g) \rightarrow 8\ CO_2(g) + 10\ H_2O(g)$

4. The formation of $NO(g)$ from $N_2(g)$ and $O_2(g)$ is endothermic.

$N_2(g) + O_2(g) \rightarrow 2\ NO(g)\ \Delta H = +180$ kJ

What is the value of ΔH for the decomposition of $NO(g)$ to its elements?

$2\ NO(g) \rightarrow N_2(g) + O_2(g)\ \Delta H = ?$

Answer: $\Delta H = -180$ kJ. The endothermic equation is reversed and becomes an exothermic equation. The sign of ΔH changes, though the amount of heat energy does not.

Work Problems

1. Classify each reaction as either exothermic or endothermic.

(a) $2\ Na(s) + Cl_2(g) \rightarrow 2\ NaCl(s)\ \Delta H = -822$ kJ

(b) $56\ kJ + PCl_5(g) \rightarrow PCl_3(g) + Cl_2(g)$

(c) $4\ NH_3(g) + 5\ O_2(g) \rightarrow 4\ NO(g) + 6\ H_2O(g) + 906$ kJ

2. The balanced equation for the synthesis of ammonia, $NH_3(g)$, from its elements is:

$N_2(g) + 3\ H_2(g) \rightarrow 2\ NH_3(g)\ \Delta H = -91.8$ kJ

What amount of heat energy is produced for each mole of ammonia produced?

3. Write the balanced equation for the combustion of propane, $C_3H_8(g)$.

4. Knowing that $H_2(g) + I_2(s) \rightarrow 2\ HI(g)\ \Delta H = +53$ kJ, predict the heat of reaction for $2\ HI(g) \rightarrow H_2(g) + I_2(s)$.

Worked Solutions

1. **(a) The negative sign of the heat of reaction indicates the reaction is exothermic.**

(b) The decomposition of phosphorus pentachloride is endothermic.

(c) The combustion of ammonia is exothermic.

2. **45.9 kJ.** The formation of 2 moles of ammonia produces 91.8 kJ of heat energy; the production of 1 mole produces just half that, 45.9 kJ.

3. $\mathbf{C_3H_8(g) + 5\ O_2(g) \rightarrow 3\ CO_2(g) + 4\ H_2O(g)}$

4. **$\Delta H = -53$ kJ, reversing the reaction changes the sign of ΔH, not the size of ΔH.**

Chapter Problems and Answers

Problems

For Questions 1 through 5, convert the word equations into balanced formula equations. You do not need to indicate the physical state, *(s)*, *(l)*, *(g)*, or *(aq)*, of the reactants and products.

1. silver oxide → silver + oxygen

2. iron(III) chloride + potassium hydroxide → iron(III) hydroxide + potassium chloride

3. calcium carbonate + hydrochloric acid → calcium chloride + carbon dioxide + water

4. aluminum + copper(II)sulfate → aluminum sulfate + copper

5. barium hydroxide + nitric acid → barium nitrate + water

For Questions 6 through 8, describe the reactions in words by giving the name of each reactant and product, their physical states, and any special conditions required for the reaction.

6. $3\ H_2(g) + Al_2O_3(s) \xrightarrow{\Delta} 2\ Al(s) + 3\ H_2O(g)$

7. $Cd(s) + 2\ AgNO_3(aq) \rightarrow Cd(NO_3)_2(aq) + 2\ Ag(s)$

8. $2\ NO_2(g) \xrightarrow[Pt]{\Delta} N_2(g) + 2\ O_2(g)$

For Questions 9 through 22, correctly balance the equations.

9. $CO(g) + H_2(g) \rightarrow CH_4(g) + H_2O(l)$

10. $Na(s) + H_2O(l) \rightarrow NaOH(aq) + H_2(g)$

11. $Mg_3N_2(s) + H_2O(l) \rightarrow Mg(OH)_2(aq) + NH_3(aq)$

12. $C_2H_6(g) + O_2(g) \rightarrow CO_2(g) + H_2O(g)$

13. $Fe_2O_3(s) + H_2(g) \rightarrow Fe(s) + H_2O(g)$

14. $CaCl_2(aq) + H_3PO_4(aq) \rightarrow Ca_3(PO_4)_2(s) + HCl(aq)$

15. $AgNO_3(aq) + K_2CrO_4(aq) \rightarrow Ag_2CrO_4(s) + KNO_3(aq)$

16. $NH_3(g) + O_2(g) \rightarrow NO(g) + H_2O(l)$

17. $C(s) + Fe_2O_3(s) \rightarrow Fe(s) + CO_2(g)$

18. $Hg_2(NO_3)_2(aq) + NaCl(aq) \rightarrow NaNO_3(aq) + Hg_2Cl_2(s)$

19. $C_3H_8(g) + O_2(g) \rightarrow CO_2(g) + H_2O(g)$

20. $CH_4O(l) + O_2(g) \rightarrow CO_2(g) + H_2O(g)$

21. $C_8H_{18}(l) + O_2(g) \rightarrow CO_2(g) + H_2O(l)$

22. $H_2S(g) + Fe(OH)_3(s) \rightarrow Fe_2S_3(s) + H_2O(g)$

For Questions 23 through 31, classify each reaction as a synthesis, decomposition, single-replacement reaction, or double-replacement reaction.

23. $2\ Fe(s) + 6\ HNO_3(aq) \rightarrow 2\ Fe(NO_3)_3(aq) + 3\ H_2(g)$

24. $2\ AgNO_3(aq) + K_2CrO_4(aq) \rightarrow Ag_2CrO_4(s) + 2\ KNO_3(aq)$

25. $Br_2(l) + 2\ H_2(g) \rightarrow 2\ HBr(g)$

26. $Sr(OH)_2(aq) + 2\ HC_2H_3O_2(aq) \rightarrow Sr(C_2H_3O_2)_2(aq) + 2\ H_2O(l)$

27. $2\ NH_3(g) + H_2SO_4(aq) \rightarrow (NH_4)_2SO_4(aq)$

28. $K_2O(s) + 2\ HNO_3(aq) \rightarrow 2\ KNO_3(aq) + H_2O(l)$

29. $Mg(s) + CuSO_4(aq) \rightarrow MgSO_4(aq) + Cu(s)$

30. $MgCO_3(s) + 2\ HCl(aq) \rightarrow MgCl_2(aq) + H_2O + CO_2(g)$

31. $2\ Ag_2O(s) \rightarrow 4\ Ag(s) + O_2(g)$

32. Each class of reaction has three subclasses of reaction.

    (a) What are the three classes of double-replacement reaction?

    (b) What are the three classes of decomposition reaction?

For Questions 33 through 36, complete and balance each single-replacement equation.

33. $Cl_2(aq)$ + _____ $\rightarrow CuCl_2(aq) + I_2(aq)$

34. $K(s)$ + _____ $\rightarrow KOH(aq) + H_2(g)$

35. _____ + $CdCl_2(aq) \rightarrow CrCl_3(aq) + Cd(s)$

36. $Ni(s) + Pb(NO_3)_2(aq) \rightarrow Pb(s)$ + _____

For Questions 37 through 40, complete and balance the double-replacement equation.

37. $KI(aq)$ + _____ $\rightarrow KNO_3(aq) + PbI_2(s)$

38. $LiOH(aq) + H_3PO_4(aq) \rightarrow$ _____ + _____

39. $H_2S(aq) + NiCl_2(aq) \rightarrow$ _____ + $HCl(aq)$

40. $K_2CO_3(s)$ + _____ $\rightarrow KNO_3(aq)$ + _____ + _____

For Questions 41 through 43, identify the oxidizing agent and reducing agent in the equations.

41. $Mg(s) + Br_2(l) \rightarrow Mg^{2+}(aq) + 2\ Br^-(aq)$

42. $Hg^{2+}(aq) + Pb(s) \rightarrow Hg(l) + Pb^{2+}(aq)$

43. $Cr^{2+}(aq) + Fe^{3+}(aq) \rightarrow Cr^{3+}(aq) + Fe^{2+}(aq)$

For Questions 44 through 47, classify the reaction as exothermic or endothermic.

44. $67.8\ kJ + N_2(g) + 2\ O_2(g) \rightarrow 2\ NO_2(g)$

45. $N_2O_5(g) + H_2O(l) \rightarrow HNO_3(aq)\ \Delta H = -77\ kJ$

46. $2\ Al_2O_3(s) + 3\ C(s) \rightarrow 4\ Al(s) + 3\ CO_2(g)\ \Delta H = +1630\ kJ$

47. $S(s) + O_2(g) \rightarrow SO_2(g) + 297\ kJ$

48. The neutralization of 1 mole of $HCl(aq)$ with $NaOH(aq)$ produces 56.2 kJ of heat. How much heat energy is produced if 4 moles of $HCl(aq)$ are neutralized? If 0.3 mole of $HCl(aq)$ is neutralized?

49. Knowing that $2\ SO_3(g) \rightarrow 2\ SO_2(g) + O_2(g)\ \Delta H = +197\ kJ$, how much heat energy is required if 3 moles of $SO_3(g)$ are consumed?

50. If 177.8 kJ of heat energy are required to decompose 1 mole of calcium carbonate, what is the heat of the reaction, ΔH, for the synthesis of calcium carbonate in the reverse reaction?

51. Write the balanced equation for the combustion of pentane, $C_5H_{12}(l)$.

Answers

1. $\mathbf{2\ Ag_2O \rightarrow 4\ Ag + O_2}$

2. $\mathbf{FeCl_3 + 3\ KOH \rightarrow Fe(OH)_3 + 3\ KCl}$

3. $\mathbf{CaCO_3 + 2\ HCl \rightarrow CaCl_2 + CO_2 + H_2O}$

4. $\mathbf{2\ Al + 3\ CuSO_4 \rightarrow Al_2(SO_4)_3 + 3\ Cu}$

5. **$Ba(OH)_2 + 2\ HNO_3 \rightarrow Ba(NO_3)_2 + 2\ H_2O$**

6. **Heating aluminum oxide with hydrogen gas produces solid aluminum and gaseous water.**

7. **Adding solid cadmium metal to a solution of silver nitrate produces a solution of cadmium nitrate and solid silver.**

8. **Nitrogen dioxide gas, when heated in the presence of a platinum catalyst, produces nitrogen and oxygen gas (or, dinitrogen and dioxygen gas).**

9. **$CO(g) + 3\ H_2(g) \rightarrow CH_4(g) + H_2O(l)$**

   C and O are balanced. Place a coefficient of 3 in front of H_2 to balance H.

10. **$2\ Na(s) + 2\ H_2O(l) \rightarrow 2\ NaOH(aq) + H_2(g)$**

    Inspecting the unbalanced equation shows 2 H on the left and 3 H on the right. Balance H by placing the coefficient 1/2 in front of H_2. Then multiply all coefficients by 2 to remove the fractional coefficient.

11. **$Mg_3N_2(s) + 6\ H_2O(l) \rightarrow 3\ Mg(OH)_2(aq) + 2\ NH_3(aq)$**

    Inspecting the unbalanced equation suggests balancing Mg and N first. Balance Mg and N by placing a 3 in front of $Mg(OH)_2$ and a 2 in front of NH_3. With 6 O now on the right and only 1 on the left, balance O by placing a 6 in front of H_2O. This also balances H.

12. **$2\ C_2H_6(g) + 7\ O_2(g) \rightarrow 4\ CO_2(g) + 6\ H_2O(g)$**

    Inspecting the unbalanced equation shows that C and H can be quickly balanced by placing a 2 in front of CO_2 and a 3 in front of H_2O. Now there are 7 O on the right side and 2 O on the left. Place a coefficient of 7/2 in front of O_2 ($7/2 \times O_2 = 7$ O) to balance O. Then remove the fractional coefficient by multiplying all coefficients in the equation by 2.

13. **$Fe_2O_3(s) + 3\ H_2(g) \rightarrow 2\ Fe(s) + 3\ H_2O(g)$**

    Fe and O can be quickly balanced by placing a 2 in front of Fe and a 3 in front of H_2O. Now there are 6 H on the right that can be balanced on the left with a coefficient of 3 in front of H_2.

14. **$3\ CaCl_2(aq) + 2\ H_3PO_4(aq) \rightarrow Ca_3(PO_4)_2(s) + 6\ HCl(aq)$**

    Balance Ca and the phosphate ion, PO_4^{3-}, by placing a 3 in front of $CaCl_2$ and a 2 in front of H_3PO_4. This now gives 6 Cl and 6 H on the left side that can be balanced with a coefficient of 2 in front of HCl.

15. **$2\ AgNO_3(aq) + K_2CrO_4(aq) \rightarrow Ag_2CrO_4(s) + 2\ KNO_3(aq)$**

    Begin by balancing Ag by placing a 2 in front of $AgNO_3$. Then balance the nitrate ion, NO_3^-, with a 2 in front of KNO_3. This also balances K.

16. **$4\ NH_3(g) + 5\ O_2(g) \rightarrow 4\ NO(g) + 6\ H_2O(l)$**

Hydrogen is in one compound on the left and one on the right, so let's start by balancing H. Because there are 3 H on the left and 2 on the right, balance H by placing a 2 in front of NH_3 and a 3 in front of H_2O, giving 6 H on both sides. Balance N by placing a 2 in front of NO. At this point you have 5 O on the right side and 2 O on the left. Place a coefficient of 5/2 in front of O_2 to balance O ($5/2 \times O_2 = 5$ O). Multiply through by 2 to remove the fractional coefficient.

17. **$3\ C(s) + 2\ Fe_2O_3(s) \rightarrow 4\ Fe(s) + 3\ CO_2(g)$**

One compound on each side of the arrow contains O. Begin by balancing O by placing a 2 in front of Fe_2O_3 and a 3 in front of CO_2. This gives 6 O on both sides. Now balance Fe and C with a 4 in front of Fe and a 3 in front of C.

18. **$Hg_2(NO_3)_2(aq) + 2\ NaCl(aq) \rightarrow 2\ NaNO_3(aq) + Hg_2Cl_2(s)$**

Balance the nitrate ion, NO_3^-, by placing a 2 in front of $NaNO_3$. Placing a 2 in front of NaCl balances both Na and Cl.

19. **$C_3H_8(g) + 5\ O_2(g) \rightarrow 3\ CO_2(g) + 4\ H_2O(g)$**

Balance C and H by placing a 3 in front of CO_2 and a 4 in front of H_2O. This results in 10 O on the right side of the equation that can be balanced with a 5 in front of O_2.

20. **$2\ CH_4O(l) + 3\ O_2(g) \rightarrow 2\ CO_2(g) + 4\ H_2O(g)$**

Since everything has oxygen in it, begin by balancing H by adding a 2 in front of H_2O. This gives 4 O on the right and 3 O on the left. To add one more O to the left without upsetting the hydrogen balance, place a coefficient of 3/2 before O_2. Then, multiply all coefficients by 2 to remove the fractional coefficient and the equation is correctly balanced.

21. **$2\ C_8H_{18}(l) + 25\ O_2(g) \rightarrow 16\ CO_2(g) + 18\ H_2O(l)$**

Start by using a coefficient of 8 for CO_2 and 9 for H_2O. This gives 25 O on the right side, so use a coefficient of 25/2 with O_2. Multiply through by 2 to remove the fractional coefficient.

22. **$3\ H_2S(g) + 2\ Fe(OH)_3(s) \rightarrow Fe_2S_3(s) + 6\ H_2O(g)$**

First balance Fe by placing a 2 in front of $Fe(OH)_3$, then balance S by placing a 3 in front of H_2. A coefficient of 6 for H_2O balances the equation.

23. **single-replacement reaction**

24. **double-replacement reaction**

25. **synthesis reaction**

26. **double-replacement reaction**

27. **synthesis reaction**

28. **double-replacement reaction**

29. **single-replacement reaction**

30. **double-replacement reaction**

31. **decomposition reaction**

32. **(a) The three kinds of double-replacement reactions are precipitate formation, gas formation, and water formation.**

    (b) The three kinds of decomposition reactions are decomposition forming only elements, decomposition forming elements and compounds, and decomposition forming only compounds.

33. $Cl_2(aq) + CuI_2(aq) \rightarrow CuCl_2(aq) + I_2(aq)$

34. $2\ K(s) + H_2O(l) \rightarrow 2\ KOH(aq) + H_2(g)$

35. $2\ Cr(s) + 3\ CdCl_2(aq) \rightarrow 2\ CrCl_3(aq) + 3\ Cd(s)$

36. $Ni(s) + Pb(NO_3)_2(aq) \rightarrow Pb(s) + Ni(NO_3)_2(aq)$

37. $2\ KI(aq) + Pb(NO_3)_2(aq) \rightarrow 2\ KNO_3(aq) + PbI_2(s)$

38. $3\ LiOH(aq) + H_3PO_4(aq) \rightarrow Li_3PO_4(aq) + 3\ H_2O(l)$

39. $H_2S(aq) + NiCl_2(aq) \rightarrow NiS(s) + 2\ HCl(aq)$

40. $K_2CO_3(s) + 2\ HNO_3(aq) \rightarrow 2\ KNO_3(aq) + H_2O(l) + CO_2(g)$

41. **Mg is oxidized and is the reducing agent. Br_2 is reduced and is the oxidizing agent.**

42. **Pb is oxidized and is the reducing agent. Hg^{2+} is reduced and is the oxidizing agent.**

43. **Cr^{2+} is oxidized and is the reducing agent. Fe^{3+} is reduced and is the oxidizing agent.**

44. **Endothermic reaction; heat is absorbed as the reaction proceeds.**

45. **The sign of ΔH indicates this is an exothermic reaction.**

46. **The sign of ΔH indicates this is an endothermic reaction.**

47. **Heat is a product of an exothermic reaction.**

48. **Neutralization of 4 moles of HCl(*aq*) with NaOH(*aq*) produces (4×56.2 kJ), 225 kJ. Neutralization of 0.3 mole produces (0.3×56.2 kJ), 16.9 kJ.**

49. **If 197 kJ are required to decompose 2 moles of $SO_3(g)$, ($3/2 \times 197$ kJ), 296 kJ of heat are required to decompose 3 moles of $SO_3(g)$.**

50. **ΔH = −177.8 kJ.** If energy is required to decompose $CaCO_3$*(s)*, the process is endothermic and ΔH is positive. Reversing the reaction converts it to an exothermic process, and ΔH is negative. The quantity of heat energy stays the same.

51. $C_5H_{12}(l) + 8\ O_2(g) \rightarrow 5\ CO_2(g) + 6\ H_2O(g)$

Supplemental Chapter Problems

Problems

For Questions 1 through 3, convert the word equations into balanced formula equations. You do not need to indicate the physical state, *(s)*, *(l)*, *(g)*, or *(aq)*, of the reactants and products.

1. barium nitrate + sodium sulfate → barium sulfate + sodium nitrate

2. tetraphosphorus + dichlorine → phosphorus trichloride

3. aluminum sulfide + water → aluminum hydroxide + hydrogen sulfide

For Questions 4 through 6, describe the reactions in words by giving the name of each reactant and product, their physical states, and any special conditions required for the reaction.

4. $Zn(s) + H_2SO_4(aq) \rightarrow ZnSO_4(aq) + H_2(g)$

5. $2\ H_2O_2(aq) \xrightarrow{KI} 2\ H_2O(l) + O_2(g)$ (H_2O_2 is hydrogen peroxide.)

6. $NH_4Cl(s) \xrightarrow{\Delta} NH_3(g) + HCl(g)$

For Questions 7 through 16, correctly balance the equations.

7. $S_8(s) + O_2(g) \rightarrow SO_2(g)$

8. $Cl_2(g) + NaI(aq) \rightarrow NaCl(aq) + I_2(s)$

9. $NaOH(aq) + H_3PO_4(aq) \rightarrow Na_3PO_4(aq) + H_2O(l)$

10. $KNO_3(s) \rightarrow KNO_2(s) + O_2(g)$

11. $P_4O_{10}(s) + H_2O(l) \rightarrow H_3PO_4(aq)$

12. $NH_3(aq) + H_2SO_4(aq) \rightarrow (NH_4)_2SO_4(aq)$

13. $Mg(s) + Fe_2(SO_4)_3(aq) \rightarrow MgSO_4(aq) + Fe(s)$

14. $C_2H_6O(l) + O_2(g) \rightarrow CO_2(g) + H_2O(l)$

15. $HCN(g) + O_2(g) \rightarrow N_2(g) + CO_2(g) + H_2O(l)$

16. $Al_4C_3(s) + H_2O(l) \rightarrow Al(OH)_3(s) + CH_4(g)$

For Questions 17 through 22, classify each reaction as a synthesis, decomposition, single-replacement, or double-replacement reaction.

17. $P_2O_3(s) + O_2(g) \rightarrow P_2O_5(s)$

18. $BaCl_2(aq) + K_2CrO_4(aq) \rightarrow 2\ KCl(aq) + BaCrO_4(s)$

19. $2\ AgI(s) \rightarrow 2\ Ag(s) + I_2(s)$

20. $2\ K(s) + H_2O(l) \rightarrow 2\ KOH(aq) + H_2(g)$

21. $SO_3(g) + H_2O(l) \rightarrow H_2SO_4(aq)$

22. $2\ NaHCO_3(s) + H_2SO_4(aq) \rightarrow Na_2SO_4(aq) + 2\ H_2O(l) + 2\ CO_2(g)$

For Questions 23 and 24, identify the oxidizing agent and the reducing agent in the equations.

23. $2\ Ag^+(aq) + Cu(s) \rightarrow 2\ Ag(s) + Cu^{2+}(aq)$

24. $Br_2(aq) + 2\ I^-(aq) \rightarrow 2\ Br^-(aq) + I_2(aq)$

For Questions 25 and 26, complete and balance the single-replacement equation.

25. $Ba(s) + H_2O(l) \rightarrow$ _____ $+ H_2(g)$

26. $Mg(s) +$ _____ $\rightarrow Mg(NO_3)_2(aq) + Cu(s)$

For Questions 27 and 28, complete and balance the double-replacement equation.

27. _____ $+ ZnCl_2(aq) \rightarrow NaCl(aq) + ZnS(s)$

28. $K_3PO_4(aq) +$ _____ $\rightarrow Sr_3(PO_4)_2(s) + KNO_3(aq)$

For Questions 29 and 30, indicate whether the reaction is exothermic or endothermic.

29. $C(s) + H_2O(g) + 131\ kJ \rightarrow CO(g) + H_2(g)$

30. $H_2(g) + I_2(g) \rightarrow 2\ HI(g)\ \Delta H = -9.4\ kJ$

31. Knowing that $C(s) + 2\ H_2(g) \rightarrow CH_4(g) + 75\ kJ$, how many moles of $C(s)$ must react to produce 300 kJ of heat energy?

32. $Na_2SO_4 \cdot 10H_2O(s) \rightarrow Na_2SO_4(s) + 10\ H_2O(g)\ \Delta H = +522\ kJ$

$Na_2SO_4(s) + 10\ H_2O(g) \rightarrow Na_2SO_4 \cdot 10H_2O(s)\ \Delta H =$ _____?

33. Write the balanced equation for the combustion of hexane, $C_6H_{14}(l)$.

Answers

Word equations to formula equations (page 157):

1. $Ba(NO_3)_2 + Na_2SO_4 \rightarrow BaSO_4 + 2\ NaNO_3$

2. $P_4 + 6\ Cl_2 \rightarrow 4\ PCl_3$

3. $Al_2S_3 + 6\ H_2O \rightarrow 2\ Al(OH)_3 + 3\ H_2S$

Reading equations (page 157):

4. When solid zinc is added to a solution of sulfuric acid, it reacts to form a solution of zinc sulfate and hydrogen gas (or dihydrogen gas).

5. Potassium iodide catalyzes the decomposition of aqueous hydrogen peroxide to form water and oxygen gas (or dioxygen gas).

6. Ammonium chloride decomposes with heating to form gaseous ammonia and gaseous hydrogen chloride.

Balancing equations (page 160):

7. $S_8(s) + 8\ O_2(g) \rightarrow 8\ SO_2(g)$

8. $Cl_2(g) + 2\ NaI(aq) \rightarrow 2\ NaCl(aq) + I_2(s)$

9. $3\ NaOH(aq) + H_3PO_4(aq) \rightarrow Na_3PO_4(aq) + 3\ H_2O(l)$

10. $2\ KNO_3(s) \rightarrow 2\ KNO_2(s) + O_2(g)$

11. $P_4O_{10}(s) + 6\ H_2O(l) \rightarrow 4\ H_3PO_4(aq)$

12. $2\ NH_3(aq) + H_2SO_4(aq) \rightarrow (NH_4)_2SO_4(aq)$

13. $3\ Mg(s) + Fe_2(SO_4)_3(aq) \rightarrow 3\ MgSO_4(aq) + 2\ Fe(s)$

14. $C_2H_6O(l) + 3\ O_2(g) \rightarrow 2\ CO_2(g) + 3\ H_2O(l)$

15. $4\ HCN(g) + 5\ O_2(g) \rightarrow 2\ N_2(g) + 4\ CO_2(g) + 2\ H_2O(l)$

16. $Al_4C_3(s) + 12\ H_2O(l) \rightarrow 4\ Al(OH)_3(s) + 3\ CH_4(g)$

Classifying equations (page 165):

17. synthesis reaction

18. double-replacement reaction

19. decomposition reaction

20. single-replacement reaction

21. synthesis reaction

22. double-replacement reaction

Oxidation-reduction equations (page 169):

23. Cu is oxidized and is the reducing agent. Ag^+ is reduced and is the oxidizing agent.

24. I^- is oxidized and is the reducing agent. Br_2 is reduced and is the oxidizing agent.

Completing single-replacement equations (page 173):

25. $Ba(s) + 2\ H_2O(l) \rightarrow Ba(OH)_2(aq) + H_2(g)$

26. $Mg(s) + Cu(NO_3)_2(aq) \rightarrow Mg(NO_3)_2(aq) + Cu(s)$

Completing double-replacement equations (page 174):

27. $Na_2S(aq) + ZnCl_2(aq) \rightarrow 2\ NaCl(aq) + ZnS(s)$

28. $2\ K_3PO_4(aq) + 3\ Sr(NO_3)_2(aq) \rightarrow Sr_2(PO_4)_2(s) + 6\ KNO_3(aq)$

Endothermic and exothermic reactions (page 177):

29. endothermic reaction

30. exothermic reaction

31. Four moles of C(*s*). If 75 kJ of heat are produced when one mole of C(*g*) reacts, 300 kJ of heat, which is four times 75 kJ, requires four times the amount of C(*s*), four moles of C(*s*) (page 177).

32. $\Delta H = -522$ kJ (page 177)

33. $2\ C_6H_{14}(l) + 19\ O_2(g) \rightarrow 12\ CO_2(g) + 14\ H_2O(g)$ (page 177)

Chapter 8

Calculations Using Balanced Equations

One of the most important areas of chemical arithmetic is based on balanced chemical equations. Chemists call this area of endeavor stoichiometry (stoy-key-om'-ah-tree), which concerns the quantitative relationships between the reactants and products in chemical reactions. **Stoichiometric calculations** can be used to determine the amount of one reactant needed to completely react with another, or to determine the amount of reactant needed to produce a desired amount of product. The key to understanding how this is done is found in the way balanced chemical equations can be interpreted. So that is the place to begin learning the arithmetic of balanced chemical equations.

The Meaning of the Balanced Equation

After carefully writing an equation, making certain the formulas are correct, and seeing that it is balanced, what does it tell you? It tells a lot both qualitatively and quantitatively:

- A chemical equation will tell you the formulas and symbols of the reactants and products.
- A chemical equation can show the physical state of a substance, whether it is a solid, liquid, gas, or in solution.
- A chemical equation can show if special conditions are required for a reaction to take place, such as adding heat or using a catalyst.

  And, if the equation is balanced: The coefficient numbers in the equation show the number of molecules, formula units, or atoms of the species involved in the reaction. The coefficients also equal the number of moles of each reactant and product.

It is this last statement that reveals how an equation becomes a powerful quantitative statement. Balancing an equation generates the set of coefficients that equal the number of moles of each species. Once the mole relationships between reactants and products are known, calculations concerning amounts of reactant consumed or products formed are possible. The reaction of sodium metal with chlorine gas to form sodium chloride is interpreted in several ways in the following table. The balanced equation is:

$$\mathbf{2}\ Na(s) + Cl_2(g) \rightarrow \mathbf{2}\ NaCl(s)$$

Five Interpretations of a Balanced Equation

2 Na +	*Cl_2* →	*2 NaCl*
2 atoms Na	1 molecule Cl_2	2 formula units NaCl
2 moles Na	1 mole Cl_2	2 moles NaCl
$2(6.022 \times 10^{23})$ atoms Na	$1(6.022 \times 10^{23})$ molecules Cl_2	$2(6.022 \times 10^{23})$ formula units NaCl
2 molar mass Na	1 molar mass Cl_2	2 molar mass NaCl
2 × 22.99 g Na	70.90 g Cl_2	2 × 58.44 g NaCl
45.98 g Na	70.90 g Cl_2	116.88 g NaCl

Only a *balanced* equation is in agreement with the Law of Conservation of Mass, which means that mass is not gained or lost as the reaction takes place. The last line of data in the table shows that this is true:

$$45.98 \text{ g Na} + 70.90 \text{ g } Cl_2 = 116.88 \text{ g NaCl}$$

The reaction begins with 116.88 grams of reactant (45.98 g + 70.90 g) and ends with 116.88 g of product. Mass is conserved.

Example Problems

These problems have the answers given.

1. Interpret the following balanced equation in terms of: (a) atoms or molecules of each reactant and product, (b) moles of each reactant and product, and (c) the mass of each reactant and product. (d) Show that the balanced equation obeys the Law of Conservation of Mass.

   $H_2O(g) + C(s) \rightarrow CO(g) + H_2(g)$

   Answer: (a) 1 molecule H_2O + 1 atom C → 1 molecule CO + 1 molecule H_2

   (b) 1 mole H_2O + 1 mole C → 1 mole CO + 1 mole H_2

   (c) 18.016 g H_2O + 12.01 g C → 28.01 g CO + 2.016 g H_2

   (d) 30.03 g reactants = 30.03 g products. Mass is conserved.

2. Interpret the following balanced equation in terms of: (a) atoms, molecules, or formula units of each reactant and product, (b) moles of each reactant and product, and (c) the mass of each reactant and product. (d) Show that the balanced equation obeys the Law of Conservation of Mass.

   $4\ Al(s) + 3\ O_2(g) \rightarrow 2\ Al_2O_3(s)$

   Answer:

   (a) 4 atoms Al + 3 molecules O_2 → 2 formula units Al_2O_3

(b) 4 moles Al + 3 moles O_2 → 2 moles Al_2O_3

(c) 107.92 g Al + 96.00 g O_2 → 203.92 g Al_2O_3

(d) 203.92 g reactants = 203.92 g product. Mass is conserved.

Work Problem

Use this problem for additional practice.

Interpret the following balanced equation in terms of: (a) atoms or molecules of each reactant and product, (b) moles of each reactant and product, and (c) the mass of each reactant and product.

$CH_4(g) + 2\ O_2(g) \rightarrow CO_2(g) + 2\ H_2O(g)$

Worked Solution

(a) 1 molecule CH_4 + 2 molecules O_2 → 1 molecule CO_2 + 2 molecules H_2O

(b) 1 mole CH_4 + 2 moles O_2 → 1 mole CO_2 + 2 moles H_2O

(c) 16.042 g CH_4 + 64.00 g O_2 → 44.01 g CO_2 + 36.032 g H_2O

Mole-to-Mole Conversions

Because the coefficients of balanced equations indicate the number of moles of each species, the conversion from "moles" of one substance to "moles" of another in a reaction will be governed by the values of the coefficients. The equation for the reaction of copper metal with silver nitrate can be read in terms of moles: 2 moles of $AgNO_3$ and 1 mole of Cu react to form 2 moles of Ag and 1 mole of $Cu(NO_3)_2$.

$$2\ AgNO_3(aq) + Cu(s) \rightarrow 2\ Ag(s) + Cu(NO_3)_2(aq)$$

How many moles of Ag*(s)* would be produced if 0.295 mole of Cu*(s)* is consumed? The answer will require a mole-to-mole conversion, converting the KNOWN mole of Cu*(s)* to the SOUGHT mole of Ag*(s)*. The coefficients in the balanced equation show that for each 1 mole of copper consumed, 2 moles of silver metal are produced. The designation of SOUGHT and KNOWN species is of critical importance. The KNOWN species is the one of *known number of moles,* Cu*(s)* in this case. We are seeking the number of moles of the SOUGHT species, Ag*(s)* in this case. The conversion is done using a conversion factor that uses the coefficients of these species in the balanced equation.

$$\text{mole of SOUGHT species} = \left(\frac{\text{coefficient of SOUGHT species}}{\text{coefficient of KNOWN species}}\right) \times (\text{moles of KNOWN species})$$

The SOUGHT species is Ag*(s)*; we are seeking the mole of Ag*(s)* that will form.

The KNOWN species is Cu*(s)*; we know 0.295 mole of Cu*(s)* is being consumed.

Coefficient of SOUGHT = 2

Coefficient of KNOWN = 1

Mole of KNOWN species = 0.295 mole Cu(*s*)

Substituting these values into the conversion equation

$$\text{moles of Ag} = \left(\frac{2 \text{ moles Ag}}{1 \text{ mole Cn}}\right) \times (0.295 \text{ moles Cu}) = 0.590 \text{ moles Ag}$$

shows that 0.590 mole of Ag(*s*) will be produced if 0.295 mole Cu(*s*) is consumed. Notice in the conversion calculation that the unit of the known species, "mole Cu," cancels and the unit of the sought species, "mole Ag," is retained. This will always happen if the conversion is done correctly. The conversion factor is always in the order of SOUGHT **over** KNOWN. Mole-to-mole conversions are one-step calculations that are *always* based on *balanced* chemical equations.

The coefficients in the copper-silver nitrate equation were fairly easy to use and you may have quickly reasoned that the mole of Ag(*s*) would be twice that of Cu(*s*) because of the 1 to 2 coefficient ratio, but when the coefficients are more complicated than those in this problem, the conversion equation allows a quick route to the answer. A similar conversion equation using sought and known terms was presented in Chapter 1 to convert units of measurement, and in Chapter 5 for mole-to-mass and mass-to-mole conversions.

The following problems involve mole-to-mole conversions based on balanced equations:

Problem 1(a): How many moles of CO_2 are produced in the combustion of 0.750 mole of hexane, C_6H_{14}? The equation is:

$$2\ C_6H_{14}(l) + 19\ O_2(g) \rightarrow 12\ CO_2(g) + 14\ H_2O(l)$$

First, make certain the equation is balanced. It is in this case. Assembling the facts:

Coefficient of SOUGHT, $CO_2(g)$ = 12

Coefficient of KNOWN, $C_6H_{14}(l)$ = 2

Moles of KNOWN, $C_6H_{14}(l)$ = 0.750 mole

$$\text{moles } CO_2 = \left(\frac{12 \text{ moles } CO_2}{2 \text{ moles } C_6H_{14}}\right) \times (0.750 \text{ mole } C_6H_{14}) = 4.50 \text{ moles } CO_2$$

Answer: Combustion of 0.750 mole of C_6H_{14} produces 4.50 moles of CO_2.

Problem 1(b): How many moles of O_2 are consumed in the reaction?

The sought species is now O_2:

Coefficient of SOUGHT, $O_2(g)$ = 19

Coefficient of KNOWN, $C_6H_{14}(l)$ = 2

Moles of KNOWN, $C_6H_{14}(l)$ = 0.750 mole

$$\text{moles } O_2 = \left(\frac{19 \text{ moles } O_2}{2 \text{ moles } C_6H_{14}}\right) \times (0.750 \text{ mole } C_6H_{14}) = 7.13 \text{ moles } O_2$$

Answer: The combustion of 0.750 mole of C_6H_{14} consumes 7.13 moles of O_2.

Ammonia, $NH_3(g)$, burns in oxygen, $O_2(g)$, forming nitrogen oxide, $NO(g)$, and water, $H_2O(g)$.

$NH_3(g) + O_2(g) \rightarrow NO(g) + H_2O(g)$

Problem 2(a): How many moles of ammonia are consumed if 1.35 moles of O_2 are consumed?

Notice the equation is not balanced! Before going any further, it *must* be balanced.

$4\ NH_3(g) + 5\ O_2(g) \rightarrow 4\ NO(g) + 6\ H_2O(g)$

Sought: $NH_3(g)$, its coefficient = 4. Known: 1.35 moles of $O_2(g)$, its coefficient = 5.

$$\text{moles } NH_3 = \left(\frac{4 \text{ moles } NH_3}{5 \cancel{\text{moles } O_2}}\right) \times \left(1.35 \cancel{\text{moles } O_2}\right) = 1.08 \text{ moles } NH_3$$

Answer: 1.08 moles of NH_3 are consumed by 1.35 moles of O_2.

Problem 2(b): How many moles of water are formed in this reaction?

Sought: $H_2O(g)$, its coefficient = 6. Known: 1.35 moles of $O_2(g)$, its coefficient = 5.

$$\text{moles } H_2O = \left(\frac{6 \text{ moles } H_2O}{5 \cancel{\text{moles } O_2}}\right) \times \left(1.35 \cancel{\text{moles } O_2}\right) = 1.62 \text{ moles } H_2O$$

Answer: 1.62 moles of H_2O are produced as 1.35 moles of O_2 are consumed.

The mole-to-mole conversion is the first of three types of calculations based on balanced chemical equations. Because balanced equations are "mole statements," mole-to-mole conversions are a critical part of all calculations based on balanced equations. When doing mole-to-mole conversions:

1. Make certain the equation is balanced.
2. Identify the SOUGHT species. You want to calculate the mole quantity of this species.
3. Identify the KNOWN species. You know the number of moles of this species.
4. Set up the conversion equation using the coefficient ratio of "sought over known" times the moles of the known species. The units of the known species will divide out leaving the unit of the sought species.

Example Problems

These problems have both answers and solutions given.

1. At high temperatures, aluminum reacts with oxygen to form aluminum oxide. The balanced equation is:

   $4\ Al(s) + 3\ O_2(g) \rightarrow 2\ Al_2O_3(s)$

   (a) How many moles of O_2 are consumed for each 0.450 mole of Al?

Answer: 0.338 mole of O_2.

Sought: mole $O_2(g)$, its coefficient = 3. Known: 0.450 mole $Al(s)$, its coefficient = 4.

$$\text{moles } O_2 = \left(\frac{3 \text{ moles } O_2}{4 \text{ moles Al}}\right) \times (0.450 \text{ mole Al}) = 0.338 \text{ mole } O_2$$

0.338 mole of O_2 is consumed for each 0.450 mole of Al consumed.

(b) How many moles of Al_2O_3 will be obtained when 0.450 mole Al is consumed?

Answer: 0.225 mole of Al_2O_3.

Sought: $Al_2O_3(s)$, its coefficient = 2. Known: 0.450 mole $Al(s)$, its coefficient = 4.

$$\text{moles } Al_2O_3 = \left(\frac{2 \text{ moles } Al_2O_3}{4 \text{ moles Al}}\right) \times (0.450 \text{ mole Al}) = 0.225 \text{ mole } Al_2O_3$$

0.225 mole of Al_2O_3 will be produced for each 0.450 mole of Al consumed.

2. The equation for the combustion of propane gas, $C_3H_8(g)$, is:

$C_3H_8(g) + O_2(g) \rightarrow CO_2(g) + H_2O(g)$

(a) How many moles of O_2 would be consumed in the combustion of 0.300 mole of C_3H_8?

(b) How many moles of H_2O would be produced?

Answer: (a) 1.50 moles of O_2 consumed; (b) 1.20 moles of H_2O produced.

(a) First, the equation must be balanced.

$C_3H_8(g) + 5\ O_2(g) \rightarrow 3\ CO_2(g) + 4\ H_2O(g)$

Sought: $O_2(g)$, its coefficient = 5. Known: 0.300 mole $C_3H_8(g)$, its coefficient = 1.

$$\text{moles } O_2 = \left(\frac{5 \text{ moles } O_2}{1 \text{ mole } C_3H_8}\right) \times (0.300 \text{ mole } C_3H_8) = 1.50 \text{ moles } O_2$$

The combustion of 0.300 mole of C_3H_8 consumes 1.50 moles of O_2.

(b) Sought: $H_2O(g)$, its coefficient = 4. Known: 0.300 mole $C_3H_8(g)$, its coefficient = 1.

$$\text{moles } H_2O = \left(\frac{4 \text{ moles } H_2O}{1 \text{ mole } C_3H_8}\right) \times (0.300 \text{ mole } C_3H_8) = 1.20 \text{ moles } H_2O$$

The combustion of 0.300 mole of C_3H_8 produces 1.20 moles of H_2O.

Work Problems

Use these problems for additional practice.

1. How many moles of Al could be produced if 5.00 moles of H_2 are consumed?

$3\ H_2(g) + Al_2O_3(s) \rightarrow 2\ Al(s) + 3\ H_2O(g)$

2. How many moles of $Fe(OH)_3$ would be produced for each 0.250 mole of NaOH consumed? The equation is:

$FeCl_3(aq) + NaOH(aq) \rightarrow Fe(OH)_3(s) + NaCl(aq)$

Worked Solutions

1. **3.33 moles of Al are produced.**

The equation is balanced as written.

Sought: Al*(s)*, its coefficient = 2. Known: 5.00 moles $H_2(g)$, its coefficient = 3.

$$\text{moles Al} = \left(\frac{2 \text{ moles Al}}{3 \text{ moles } H_2}\right) \times (5.00 \text{ moles } H_2) = 3.33 \text{ moles Al}$$

3.33 moles of Al are produced for each 5.00 moles of H_2 consumed.

2. **0.0833 mole of $Fe(OH)_3$ is produced.**

First, the equation must be balanced.

$FeCl_3(aq) + 3\ NaOH(aq) \rightarrow Fe(OH)_3(s) + 3\ NaCl(aq)$

Sought: $Fe(OH)_3(s)$, its coefficient = 1. Known: 0.250 mole NaOH*(aq)*, its coefficient = 3.

$$\text{moles } Fe(OH)_3 = \left(\frac{1 \text{ mole } Fe(OH)_3}{3 \text{ moles NaOH}}\right) \times (0.250 \text{ mole NaOH}) = 0.0833 \text{ mole } Fe(OH)_3$$

0.0883 mole of $Fe(OH)_3$ is produced for each 0.250 mole of NaOH consumed.

Mole-to-Mass and Mass-to-Mole Conversions

The second level of balanced equation calculations adds a second conversion step to the mole-to-mole conversion described in the previous section. There are two variations: (a) starting with a known mass of one substance and seeking the moles of another:

mass (known) → mole (known) → mole (sought)

and (b) starting with a known number of moles of one substance and seeking the mass of another:

mole (known) → mole (sought) → mass (sought)

Two types of conversion are required: (a) converting of the mass of a species to moles using the molar mass of that species, and (b) converting the moles of a species to moles of another using the coefficients from the balanced equation. This is the mole-to-mole conversion discussed in the previous section.

Carbon, C*(s)*, reacts with sulfur dioxide, $SO_2(g)$, to form carbon disulfide, $CS_2(l)$, and carbon dioxide, $CO_2(g)$. The balanced equation is:

$$3\ C(s) + 2\ SO_2(g) \rightarrow CS_2(l) + 2\ CO_2(g)$$

Problem 1: How many grams of CO_2 would be produced if 0.852 mole of C is completely consumed by sulfur dioxide?

Two conversions are required: mole C to mole CO_2, and mole CO_2 to mass CO_2.

Step 1: Convert the known mole *C(s)* → sought mole of $CO_2(g)$. Coefficient of sought = 2, coefficient of known = 3.

$$\text{moles } CO_2 = \left(\frac{2 \text{ moles } CO_2}{3 \text{ moles C}}\right) \times (0.852 \text{ mole C}) = 0.568 \text{ mole } CO_2$$

Step 2: Convert the mole of $CO_2(g)$ → mass of $CO_2(g)$. The molar mass of CO_2 is 44.01 g.

$$\text{mass } CO_2 = (0.568 \text{ mole } CO_2) \times \left(\frac{44.01 \text{ g } CO_2}{1 \text{ mole } CO_2}\right) = 25.0 \text{ g } CO_2$$

Answer 1: If 0.852 mole of C is consumed, 25.0 g of CO_2 will be produced.

Elemental phosphorus reacts with chlorine to form phosphorus pentachloride. The balanced equation is:

$2\ P(s) + 5\ Cl_2(g) \rightarrow 2\ PCl_5(s)$

Problem 2: How many moles of Cl_2 would be consumed if 5.50 g PCl_5 are produced?

Two conversions are required: mass of PCl_5 to mole PCl_5, and mole PCl_5 to mole Cl_2.

Step 1: Convert the mass of $PCl_5(s)$ → mole $PCl_5(s)$. The molar mass of PCl_5 is 208.2 g.

$$\text{moles } PCl_5 = (5.50 \text{ g } PCl_5) \times \left(\frac{1 \text{ mole } PCl_5}{208.2 \text{ g } PCl_5}\right) = 0.0264 \text{ mole } PCl_5$$

Step 2: Convert the known mole of $PCl_5(s)$ → sought mole of $Cl_2(g)$. The necessary coefficients are given in the balanced equation.

$$\text{moles } Cl_2 = \left(\frac{5 \text{ moles } Cl_2}{2 \text{ moles } PCl_5}\right) \times (0.0264 \text{ mole } PCl_5) = 0.0660 \text{ mole } Cl_2$$

Answer 2: The formation of 5.50 g of PCl_5 will consume 0.0660 mole of Cl_2.

Once you have determined the conversions needed to solve a two-step problem, the steps can be linked into a single calculation, which might save some time. The solution to Problem 2 is calculated again, linking the two conversions. Note how units cancel, leaving the desired unit of "mole Cl_2."

mass PCl_5 → moles PCl_5

moles PCl_5 → moles Cl_2 (from the balanced equation)

$$\text{moles } Cl_2 = (5.50 \text{ g } PCl_5)\left(\frac{1 \text{ mole } PCl_5}{208.7 \text{ g } PCl_5}\right)\left(\frac{5 \text{ moles } Cl_2}{2 \text{ moles } PCl_5}\right) = 0.0660 \text{ mole } Cl_2$$

Powdered aluminum metal will react with bromine to form aluminum bromide. The equation is:

$Al(s) + Br_2(l) \rightarrow AlBr_3(s)$

Problem 3(a): What mass of Br_2 will be consumed by 0.200 mole Al?

The equation must first be balanced: $2\ Al(s) + 3\ Br_2(l) \rightarrow 2\ AlBr_3(s)$

The two steps needed to solve this problem will be combined in a single linked calculation. The first fraction converts the known mole of Al to the sought mole of Br_2. The second converts the mole of Br_2 to the mass of Br_2. The molar mass of Br_2 is 159.8 g.

$$\text{mass } Br_2 = (0.200 \cancel{\text{ mole Al}})\left(\frac{3 \cancel{\text{ moles } Br_2}}{2 \cancel{\text{ moles Al}}}\right)\left(\frac{159.8 \text{ g } Br_2}{1 \cancel{\text{ mole } Br_2}}\right) = 47.9 \text{ g } Br_2$$

Answer 3(a): 0.200 mole of Al will consume 47.9 g of Br_2.

Problem 3(b): What mass of $AlBr_3$ will be produced in this reaction?

Again, the two steps will be combined in a single calculation. The first fraction converts the known mole of Al to the sought mole of $AlBr_3$, and the second converts the mole of $AlBr_3$ to mass. The molar mass of $AlBr_3$ is 266.7 g.

$$\text{mass } AlBr_3 = (0.200 \cancel{\text{ mole Al}})\left(\frac{2 \cancel{\text{ moles } AlBr_3}}{2 \cancel{\text{ moles Al}}}\right)\left(\frac{266.7 \text{ g } AlBr_3}{1 \cancel{\text{ mole } AlBr_3}}\right) = 53.3 \text{ g } AlBr_3$$

Answer 3(b): 53.3 g of $AlBr_3$ will be produced.

Example Problems

These problems have both answers and solutions given.

1. Consider the synthesis of ammonia from its elements. The balanced equation is:

   $N_2(g) + 3\ H_2(g) \rightarrow 2\ NH_3(g)$

   What mass of ammonia, NH_3, can be prepared from 2.50 moles of N_2?

   Answer: 85.2 g of NH_3. This problem will be solved in two separate steps.

   Step 1: Convert the known mole of $N_2(g) \rightarrow$ sought mole of $NH_3(g)$. Coefficient of known = 1. Coefficient of sought = 2.

   $$\text{moles } NH_3 = \left(\frac{2 \text{ moles } NH_3}{1 \cancel{\text{ mole } N_2}}\right) \times (2.50 \cancel{\text{ moles } N_2}) = 5.00 \text{ moles } NH_3$$

   Step 2: Convert the mole of $NH_3(g) \rightarrow$ mass $NH_3(g)$. The molar mass of NH_3 is 17.03 g.

   $$\text{mass } NH_3 = (5.00 \cancel{\text{ moles } NH_3}) \times \left(\frac{17.03 \text{ g } NH_3}{1 \cancel{\text{ mole } NH_3}}\right) = 85.2 \text{ g } NH_3$$

2. Iron metal can be freed from its ore by heating it to high temperatures with carbon.

$$2\ Fe_2O_3(s) + 3\ C(s)\ 4 \xrightarrow{\Delta} 4\ Fe(s) + 3\ CO_2(g)$$

How many moles of Fe_2O_3 are required to produce 1.00×10^3 g of Fe?

Answer: 8.95 moles of Fe_2O_3. This problem will be solved using a linked calculation.

The desired mass of Fe is known. Once converted to moles of iron, the coefficients of the balanced equation will guide the determination of the required moles of Fe_2O_3. The molar mass of Fe is 55.85 g.

$$\text{moles } Fe_2O_3 = (1.00 \times 10^3 \text{ g Fe})\left(\frac{1 \text{ mole Fe}}{55.85 \text{ g Fe}}\right)\left(\frac{2 \text{ moles } Fe_2O_3}{4 \text{ moles Fe}}\right) = 8.95 \text{ moles } Fe_2O_3$$

Work Problems

Use these problems for additional practice.

1. Calcium metal reacts with cold water to produce calcium hydroxide and hydrogen gas.

   $2\ Ca(s) + 2\ H_2O(l) \rightarrow 2\ Ca(OH)_2(aq) + H_2(g)$

   What mass of $Ca(OH)_2$ will form if 0.750 mole of $H_2(g)$ is produced in the reaction?

2. Sulfur reacts with fluorine gas to form sulfur hexafluoride.

   $S(s) + 3\ F_2(g) \rightarrow SF_6(g)$

   How many moles of $F_2(g)$ are consumed when 50.0 g of $SF_6(g)$ are produced?

Worked Solutions

1. **111 g of $Ca(OH)_2$ are produced as 0.750 mole of H_2 is produced.**

   Step 1: Convert known mole $H_2(g) \rightarrow$ sought mole of $Ca(OH)_2$. Coefficient of known = 1. Coefficient of sought = 2.

   $$\text{moles } Ca(OH)_2 = \left(\frac{2 \text{ moles } Ca(OH)_2}{1 \text{ mole } H_2}\right) \times (0.750 \text{ mole } H_2) = 1.50 \text{ moles } Ca(OH)_2$$

   Step 2: Convert mole $Ca(OH)_2 \rightarrow$ mass of $Ca(OH)_2$. The molar mass of $Ca(OH)_2$ is 74.10 g.

   $$\text{mass } Ca(OH)_2 = (1.50 \text{ moles } Ca(OH)_2) \times \left|\frac{74.10 \text{ g } Ca(OH)_2}{1 \text{ mole } Ca(OH)_2}\right| = 111 \text{ g } Ca(OH)_2$$

2. **1.03 moles of F_2 is consumed as 50.0 g SF_6 is formed.**

   Both steps, converting the mass SF_6 to mole SF_6, and the mole SF_6 to mole F_2, are combined in a single linked calculation. The molar mass of SF_6 is 146.1 g.

   $$\text{mole } F_2 = (50.0 \text{ g } SF_6)\left(\frac{1 \text{ mole } SF_6}{146.1 \text{ g } SF_6}\right)\left(\frac{3 \text{ moles } F_2}{1 \text{ mole } SF_6}\right) = 1.03 \text{ moles } F_2$$

Mass-to-Mass Conversions

As the name implies, mass-to-mass conversions start with a known mass of one substance and, by use of a balanced equation, seek the mass of another in the reaction. Mass-to-mass problems involve three conversion steps:

1. Converting the mass of the known species to moles: a mass-to-mole conversion.
2. Converting the moles of the known species to moles of the sought species: a mole-to-mole conversion based on the balanced equation.
3. Converting the moles of the sought species to mass: a mole-to-mass conversion.

mass (known) → mole (known) → mole (sought) → mass (sought)

The following problem uses mass-to-mass conversions. Silver oxide, Ag_2O, reacts with hydrogen, H_2, at high temperatures to produce silver metal, Ag, and water, H_2O. The balanced equation is:

$$2\,Ag_2O(s) + H_2(g) \xrightarrow{\Delta} 2\,Ag(s) + H_2O(l)$$

Problem 1: How many grams of Ag_2O must be consumed to produce 50.0 g of silver metal? Silver metal is the known species; silver oxide, the sought. The three steps are:

mass (Ag) → mole (Ag) → mole (Ag_2O) → mass (Ag_2O)

Step 1: Convert the mass of Ag to moles of Ag. The molar mass of Ag is 107.9 g.

$$\text{moles Ag} = (50.0\ \cancel{\text{g Ag}}) \times \left(\frac{1\ \text{mole Ag}}{107.9\ \cancel{\text{g Ag}}}\right) = 0.463\ \text{mole Ag}$$

Step 2: Convert the known mole of Ag → sought mole of Ag_2O.

$$\text{moles Ag}_2\text{O} = \left(\frac{1\ \text{mole Ag}_2\text{O}}{2\ \cancel{\text{moles Ag}}}\right) \times (0.463\ \cancel{\text{mole Ag}}) = 0.232\ \text{mole Ag}_2\text{O}$$

Step 3: Convert the mole of Ag_2O to mass of Ag_2O. The molar mass of Ag_2O is 231.8 g.

$$\text{mass Ag}_2\text{O} = (0.232\ \cancel{\text{mole Ag}_2\text{O}}) \times \left(\frac{231.8\ \text{g Ag}_2\text{O}}{1\ \cancel{\text{mole Ag}_2\text{O}}}\right) = 53.7\ \text{g Ag}_2\text{O}$$

Answer 1: It requires 53.7 g of Ag_2O to produce 50.0 g silver metal, Ag.

The three steps of this calculation can be linked into one long calculation. Each conversion factor represents one of the three steps.

mass Ag → moles Ag

moles Ag → moles Ag_2O (Based on the balanced equation)

moles Ag_2O → mass Ag_2O

$$\text{mass Ag}_2\text{O} = (50.0\ \cancel{\text{g Ag}})\left(\frac{1\ \cancel{\text{mole Ag}}}{107.9\ \cancel{\text{g Ag}}}\right)\left(\frac{1\ \cancel{\text{mole Ag}_2\text{O}}}{2\ \cancel{\text{moles Ag}}}\right)\left(\frac{231.8\ \text{g Ag}_2\text{O}}{1\ \cancel{\text{mole Ag}_2\text{O}}}\right) = 53.7\ \text{g Ag}_2\text{O}$$

Once you are confident you know the reason for each step, it may be easier to do the mass-to-mass conversions by linking them together. If done correctly, units cancel, leaving the single correct unit for the answer.

Copper metal, Cu, reacts with nitric acid, $HNO_3(aq)$, to produce copper(II) nitrate, $Cu(NO_3)_2$, nitrogen dioxide, NO_2, and water, H_2O. The balanced equation is:

$Cu(s) + 4\ HNO_3(aq) \rightarrow Cu(NO_3)_2(aq) + 2\ NO_2(g) + 2\ H_2O(l)$

Problem 2: What mass of nitric acid, $HNO_3(aq)$, is required to consume just 7.50 g of copper metal, Cu?

The three steps required to solve this problem will be linked into a single calculation. The molar mass of Cu is 63.55 g, and for $HNO_3(aq)$ is 63.02 g.

mass (Cu) $\rightarrow$ mole (Cu) $\rightarrow$ mole (HNO_3) $\rightarrow$ mass (HNO_3)

$$\text{mass } HNO_3 = \left(7.50\text{ g } \cancel{Cu}\right)\left(\frac{1\ \cancel{\text{mole Cu}}}{63.55\text{ g } \cancel{Cu}}\right)\left(\frac{4\ \cancel{\text{moles } HNO_3}}{1\ \cancel{\text{mole Cu}}}\right)\left(\frac{63.02\text{ g } HNO_3}{1\ \cancel{\text{mole } HNO_3}}\right) = 29.7\text{ g } HNO_3$$

Answer 2: It requires 29.7 g $HNO_3(aq)$ to consume just 7.50 g of copper metal.

Potassium chlorate, $KClO_3$, decomposes when heated in the presence of a catalyst to form potassium chloride, KCl, and oxygen gas, O_2. The equation is:

$KClO_3(s) \xrightarrow[MnO_2]{\Delta} KCl(s) + O_2(g)$

Problem 3: What mass of $KClO_3$ must be decomposed to obtain 1.75 g of O_2?

Did you notice the equation was not balanced? The balanced equation is:

$2\ KClO_3(s) \rightarrow 2\ KCl(s) + 3\ O_2(g)$

The three steps required to answer this question are linked into a single calculation.

mass (O_2) $\rightarrow$ mole (O_2) $\rightarrow$ mole ($KClO_3$) $\rightarrow$ mass ($KClO_3$)

The molar mass of O_2 is 32.00 g; that of $KClO_3$ is 122.6 g.

$$\text{mass } KClO_3 = \left(1.75\text{ g } \cancel{O_2}\right)\left(\frac{1\ \cancel{\text{mole } O_2}}{32.00\text{ g } \cancel{O_2}}\right)\left(\frac{2\ \cancel{\text{moles } KClO_3}}{3\ \cancel{\text{moles } O_2}}\right)\left(\frac{122.6\text{ g } KClO_3}{1\ \cancel{\text{mole } KClO_3}}\right) = 4.47\text{ g } KClO_3$$

Answer 3: 4.47 g of $KClO_3$ must be decomposed to produce 1.75 g of O_2.

Example Problems

These problems have both answers and solutions given.

1. What mass of CuO would be produced if 1.00×10^2 g of Cu_2O is consumed?

   $2\ Cu_2O(s) + O_2(g) \rightarrow 4\ CuO(s)$

   The molar mass of Cu_2O is 143.1 g; that of CuO is 79.55 g.

   Answer: 111 g of CuO is produced.

This problem will be solved using three individual steps.

mass (Cu_2O) → mole (Cu_2O) → mole (CuO) → mass (CuO)

Step 1: Converting the mass of Cu_2O to mole.

$$\text{moles } Cu_2O = (1.00 \times 10^2 \text{ g } \cancel{Cu_2O}) \times \left(\frac{1 \text{ mole } Cu_2O}{143.1 \text{ g } \cancel{Cu_2O}}\right) = 0.699 \text{ mole } Cu_2O$$

Step 2: Converting the known mole of Cu_2O to the sought mole of CuO.

$$\text{moles CuO} = \left(\frac{4 \text{ moles CuO}}{2 \cancel{\text{moles } Cu_2O}}\right) \times (0.699 \cancel{\text{mole } Cu_2O}) = 1.40 \text{ moles CuO}$$

Step 3: Converting the moles of CuO to mass of CuO.

$$\text{mass CuO} = (1.40 \cancel{\text{moles CuO}}) \times \left(\frac{79.55 \text{ g CuO}}{1 \cancel{\text{mole CuO}}}\right) = 111.4 \text{ g} = 111 \text{ g CuO}$$

2. What mass of H_2 will be produced if 5.00 g of Na are consumed in a reaction with water?

$2\ Na(s) + 2\ H_2O(l) \rightarrow 2\ NaOH(aq) + H_2(g)$

The molar mass of Na is 22.99 g; that of H_2 is 2.016 g.

Answer: 0.219 g H_2 is produced.

This problem will be solved linking the three conversion steps in a single equation.

$$\text{mass } H_2 = (5.00 \text{ g } \cancel{Na})\left(\frac{1 \cancel{\text{mole Na}}}{22.99 \text{ g } \cancel{Na}}\right)\left(\frac{1 \cancel{\text{mole } H_2}}{2 \cancel{\text{moles Na}}}\right)\left(\frac{2.016 \text{ g } H_2}{1 \cancel{\text{mole } H_2}}\right) = 0.219 \text{ g } H_2$$

Work Problems

Use these problems for additional practice.

1. What mass of $FeCl_3$ will be produced if 25.0 g of Cl_2 reacts with iron metal? The balanced equation is:

   $2\ Fe(s) + 3\ Cl_2(g) \xrightarrow{\Delta} 2\ FeCl_3(s)$

   Atomic masses (amu): Cl = 35.45; Fe = 55.85. You will need to calculate molar masses.

2. What mass of KO_2 must be used to prepare exactly 50.0 g of O_2? The balanced equation is:

   $4\ KO_2(s) + 2\ H_2O(l) \rightarrow 4\ KOH(aq) + 3\ O_2(g)$

   Atomic masses (amu): K = 39.10; O = 16.00

Worked Solutions

1. **38.1 g of $FeCl_3$ will be produced as 25.0 g of Cl_2 is consumed.**

   The three-step conversion will be linked in a single calculation.

   mass (Cl_2) → mole (Cl_2) → mole ($FeCl_3$) → mass ($FeCl_3$)

   The molar mass of Cl_2 is 70.90 g; that of $FeCl_3$ is 162.2 g.

$$\text{mass } FeCl_3 = \left(25.0 \text{ g } \cancel{Cl_2}\right)\left(\frac{1 \text{ } \cancel{\text{mole } Cl_2}}{70.90 \text{ } \cancel{\text{g } Cl_2}}\right)\left(\frac{2 \text{ } \cancel{\text{moles } FeCl_3}}{3 \text{ } \cancel{\text{moles } Cl_2}}\right)\left(\frac{162.2 \text{ g } FeCl_3}{1 \text{ } \cancel{\text{mole } FeCl_3}}\right) = 38.1 \text{ g } FeCl_3$$

2. **148 g of KO_2 must be used to prepare 50.0 g of O_2.**

   The three-step conversion will be linked in a single calculation.

   mass (O_2) → mole (O_2) → mole (KO_2) → mass (KO_2)

   The molar mass of KO_2 is 71.10 g; that of O_2 is 32.00 g.

$$\text{mass } KO_2 = \left(50.0 \text{ g } \cancel{O_2}\right)\left(\frac{1 \text{ } \cancel{\text{mole } O_2}}{32.00 \text{ } \cancel{\text{g } O_2}}\right)\left(\frac{4 \text{ } \cancel{\text{moles } KO_2}}{3 \text{ } \cancel{\text{moles } O_2}}\right)\left(\frac{71.10 \text{ g } KO_2}{1 \text{ } \cancel{\text{mole } KO_2}}\right) = 148 \text{ g } KO_2$$

The Limiting Reactant

In chemical reactions with two or more reactants, the reaction will continue until one of the reactants is used up. When that reactant is gone, the reaction stops. No more product can form. A **limiting reactant** (or limiting reagent) is used up first. It is the reactant that limits the amount of product that can be made. Consider the reaction of sodium metal with water to form hydrogen gas:

$$2 \text{ Na}(s) + 2 \text{ H}_2\text{O}(l) \rightarrow 2 \text{ NaOH}(aq) + \text{H}_2(g)$$

If 1.0 gram of sodium is dropped into a bathtub full of water, it will react vigorously, producing hydrogen gas. The reaction will continue until the sodium is completely consumed. At that point no more hydrogen gas is produced. Sodium metal is the limiting reactant; its disappearance limits the amount of hydrogen gas produced. A 1.0-gram sample of sodium requires less than 0.8 gram of water to be completely consumed. A bathtub full of water is immensely more than the 0.8 gram of water sodium requires. Water is the **excess reactant.**

In the reaction of sodium hydroxide, NaOH, with hydrochloric acid, HCl(*aq*), the balanced equation shows that each mole of NaOH reacts with the exact same number of moles of HCl(*aq*). They react in a 1 to 1 mole ratio, the number of moles of both being identical.

$$\text{NaOH}(aq) + \text{HCl}(aq) \rightarrow \text{NaCl}(aq) + \text{H}_2\text{O}(l)$$

If NaOH and HCl(*aq*) are combined in anything other than a 1 to 1 mole ratio, one reactant will be in excess. The other will be the limiting reactant. Whenever starting amounts of two reactants are given in a stoichiometry problem, before you can calculate amounts of product you will first have to learn which reactant is the limiting reactant. It will govern how much product forms.

To show how to analyze a limiting reactant problem, three different starting mixtures of NaOH and HCl(*aq*) follow. See how the limiting reactant is determined in each mixture.

1. 0.35 mole of NaOH is combined with 0.50 mole of HCl*(aq)*.

   To determine which is the limiting reactant, choose one of the reactants and determine how many moles of the other it will consume. Let's choose NaOH and ask the question: How many moles of HCl will be consumed by 0.35 mole of NaOH? Even without doing a mole-to-mole conversion, the 1-to-1 mole ratio in the balanced equation indicates that 0.35 mole of NaOH will consume 0.35 mole of HCl, a smaller amount of HCl than we started with. There is an excess of HCl. The reaction started with more HCl than NaOH can consume. *Knowing that HCl is in excess indicates that NaOH is the limiting reactant.* The reaction will stop once the NaOH is consumed, and 0.35 mole of both NaCl and H_2O will be produced. The excess HCl (0.15 mole HCl) remains unreacted (0.50 mole at the start– 0.35 mole consumed by NaOH = 0.15 mole HCl left over).

2. 0.35 mole of NaOH is combined with 0.25 mole of HCl*(aq)*.

   As before, start by asking the question: How many moles of HCl would be consumed by 0.35 mole of NaOH? The answer is that 0.35 mole of NaOH would consume 0.35 mole of HCl, but only 0.25 mole of acid is present. There is not enough HCl to consume all the NaOH; there is an excess of NaOH and that means that HCl is the limiting reactant. The reaction will stop once the HCl is consumed and the amount of product that forms will depend on the starting number of moles of HCl, not NaOH.

3. 0.35 mole of NaOH is combined with 0.35 mole of HCl*(aq)*.

   These mole quantities of the base and acid are in the 1-to-1 stoichiometric ratio of the balanced equation. Neither is in excess. 0.35 mole of NaOH will just exactly consume 0.35 mole of HCl. In a sense, they are both limiting reactants.

Remember, if you are given starting quantities of two reactants in a balanced-equation problem, always check for the limiting reactant first. The limiting reactant will govern the amount of product formed. In all the reactions studied up to this point, the amount of only one reactant was given and the other reactant was considered to be in excess.

Guidelines for determining the limiting reactant:

1. To check for a limiting reactant, the equation must be balanced and the starting amounts of both reactants must be in moles. You may need to do mass-to-mole conversions.

2. Referring to the reactants as A and B, to determine if reactant A is the limiting reactant, calculate the number of moles of B it will consume (a mole-to-mole conversion).

   If you start with more B than needed, B is in excess and A is the limiting reactant.

   If you start with less B than needed, B is the limiting reactant.

   If you start with the same amount of B as calculated, both reactants will be completely consumed and neither is in excess.

3. If one reactant is in excess, the other is the limiting reactant, and vice versa.

4. Use the number of moles of the limiting reactant to calculate the moles of product formed.

Problem 1: 1.0 mole of Cr is mixed with 1.0 mole of O_2, and the heated mixture is allowed to react. How many moles of Cr_2O_3*(s)* will be produced? The balanced equation is:

$$4\,Cr(s) + 3\,O_2(g) \xrightarrow{\Delta} \rightarrow 2\,Cr_2O_3(s)$$

Because starting amounts of both reactants are given, first check for the limiting reactant. Choose one reactant (either one is fine) and calculate the number of moles of the other it would consume. Let's choose Cr and ask the question: How many moles of O_2 would be consumed by1.0 mole of Cr? This is a mole-to-mole conversion based on the balanced equation. Cr is the known species; O_2, the sought.

$$\text{moles } O_2 \text{ consumed by } 1.0 \text{ mole Cr} = \left(\frac{3 \text{ moles } O_2}{4 \text{ moles Cr}}\right) \times (1.0 \text{ mole Cr}) = 0.75 \text{ mole } O_2$$

We learn that 1.0 mole of Cr will consume 0.75 mole of O_2, less than the amount of O_2 in the starting mixture. There is in excess of O_2, indicating Cr is the limiting reactant. All the chromium will be consumed, leaving the excess of O_2. The number of moles of Cr consumed determines the number of moles of Cr_2O_3 produced.

$$\text{moles } Cr_2O_3 = \left(\frac{2 \text{ moles } Cr_2O_3}{4 \text{ moles Cr}}\right) \times (1.00 \text{ mole Cr}) = 0.500 \text{ mole } Cr_2O_3$$

Answer: Cr is the limiting reactant. 1.00 mole of Cr will produce 0.500 mole of Cr_2O_3.

Problem 2(a): 12.0 moles of N_2 and 85.0 grams of H_2 are reacted to produce ammonia, NH_3. The balanced equation is:

$N_2(g) + 3\ H_2(g) \rightarrow 2\ NH_3(g)$

Which is the limiting reactant? First, the starting amount of each reactant must be in moles, not grams. The 85.0 g mass of H_2 is 42.2 moles H_2. The molar mass of H_2 is 2.016 g.

$$\text{moles } H_2 = (85.0 \text{ g } H_2) \times \left(\frac{1 \text{ mole } H_2}{2.016 \text{ g } H_2}\right) = 42.2 \text{ moles } H_2$$

In terms of moles, the starting mixture contains 42.2 moles H_2 and 12.0 moles N_2. The balanced equation shows that 1 mole of N_2 reacts with exactly 3 moles of H_2, a 1 to 3 ratio. It follows, then, that 12.0 moles of $N_2(g)$ would consume exactly 3×12.0 moles, or 36.0 moles of $H_2(g)$. The mole-to-mole calculation verifies this.

$$\text{moles } H_2 = \left(\frac{3 \text{ moles } H_2}{1 \text{ mole } N_2}\right) \times (12.0 \text{ moles } N_2) = 36.0 \text{ moles } H_2$$

12.0 moles of N_2 requires 36.0 moles of H_2, a smaller amount of hydrogen than in the starting mixture. H_2 is in excess by 6.2 moles, (42.2 moles - 36.0 moles = 6.2 moles), and if H_2 is the excess, N_2 is the limiting reactant.

Answer 2(a): N_2 is the limiting reactant.

Problem 2(b): How many grams of ammonia, NH_3, will be produced? The molar mass of NH_3 is 17.03 g.

The amount of ammonia produced is governed by the number of moles of N_2 present at the start of the reaction, since it is the limiting reactant. The number of moles of NH_3 produced by 12.0 moles of N_2 is:

$$\text{moles } NH_3 = (12.0 \text{ moles } N_2) \times \left(\frac{2 \text{ moles } NH_3}{1 \text{ mole } N_2}\right) = 24.0 \text{ moles } NH_3$$

The mass of 24.0 moles of NH_3 is:

$$\text{moles } NH_3 = (24.0 \text{ moles } NH_3) \times \left(\frac{17.03 \text{ g } NH_3}{1 \text{ mole } NH_3}\right) = 408.7 \text{ g} = 409 \text{ g } NH_3$$

Answer 2(b): 409 g of NH_3 is produced.

Example Problems

These problems have both answers and solutions given:

1. Looking again at the blast furnace reaction: 2 Fe_2O_3*(s)* + 3 C*(s)* $\xrightarrow{\Delta}$ 4 Fe*(s)* + 3 CO_2*(g)*

Three starting mixtures of Fe_2O_3 and C are listed. Which is the limiting reactant in each mixture?

(a) 2.0 moles Fe_2O_3 + 4.0 moles C

(b) 2.0 moles Fe_2O_3 + 3.0 moles C

(c) 2.0 moles Fe_2O_3 + 2.0 moles C

Answer: (a) Fe_2O_3 is the limiting reactant. (b) Both are in the 2-to-3 stoichiometric ratio and will be completely consumed. (c) C is the limiting reactant.

Because each of the three starting mixtures has 2.0 moles of Fe_2O_3, the number of moles of C it would consume needs to be calculated only once. Then compare the starting moles of C given in each mixture with the moles of C consumed by 2.0 moles of Fe_2O_3 to find the limiting reactant.

$$\text{moles C} = \left(\frac{3 \text{ moles C}}{2 \text{ moles } Fe_2O_3}\right) \times (2.0 \text{ moles } Fe_2O_3) = 3.0 \text{ moles C}$$

(a) 2.0 moles of Fe_2O_3 will consume 3.0 moles of C. The starting mixture has 4.0 moles of C, an excess amount. If C is in excess, Fe_2O_3 is the limiting reactant.

(b) 2.0 moles of Fe_2O_3 will consume 3.0 moles of C, which is the exact composition of the starting mixture. In this stoichiometric mixture, both reactants will be completely consumed.

(c) 2.0 moles of Fe_2O_3 will consume 3.0 moles of C. With only 2.0 moles of C in the starting mixture, C is the limiting reactant. There is an excess of Fe_2O_3.

2. 6.00 moles of Al are combined with 6.00 moles of S and heated to form Al_2S_3.

2 Al*(s)* + 3 S*(s)* $\xrightarrow{\Delta}$ → Al_2S_3*(s)*

How many moles of Al_2S_3 will be produced?

Answer: 2.00 moles of Al_2S_3 will be produced.

Check if Al is the limiting reactant. Calculate the moles of S consumed by 6.00 moles of Al.

$$\text{moles S} = \left(\frac{3\ \text{moles S}}{2\ \cancel{\text{moles Al}}}\right) \times \left(6.00\ \cancel{\text{moles Al}}\right) = 9.00\ \text{moles S}$$

6.00 moles of Al will consume 9.00 moles of S, *more* than is present in the starting reaction mixture. This means S is the limiting reactant. The amount of Al_2S_3 will be governed by the starting amount of sulfur. The moles of Al_3S_3 produced in the reaction are calculated from the initial number of moles of S:

$$\text{moles } Al_2S_3 = \left(\frac{1\ \text{mole } Al_2S_3}{3\ \cancel{\text{moles S}}}\right) \times \left(6.00\ \cancel{\text{moles S}}\right) = 2.00\ \text{moles } Al_2S_3$$

Work Problems

Use these problems for additional practice.

1. $4\ NH_3(g) + 5\ O_2(g) \rightarrow 4\ NO(g) + 6\ H_2O(l)$

   Three starting mixtures of $NH_3(g)$ and $O_2(g)$ are listed. Which is the limiting reactant in each mixture?

   (a) 2.5 moles NH_3 + 3.0 moles O_2

   (b) 2.0 moles NH_3 + 3.0 moles O_2

   (c) 2.4 moles NH_3 + 3.0 moles O_2

2. $2\ C_2H_6(g) + 7\ O_2(g) \rightarrow 4\ CO_2(g) + 6\ H_2O(g)$

   4.00 moles of C_2H_6 and 12.0 moles of O_2 are reacted. How many moles of CO_2 will form?

Worked Solutions

1. **(a) O_2 is the limiting reactant. (b) NH_3 is the limiting reactant. (c) Neither reactant is in excess; both reactants will be completely consumed.**

   Each starting mixture contains 3.0 moles of O_2. How many moles of NH_3 would be consumed by 3.0 moles of O_2?

   $$\text{moles } NH_3 = \left(\frac{4\ \text{moles } NH_3}{5\ \cancel{\text{moles } O_2}}\right) \times \left(3.0\ \cancel{\text{moles } O_2}\right) = 2.4\ \text{moles } NH_3$$

   (a) The starting number of moles of $NH_3(g)$ is greater than 2.4 moles. An excess of NH_3 is present; therefore, O_2 is the limiting reactant.

   (b) Only 2.0 moles of NH_3 are present, less than what is needed to completely consume 3.0 moles of O_2. This means NH_3 is the limiting reactant.

   (c) There is exactly enough NH_3 to consume all the O_2. Neither is in excess and both will be completely consumed.

2. **6.86 moles of CO_2 will be produced.**

To check if C_2H_6 is the limiting reactant, the number of moles of O_2 consumed by 4.00 moles of C_2H_6 is determined.

$$\text{moles } O_2 = \left(\frac{7 \text{ moles } O_2}{2 \text{ moles } C_2H_6}\right) \times (4.00 \text{ moles } C_2H_6) = 14.0 \text{ moles } O_2$$

4.00 moles of C_2H_6 will consume 14.00 moles of O_2, more than is present in the starting mixture. An insufficient amount of O_2 indicates it is the limiting reactant. The number of moles of CO_2 produced will depend on the starting number of moles of O_2.

$$\text{moles } CO_2 = \left(\frac{4 \text{ moles } CO_2}{7 \text{ moles } O_2}\right) \times (12.0 \text{ moles } O_2) = 6.86 \text{ moles } CO_2$$

Percent Yield of Reactions

The **theoretical yield** of a reaction is the calculated amount of product that should be obtained in a reaction. But in practice, something less than the theoretical yield, the **actual yield,** is obtained. For many reasons—impure reagents, solubility problems, decomposition, or spillage—the amount of product isolated from a reaction is frequently less than that calculated from the balanced equation. The fraction of the theoretical yield actually obtained in a reaction is expressed as the **percent yield** of the reaction.

$$\% \text{ yield} = \left(\frac{\text{actual yield}}{\text{theoretical yield}}\right) \times 100\%$$

The actual yield and theoretical yield should be given in the same units: grams, moles, and so on. Suppose in a synthesis reaction you calculated that you should obtain 50.0 grams of product, the theoretical yield. But when you did the reaction you only obtained 45.5 grams. The percent yield would be:

$$\% \text{ yield} = \left(\frac{45.5 \text{ g}}{50.0 \text{ g}}\right) \times 100\% = 91.0\% \text{ yield}$$

Problem: 0.800 mole of phosphorus was reacted with an excess of sulfur to form diphosphorus pentasulfide, P_2S_5.

$2\ P(s) + 5\ S(s) \xrightarrow{\Delta} \rightarrow P_2S_5(s)$

At the end of the reaction, the actual yield was 82.7 g of P_2S_5. What is the percent yield of the reaction? The molar mass of P_2S_5 is 222.3 g.

First, calculate the mass of P_2S_3 you should obtain from 0.400 mole of phosphorus. This is the theoretical yield of the synthesis. A linked calculation is used.

$$\text{mass } P_2S_5 = (0.800 \text{ mole P})\left(\frac{1 \text{ mole } P_2S_5}{2 \text{ moles P}}\right)\left(\frac{222.3 \text{ g } P_2S_5}{1 \text{ mole } P_2S_5}\right) = 88.9 \text{ g } P_2S_5$$

Knowing the actual yield is 82.7 g and the theoretical yield is 88.9 g, the percent yield is:

$$\% \text{ yield} = \left(\frac{82.7 \text{ g } P_2S_5}{88.9 \text{ g } P_2S_5}\right) \times 100\% = 93.0\%$$

Answer: The percent yield is 93.0%.

Example Problem

This problem has both the answer and solution given.

When solutions of calcium chloride and silver nitrate are mixed, a precipitate of silver chloride, AgCl, forms.

$CaCl_2(aq) + 2\ AgNO_3(aq) \rightarrow 2\ AgCl(s) + Ca(NO_3)_2(aq)$

A solution containing 0.320 mole of $AgNO_3$ was treated with an excess amount of $CaCl_2$. After removing and drying the precipitate, 44.1 g of AgCl was obtained. What is the percent yield of the reaction? The molar mass of AgCl is 143.4 g.

Answer: The percent yield is 96.1%.

The actual yield is 44.1 g of AgCl. Calculating the theoretical yield:

$$\text{mass AgCl} = (0.320 \text{ mole } AgNO_3)\left(\frac{2 \text{ moles AgCl}}{2 \text{ moles } AgNO_3}\right)\left(\frac{143.4 \text{ g AgCl}}{1 \text{ mole AgCl}}\right) = 45.9 \text{ g AgCl}$$

The percent yield is:

$$\% \text{ yield} = \left(\frac{44.1 \text{ g AgCl}}{45.9 \text{ g AgCl}}\right) \times 100\% = 96.1\%$$

Work Problem

Use this problem for additional practice.

Mercury(II) oxide decomposes with heating to form elemental mercury, Hg, and oxygen gas, O_2.

$2\ HgO(s) \xrightarrow{\Delta} 2\ Hg(l) + O_2(g)$

The decomposition of 0.0580 mole of HgO produced 0.724 g of O_2. What is the percent yield of this reaction? The molar mass of O_2 is 32.00 g.

Worked Solution

The percent yield is 78.0%.

The theoretical yield of oxygen gas produced in the decomposition of 0.0580 mole of HgO is:

$$\text{mass } O_2 = (0.0580 \text{ mole HgO})\left(\frac{1 \text{ mole } O_2}{2 \text{ moles HgO}}\right)\left(\frac{32.00 \text{ g } O_2}{1 \text{ mole } O_2}\right) = 0.928 \text{ g } O_2$$

The percent yield is:

$$\% \text{ yield} = \left(\frac{0.724 \text{ g } O_2}{0.928 \text{ g } O_2}\right) \times 100\% = 78.0\%$$

Chapter Problems and Answers

Problems

For Questions 1 through 3, interpret the balanced equation in terms of: (a) atoms or molecules of each reactant and product, (b) moles of each reactant and product, and (c) the mass of each reactant and product. (d) Is the balanced equation consistent with the Law of Conservation of Mass?

1. $N_2(g) + 2\ O_2(g) \rightarrow 2\ NO_2(g)$

2. $NH_4NO_3(s) \rightarrow N_2O(g) + 2\ H_2O(l)$

3. $CS_2(l) + 3\ O_2(g) \rightarrow CO_2(g) + 2\ SO_2(g)$

Questions 4 through 6 concern mole-to-mole problems based on balanced chemical equations.

4. $P_4O_6(s) + 6\ H_2O(l) \rightarrow 4\ H_3PO_3(aq)$

   (a) How many moles of H_3PO_3 could be prepared from 0.650 mole of P_4O_6?

   (b) If 0.750 mole of H_2O is consumed, how many moles of H_3PO_3 will form?

5. $Al_2S_3(s) + 6\ H_2O(l) \rightarrow 2\ Al(OH)_3(s) + 3\ H_2S(g)$

   (a) If 2.30 moles of H_2S are produced, how many moles of Al_2S_3 are consumed?

   (b) What number of moles of $Al(OH)_3$ would be produced along with 2.30 moles of H_2S?

6. $8\ BF_3(g) + 6\ NaH(s) \rightarrow 6\ NaBF_4(s) + B_2H_6(g)$

   (a) How many moles of BF_3 are required to produce 0.520 mole of B_2H_6?

   (b) How many moles of NaH are required to produce 0.520 mole of B_2H_6?

Questions 7 through 10 concern mole-to-mass and mass-to-mole problems based on balanced equations.

7. Phosphorus pentachloride reacts with water to produce phosphoric acid and hydrochloric acid.

   $PCl_5(s) + H_2O(l) \rightarrow H_3PO_4(aq) + HCl(aq)$

   (a) What mass of HCl is produced if 0.250 mole of PCl_3 is consumed?

   (b) What mass of H_2O is consumed if 0.800 mole of H_3PO_4 is formed?

The molar mass of HCl is 36.46 g; that of H_2O is 18.02 g.

8. If a mixture of $MgO(s)$ and Fe is heated, Fe_2O_3 and molten magnesium, Mg, is formed.

$3\ MgO(s) + 2\ Fe(s) \rightarrow Fe_2O_3(s) + 3\ Mg(l)$

Atomic masses (amu): Fe = 55.85; Mg = 24.31

(a) What mass of Fe is required to just completely react with 0.850 mole of MgO?

(b) How many moles of Fe are required to produce 100.0 g of Mg?

9. $2\ Na_3PO_4(aq) + 3\ Ni(NO_3)_2(aq) \rightarrow Ni_3(PO_4)_2(s) + 6\ NaNO_3$

(a) What mass of $Ni_3(PO_4)_2$ forms if 0.350 mole of Na_3PO_4 is consumed?

(b) How many moles of $Ni(NO_3)_2$ are required to consume exactly 250.0 g of Na_3PO_4?

Atomic masses (amu): Na = 22.99; P = 30.97; Ni = 58.69; O = 16.00

Questions 10 through 13 are mass-to-mass problems based on balanced chemical equations.

10. $Cl_2(aq) + 2\ NaBr(aq) \rightarrow 2\ NaCl(aq) + Br_2(aq)$

(a) How many grams of NaBr are needed to prepare 50.0 g of Br_2?

(b) How many grams of Cl_2 are consumed?

Molar masses: Cl_2 = 70.90 g; NaBr = 102.9 g; Br_2 = 159.8 g

11. $CdCl_2(aq) + NaOH(aq) \rightarrow Cd(OH)_2(s) + NaCl(aq)$

What mass of $Cd(OH)_2$ would form if exactly 1.00 g of NaCl is produced?

Molar masses: $Cd(OH)_2$ = 146.4 g; NaCl = 58.44 g

12. $CaF_2(s) + H_2SO_4(l) \rightarrow 2\ HF(g) + CaSO_4(s)$

What mass of CaF_2 must be consumed to make 25.0 g HF?

Atomic masses (amu): Ca = 40.08; H = 1.008; F = 19.00

13. $C_6H_{12}O_6(s) + O_2(g) \rightarrow CO_2(g) + H_2O(l)$

(a) What mass of glucose, $C_6H_{12}O_6$, would produce 5.50 g of CO_2?

(b) What mass of O_2 would be consumed in the process?

Molar masses: $C_6H_{12}O_6$ = 180.2 g; CO_2 = 44.01 g; O_2 = 32.00 g

Questions 14 through 16 are limiting reactant problems.

14. Consider the reaction in which a precipitate of lead (II) chloride is formed:

$2\ NaCl(aq) + Pb(NO_3)_2(aq) \rightarrow PbCl_2(s) + 2\ NaNO_3(aq)$

Three starting mixtures for this reaction are listed below. Identify the limiting reactant in each one.

(a) 0.30 mole NaCl 0.20 mole $Pb(NO_3)_2$

(b) 0.15 mole NaCl 0.30 mole $Pb(NO_3)_2$

(c) 0.20 mole NaCl 0.05 mole $Pb(NO_3)_2$

15. The equation for the reaction in certain rocket engines is:

$C_2H_8N_2(l) + N_2O_4(g) \rightarrow N_2(g) + CO_2(g) + H_2(g)$

(a) 45.0 g of $C_2H_8N_2$ and 75.0 g of N_2O_4 are mixed and reacted. Which is the limiting reactant?

(b) What mass of CO_2 will be produced in the reaction?

Molar masses: $C_2H_8N_2$ = 60.10 g; N_2O_4 = 92.02 g; CO_2 = 44.01 g

16. $NiCl_2(aq) + (NH_4)_2S(aq) \rightarrow NiS(s) + 2\ NH_4Cl(aq)$

One solution containing 31.0 grams of $NiCl_2$ and another containing 22.0 grams of $(NH_4)_2S$ are mixed, producing a precipitate of NiS. What mass of NiS will form?

Atomic masses (amu): Ni = 55.69; Cl = 35.45; S = 32.07; H = 1.008; N = 14.01

Questions 17 and 18 are percent yield problems.

17. Calcium oxalate precipitates when solutions of calcium hydroxide and oxalic acid are mixed:

$Ca(OH)_2(aq) + H_2C_2O_4(aq) \rightarrow CaC_2O_4(s) + 2\ H_2O(l)$

When 0.650 mole of calcium hydroxide was mixed with an excess of oxalic acid, 78.4 g of dried calcium oxalate was isolated. What is the percent yield of the reaction? The molar mass of calcium oxalate is 128.1 g.

18. Aluminum hydroxide is converted to aluminum oxide at high temperatures:

$2\ Al(OH)_3(s) \rightarrow Al_2O_3(s) + 3\ H_2O(g)$

3.50 g of $Al(OH)_3$ produced 1.96 g of Al_2O_3 after intense heating. What is the percent yield of the reaction? The molar mass of $Al(OH)_3$ is 78.00 g; that of Al_2O_3 is 101.96 g.

Answers

1. **(a) 1 molecule N_2 + 2 molecules $O_2 \rightarrow$ 2 molecules NO_2**

   (b) 1 mole N_2 + 2 moles $O_2 \rightarrow$ 2 moles NO_2

   (c) 28.02 g N_2 + 64.00 g $O_2 \rightarrow$ 92.02 g NO_2

   (d) Yes. 92.02 g reactants = 92.02 g product

2. **(a) 1 formula unit $NH_4NO_3 \rightarrow$ 1 molecule N_2O + 2 molecules H_2O**

   (b) 1 mole $NH_4NO_3 \rightarrow$ 1 mole N_2O + 2 moles H_2O

   (c) 80.05 g $NH_4NO_3 \rightarrow$ 44.02 g N_2O + 36.03 g H_2O

   (d) Yes. 80.05 g reactant = 80.05 g products

3. **(a) 1 molecule CS_2 + 3 molecules $O_2 \rightarrow$ 1 molecule CO_2 + 2 molecules SO_2**

   (b) 1 mole CS_2 + 3 moles $O_2 \rightarrow$ 1 mole CO_2 + 2 moles SO_2

   (c) 76.15 g CS_2 + 96.00 g $O_2 \rightarrow$ 44.01 g CO_2 + 172.15 g SO_2

   (d) Yes. 172.15 g reactants = 172.15 g products

4. **(a) 2.60 moles of H_3PO_3 will be produced.**

$$\text{moles } H_3PO_3 = \left(\frac{4 \text{ moles } H_3PO_3}{1 \text{ } \cancel{\text{mole } P_4O_6}}\right) \times \left(0.650 \text{ } \cancel{\text{mole } P_4O_6}\right) = 2.60 \text{ moles } H_3PO_3$$

   (b) 0.500 mole of H_3PO_3 will be produced.

$$\text{moles } H_3PO_3 = \left(\frac{4 \text{ moles } H_3PO_3}{6 \text{ } \cancel{\text{moles } H_2O}}\right) \times \left(0.750 \text{ } \cancel{\text{mole } H_2O}\right) = 0.500 \text{ mole } H_3PO_3$$

5. **(a) 0.767 mole of Al_2S_3 is consumed.**

$$\text{moles } Al_2S_3 = \left(\frac{1 \text{ mole } Al_2S_3}{3 \text{ } \cancel{\text{moles } H_2S}}\right) \times \left(2.30 \text{ } \cancel{\text{moles } H_2S}\right) = 0.767 \text{ mole } Al_2S_3$$

   (b) 1.53 moles of $Al(OH)_3$ are also formed.

$$\text{moles } Al(OH)_3 = \left(\frac{2 \text{ moles } Al(OH)_3}{3 \text{ } \cancel{\text{moles } H_2S}}\right) \times \left(2.30 \text{ } \cancel{\text{moles } H_2S}\right) = 1.53 \text{ moles } Al(OH)_3$$

6. **(a) 4.16 moles of BF_3 are required.**

$$\text{moles } BF_3 = \left(\frac{8 \text{ moles } BF_3}{1 \text{ } \cancel{\text{mole } B_2H_6}}\right) \times \left(0.520 \text{ } \cancel{\text{mole } B_2H_6}\right) = 4.16 \text{ moles } BF_3$$

   (b) 3.12 moles of NaH are required.

$$\text{moles NaH} = \left(\frac{6 \text{ moles NaH}}{1 \text{ } \cancel{\text{mole } B_2H_6}}\right) \times \left(0.520 \text{ } \cancel{\text{mole } B_2H_6}\right) = 3.12 \text{ moles NaH}$$

7. **The equation must first be balanced.** $PCl_5(s) + 4\ H_2O(l) \rightarrow H_3PO_4(aq) + 5\ HCl(aq)$

(a) 45.6 g of HCl(*aq*) will be produced. The linked calculation to convert the moles of $PCl_5(s)$ to moles of HCl(*aq*), then to mass of HCl(*aq*), is:

$$\text{mass HCl} = \left(0.250 \text{ mole } PCl_5\right)\left(\frac{5 \text{ moles HCl}}{1 \text{ mole } PCl_5}\right)\left(\frac{36.46 \text{ g HCl}}{1 \text{ mole HCl}}\right) = 45.6 \text{ HCl}$$

(b) 57.7 g of water will be consumed.

$$\text{mass } H_2O = \left(0.800 \text{ mole } H_3PO_4\right)\left(\frac{4 \text{ moles } H_2O}{1 \text{ mole } H_3PO_4}\right)\left(\frac{18.02 \text{ g } H_2O}{1 \text{ mole } H_2O}\right) = 57.7 \text{ g } H_2O$$

8. **(a) 31.6 g of Fe are required.**

$$\text{mass Fe} = \left(0.850 \text{ mole MgO}\right)\left(\frac{2 \text{ moles Fe}}{3 \text{ moles MgO}}\right)\left(\frac{55.85 \text{ g Fe}}{1 \text{ mole Fe}}\right) = 31.6 \text{ g Fe}$$

(b) 2.74 moles of Fe are required.

$$\text{moles Fe} = \left(100.0 \text{ g Mg}\right)\left(\frac{1 \text{ mole Mg}}{24.31 \text{ g Mg}}\right)\left(\frac{2 \text{ moles Fe}}{3 \text{ moles Mg}}\right) = 2.74 \text{ moles Fe}$$

9. **(a) 64.1 grams of $Ni_3(PO_4)_2$ are formed.**

From the atomic masses the molar masses are: $Ni_3(PO_4)_2$ = 366.0 g; Na_3PO_4 = 163.9 g

$$\text{mass } Ni_3(PO_4)_2 = \left(0.350 \text{ mole } Na_3PO_4\right)\left(\frac{1 \text{ mole } Ni_3(PO_4)_2}{2 \text{ moles } Na_3PO_4}\right)\left(\frac{366.0 \text{ g } Ni_3(PO_4)_2}{1 \text{ mole } Ni_3(PO_4)_2}\right) = 64.1 \text{ g Ni}(PO_4)_2$$

(b) 2.29 moles of $Ni(NO_3)_2$ are required.

$$\text{moles Ni}(NO_3)_2 = \left(250.0 \text{ g } Na_3PO_4\right)\left(\frac{1 \text{ mole } Na_3PO_4}{163.9 \text{ g } Na_3PO_4}\right)\left(\frac{3 \text{ moles Ni}(NO_3)_2}{2 \text{ moles } Na_3PO_4}\right) = 2.29 \text{ moles Ni}(NO_3)_2$$

10. **(a) 64.4 grams of NaBr are required.**

$$\text{mass NaBr} = 50.0 \text{ g } Br_2\left(\frac{1 \text{ mole } Br_2}{159.8 \text{ g } Br_2}\right)\left(\frac{2 \text{ moles NaBr}}{1 \text{ mole } Br_2}\right)\left(\frac{102.9 \text{ g NaBr}}{1 \text{ mole NaBr}}\right) = 64.4 \text{ g NaBr}$$

(b) 22.2 grams of Cl_2 are consumed.

$$\text{mass } Cl_2 = \left(50.0 \text{ g } Br_2\right)\left(\frac{1 \text{ mole } Br_2}{159.8 \text{ g } Br_2}\right)\left(\frac{1 \text{ mole } Cl_2}{1 \text{ mole } Br_2}\right)\left(\frac{70.90 \text{ g } Cl_2}{1 \text{ mole } Cl_2}\right) = 22.2 \text{ g } Cl_2$$

11. **1.25 g of $Cd(OH)_2$ would form if 1.00 g of NaCl forms.**

The equation as given is not balanced. Once balanced the equation is:

$CdCl_2(aq) + 2\ NaOH(aq) \rightarrow Cd(OH)_2(s) + 2\ NaCl(aq)$

$$\text{mass Cd}(OH)_2 = \left(1.00 \text{ g NaCl}\right)\left(\frac{1 \text{ mole NaCl}}{58.44 \text{ g NaCl}}\right)\left(\frac{1 \text{ mole Cd}(OH)_2}{2 \text{ moles NaCl}}\right)\left(\frac{146.4 \text{ g Cd}(OH)_2}{1 \text{ mole Cd}(OH)_2}\right) = 1.25 \text{ g Cd}(OH)_2$$

12. **48.8 grams of CaF_2 are required.**

From the atomic masses the required molar masses are: HF = 20.01 g; CaF_2 = 78.08 g

$$\text{mass } CaF_2 = (25.0 \text{ g HF})\left(\frac{1 \text{ mole HF}}{20.01 \text{ g HF}}\right)\left(\frac{1 \text{ mole } CaF_2}{2 \text{ moles HF}}\right)\left(\frac{78.08 \text{ g } CaF_2}{1 \text{ mole } CaF_2}\right) = 48.8 \text{ g } CaF_2$$

13. **First, balance the equation: $C_6H_{12}O_6(s) + 6\ O_2(g) \rightarrow 6\ CO_2(g) + 6\ H_2O(l)$**

(a) 3.75 g of $C_6H_{12}O_6$ would produce 5.50 g of CO_2.

$$\text{g } C_6H_{12}O_6 = (5.50 \text{ g } CO_2)\left(\frac{1 \text{ mole } CO_2}{44.01 \text{ g } CO_2}\right)\left(\frac{1 \text{ mole } C_6H_{12}O_6}{6 \text{ moles } CO_2}\right)\left(\frac{180.2 \text{ g } C_6H_{12}O_6}{1 \text{ mole } C_6H_{12}O_6}\right) = 3.75 \text{ g } C_6H_{12}O_6$$

(b) 4.00 g of O_2 would be consumed.

$$\text{mass } O_2 = (5.50 \text{ g } CO_2)\left(\frac{1 \text{ mole } CO_2}{44.01 \text{ g } CO_2}\right)\left(\frac{6 \text{ moles } O_2}{6 \text{ moles } CO_2}\right)\left(\frac{32.00 \text{ g } O_2}{1 \text{ mole } O_2}\right) = 4.00 \text{ g } O_2$$

14. **In (a) and (b), NaCl is the limiting reactant. In (c), $Pb(NO_3)_2$ is the limiting reactant.**

To check for the limiting reactant, calculate the number of $Pb(NO_3)_2$ consumed by NaCl and compare the result with the number of moles of $Pb(NO_3)_2$ present at the start of the reaction.

(a) How many moles of $Pb(NO_3)_2$ will 0.30 mole of NaCl consume?

$$\text{moles } Pb(NO_3)_2 = \left(\frac{1 \text{ mole } Pb(NO_3)_2}{2 \text{ moles NaCl}}\right) \times (0.30 \text{ mole NaCl}) = 0.15 \text{ mole } Pb(NO_3)_2$$

There is more than 0.15 mole of $Pb(NO_3)_2$ at the start of the reaction, indicating that $Pb(NO_3)_2$ is in excess. This means NaCl must be the limiting reactant.

(b) From a calculation similar to that used in (a), 0.15 mole of NaCl would consume 0.075 mole of $Pb(NO_3)_2$. But starting with 0.30 mole of $Pb(NO_3)_2$ shows it is clearly in excess, making NaCl the limiting reactant.

(c) Again, using a calculation similar to that in (a), 0.20 mole of NaCl would consume 0.10 mole of $Pb(NO_3)_2$, a greater amount than that present at the start of the reaction, 0.05 mole. With insufficient $Pb(NO_3)_2$ present to consume all the NaCl, $Pb(NO_3)_2$ is the limiting reactant.

15. **$C_2H_8N_2$ is the limiting reactant.**

(a) First, balance the equation: $C_2H_8N_2(l) + N_2O_4(g) \rightarrow 2\ N_2(g) + 2\ CO_2(g) + 4\ H_2(g)$

The amounts of both reactants must be in moles to determine the limiting reactant.

$$\text{moles } C_2H_8O_4 = (45 \text{ g } C_2H_8N_2)\left(\frac{1 \text{ mole } C_2H_8N_2}{60.10 \text{ g } C_2H_8N_2}\right) = 0.750 \text{ mole } C_2H_8N_2$$

$$\text{moles } N_2O_4 = (75.0 \text{ g } N_2O_4)\left(\frac{1 \text{ mole } N_2O_4}{92.02 \text{ g } N_2O_4}\right) = 0.815 \text{ mole } N_2O_4$$

The balanced equation clearly shows a 1-to-1 mole ratio between $C_2H_8N_2$ and N_2H_4. 0.750 mole of $C_2H_8N_2$ will consume 0.750 mole of N_2H_4, a smaller amount than that present at the beginning of the reaction. N_2H_4 is in excess and $C_2H_8N_2$ is the limiting reactant.

(b) 66.0 g of CO_2 will be produced in the reaction.

The mass of CO_2 produced will be limited by $C_2H_8N_2$, so the calculation will be based on the starting number of moles of the limiting reactant, 0.750 mole of $C_2H_8N_2$.

$$\text{mass } CO_2 = (0.750 \text{ mole } C_2H_8N_2)\left(\frac{2 \text{ moles } CO_2}{1 \text{ mole } C_2H_8N_2}\right)\left(\frac{44.01 \text{ g } CO_2}{1 \text{ mole } CO_2}\right) = 66.0 \text{ g } CO_2$$

16. **$NiCl_2$ is the limiting reactant. 21.7 g of NiS will be produced.**

The equation is balanced and starting amounts of two reactants are given. Determine the limiting reactant, then calculate the mass of NiS produced based on the limiting reactant. The necessary molar masses are calculated: $NiCl_2$ = 129.6 g; $(NH_4)_2S$ = 68.15 g; NiS = 90.76 g.

Using the molar mass of $NiCl_2$, 31.0 g of $NiCl_2$ is 0.239 mole of $NiCl_2$.

Using the molar mass of $(NH_4)_2S$, 22.0 g of $(NH_4)_2S$ is 0.323 mole of $(NH_4)_2S$.

The 1-to-1 mole ratio of $NiCl_2$ and $(NH_4)_2S$ indicates that 0.239 mole of $NiCl_2$ will consume 0.239 mole of $(NH_4)_2S$, less than that present at the start of the reaction. $(NH_4)_2S$ is in excess. $NiCl_2$ must be the limiting reactant. The mass of NiS produced will be based on 0.239 mole of $NiCl_2$.

$$\text{mass NiS} = (0.239 \text{ mole } NiCl_2)\left(\frac{1 \text{ mole NiS}}{1 \text{ mole } NiCl_2}\right)\left(\frac{90.76 \text{ g NiS}}{1 \text{ mole NiS}}\right) = 21.7 \text{ g NiS}$$

17. **The percent yield is 94.1%.**

The equation is: $Ca(OH)_2(aq) + H_2C_2O_4(aq) \rightarrow CaC_2O_4(s) + 2\ H_2O(l)$

The 1-to-1 mole relationship indicates that 0.650 mole of $Ca(OH)_2$ will produce 0.650 mole of CaC_2O_4, which has a mass of:

$$\text{mass } CaC_2O_4 = (0.650 \text{ mole } CaC_2O_4) \times \left(\frac{128.1 \text{ g } CaC_2O_4}{1 \text{ mole } CaC_2O_4}\right) = 83.3 \text{ g } CaC_2O_4$$

The theoretical yield of CaC_2O_4 is 83.3 g. The actual yield is 78.4 g. The percent yield is:

$$\%\text{ yield} = \left(\frac{78.4 \text{ g}}{83.3 \text{ g}}\right) \times 100\% = 94.1\%$$

18. **The percent yield is 85.6%.**

The equation is: $2\ Al(OH)_3(s) \rightarrow Al_2O_3(s) + 3\ H_2O(g)$

The starting mass of $Al(OH)_3$, 3.50 g, equals 0.0449 mole of $Al(OH)_3$. The expected number of mole of Al_2O_3 produced is:

$$\text{moles } Al_2O_3 = \left[\frac{1 \text{ mole } Al_2O_3}{2 \text{ moles } Al(OH)_3}\right] \times (0.0449 \text{ mole } Al(OH)_3) = 0.0225 \text{ mole } Al_2O_3$$

0.0225 mole of Al_2O_3 has a mass of 2.29 g, the theoretical yield, which is larger than the actual yield of 1.96 g. Knowing the actual and theoretical yields, the percent yield is:

$$\%\text{ yield} = \left(\frac{1.96 \text{ g}}{2.29 \text{ g}}\right) \times 100\% = 85.6\%$$

Supplemental Chapter Problems

Problems

1. Interpret the balanced equations in terms of: (a) atoms or molecules of each reactant and product, (b) moles of each reactant and product, and (c) the mass of each reactant and product.

   (a) $2\ Na(s) + 2\ H_2O(l) \rightarrow 2\ NaOH(aq) + H_2(g)$

   (b) $2\ Al(s) + 3\ H_2SO_4(aq) \rightarrow Al_2(SO_4)_3(aq) + 3\ H_2(g)$

2. 0.540 mole of Cl_2 is reacted with an excess of CS_2. How many moles of CCl_4 are produced?

   $CS_2(l) + Cl_2(g) \rightarrow CCl_4(l) + S_2Cl_2(l)$

3. Octane, a component of gasoline, is burned in air. If 0.800 mole of octane, C_8H_{18}, is consumed, how many moles of O_2 are consumed? How many moles of CO_2 are produced?

   $C_8H_{18}(l) + O_2(g) \rightarrow CO_2(g) + H_2O(l)$

4. $2\ Al(s) + 3\ I_2(s) \rightarrow 2\ AlI_3(s)$

   If 30.0 g of I_2 were consumed, how many moles of AlI_3 would be produced? The atomic masses (amu): Al = 26.98; I = 126.9.

5. Lithium reacts with oxygen to produce lithium oxide. If 0.500 mole of Li is consumed, what mass of Li_2O is formed? Atomic masses (amu): Li =6.941; O = 16.00.

   $Li(s) + O_2(g) \rightarrow Li_2O(s)$

6. $3\ Fe_3O_4(s) + 8\ Al(s) \rightarrow 4\ Al_2O_3(s) + 9\ Fe(s)$

   2.40 moles of Fe_3O_4 are consumed. What mass of Al_2O_3 is formed? What mass of Fe is formed?

   The molar mass of Al_2O_3 is 101.96 g; that of Fe is 55.85 g.

7. What mass of water can be produced from 1.00 g of glucose, $C_6H_{12}O_6$?

   $C_6H_{12}O_6(s) + 6\ O_2(g) \rightarrow 6\ CO_2(g) + 6\ H_2O(l)$

   The molar mass of glucose is 180.2 g; that of water is 18.02 g.

8. What mass of CO_2 can react with 1.00 g of lithium hydroxide, LiOH?

   $LiOH(s) + CO_2(g) \rightarrow Li_2CO_3(s) + H_2O(l)$

   The molar mass of LiOH is 23.95 g; that of CO_2 is 44.01 g.

9. How many grams of oxygen gas can be obtained from the decomposition of 4.50 g of $KClO_3$?

$2\ KClO_3(s) \xrightarrow{\Delta} 2\ KCl(s) + 3\ O_2(g)$

The molar mass of $KClO_3$ is 122.55 g; that of O_2 is 32.00 g.

10. 0.450 mole of $Al(OH)_3$ and 0.550 mole of H_2SO_4 were reacted. How many moles of $Al_2(SO_4)_3$ are produced?

$2\ Al(OH)_3(s) + 3\ H_2SO_4(aq) \rightarrow Al_2(SO_4)_3(aq) + 6\ H_2O(l)$

11. 6.50 g of sodium carbonate and 7.00 g of silver nitrate were reacted in solution. What mass of silver carbonate precipitated?

$Na_2CO_3(aq) + AgNO_3(aq) \rightarrow Ag_2CO_3(s) + NaNO_3(aq)$

Molar masses: Na_2CO_3 = 106.0 g; $AgNO_3$ = 169.9 g; Ag_2CO_3 = 275.8 g

12. 2.00 g of zinc and 2.50 g of silver nitrate were combined. Which is the limiting reactant and what mass of silver metal will be produced? What mass of $Zn(NO_3)_2$ will be produced?

$Zn(s) + 2\ AgNO_3(aq) \rightarrow 2\ Ag(s) + Zn(NO_3)_2(aq)$

Molar masses: Zn = 65.39 g; $AgNO_3$ = 169.9 g; Ag = 107.9 g; $Zn(NO_3)_2$ = 189.4 g

13. 30.0 g of C_6H_6 and 65.0 g of Br_2 were combined and reacted. The actual yield of C_6H_5Br is 56.7 g. What is the theoretical yield of the reaction? What is the percent yield?

$C_6H_6(l) + Br_2(l) \rightarrow C_6H_5Br(l) + HBr(g)$

Atomic masses (amu): C = 12.01; H = 1.008; Br = 79.90

14. $Fe_2O_3(s) + 3\ CO(g) \rightarrow 2\ Fe(s) + 3\ CO_2(g)$

150.0 grams of Fe_2O_3 were reacted with an excess of CO. The actual yield of Fe was 87.9 g. What is the theoretical yield of Fe and the percent yield of the reaction?

Atomic masses (amu): Fe = 55.85; O = 16.00

Answers

Interpreting equations (page 191):

1. **(a) 2 atoms Na + 2 molecules H_2O → 2 formula units NaOH + 1 molecule H_2**

2 moles Na + 2 moles H_2O → 2 moles NaOH + 1 mole H_2

45.98 g Na + 36.03 g H_2O → 80.00 NaOH + 2.02 H_2

(b) 2 atoms Al + 3 molecules $H_2SO_4 \rightarrow$ 1 formula unit $Al_2(SO_4)_3$ + 3 molecules H_2

2 moles Al + 3 moles $H_2SO_4 \rightarrow$ 1 mole $Al_2(SO_4)_3$ + 3 moles H_2

53.96 g Al + 294.26 g $H_2SO_4 \rightarrow$ 342.17 g $Al_2(SO_4)_3$ + 6.05 g H_2

Mole-to-mole problems (page 193):

2. **0.180 mole of CCl_4 is produced.**

   The equation must be balanced: $CS_2(l) + 3\ Cl_2(g) \rightarrow CCl_4(l) + S_2Cl_2(l)$

3. **10.0 moles of O_2 will be consumed, 6.40 moles of CO_2 will be produced.**

   The equation must be balanced: $2\ C_8H_{18}(l) + 25\ O_2(g) \rightarrow 16\ CO_2(g) + 18\ H_2O(l)$

Mole-to-mass and mass-to-mole problems (page 197):

4. **0.0788 mole of AlI_3 is produced.**

5. **The molar mass of Li_2O is 29.88 g. 7.741 g of Li_2O (0.250 mole) will be produced.**

   The equation must be balanced: $4\ Li(s) + O_2(g) \rightarrow 2\ Li_2O(s)$

6. **2.40 moles of Fe_3O_4 will produce 326.3 g of Al_2O_3 and 402.1 g of Fe.**

Mass-to-mass problems (page 201):

7. **0.600 g of water is obtained from 1.00 g of glucose.**

8. **0.919 g of CO_2 can be consumed by 1.00 g of LiOH.**

   The equation needs to be balanced: $2\ LiOH(s) + CO_2(g) \rightarrow Li_2CO_3(s) + H_2O(l)$

9. **1.77 g of O_2 is obtained in the decomposition of 4.50 g of $KClO_3$.**

Limiting reactant problems (page 204):

10. **H_2SO_4 is the limiting reactant. 0.183 mole of $Al_2(SO_4)_3$ is produced.**

11. **Balance the equation. The limiting reactant is $AgNO_3$. 5.68 g of Ag_2CO_3 will be produced.**

12. **The limiting reactant is $AgNO_3$. 1.59 g of Ag and 1.39 g of $Zn(NO_3)_2$ will be produced.**

Theoretical yield, actual yield, and percent yield (page 209):

13. **C_6H_6 is the limiting reactant. Theoretical yield of C_6H_5Br = 60.3 g. Percent yield = 94.0%.**

14. **The theoretical yield is 104.9 g of Fe. The percent yield is 83.8%.**

Chapter 9

Atoms II—Atomic Structure and Periodic Properties

The composition of atoms in terms of the numbers of neutrons (n), protons (p^+), and electrons (e^-) that make them up was described in Chapter 3. The number of each subatomic particle could be determined from the atomic number and mass number of the atom.

$$^{27}_{13}Al$$

- Atomic number = number of p^+ in the nucleus, and the number of e^- about the nucleus in the neutral atom
- Mass number = number of p^+ + number of neutrons in the nucleus
- Number of neutrons in the nucleus = mass number − atomic number

The chemical properties of an element are determined by the number and arrangement of electrons about the nucleus. The arrangement of electrons in an atom is called its **electronic configuration.** If we want to know why sodium, Na, forms an ion with a single positive charge, Na^+, and why the other metals with sodium in Group IA do the same thing, then we need to know the electronic structure of these elements. The electronic structures will give us the answer. The electronic structures of oxygen and hydrogen can tell us why water is H_2O and not HO or HO_2.

The way electrons behave in an atom is unlike anything in our normal experience. The negative electron is strongly attracted to the positive charge of the nucleus, yet it doesn't fall into the nucleus. Instead, it resides outside the nucleus in one of several fixed energy levels, called *quantized states.* Electrons can move from one quantized state (energy level) to another with the absorption or emission of energy, a transition called a *quantum jump.* A "jump" from a lower to a higher energy state requires the absorption of a specific amount of energy. A "jump" from a higher to a lower energy state emits a specific amount of energy, and that energy is emitted as light energy.

Electrons in motion about the nucleus also exhibit a dual personality. They can be regarded as *particles* moving about a nucleus or they can be treated as *waves* existing about a nucleus. There is nothing in our macroscopic world that parallels this dual nature, yet we constantly find ourselves thinking of the electron as a particle (which is not wrong) while writing electronic configurations using orbitals that are the product of the wave nature of electrons (which is also not wrong). Our understanding of the electronic structure of atoms relies on the tools of *quantum mechanics,* a special kind of mathematics developed early in the 20th century. We won't do the math here.

The first person to offer a reasonable explanation of the electronic structure of an atom was Niels Bohr, who in 1913 quantitatively explained the source of the lines of light emitted from "excited" hydrogen atoms. His explanation gave the world the first workable model of the hydrogen atom.

The Bohr Model of the Hydrogen Atom

The key used to unlock the mystery of the structure of the atom is energy, specifically light energy. Before we can understand how Bohr used light to solve this mystery, a little light background is necessary.

The Nature of Light

Visible light is a type of *electromagnetic radiation,* a form of energy that travels through space at the "speed of light," 3.00×10^{10} meters/second. The energy of light is related to its **wavelength,** λ (lambda), the distance between two adjacent crests of a wave. The shorter the wavelength, the higher the energy of the light. The longer the wavelength, the lower the energy.

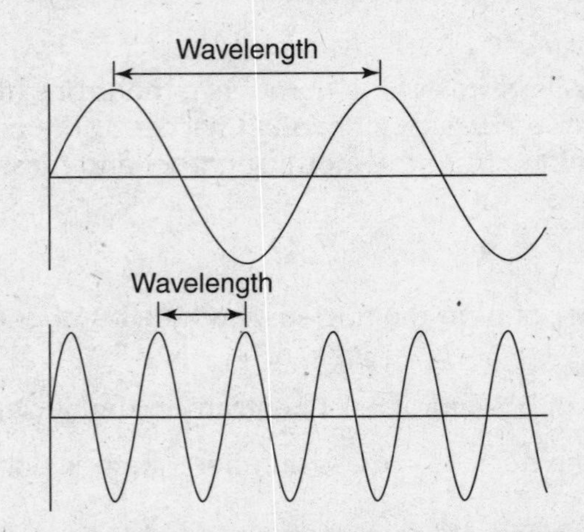

The wavelength of a light wave is the distance between two adjacent crests of the wave.

The number of waves of light that pass a point in one second is its **frequency,** ν (nu). If the wavelength is long, the frequency is low. If the wavelength is short, the frequency is high.

- High-energy light has short wavelengths and high frequencies.
- Low-energy light has long wavelengths and low frequencies.

The relationship between wavelength and frequency is $\lambda\nu$ = speed of light. The equation shows that wavelength and frequency are *inversely* related; that is, if wavelength doubles, the frequency is cut in half, and if wavelength is cut in half, frequency doubles. We know that light is a form of energy because it can expose film (it causes a chemical change) and will tan our skin in the summer (also a chemical change).

The *electromagnetic spectrum* ranges from very high-energy gamma and x-rays (short λ, high ν) to very low-energy radio and TV waves (long λ, low ν). The visible spectrum, that part we can see with our eyes, spans a narrow region between these two extremes.

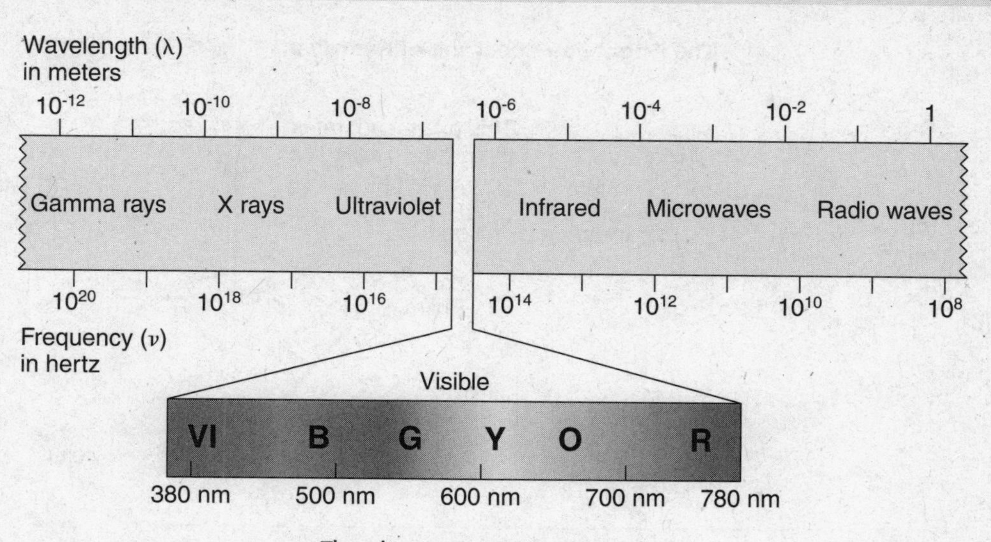

The electromagnetic spectrum.

When white light is passed through a prism, it separates into the familiar continuous rainbow of colors: red, orange, yellow, green, blue, indigo, and violet. The wavelength of red light (about 700 nm) is longer than that of violet light (about 400 nm). Moving across the visible spectrum from red to violet is moving in the direction of increasing energy as the wavelengths get shorter. Violet light has a higher energy than red light.

When a confined sample of hydrogen gas is heated to a high temperature or subjected to an electrical discharge, it glows with a red-violet color. As this light is passed through a prism, it produces a set of brightly colored lines in the visible region, called a bright-line spectrum. Each line is a specific color, with a specific wavelength and a specific energy. Each line represents an energy associated with a specific process in the hydrogen atom. Lines are also produced in the infrared and ultraviolet regions.

The Bright-line Spectrum of Hydrogen

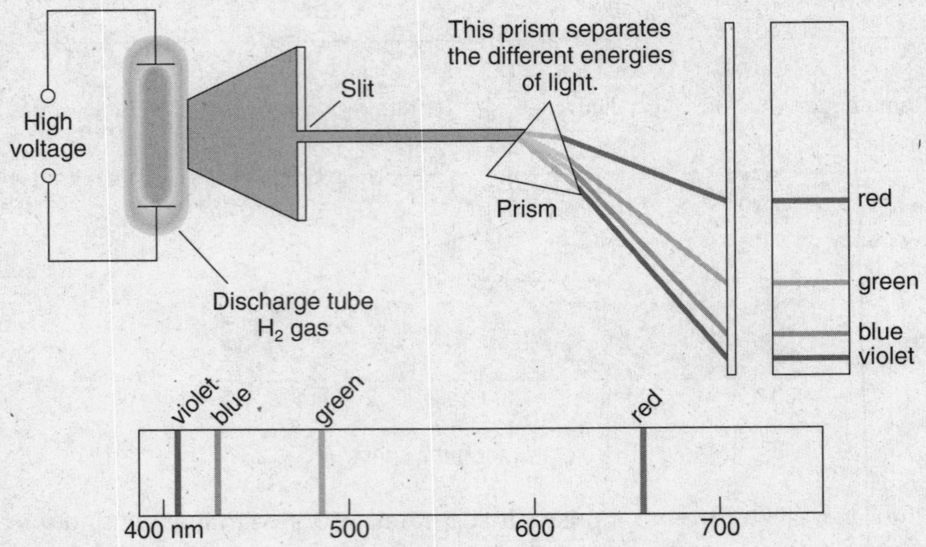

Light, Electrons, and Niels Bohr

One of the greatest discoveries of the early 20th century was the model of the hydrogen atom put forth in 1913 by Danish physicist Niels Bohr (1885–1962). His model explained the bright lines in the visible spectrum of hydrogen. In Bohr's model of the hydrogen atom, the single electron revolves about the nucleus in one of several allowed circular orbits, like the earth revolves around the sun. Here is a summary of the Bohr model of the hydrogen atom:

1. There are several orbits in which the electron can travel, and each orbit has a specific radius and energy. Each orbit, or energy level, is identified with a number, **n,** called the **principle quantum number.** The values of **n** are positive, whole numbers 1, 2, 3, etc. The orbit closest to the nucleus, the **n** = 1 principle level, has the lowest energy and the smallest radius. The **n** = 2 orbit has a higher energy and larger radius. As **n** gets larger, the energy *and* radius of the orbit increases. Opposite charges attract, so it requires energy to move a *negative* electron away from a *positive* nucleus. As the electron moves from a smaller orbit (a smaller value of **n**) to a larger orbit (a larger value of **n**), it is moving farther away from the nucleus. Increasing the separation of opposite charges requires energy. Therefore, as the value of **n** increases, the energy of the electron increases (the potential energy of the electron is increasing, though it is often referred to simply as the energy of the electron). The electron can only be in *one* orbit (energy level) at a time. It cannot reside between orbits but it can move from one orbit to another.

   Before going any further, let's make certain you understand the terms. An energy *level* and an energy *state* are the same thing. Both indicate the energy of the electron. Both the energy state and the associated orbit are identified with a quantum number, **n:** the larger the value of **n,** the higher the energy, the larger the orbit.

2. The most stable location for the electron is in the n = 1 level, which is called the **ground state.** The electron can be *excited* to a higher energy level (n = 2, 3, or a higher number) with the *absorption* of energy. *The energy absorbed exactly equals the energy difference between the two states,* and the transition is described as a quantum jump. When the electron is not in its ground state, it is in an **excited state.** An electrical discharge or high temperatures can provide the energy to excite an electron to an excited state.

3. An electron in a higher energy state (larger **n**) will spontaneously return to a lower energy state (smaller **n**) with the emission of energy. The energy given off is in the form of light. *The energy of the emitted light exactly equals the energy difference between the higher*

and lower energy states. The greater the energy separation of the two states, the greater the energy of the emitted light (the shorter the wavelength). The smaller the energy separation, the lower the energy of the emitted light (the longer its wavelength). The transitions of electrons from higher to lower energy levels produce the different lines of light that make up the bright line spectrum of hydrogen.

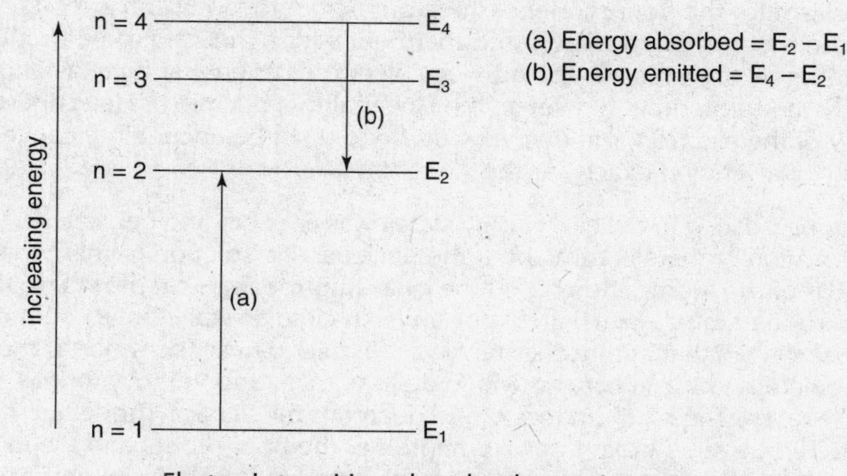

Electronic transitions that absorb or emit energy.

Bohr realized the energy of each line in the hydrogen spectrum represented a "difference in energy" between two energy levels. Bohr's impact on the growing knowledge of atomic structure was the idea that an electron in an atom could exist only in definite, fixed energy levels. The electron can change its energy by moving from one level to another but can *never* have an energy between adjacent levels. Think of the energy levels as stair steps and the electron as a rubber ball. The ball can come to rest on one step or another, but never between two steps. The following figure shows how electronic transitions from higher energy levels to the n = 2 level emit the specific wavelengths of light that produce the bright-line emission spectrum of hydrogen.

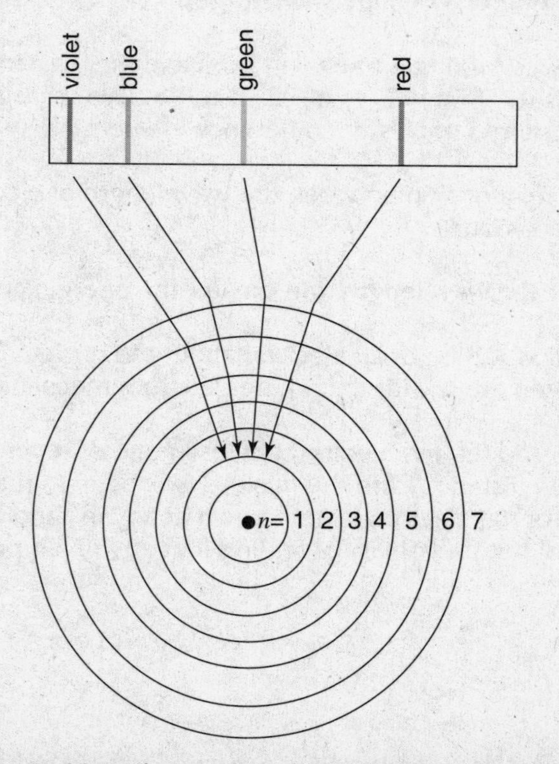

The electronic transitions that produce the bright-line spectrum of hydrogen.

Every element produces a unique bright-line spectrum. No two elements have the same spectrum. The emission spectrum of an element is like a fingerprint of that element. That every element can produce a bright-line spectrum proves that electrons reside only in definite energy states in *all* elements, and that each element has its own set of allowed energy states that are unlike those of any other element.

The initial enthusiasm for the Bohr model of the atom began to fade soon after its introduction. New discoveries in modern physics indicated that Bohr's ideas needed to be modified. The Bohr model of the hydrogen atom pinpointed both the electron's energy and its distance from the nucleus at exactly the same time. However, the Uncertainty Principle of Heisenberg required that if the energy of the electron is known exactly there must be uncertainty concerning its location. Both could not be known exactly at the same time.

The idea of electrons existing in definite energy states was fine, but another way had to be devised to describe the location of the electron about the nucleus. The solution to this problem produced the modern model of the atom, often called the **quantum mechanical model.** In this new model of the hydrogen atom, electrons do not travel in circular orbits but exist in orbitals with three-dimensional shapes that are inconsistent with circular paths. The modern model of the atom treats the electron not as a particle with a definite mass and velocity, but as a wave with the properties of waves. The mathematics of the quantum mechanical model are much more complex, but the results are a great improvement over the Bohr model and are in better agreement with what we know about nature. In the quantum mechanical model of the atom, the location of an electron about the nucleus is described in terms of probability, not paths, and these volumes where the probability of finding the electron is high are called orbitals.

Example Problems

These problems have both answers and solutions given.

1. Considering the Bohr model of the hydrogen atom, what process is taking place in the atom when it emits light of a definite wavelength?

   Answer: An electron is making a transition from a higher to a lower principle energy level, such as from $\mathbf{n} = 4 \rightarrow \mathbf{n} = 2$, or $\mathbf{n} = 2 \rightarrow \mathbf{n} = 1$. Each energy level is fixed, and the energy of the light emitted equals the difference in energy between the two levels.

2. What is the general relationship between the wavelength of electromagnetic radiation and the energy of that radiation?

   Answer: The shorter the wavelength, the greater the energy, and vice versa.

3. Compare the way the location of an electron about the nucleus is described in the Bohr model of the hydrogen atom with that in the quantum mechanical model.

   Answer: In the Bohr model, the electron travels in one of several allowed circular paths about the nucleus. Its distance from the nucleus is known. In the quantum mechanical model, the location of the electron is not known at any instant; rather, its location is described in terms of the probability of finding it at any given point around the nucleus.

Work Problems

Use these problems for additional practice.

1. Explain how the bright-line spectrum of hydrogen is consistent with Bohr's model of quantized energy states for the electrons in the hydrogen atom.

2. Arrange the following types of electromagnetic radiation in order of increasing energy. The wavelength of each is given: blue light (460 nm), infrared light (0.10 mm), red light (680 nm), ultraviolet light (110 nm), and yellow light (585 nm).

3. Is energy emitted or absorbed in the following electronic transitions of the hydrogen atom?

   (a) from $\mathbf{n} = 2 \rightarrow \mathbf{n} = 5$

   (b) from an orbit of radius 2.1×10^{-10} m to an orbit of radius 8.5×10^{-10} m

   (c) from $\mathbf{n} = 2 \rightarrow \mathbf{n} = 1$

Worked Solutions

1. **Since electrons can exist only in definite, fixed energy levels (n = 1, n = 2, etc.), transitions from higher to lower energy levels will be accompanied by a loss of energy by the electron.** This energy is emitted as light with an energy exactly equal to the energy difference between the two levels.

2. **As the wavelength gets shorter, the energy increases in this sequence: infrared light, red light, yellow light, blue light, and ultraviolet light.**

3. **(a) Energy is absorbed moving from a lower to a higher principle energy level.**

   (b) Energy is absorbed. The larger orbit is the one of larger **n** value. Moving from a smaller orbit to a larger orbit is moving from a smaller **n** value to a larger **n** value.

   (c) Energy is emitted in a transition from a larger n value (2) to a smaller n value (1).

The Quantum Mechanical Atom: Principal Shells, Subshells, and Orbitals

An **orbital** is a volume of space about the nucleus where the probability of finding an electron is high. Unlike orbits that are easy to visualize, orbitals have shapes that do not resemble the circular paths of orbits. In the quantum mechanical model of the hydrogen atom, the energy of the electron is accurately known but its location about the nucleus is not known with certainty at any instant. The three-dimensional volumes that represent the orbitals indicate where an electron will *likely* be at any instant. This uncertainty in location is a necessity of physics.

The quantum mechanical description of the hydrogen atom is more complex than Bohr's picture, but it is a better picture. In the quantum mechanical model, there are several principal shells (energy levels) about the nucleus (these parallel but are not the same as Bohr's orbits), and each principal shell is divided into 1, 2, 3, or 4 subshells. Each subshell is composed of 1 or more orbitals, and each orbital can hold 1 or 2 electrons.

Principal shells → Subshells → Orbitals

To help you get a better idea of this nested structure, the general characteristics of principal shells, subshells, and orbitals are described below.

1. The **principal shells** are the major energy levels within the atom. Each principal shell is identified with a **quantum number, n,** a positive, whole number, 1 or greater. The **n** = 1 principal shell is closest to the nucleus and lowest in energy. As **n** increases, 2, 3, 4, and so on, the size *and* energy of the principal shell increases. The quantum number used to label each principal shell is similar to the quantum number used to identify the orbits in Bohr's model. But a principal shell is not an orbit, though both energy and size increase as **n** gets larger.
2. Within each principal shell are **subshells.** The number of subshells in a principal shell equals the value of its quantum number, **n.** Each subshell is identified with a one-letter label: s, p, d, or f. Within known elements, the s-, p-, and d-subshells are encountered most often.

   The **n** = 1 principal shell contains one subshell, an s-subshell. It is identified as the 1s-subshell.

   The **n** = 2 principal shell contains two subshells, an s-subshell and a p-subshell. They are identified as the 2s-subshell and the 2p-subshell.

   The **n** = 3 principal shell contains three subshells, an s-, p-, and a d-subshell. They are called the 3s-subshell, 3p-subshell, and the 3d-subshell.

   The **n** = 4 principal shell contains four subshells, an s-, p-, d-, and an f-subshell. As before, they are known as the 4s-subshell, 4p-subshell, 4d-subshell, and the 4f-subshell.
3. The **orbital** is a region of space where an electron assigned to that orbital is most likely to be found. Each orbital can hold a maximum of two electrons. The subshells are composed of one or more orbitals. There is one orbital in an s-subshell. There are three orbitals in a p-subshell, five in a d-subshell, and seven in an f-subshell. The number of orbitals in a principal shell equals $\mathbf{n}^2$. There are nine orbitals in the **n** = 3 principal shell, $3^2 = 9$.

   The following figure shows the hierarchy of principal shells, subshells, and the orbitals in each subshell. The orbitals are represented by small boxes.

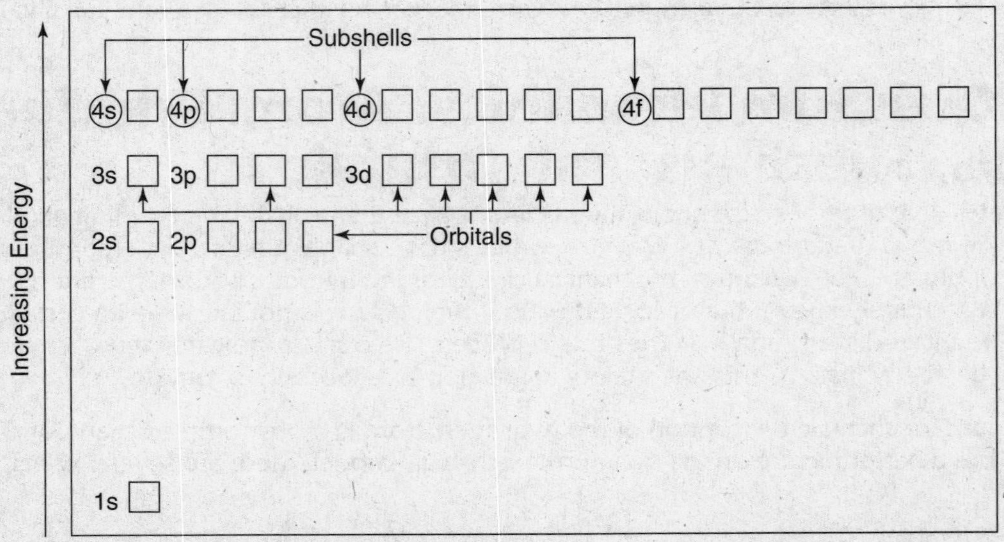

Subshells
4s 4p 4d 4f
3s 3p 3d
2s 2p Orbitals
1s
Increasing Energy

The orbitals in each of the four subshells have characteristic shapes. The orbital in an s-subshell does not look like the orbitals in a p-subshell, and so forth. The diagrams of orbitals shown below indicate where an electron will be 95 percent of the time. They are fuzzy, probability pictures. The nucleus is at the center of each orbital. Since subshells and orbitals are closely linked, let's look at the orbitals in terms of the subshell they are in.

s-orbitals:

There is *one* s-orbital in each s-subshell.

Every principal shell contains one s-orbital in an s-subshell. The labels used to identify the s-subshells, 1s, 2s, 3s, etc., are also used to identify the s-orbital they contain. The 1s-orbital is in the 1s-subshell, and so forth. An s-orbital can hold a minimum of 2 electrons.

Note that s-orbitals have spherical shapes.

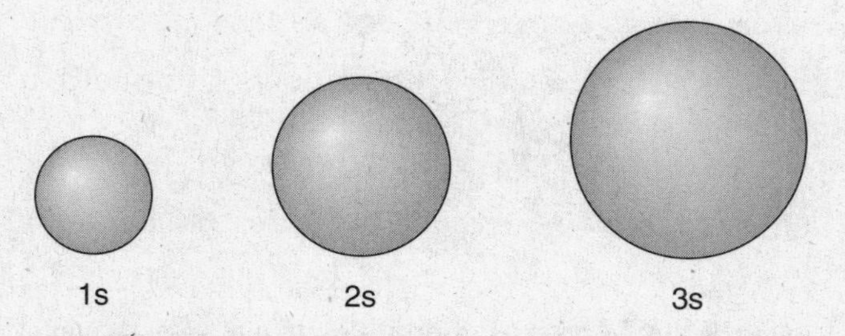

The spherical s-orbitals get larger as **n** increases.
The 1s-orbital is nested within the 2s-orbital, which is nested within the 3s-orbital. The nucleus is at the center of the orbitals.

p-orbitals:

p-orbitals come in sets of *three* within each p-subshell. The three orbitals are oriented at right angles to one another and are often pictured aligned with x-, y-, and z-axes. For that reason each of the three orbitals has its own label: p_x, p_y, p_z.

p-subshells first appear in the **n** = 2 principal shell and are in all higher principal shells. So it is not surprising that p-orbitals first appear in the **n** = 2 principal shell. The labels used to identify the p-subshells, 2p, 3p, 4p, etc., also identify the sets of p-orbitals. The p-orbitals in the 2p-subshell are referred to as the 2p-orbitals, $2p_x$, $2p_y$, and $2p_z$. p-orbitals are always kept together as a set, and the three orbitals always have the same energy. A p-subshell can contain a maximum of six electrons, two in each of the three p-orbitals.

p-orbitals have two-lobe, dumbbell shapes. The nucleus is at the point where the two lobes meet.

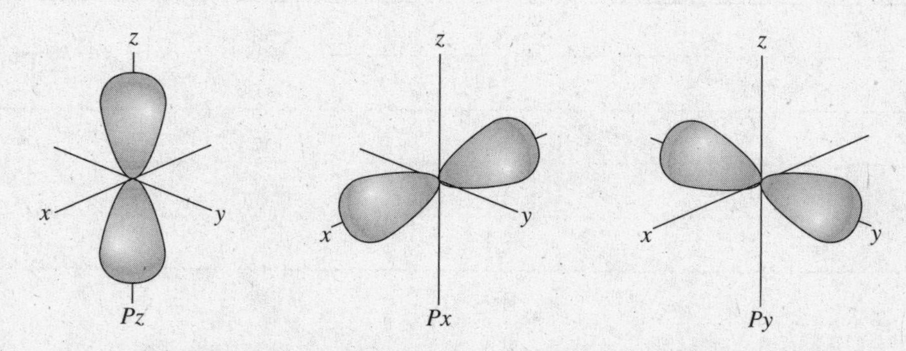

p-orbitals come in sets of three oriented at right angles to one another.

d-orbitals:

d-orbitals come in sets of *five* within each d-subshell.

Since d-subshells first appear in the **n** = 3 principal shell, d-orbitals are first encountered in the **n** = 3 principal shell. The d-orbitals in the 3d-subshell are referred to as the 3d-orbitals. d-orbitals are always kept together as a set and they all have the same energy. A d-subshell can contain a maximum of ten electrons, two in each of the five d-orbitals.

One of the 3d-orbitals that has a four-lobe, four-leaf clover shape is shown in the following figure.

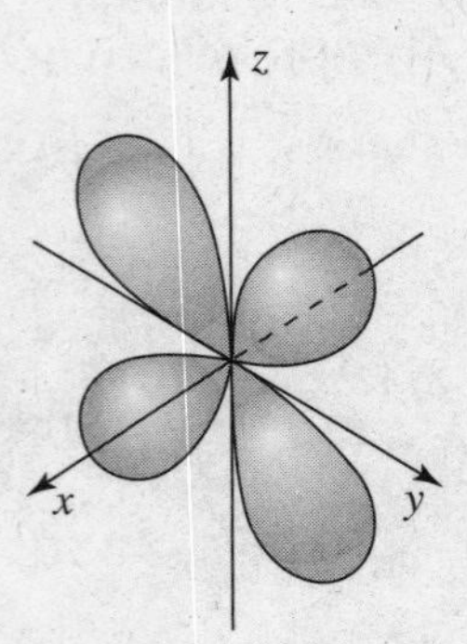

One of the 3d-orbitals; d-orbitals come in sets of five.

f-orbitals:

f-orbitals come in sets of *seven* within an f-subshell.

f-subshells first appear in the **n** = 4 principal shell. The orbitals in the 4f-subshell are referred to as 4f-orbitals. All the orbitals in an f-subshell stay together and have the same energy. An f-subshell can hold a maximum of 14 electrons, 2 in each of the 7 orbitals.

f-orbitals have complex eight-lobe shapes and will not be pictured here.

The relationship among principal shells, subshells, and orbitals is summarized in the following table.

Principal Shells, Subshells, and Orbitals

Principal Shell (n)	*Number of Subshells in the Principal Shell*	*Identity of Subshells*	*Number of Orbitals in the Subshell*
1	1	1s	one
2	2	2s	one
		2p	three
3	3	3s	one
		3p	three
		3d	five
4	4	4s	one
		4p	three
		4d	five
		4f	seven

The symbols used to indicate the number of electrons in a subshell include the number of the principal shell (1, 2, 3...), the letter designation of the subshell (s, p, d, f), with a superscript number indicating the number of electrons.

One electron in the s-orbital in the 2s-subshell: $2s^1$

Four electrons in a set of p-orbitals in the 2p-subshell: $2p^4$

Six electrons in a set of d-orbitals in the 3d-subshell: $3d^6$

Electron Spin

Electrons behave as if they are spinning on an axis much as the earth spins on its axis. A spinning electron acts like a very small bar magnet with north and south poles. Small arrows pointing upward, ↑, or downward, ↓, are used to indicate the two orientations of spin. Electron spin is important because two electrons in the same orbital must spin in opposite directions, ↑ ↓, a fact stated below in the Pauli principle.

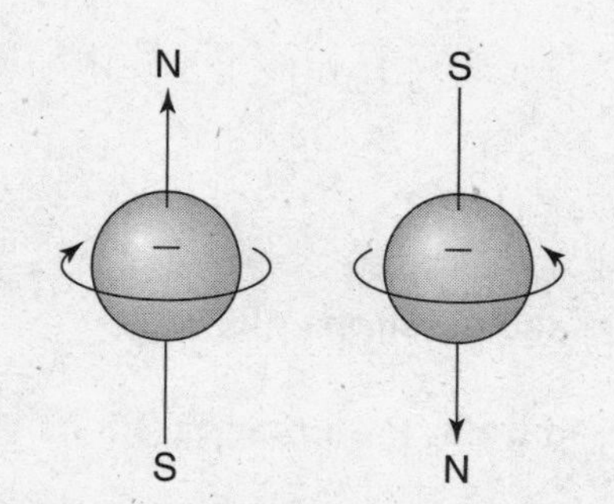

Two electrons spinning in opposite direction act like two small bar magnets with opposite poles aligned. Upward and downward arrows are used to show the two directions of spin.

There are two consequences of electron spin:

- The **Pauli principle:** No more than two electrons can be placed in a single orbital. Two electrons in the same orbital must have spins in opposite directions, ↑ ↓. Two electrons with opposite spins in an orbital are said to be "paired."
- **Hund's rule:** Electrons entering a set of equal-energy orbitals will fill them singly before any electrons are paired. Hund's rule concerns the way electrons fill the sets of orbitals in p-, d-, and f-subshells. For example, the three orbitals in a p-subshell have the same energy. If three electrons are in a set of p-orbitals, they will arrange themselves so that each orbital has one electron. If a fourth electron is added, it will "pair up" with one of the single electrons. It makes no difference which one. *Remember, orbitals of the same energy fill singly before any pairing occurs.*

In the following orbital diagram, a total of six electrons is added, one at a time, to the three orbitals in a p-subshell. Each orbital is represented as a box, and each electron as an arrow. Notice how the Pauli principle and Hund's rule are followed.

	Px	Py	Pz
$2p^1$	↑		
$2p^2$	↑	↑	
$2p^3$	↑	↑	↑
$2p^4$	↑↓	↑	↑
$2p^5$	↑↓	↑↓	↑
$2p^6$	↑↓	↑↓	↑↓

Electrons will fill the orbitals in a p-subshell singly before any electrons are paired.

The same holds true for the five orbitals in a d-subshell. Adding three electrons to a d-subshell will place one electron in three of the five orbitals, leaving two orbitals empty. No pairing would take place until more than five electrons are added to the d-subshell.

$3d^3$ [↑] [↑] [↑] [ ] [ ]

Example Problems

These problems have both answers and solutions given.

1. (a) How many subshells are in the **n** = 3 principal shell? What subshells are they?

   (b) What would be the maximum number of orbitals of all types in the **n** = 3 principal shell?

   Answer: (a) The number of subshells equals the value of **n.** If **n** = 3, there are three subshells, the 3s-, 3p-, and 3d-subshells.

   (b) The number of orbitals equals the square of the principal quantum number. There are nine orbitals (3^2) in the n = 3 principal shell: one 3s-orbital, three 3p-orbitals, and five 3d-orbitals.

2. Which of the following electron arrangements are allowed and which are not allowed according to the Pauli principle and Hund's rule?

   (a) $2p^4$ $2p^4$ [↑↓] [↑↓] [ ]

   (b) $2s^2$ $2s^2$ [↑↓]

   (c) $3p^7$

   (d) $3d^8$ $3d^8$ [↑↑] [↑↑] [↑↑] [↑] [↑]

   Answer: (a) Not allowed. Violates Hund's rule. The p-orbitals fill singly before any pairing occurs, as shown here:

   $2p^4$ [↑↓] [↑] [↑]

(b) Allowed, the arrangement agrees with both the Pauli principle and Hund's rule.

(c) Not allowed; violates the Pauli principle. A set of p-orbitals can hold a maximum of six electrons. If seven electrons were present, one orbital would have more than two electrons in it.

(d) Not allowed; violates the Pauli principle. If an orbital contains two electrons, the spins must be in opposite direction, as shown here:

$3d^8$ [↑↓] [↑↓] [↑↓] [↑] [↑]

Work Problems

Use these problems for additional practice.

1. (a) What is the maximum number of electrons that can be in the **n** = 2 principal shell?

   (b) A principal shell contains 16 orbitals. What is the number of this principal shell?

2. Comment on each of the following arrangements of electrons within a subshell.

   (a) $3p^3$ $3p^3$ [↑] [↑] [↓]

   (b) $3p^3$ $3p^3$ [↑↓] [↑] [ ]

   (c) $3p^3$ $3p^3$ [↑] [↑] [↑]

   (d) $3p^3$ $3p^3$ [↑↓↑] [ ] [ ]

Worked Solutions

1. **(a) A maximum of eight electrons: two electrons in the 2s-subshell and six electrons in the 2p-subshell.**

   (b) It is the n = 4 shell: 4s (1 orbital) + 4p (3 orbitals) + 4d (5 orbitals) + 4f (7 orbitals).

2. **(a) Violates Hund's rule.** Placing one electron in each orbital is right, but they should all be spinning in the same direction. Here, two are spinning up and one is spinning down.

   (b) Violates Hund's rule. A set of p-orbitals fills singly before any electrons pair up. There should be one electron in each orbital, all spinning in the same direction.

   (c) Agrees with both the Pauli principle and Hund's rule.

   (d) Violates the Pauli principle. There can be no more than two electrons in one orbital. The arrangement also violates Hund's rule.

The Electronic Configuration of Atoms

The **electronic configuration** of an atom shows the way the electrons are distributed among its subshells and orbitals. Though there are countless ways in which electrons could be arranged about the nucleus, the *most stable* arrangement of electrons is the one in which the electrons are in the lowest energy subshells possible, and this arrangement is called the **ground state electronic configuration** of the atom. The term "ground state" indicates the most stable, lowest energy electronic arrangement. Just as water seeks its lowest, most stable level, electrons seek their lowest, most stable arrangement. Any other arrangement of electrons would be an "excited" state, one less stable than the ground state. The chemical properties of an element are largely determined by its ground state electronic configuration.

In the Bohr model of the hydrogen atom, the electron's energy was a function of its principle quantum number only. This was true because there is only one electron in hydrogen. When the quantum mechanical model of the hydrogen atom is extended beyond hydrogen to *many-electron atoms*, the energy of an electron is determined by both the energy of the principal shell (n) and the subshell it is in. Within a given principal shell, the s-subshell is a bit more stable (lower energy) than the p-subshell, and the p-subshell is a bit more stable than the d-subshell.

The ground state configuration of electrons for a many-electron element is determined using a building-up process in which electrons are added to subshells in a specific sequence starting with the 1s-subshell, the most stable subshell, and continuing in order of increasing subshell energies. The sequence of increasing energies is: 1s $\rightarrow$ 2s $\rightarrow$ 2p $\rightarrow$ 3s $\rightarrow$ 3p $\rightarrow$ 4s $\rightarrow$ 3d $\rightarrow$ 4p $\rightarrow$ 5s $\rightarrow$ 4d, etc. You need not memorize this sequence because it is laid out below in a two-dimensional array that is easy to reproduce. For years, this building-up principle has been known by its German name, the **Aufbau principle.**

Things to keep in mind when writing ground state electronic configurations:

- The number of electrons in a neutral atom equals its atomic number.
- The lowest energy subshell is filled first, then the next lowest, following the sequence of increasing subshell energies. The increasing subshell energies are shown in the two-dimensional array given below.
- No more than two electrons can be placed in an orbital. If you are writing the electronic configuration using orbital diagrams (boxes representing the orbitals), two electrons in one orbital must have their spins paired, $\uparrow\downarrow$. This is a requirement of the Pauli principle.
- In p-, d-, and f-subshells that have three, five, or seven orbitals, each orbital is filled singly before any orbital contains two electrons. This is Hund's rule.
- The orbitals in p-, d-, or f-subshells must be completely filled with electrons before moving to the next higher subshell. A p-subshell can hold a maximum of six electrons; a d-subshell, ten electrons; an f-subshell, fourteen.

The ground state electronic configuration of an atom is developed by adding electrons to subshells, following the above rules, until all electrons are used. The filling order of the subshells is shown in the following figure. The lowest energy subshell is the 1s-subshell. The arrow moves back and forth tracing the sequence of *increasing* subshell energies. It is much easier to do than it sounds.

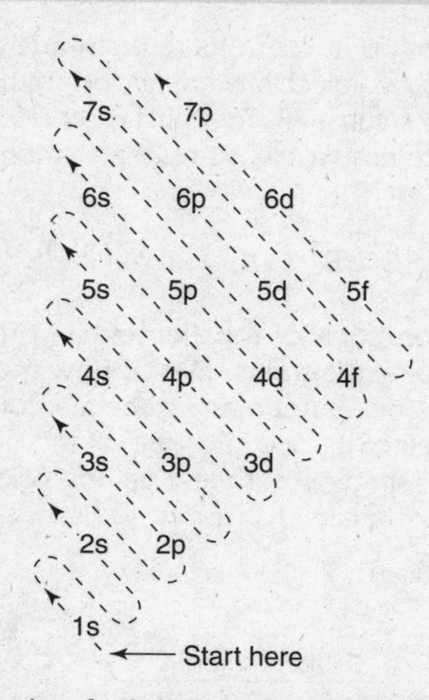

The building-up order of subshells is obtained by following the arrow through this array of subshells. The 1s-subshell is filled first, and electrons are added to subsequent subshells as traced by the arrow.

In the following examples, the atomic number of each element will accompany its symbol. Later on you will need to find this on a periodic table to learn the number of electrons possessed by an element.

To derive the ground state electronic configuration for hydrogen (atomic number 1), $_1H$, move up the bottom arrow in the previous figure until the 1s-subshell is reached. The single electron of hydrogen will go into the orbital in the 1s-subshell, giving a configuration of $1s^1$. The electronic configuration can also be written using an orbital diagram that allows you to show the spin of the electron, ↑. Both represent the ground state electronic configuration of hydrogen and would be read or spoken as "one-s-one."

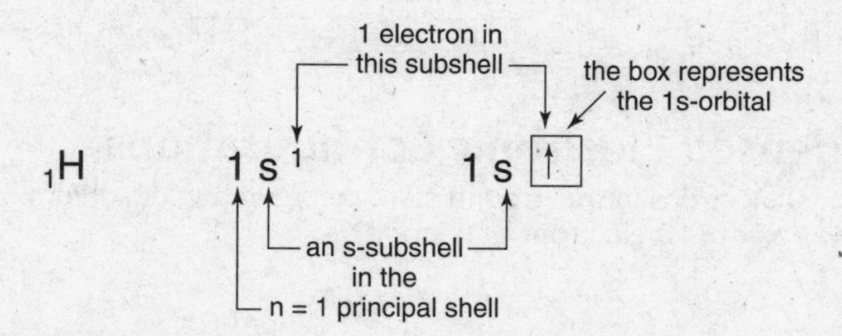

The ground state electronic configuration of helium, $_2He$, completely fills the 1s-orbital with two electrons. The spins of the electrons are "paired." The configuration for helium would be read as "one-s-two" (not one-s-squared).

$_2He$ $1s^2$ or 1s ↑↓

Lithium (atomic number 3), $_3Li$, has three electrons to distribute among the subshells. Following the first arrow in the filling diagram, two electrons are added to the 1s-orbital, filling it completely. The next higher arrow leads to the 2s-subshell. The third electron is placed in the 2s-orbital in the 2s-subshell. The configuration of lithium would be read as "one-s-two, two-s-one."

$$_3Li \quad 1s^2 2s^1 \quad \text{or} \quad 1s \; [\uparrow\downarrow] \quad 2s \; [\uparrow]$$

The ground state electronic configurations for the elements, from $_3Li$ to $_{13}Al$, are given below in both conventional notation and orbital diagrams. Notice how the configuration of each element builds on the one before it. For instance, after placing the first four electrons of boron in the 1s- and 2s-subshells, the fifth and last electron occupies one of the orbitals in the 2p-subshell. From B to Ne, notice how the p-orbitals fill singly before any pairing of electrons occurs. Once the 2p-subshell is filled with six electrons, the last electron in sodium, $_{11}Na$, is placed in the next higher energy subshell, the 3s-subshell.

Element	Total Electrons	Orbital Diagram					Electron Configuration
		1s	2s	2p	3s	3p	
Li	3	[↑↓]	[↑]	[ ][ ][ ]	[ ]		$1s^2 2s^1$
Be	4	[↑↓]	[↑↓]	[ ][ ][ ]	[ ]		$1s^2 2s^2$
B	5	[↑↓]	[↑↓]	[↑][ ][ ]	[ ]		$1s^2 2s^2 2p^1$
C	6	[↑↓]	[↑↓]	[↑][↑][ ]	[ ]		$1s^2 2s^2 2p^2$
N	7	[↑↓]	[↑↓]	[↑][↑][↑]	[ ]		$1s^2 2s^2 2p^3$
O	8	[↑↓]	[↑↓]	[↑↓][↑][↑]	[ ]		$1s^2 2s^2 2p^4$
F	9	[↑↓]	[↑↓]	[↑↓][↑↓][↑]	[ ]		$1s^2 2s^2 2p^5$
Ne	10	[↑↓]	[↑↓]	[↑↓][↑↓][↑↓]	[ ]		$1s^2 2s^2 2p^6$
Na	11	[↑↓]	[↑↓]	[↑↓][↑↓][↑↓]	[↑]		$1s^2 2s^2 2p^6 3s^1$
Mg	12	[↑↓]	[↑↓]	[↑↓][↑↓][↑↓]	[↑↓]		$1s^2 2s^2 2p^6 3s^2$
Al	13	[↑↓]	[↑↓]	[↑↓][↑↓][↑↓]	[↑↓]	[↑][ ][ ]	$1s^2 2s^2 2p^6 3s^2 3p^1$

Writing Condensed Electronic Configurations

Notice how the first 10 electrons in the ground state configuration of sodium, $_{11}Na$, are arranged in the exact same way as the 10 electrons of neon, $_{10}Ne$.

$$_{10}Ne \; 1s^2 2s^2 2p^6$$

$$_{11}Na \; [1s^2 2s^2 2p^6] 3s^1$$

A condensed ground state configuration for sodium can be written using [Ne] to represent the first 10 electrons of sodium that have the identical configuration as neon.

$$_{11}Na \; [Ne] 3s^1$$

Because the elements on the far right edge of the periodic table, Group VIIIA, the noble gases, have electronic configurations that are uniquely stable, they are the *only* elements whose symbols are used in writing condensed formulas. Knowing that the configuration of $_2He$ is $1s^2$, the configuration of $_3Li$ can be written as $1s^22s^1$ or in condensed form as $[He]2s^1$. The ground state configurations of the first four noble gases are:

$_2He$	$[He] = 1s^2$	(2 electrons)
$_{10}Ne$	$[Ne] = 1s^22s^22p^6$	(10 electrons)
$_{18}Ar$	$[Ar] = 1s^22s^22p^63s^23p^6$	(18 electrons)
$_{36}Kr$	$[Ar] = 1s^22s^22p^63s^23p^64s^23d^{10}4p^6$	(36 electrons)

The electronic configuration of calcium, $_{20}Ca$, can be written as $1s^22s^22p^63s^23p^64s^2$ or as $[Ar]4s^2$. Condensed electronic configuration will be extensively used in the discussion of valence electrons later in this chapter.

The Transition Metals

The ten elements that follow calcium in the fourth period of the periodic table are the metals of the first transition series, scandium, $_{21}Sc$, through zinc, $_{30}Zn$. You may want to verify this on a periodic table. After filling the 4s-subshell (calcium), the next electron enters the 3d-subshell with its five equivalent orbitals. Notice that the 4s-subshell is more stable than the 3d-subshell. The **n** = 3 and **n** = 4 principal shells overlap at this point. The 3d-subshell can hold a total of ten electrons, which is why there are ten transition metals in the fourth and subsequent periods. The **transition metals** are the elements that are built by filling a d-subshell. The ground state electronic configurations of three transition metals, $_{21}Sc$, $_{25}Mn$, and $_{28}Ni$, are shown below. Notice how the five equivalent 3d-orbitals fill singly before any pairing occurs (Hund's rule).

$_{21}Sc$	$[Ar]4s^23d^1$	$_{21}Sc$ [Ar] 4s [↑↓] 3d [↑] [ ] [ ] [ ] [ ]
$_{25}Mn$	$[Ar]4s^23d^5$	$_{25}Mn$ [Ar] 4s [↑↓] 3d [↑] [↑] [↑] [↑] [↑]
$_{28}Ni$	$[Ar]4s^23d^8$	$_{28}Ni$ [Ar] 4s [↑↓] 3d [↑↓] [↑↓] [↑↓] [↑] [↑]

At the very bottom of the periodic table are two rows of 14 elements. The elements in the first row of 14 ($_{58}Ce$ to $_{71}Lu$) have atomic numbers that come after that of lanthanum, $_{57}La$. The atomic numbers of those in the second row of 14 ($_{90}Th$ to $_{103}Lr$) come immediately after actinium, $_{89}Ac$. The metals in these rows are referred to as the *lanthanides* and *actinides* and together they comprise the **inner transition** metals. The reason there are 14 elements in each is because the 7 orbitals in the f-subshells are each filling one at a time with 14 electrons. The inner transition metals are relatively rare and all those in the actinide series are radioactive.

Example Problems

These problems have both answers and solutions given.

1. Write out the ground state electronic configurations for (a) phosphorus, (b) sodium, and (c) titanium. You can find the atomic numbers of these elements on a periodic table.

Answer: (a) 15 electrons: $P = 1s^22s^22p^63s^23p^3$

(b) 11 electrons: $Na = 1s^22s^22p^63s^1$

(c) 22 electrons: $Ti = 1s^22s^22p^63s^23p^64s^23d^2$

2. Identify the element represented by each ground state electronic configuration.

(a) $1s^2 2s^2 2p^6 3s^2 3p^6$

(b) $[He]2s^2 2p^4$

(c) $[Ar]4s^2 3d^{10} 4p^4$

Answer: (a) The number of electrons, 18, equals the atomic number of the element, argon, Ar

(b) 8 electrons, the element is oxygen, O

(c) 34 electrons, the element is selenium, Se

3. Using boxes for the orbitals, write out the orbital diagrams for (a) nitrogen, (b) fluorine, and (c) vanadium.

Answer: (a) $_7N$ 1s [↑↓] 2s [↑↓] 2p [↑] [↑] [↑]

(b) $_9F$ 1s [↑↓] 2s [↑↓] 2p [↑↓] [↑↓] [↑]

(c) $_{23}V$ 1s [↑↓] 2s [↑↓] 2p [↑↓] [↑↓] [↑↓] 3s [↑↓] 3p [↑↓] [↑↓] [↑↓] 4s [↑↓] 3d [↑] [↑] [↑] [ ] [ ]

4. What is wrong with the following ground state electronic configurations? Write the correct configuration for each species.

(a) C $1s^2 2s^2 3s^2$

(b) Ti $1s^2 2s^2 2p^6 3s^2 3p^6 4s^2 3d^3$

(c) Mn $[Ar]4s^2 4p^5$

Answer: (a) The 2p-subshell comes before the 3s. The correct configuration is C = $1s^2 2s^2 2p^2$.

(b) Titanium has 22 electrons, not 23. Ti = $1s^2 2s^2 2p^6 3s^2 3p^6 4s^2 3d^2$.

(c) The 3d-subshell comes before the 4p. Mn = $[Ar]4s^2 3d^5$.

Work Problems

Use these problems for additional practice.

1. Write out the ground state electronic configurations for (a) magnesium, (b) silicon, and (c) nickel.

2. Identify the element represented by each ground state electronic configuration.

(a) $1s^2 2s^2 2p^1$

(b) $[Kr]5s^1$

(c) $[Ne]3s^2 3p^4$

3. Using boxes for the orbitals, write out the orbital diagrams for (a) magnesium, (b) phosphorus, and (c) carbon.

4. What is wrong with the following ground state electronic configurations? Write the correct configuration for each species.

(a) $_{13}$Al $1s^2 2s^2 2p^6 3s^3$

(b) $_{19}$K $1s^2 2s^2 2p^6 3s^2 3p^6 4s^2$

(c) $_{16}$S $1s^2 2s^2 2p^6 3p^6$

Worked Solutions

1. **(a) 12 electrons: Mg = $1s^2 2s^2 2p^6 3s^2$**

(b) 14 electrons: Si = $1s^2 2s^2 2p^6 3s^2 3p^2$

(c) 28 electrons: Ni = $1s^2 2s^2 2p^6 3s^2 3p^6 4s^2 3d^8$

2. **(a) 5 electrons, the element is boron.**

(b) 37 electrons, the element is rubidium, Rb.

(c) 16 electrons, the element is sulfur.

3. **(a)** $_{12}$Mg 1s [↑↓] 2s [↑↓] 2p [↑↓] [↑↓] [↑↓] 3s [↑↓]

(b) $_{15}$P 1s [↑↓] 2s [↑↓] 2p [↑↓] [↑↓] [↑↓] 3s [↑↓] 3p [↑] [↑] [↑]

(c) $_{6}$C 1s [↑↓] 2s [↑↓] 2p [↑] [↑] [ ]

4. **(a) Too many electrons in the 3s-subshell.** Al = $1s^2 2s^2 2p^6 3s^2 3p^1$.

(b) The configuration shows 20 electrons; only 19 are needed. K = $1s^2 2s^2 2p^6 3s^2 3p^6 4s^1$.

(c) The 3s-subshell is missing. S = $1s^2 2s^2 2p^6 3s^2 3p^4$.

Electronic Configuration and the Periodic Table

In Chapter 2, you learned that the elements in the periodic table are arranged in order of increasing atomic number, but that doesn't explain the unusual shape of the table. The shape is determined by the sequence used to fill the subshells with electrons. In the following figure, the periodic table is divided into sections based on the type of subshell (s, p, d, or f) receiving the *last electron* in the building-up process.

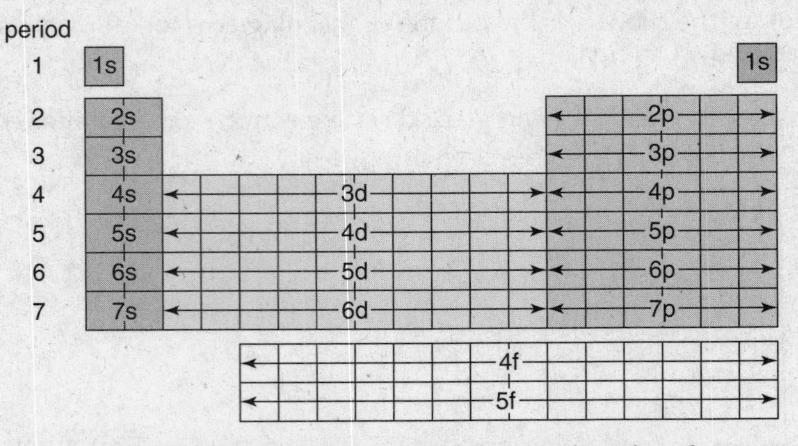

The periodic table divided into blocks of elements based on the type of subshell receiving the last electron in the building-up process.

On the left side of the periodic table is the s-block, two groups of elements that, in the building-up process, gain their last electron in an s-subshell. These are the alkali metals in Group IA and the alkaline earths in Group IIA. Moving down Group IA, notice that the principle quantum number, **n** of each s-subshell (1s, 2s, 3s, etc.), equals the number of the period. On the right side of the table are six columns (Groups IIIA through VIIIA) in which the last electron to complete their configuration enters a p-subshell. This is the p-block. The elements in the A-groups, the s- and p-blocks, are classed as the **main-group elements** or representative elements. In the middle of the table is a section of ten shorter columns in which the d-subshells are being filled. These are the **transition metals,** the elements of the d-block. Beneath the main body of the table are two rows of 14 elements, the **inner-transition metals,** in which the f-subshells are being filled, the f-block elements. Because the 1s-, 2s-, 2p-, and 3p-subshells fill before the 3d-subshell, the groups on the left and right sides of the periodic table become populated with 20 elements before the first transition metal is encountered.

The Valence Electrons

The **valence electrons** are the most chemically important electrons in an atom. They reside in the highest, occupied principal shell of an atom (largest **n**), and are the farthest away from the nucleus. They can also be described as the electrons *outside* the stable, filled inner core of electrons. Either way, the valence electrons determine the chemical properties of an element.

The valence electrons in the main-group elements are in the outermost s- and p-subshells. The Roman numeral at the top of each column equals the number of valence electrons in each member of that group (note that some texts use Arabic numerals instead of Roman numerals to identify the main-group elements). The elements in Group IA, the alkali metals, have one valence electron. Those in Group IIA have two valence electrons. Those in Group VIIA have seven, and so forth. The condensed ground state electronic configurations of the first four elements in Groups IA, IIA, IIIA, and VIIA are listed below. The valence electron(s) in each element are underlined. You should derive these configurations to verify that they are correct.

	IA		*IIA*		*IIIA*		*VIIA*
$_3$Li	[He]$\underline{2s^1}$	$_4$Be	[He]$\underline{2s^2}$	$_5$B	[He]$\underline{2s^2 2p^1}$	$_9$F	[He]$\underline{2s^2 2p^5}$
$_{11}$Na	[Ne]$\underline{3s^1}$	$_{12}$Mg	[Ne]$\underline{3s^2}$	$_{13}$Al	[Ne]$\underline{3s^2 3p^1}$	$_{17}$Cl	[Ne]$\underline{3s^2 3p^5}$
$_{19}$K	[Ar]$\underline{4s^1}$	$_{20}$Ca	[Ar]$\underline{4s^2}$	$_{31}$Ga	[Ar]$3d^{10}\underline{4s^2 4p^1}$	$_{35}$Br	[Ar]$3d^{10}\underline{4s^2 4p^5}$
$_{37}$Rb	[Kr]$\underline{5s^1}$	$_{38}$Sr	[Kr]$3d^{10}\underline{5s^2}$	$_{49}$In	[Kr]$4d^{10}\underline{5s^2 5p^1}$	$_{53}$I	[Kr]$4d^{10}\underline{5s^2 5p^5}$

Notice that each metal in Group IA has one electron in an s-subshell. This single s-electron is the valence electron. All other electrons reside in a very stable noble gas core. Because each IA metal has this similarity in their electronic configurations (one s-electron outside a noble gas core), they also have similar chemical properties. Metals tend to lose electrons in chemical reactions, and all the alkali metals lose *one* and only *one* electron in chemical reactions. Losing the valence electron forms an ion with a single positive charge. The noble gas core of each element remains intact. The reason each IA-metal loses only *one* electron is clear from its electronic configuration, and the resulting 1+ ion has the configuration of a noble gas.

Li	$[He]2s^1 \rightarrow Li^+$	$[He] + 1\ e^-$
Na	$[Ne]3s^1 \rightarrow Na^+$	$[Ne] + 1\ e^-$
K	$[Ar]4s^1 \rightarrow K^+$	$[Ar] + 1\ e^-$
Rb	$[Kr]5s^1 \rightarrow Rb^+$	$[Kr] + 1\ e^-$

Before going any further, something needs to be clarified. The ions of the alkali metals have electronic configurations exactly the same as the nearest noble gas. Though Na^+ and Ne have the same number of electrons and identical electronic configurations, they *do not* have identical chemical properties. This is because one species is an ion, Na^+, and the other is neutral, Ne. Na^+ is attracted to negative charges, Ne is not. The fact that they have the same number of electrons is overridden by the fact that one bears a charge while the other does not.

The metals in Group IIA have two valence electrons, both in an s-subshell outside the stable noble gas core. It should not be surprising that each forms an ion with a 2+ charge, the result of losing both valence electrons. Be^{2+}, Mg^{2+}, and Ca^{2+} each have the stable electronic configuration of a noble gas. The $\{[Kr]3d^{10}\}$ configuration of the strontium ion, Sr^{2+}, with a *completely filled* 3d-subshell outside the [Kr] noble gas core, is called a **pseudo-noble gas** core. Pseudo-noble gas cores are very stable, too. Don't be confused by this. Focus on the valence electrons, those in the highest occupied principal shell, the **n** = 5 shell in strontium. They are the chemically important electrons.

Be	$[He]2s^2 \rightarrow Be^{2+}$	$[He] + 2\ e^-$
Mg	$[Ne]3s^2 \rightarrow Mg^{2+}$	$[Ne] + 2\ e^-$
Ca	$[Ar]4s^2 \rightarrow Ca^{2+}$	$[Ar] + 2\ e^-$
Sr	$\{[Kr]3d^{10}\}5s^2 \rightarrow Sr^{2+}$	$\{[Kr]3d^{10}\} + 2\ e^-$

Aluminum, a chemically reactive metal in Group IIIA, has three valence electrons and forms a very stable 3+ ion. The ten electrons that remain in the Al^{3+} ion have the neon configuration. The other metals in Group IIIA behave similarly.

$$Al\ [Ne]3s^23p^1 \rightarrow Al^{3+}\ [Ne] + 3\ e^-$$

The elements of Group VIIA, the halogens, have seven valence electrons: two in an s-subshell and five in a p-subshell. The halogens are nonmetals and readily *gain* one electron to form ions bearing a single negative charge. Each ion has the electronic configuration of the next higher noble gas.

F	$[He]2s^22p^5 + 1\ e^-$	$\rightarrow F^-\ [He]2s^22p^6$	$\rightarrow F^-\ [Ne]$
Cl	$[Ne]3s^23p^5 + 1\ e^-$	$\rightarrow Cl^-\ [Ne]3s^23p^6$	$\rightarrow Cl^-\ [Ar]$
Br	$[Ar]3d^{10}4s^24p^5 + 1\ e^-$	$\rightarrow Br^-\ [Ar]3d^{10}4s^24p^6$	$\rightarrow Br^-\ [Kr]$
I	$[Kr]4d^{10}5s^25p^5 + 1\ e^-$	$\rightarrow I^-\ [Kr]4d^{10}5s^25p^6$	$\rightarrow I^-\ [Xe]$

Each element in Group VIIA readily gains *one* electron because each is only *one* electron away from achieving the uniquely stable electron configuration of a noble gas.

The unique chemical stability of the noble gas configurations cannot be overemphasized. With the exception of helium, each noble gas has eight electrons in its outermost principal shell, ns^2np^6, as seen earlier. The ions formed by the main-group elements, whether positive or negative, all have the very stable noble gas or pseudo-noble gas configurations. The metals in Groups IA, IIA, and IIIA *lose* one, two, or three electron(s), respectively, to achieve these configurations. The nonmetals in Groups VA, VIA, and VIIA *gain* three, two, or one electron(s), respectively, to form these configurations.

Group VA	P $[Ne]3s^23p^5$	+	3 e^-	→	P^{3-} [Ne]	$3s^23p^6$	→	P^{3-} [Ar]
Group VIA	S $[Ne]3s^23p^6$	+	2 e^-	→	S^{2-} [Ne]	$3s^23p^6$	→	S^{2-} [Ar]
Group VIIA	Cl $[Ne]3s^23p^5$	+	1 e^-	→	Cl^- [Ne]	$3s^23p^6$	→	Cl^- [Ar]

The elements in Group VIA (C, Si, etc.) cannot form stable ions by gaining or losing four electrons. There are no stable ions with a 4+ or 4– charge.

The metals of the first transition series have a partially filled 3d-subshell and a filled 4s-subshell outside the noble gas core of argon. Because the 3d-subshell is *partially* filled, both the 3d- and 4s-electrons are involved in the chemistry of these elements. The electronic configurations of vanadium, iron, and nickel are shown below. Each forms a 2+ ion by losing two outermost 4s-electrons. Vanadium and iron can also form a 3+ ion by losing an additional 3d-electron.

${}_{23}V$ $[Ar]4s^23d^3$	${}_{26}Fe$ $[Ar]4s^23d^6$	${}_{28}Ni$ $[Ar]4s^23d^8$
${}_{23}V^{2+}$ $[Ar]3d^3$	${}_{26}Fe^{2+}$ $[Ar]3d^6$	${}_{28}Ni^{2+}$ $[Ar]3d^8$
${}_{23}V^{3+}$ $[Ar]3d^2$	${}_{26}Fe^{3+}$ $[Ar]3d^5$	

Because the 4s-electrons are lost first when transition metals form ions, you may prefer to write the ground state electronic configuration of a transition metal with the 4s-subshell coming after the 3d-subshell, as shown below for vanadium, iron, and nickel. This is acceptable, but remember, the 4s-subshell fills before the 3d-subshell in the building-up process.

- 3d-subshell after 4s: ${}_{23}V$ $[Ar]4s^23d^3$ ${}_{26}Fe$ $[Ar]4s^23d^6$ ${}_{28}Ni$ $[Ar]4s^23d^8$
- 4s-subshell after 3d: ${}_{23}V$ $[Ar]\ 3d^34s^2$ ${}_{26}Fe$ $[Ar]\ 3d^64s^2$ ${}_{28}Ni$ $[Ar]\ 3d^84s^2$

The chemistry of the transition metals can be more complicated than that of the main-group elements and no more will be said about them here.

The unique stability of the noble gas configurations will play an important role in the next chapter, which deals with chemical bonding and the formation of compounds.

Valence Electrons and Lewis Symbols

Lewis symbols are named after G. N. Lewis (1875–1946), an American chemist who contributed greatly to our understanding of chemical bonding. Lewis symbols, or electron-dot symbols, show the number of valence electrons possessed by an element. They are used only for the main-group elements. Here is how Lewis symbols are written:

1. **Write the symbol of the element.** The symbol will represent all the electrons in the atom except the valence electrons.
2. **Determine the number of valence electrons (the group number).** Starting from either side, move around the symbol making one "dot" for each valence electron. Once one "dot" is made on all four sides (top, bottom, left, and right), begin to pair them up (two dots together) until all the electrons are used. If there are four valence electrons, there

will be four single dots. If there are five, there will be three single dots and one pair of dots. The single dots and the paired dots do not have to be arranged in any particular way around the symbol. These are all acceptable Lewis symbols for oxygen: ·Ö· :Ö· :Ȯ· :Ȯ: ·Ȯ:

3. **Two dots together represent two paired electrons.** A single dot represents an unpaired electron.

Notice how the Lewis symbols are presented in the following figure, and how the elements in each group have the same arrangement of valence electrons. The noble gases, except helium, have eight valence electrons, an "octet" of electrons.

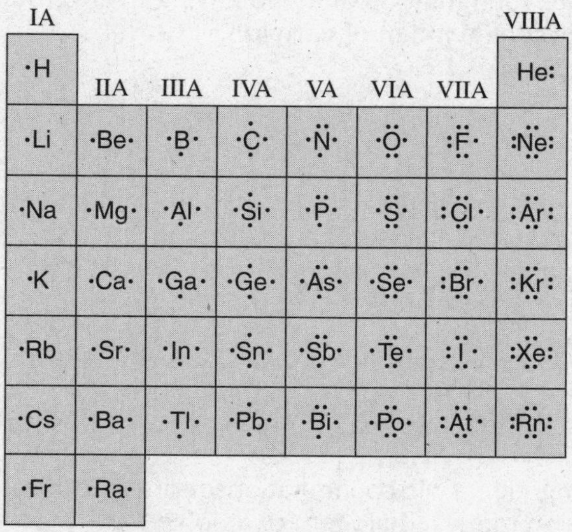

IA	IIA	IIIA	IVA	VA	VIA	VIIA	VIIIA
·H							He:
·Li	·Be·	·B·	·C·	·N·	·O·	:F·	:Ne:
·Na	·Mg·	·Al·	·Si·	·P·	·S·	:Cl·	:Ar:
·K	·Ca·	·Ga·	·Ge·	·As·	·Se·	:Br·	:Kr:
·Rb	·Sr·	·In·	·Sn·	·Sb·	·Te·	:I·	:Xe:
·Cs	·Ba·	·Tl·	·Pb·	·Bi·	·Po·	:At·	:Rn:
·Fr	·Ra·						

The Lewis symbols of the main-group elements.

Lewis symbols are useful because they emphasize the valence electrons. We will see more of them in the next chapter.

Example Problems

These problems have both answers and solutions given.

1. There are six ground state electronic configurations for main-group elements given below. Identify the valence electrons in each and then group them in pairs that would be expected to have similar chemical properties. Hint: Rewrite each as a condensed electronic configuration.

   (a) $1s^2 2s^2 2p^6 3s^1$

   (b) $1s^2 2s^2 2p^6 3s^2 3p^6 4s^2 3d^{10} 4p^3$

   (c) $1s^2 2s^2 2p^6 3s^2 3p^6 4s^1$

   (d) $1s^2 2s^2 2p^6 3s^2$

   (e) $1s^2 2s^2 2p^6 3s^2 3p^3$

   (f) $1s^2 2s^2$

Answer: Similar valence electron arrangements predict similar chemical properties: (a) and (c) are both ns^1; (b) and (e) are both ns^2np^3; (d) and (f) are both ns^2. Rewriting each as a condensed electronic configuration lets you focus on the valence electrons.

(a) $[Ne]3s^1$

(b) $\{[Ar]\ 3d^{10}\}4s^24p^3$

(c) $[Ar]4s^1$

(d) $[Ne]3s^2$

(e) $[Ne]3s^23p^3$

(f) $[He]2s^2$

2. Write the ground state configuration of the following main-group elements and their ions. Identify the noble gas configuration of each ion.

(a) N and N^{3-}

(b) S and S^{2-}

(c) Rb and Rb^+

Answer: (a) $N = 1s^22s^22p^3$; $N^{3-} = 1s^22s^22p^6 = [Ne]$

(b) $S = 1s^22s^22p^63s^23p^4$; $S^{2-} = 1s^22s^22p^63s^23p^6 = [Ar]$

(c) $Rb = 1s^22s^22p^63s^23p^64s^23d^{10}4p^65s^1$; $Rb^+ = 1s^22s^22p^63s^23p^64s^23d^{10}4p^6 = [Kr]$

3. Which of the following electronic configurations represents an alkali metal, an alkaline earth metal, a transition metal, a halogen, or a noble gas?

(a) $[Ar]4s^23d^3$

(b) $[Ne]3s^2$

(c) $1s^22s^22p^63s^23p^5$

(d) $1s^22s^22p^63s^23p^64s^23d^{10}4p^65s^24d^{10}5p^6$

(e) $[He]2s^1$

Answer: (a) Transition metal, the last electron enters the 3d-subshell. (b) Alkaline earth, Group IIA, the valence electron configuration is $3s^2$. (c) Halogen, Group VIIA, the valence electron configuration is $3s^23p^5$. (d) Noble gas, Xe, the valence electron configuration is $5s^25p^6$. (e) Alkali metal, Group IA, the valence electron configuration is $2s^1$.

4. Write the Lewis symbols for potassium, oxygen, silicon, and chlorine.

Answer: K· ·Ö: ·Ṣi· :C̤̈l·

Work Problems

Use these problems for additional practice.

1. There are six ground state electronic configurations for main-group elements given below. Identify the valence electrons and group them in pairs that would be expected to have similar chemical properties.

(a) $1s^2 2s^2 2p^2$

(b) $1s^2 2s^2 2p^6 3s^2 3p^6 4s^2 3d^{10} 4p^4$

(c) $1s^2 2s^2 2p^6 3s^2 3p^6 4s^2$

(d) $1s^2 2s^2 2p^6 3s^2 3p^4$

(e) $1s^2 2s^2 2p^6 3s^2$

(f) $1s^2 2s^2 2p^6 3s^2 3p^6 4s^2 3d^{10} 4p^2$

2. Write the ground state electronic configuration for the following main-group elements and their ions. Identify the noble gas configuration of each ion.

(a) K and K^+

(b) P and P^{3-}

(c) Br and Br^-

3. Which of the following electronic configurations represents an alkali metal, an alkaline earth metal, a transition metal, a halogen, or a noble gas?

(a) $1s^2 2s^2 2p^6 3s^2 3p^6 4s^2$

(b) $[Ar]4s^1$

(c) $[Ar]4s^2 3d^{10} 4p^6$

(d) $1s^2 2s^2 2p^6 3s^2 3p^5$

(e) $1s^2 2s^2 2p^6 3s^2 3p^6 4s^2 3d^1$

4. Write the Lewis symbol for selenium, aluminum, iodine, and magnesium.

Worked Solutions

1. **(a) and (f) both have valence electron configurations of ns^2np^2.**

(b) and (d) both have valence electron configurations of ns^2np^4.

(c) and (e) both have valence electron configurations of ns^2.

2. **(a) K = $1s^2 2s^2 2p^6 3s^2 3p^6 4s^1$; K^+ = $1s^2 2s^2 2p^6 3s^2 3p^6$ = [Ar]**

(b) P = $1s^2 2s^2 2p^6 3s^2 3p^3$; P^{3-} = $1s^2 2s^2 2p^6 3s^2 3p^6$ = [Ar]

(c) Br = $1s^2 2s^2 2p^6 3s^2 3p^6 4s^2 3d^{10} 4p^5$; Br^- = $1s^2 2s^2 2p^6 3s^2 3p^6 4s^2 3d^{10} 4p^6$ = [Kr]

3. **(a) Alkaline earth,** Group IIA, the valence electron configuration is $4s^2$.

(b) Alkali metal, Group IA, the valence electron configuration is $4s^1$.

(c) Noble gas, Group VIIIA, the valence electron configuration is $4s^2 4p^6$.

(d) Halogen, Group VIIA, the valence electron configuration is $3s^2 3p^5$.

(e) Transition metal, the last electron enters the 3d-subshell.

4. ·S̤̈e· ·Al· :Ï· ·Mg·

Properties of Atoms and the Periodic Table

The periodic table lists the elements in order of increasing atomic number and arranges them in groups with similar chemical properties. The fact that similar elements arise in a repeating pattern is because of a repeating similarity in electronic configuration. In the main-group elements, every eighth element has similar properties. The eighth element after lithium is sodium, and eight elements beyond sodium is potassium, all in Group IA and all with similar properties. Elements in the same group have the same number of valence electrons in the same subshell pattern. This repeating behavior, called a *periodic* behavior, makes the periodic table a tremendous tool in science because it allows prediction. If sodium reacts with water to produce hydrogen gas, it is reasonable to predict that potassium, the element right below it, will do the same.

Across a period, the elements become less metallic and more nonmetallic with corresponding changes in chemical properties. The arrangement of the elements in the periodic table makes it easier to see trends in their properties within groups and across periods. Two important properties of elements are the size of their atoms and the ease (or lack of ease) with which they lose an electron. Both are functions of the periodic similarities of electronic configuration, causing both size and ease of electron loss to be periodic properties of the elements.

The Size of Atoms—Trends in Atomic Radius

Atoms are regarded as being spherical bodies with blurry edges. They are not hard spheres like baseballs but rather somewhat soft and squeezable like a foam ball. Yet there are several ways to get a good estimate of the size of atoms. The size of an atom is usually expressed as its **atomic radius,** the radius of the spherical body. The atomic radius is strongly dependent on the electronic configuration of atoms that gives rise to the trends in size within groups and across periods.

- Within **groups,** the atomic radius increases with the period number from top to bottom. This is because the principal shell containing the valence electrons increases in size (larger **n**) with each element. Moving down a group, the valence electrons are increasingly farther away from the nucleus, causing the atoms to become increasingly larger.
- Across **periods,** the atomic radius decreases from left to right with increasing atomic number. Within a period, the largest atom is in Group IA; the smallest, in Group VIIIA. Moving across a period, each element has one more electron added to the same principal shell (same **n**) as the element before it. As each electron is added, the positive charge on the nucleus also increases. The increase in nuclear charge pulls the outermost electrons a bit closer to the nucleus each time. Because the electrons are entering the *same* principal shell, the size of the atoms decreases because of the increasing positive charge on the nucleus.

We are more interested in the trends in atomic radii within groups and periods as opposed to the actual numbers. The trends in atomic radii are easily seen in the following figure.

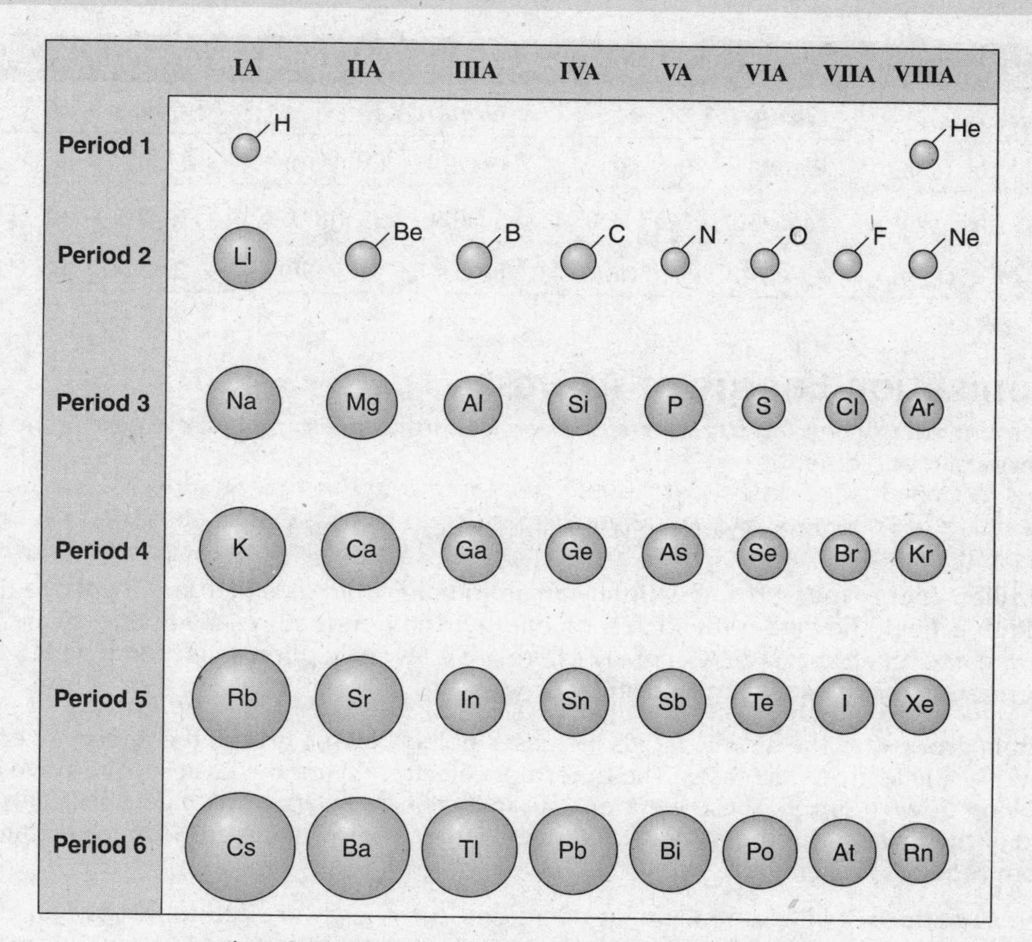

The relative atomic radii of the main-group elements.

Trends in the Sizes of Ions

The formation of positive ions requires the loss of one or more valence electrons, those that, on average, are farthest away from the nucleus. The loss of the outermost electron(s) causes a loss in size. In addition, the electrons that remain are held more tightly by the nucleus. The result is that *positive ions, cations, are always smaller than their parent atoms.*

The formation of negative ions involves the addition of one or more electrons to the valence shell of an atom. This occurs without an increase in nuclear charge to offset that of the added electron. The result is that *negative ions, anions, are always larger than their parent atoms.*

For ions of the *same* charge, the size, *ionic radius,* increases going down a group, paralleling the increase in size of their parent atoms. In the following table, the atoms and ions of the first three elements in Groups IA, IIA, VIA, and VIIA are listed. Atomic radii and ionic radii are given in picometers (pm). When studying this table, first compare the size of each ion with its parent atom. Then trace each group from top to bottom to see how both the atoms and ions increase in size.

Atomic and Ionic Radii of Several Elements (All Radii in pm)							
Group IA		*Group IIA*		*Group VIA*		*Group VIIA*	
Li (134)	Li^+ (68)	Be (90)	Be^{2+} (31)	O (73)	O^{2-} (140)	F (71)	F^- (136)
Na (154)	Na^+ (95)	Mg (130)	Mg^{2+} (65)	S (102)	S^{2-} (184)	Cl (99)	Cl^- (181)
K (196)	K^+ (133)	Ca (174)	Ca^{2+} (106)	Se (116)	Se^{2-} (198)	Br (114)	Br^- (195)

The Ionization Energy of Atoms

The minimum energy needed to remove one electron from an atom of an element is the **ionization energy** of that element.

$$\text{Ionization energy (IE)} + M(g) \rightarrow M^+(g) + e^-$$

The ionization energy measures how tightly an atom holds onto its electrons. The more tightly the electron is held, the higher the ionization energy. The trends in ionization energy are just the reverse of those for atomic radii. Generally (there are a few exceptions), as atomic radii get larger, ionization energies get smaller, and vice versa.

- Within **groups,** as the atomic radius increases going down a group, the energy needed to remove an electron decreases. The outermost electron(s) are the easiest to remove and, moving down a group, these electrons are in increasingly larger principal shells (larger **n**) and farther away from the nucleus. As **n** gets larger, removing an electron from a neutral atom requires less energy.
- Across **periods,** as the atomic radius decreases, the energy needed to remove an electron increases. The metals on the left end of a period have larger radii and lower ionization energies. They readily lose electrons to form positive ions. The nonmetals on the right end of a period have smaller radii and higher ionization energies. The nonmetals hold their valence electrons so tightly they gain, not lose, electrons in chemical reactions.

The ionization energies in kJ/mole of the main-group elements are shown in the following figure. Notice the general decrease in ionization energy going down each group and the general increase in ionization energy moving from left to right across each period. The metals to the left of the stair-step have ionization energies considerably lower than the nonmetals to the right. The trend in ionization energy across a period parallels the metallic to nonmetallic trend in chemical properties of the elements in that period.

Period	IA	IIA	IIIA	IVA	VA	VIA	VIIA	VIIIA
1	H 1312							He 2372
2	Li 520	Be 899	B 801	C 1086	N 1402	O 1314	F 1681	Ne 2081
3	Na 496	Mg 738	Al 578	Si 786	P 1012	S 1000	Cl 1251	Ar 1521
4	K 419	Ca 590	Ga 579	Ge 762	As 947	Se 941	Br 1140	Kr 1351
5	Rb 403	Sr 549	In 558	Sn 709	Sb 834	Te 869	I 1008	Xe 1170
6	Cs 376	Ba 503	Tl 589	Pb 716	Bi 703	Po 812	At ---	Rn 1037

The ionization energies of the main-group elements (kJ/mole).

Example Problems

These problems have both answers and solutions given.

1. Consulting a periodic table, arrange the elements in each set in order of decreasing atomic radius.

   (a) S, Na, K, O

   (b) Cl, Li, Al, B

   Answer: Atomic radii increase going down a group and across a period. Trace a path through the elements on a periodic table from lower left to upper right. (a) K > Na > S > O: potassium is below sodium in Group IA, sulfur is to the right of sodium in the third period, and oxygen is above sulfur in Group VIA. (b) Li > B > Al > Cl, lithium is to the left of boron in the second period. Aluminum is below boron in Group IIIA, and chlorine is to the right of aluminum in the third period.

2. Consulting a periodic table, choose the element in each pair you would expect to have the larger ionization energy.

   (a) C or Si

   (b) Ca or Br

   Answer: Ionization energies tend to decrease going down a group and increase going left to right across a group. (a) C > Si: carbon is above silicon in Group IVA, and carbon would be expected to have the greater ionization energy. (b) Br > Ca: bromine is far to the right of calcium in the fourth period and would be expected to have the greater ionization energy.

Work Problems

Use these problems for additional practice.

1. Consulting a periodic table, arrange the elements in each set in order of decreasing atomic radius.

   (a) P, Ge, As, S

   (b) Br, Cs, Sn, Sr

2. Consult a periodic table to help you decide which element in each pair has the greater ionization energy.

   (a) K or Br

   (b) S or Se

Worked Solutions

1. **(a) Ge > As > P > S**

   (b) Cs > Sr > Sn > Br

   For both sets of elements, trace the larger to smaller path of the elements from lower left to upper right on a periodic table.

2. **(a) Br > K Ionization energies tend to increase across a period.**

   (b) S > Se Ionization energies tend to decrease going down a group.

Chapter Problems and Answers

Problems

1. What is the relationship between the wavelength of light and its energy? What is the relationship between the frequency of light and its energy?

2. Arrange the following types of electromagnetic radiation in order of increasing energy. The wavelength of each type of light is given: red light (710 nm), green light (530 nm), x-rays (1 nm), infrared light (10 m), blue light (450 nm).

3. Hydrogen gas glows when subjected to an electric discharge similar to a neon sign. How does the light emitted from hydrogen differ from white light when passed through a prism?

4. In atoms, which transition emits light of a greater energy? Which transition emits light of longer wavelength? $\mathbf{n} = 4 \rightarrow \mathbf{n} = 3$ or $\mathbf{n} = 4 \rightarrow \mathbf{n} = 2$

5. In atoms, which energy level has the greater energy, $\mathbf{n} = 5$ or $\mathbf{n} = 2$? In which energy level are the electrons closer to the nucleus?

6. What is the difference between the "orbit" of a Bohr atom and the "orbital" of the quantum mechanical atom?

7. (a) In the Bohr model of the hydrogen atom, which orbit has the greater radius, $\mathbf{n} = 3$ or $\mathbf{n} = 4$?

   (b) Would an electronic transition in the hydrogen atom from the $\mathbf{n} = 3$ to the $\mathbf{n} = 4$ energy level absorb or release energy? Would the size of the atom increase or decrease?

8. What is the principal difference between the Bohr model of the hydrogen atom and the quantum mechanical model in terms of the location of an electron in a particular energy level?

9. What is the relationship among principal shell, subshell, and orbital? Use the $\mathbf{n} = 2$ principal shell to show the relationships.

10. What is the meaning of the following notations: 1s, 3p, 4d?

11. How many orbitals are in a(n): (a) s-subshell; (b) p-subshell; (c) d-subshell?

12. (a) How many orbitals are in the **n** = 1 principal shell? (b) In the **n** = 3 principal shell?

13. What is the Pauli principle as it concerns two electrons in the same orbital?

14. (a) According to the Pauli principle, what is always true for: two electrons in a 2s-orbital; (b) In what ways does the electronic configuration, $1s^2 2s^2 2p^7$, violate the Pauli principle?

15. What is Hund's rule? Does Hund's rule apply to the orbital in an s-subshell?

16. How would Hund's rule govern the placement of electrons in a $2p^4$ configuration?

17. Write the ground state electronic configurations for (a) O, (b) Ca, and (c) P.

18. Write the ground state electronic configurations for (a) O^{2-}, (b) Ca^{2+}, and (c) P^{3-}.

19. What is wrong with each of the following ground state electronic configurations? Write the correct electronic configuration using the same number of electrons.

    (a) $1s^2 2s^2 2p^6 3s^3$

    (b) $1s^2 2s^2 2p^6 3s^2 3p^6 4s^2 4p^3$

    (c) $1s^2 2s^2 2p^6 3s^2 3p^6 4s^2 3d^4 4p^2$

    (d) $1s^2 2s^2 2p^6 3s^2 3d^{10} 3p^6$

20. Examine each pair of electronic configurations and choose which represents an excited state and which represents a ground state configuration.

    (a) Ni = $1s^2 2s^2 2p^6 3s^2 3p^6 4s^1 3d^9$; Ge = $1s^2 2s^2 2p^6 3s^2 3p^6 4s^2 3d^{10} 4p^2$

    (b) S = $1s^2 2s^2 2p^6 3s^2 3p^4$; S = $1s^2 2s^2 2p^6 3s^2 3p^3 4s^1$

21. How many valence electrons are in the following electronic configurations?

    (a) $1s^2 2s^2 2p^6 3s^2 3p^4$

    (b) $1s^2 2s^2 2p^6 3s^2 3p^6 4s^2 3d^{10} 4p^1$

    (c) $1s^2 2s^2 2p^6$

22. Write out the ground state electronic configuration for O, S, and Se and show why it is correct to place them in the same group.

23. Which of the following electronic configurations represent a(n) noble gas, transition metal, alkali metal, halogen, alkaline earth?

(a) $1s^22s^22p^63s^2$

(b) $1s^22s^22p^63s^23p^64s^23d^{10}4p^5$

(c) $1s^22s^22p^63s^23p^64s^23d^{10}4p^6$

(d) $1s^22s^22p^63s^23p^64s^23d^3$

(e) $1s^22s^22p^63s^23p^64s^1$

24. Write the condensed ground state electronic configuration for iodine, manganese, and barium.

25. What are the general trends in atomic radii on the periodic table?

26. Using a periodic table to guide you, arrange the following elements in order of *decreasing* atomic radius.

(a) Si, Mg, Na, and Cl

(b) S, Sn, P, and Sb

27. Define ionization energy. What are the general trends in ionization energy on the periodic table?

28. Using a periodic table to guide you, arrange the following elements in order of *increasing* ionization energy.

(a) Ar, Na, and P

(b) Ba, Mg, and Ca

29. What is meant by the term "valence electrons?"

30. Write the Lewis symbols for carbon, magnesium, chlorine, and arsenic.

31. (a) How does the size of a positive ion compare with the size of the neutral atom of the same element?

(b) How does the size of a negative ion compare with the size of the neutral atom of the same element?

Answers

1. **The shorter the wavelength, the higher the energy of light, and vice versa. The higher the frequency of light (the shorter the wavelength), the higher the energy of light, and vice versa.**

2. **The order of increasing energy is the order of decreasing wavelength. From lowest to highest energy: infrared light, red light, green light, blue light, and x-rays are the highest energy.**

3. **The light from the hydrogen atom will form lines of light while white light will produce the continuous rainbow of colors.**

4. **The n = 4 → n = 2 transition spans a greater difference in energy and will emit light of greater energy and shorter wavelength than the n = 4 → n = 3 transition. The latter transition will emit light of longer wavelength (lower energy).**

5. **The n = 5 level is higher energy than the n = 2 level. The electrons are closer to the nucleus in the n = 2 level.**

6. **An orbit in the Bohr atom is the circular path of fixed radius taken by the electron around the nucleus. An orbital is a region of space where the probability of finding a particular electron is high. Orbitals can have various shapes.**

7. **(a) The n = 4 orbit has the greater radius.**

   (b) An electronic transition from a lower energy level, n = 3, to a higher energy level, n = 4, will absorb energy and the larger radius of the n = 4 orbit increases the size of the atom.

8. **In the Bohr model of the hydrogen atom, the electron can reside in orbits of known, fixed radius. In the quantum mechanical model, the location of the electron is not known exactly; rather, its location is described in terms of the probability of being at a given point about the nucleus.**

9. **Orbitals are contained within subshells and subshells are contained within principal shells. The n = 2 principal shell contains two subshells, an s-subshell with one orbital and a p-subshell with three orbitals.**

10. **1s = an s-subshell in the n = 1 principal shell. 3p = a p-subshell in the n = 3 principal shell. 4d = a d-subshell in the n = 4 principal shell.**

11. **(a) An s-subshell contains one orbital. (b) A p-subshell contains three orbitals. (c) A d-subshell contains five orbitals.**

12. **The number of orbitals in a principal shell, n, equals n^2. (a) There is one orbital in the n = 1 principal shell, and (b) nine orbitals in the n = 3 principal shell.**

13. **A consequence of the Pauli principle is that two electrons in the same orbital must have spins opposite one another; that is, the spins must be paired.**

14. **(a) Two electrons in a 2s-orbital must be paired. (b) The Pauli principle states that there can be no more than two electrons in any orbital. A $2p^7$ configuration would require one of the three p orbitals to contain three electrons.**

15. **Hund's rule states that electrons will fill a set of orbitals of the same energy singly before any electrons will pair. Unpaired electrons will spin in the same direction. It does not apply to the one s-orbital in an s-subshell, only to subshells with more than one orbital.**

16. **Two of the orbitals would contain one electron and one orbital would contain a pair of electrons. The two unpaired electrons would have the same spin.**

17. **(a) O = $1s^22s^22p^4$; (b) Ca = $1s^22s^22p^63s^23p^64s^2$; (c) P = $1s^22s^22p^63s^23p^3$**

18. **(a) O^{2-} = $1s^22s^22p^6$ = [Ne]; (b) Ca^{2+} = $1s^22s^22p^63s^23p^6$ = [Ar]; (c) P^{3-} = $1s^22s^22p^63s^23p^6$ = [Ar]**

19. **(a) Too many electrons are in the 3s-subshell. The correct ground state configuration for 13 electrons would be $1s^22s^22p^63s^23p^1$.**

(b) The 3d-subshell begins to fill before the 4p-subshell. The correct configuration for 23 electrons would be $1s^22s^22p^63s^23p^64s^23d^3$.

(c) Electrons should not enter the 4p-subshell until the 3d-subshell is completely filled with ten electrons. The correct configuration for 26 electrons would be $1s^22s^22p^63s^23p^64s^23d^6$.

(d) The 4s-subshell was skipped. The correct configuration for 28 electrons would be $1s^22s^22p^63s^23p^64s^23d^8$.

20. **(a) Ni = $1s^22s^22p^63s^23p^64s^13d^9$ is an excited state; an electron is taken from the 4s-subshell and placed in the higher-energy 3d-subshell. Ge = $1s^22s^22p^63s^23p^64s^23d^{10}4p^2$ is a ground state configuration.**

(b) S = $1s^22s^22p^63s^23p^4$ is a ground state configuration. S = $1s^22s^22p^63s^23p^34s^1$ is an excited state with an electron promoted from the 3p-subshell to the 4s-subshell.

21. **(a) The highest occupied principal shell is the n = 3 shell, which contains six valence electrons.**

(b) The highest occupied principal shell is the n = 4 shell, which contains three valence electrons.

(c) The highest occupied principal shell is the n = 2 shell, which contains eight valence electrons.

22. **O = $1s^22s^22p^4$; S = $1s^22s^22p^63s^23p^4$; Se = $1s^22s^22p^63s^23p^64s^23d^{10}4p^4$. All three configurations have the same configuration in the highest occupied principal shell, ns^2np^4. They all have six valence electrons.**

23. **(a) $1s^22s^22p^63s^2$ is an alkaline earth, Mg. The last electron completes the $3s^2$ configuration, [Ne]$3s^2$.**

(b) $1s^22s^22p^63s^23p^64s^23d^{10}4p^5$ is a halogen, Br, with seven valence electrons, $4s^24p^5$, {[Ar]$3d^{10}$}$4s^24p^5$.

(c) $1s^22s^22p^63s^23p^64s^23d^{10}4p^6$ is a noble gas, Kr, with eight valence electrons, $4s^24p^6$, [Kr].

(d) $1s^22s^22p^63s^23p^64s^23d^3$ is a transition metal, V. The last electron enters the 3d-subshell.

(e) $1s^22s^22p^63s^23p^64s^1$ is an alkali metal, K, [Ar]$4s^1$.

24. **$_{53}I$ = [Kr]$5s^24^{10}5p^5$; $_{25}Mn$ = [Ar]$4s^23d^5$; $_{56}Ba$ = [Xe]$6s^2$**

25. **Atomic radii increase going down a group and decrease from left to right across a period.**

26. **(a) Order of decreasing atomic radius: Na > Mg > Si > Cl, across the third period**

    (b) Order of decreasing atomic radius: Sn > Sb > P > S

27. **The ionization energy is the minimum energy needed to remove an electron from an atom. The general trends in ionization energy are decreasing down a group, increasing left to right across a period.**

28. **(a) From lowest to highest ionization energy: Na < P < Ar, across the second period**

    (b) From lowest to highest ionization energy: Ba < Ca < Mg, moving up Group IIA

29. **Valence electrons are those in the highest occupied energy level of an atom that are directly involved in the chemistry of the element. Alternatively, they are the electrons outside the noble gas core of the atom.**

30. ·Ċ̣· ·Mg· :C̤̈l· ·Ạ̈s·

31. **(a) Positive ions are always smaller in size than the neutral atom of the same element.**

    (b) Negative ions are always larger in size than the neutral atom of the same element.

Supplemental Chapter Problems

Problems

1. Arrange the following types of light in order of increasing energy: violet light (390 nm), green light (540 nm), microwaves (10 mm), yellow light (585 nm), and ultraviolet light (200 nm).

2. What do the bright lines of light represent in the visible spectrum of the hydrogen atom?

3. In what way are the 1s-, 2s-, and 3s-orbitals similar? In what way are they different?

4. What is the maximum number of electrons that can be in a 3p-subshell? A 3d-subshell?

5. List the following subshells in order of increasing energy in multi-electron atoms: 3s, 4p, 1s, 3d, 5p, 2p, 2s.

6. How are the three orbitals in a p-subshell oriented relative to one another?

7. Write the ground state electronic configurations for N, Ti, and Cl.

8. Write ground state electronic configurations for Mg^{2+}, S^{2-}, and Br^{-}.

9. Draw the condensed ground state configurations for O, Ti, K, Cl.

10. Looking at a periodic table, how can you quickly determine the number of valence electrons for a main-group element?

11. Three supposed ground state electronic configurations are given. Tell what is wrong with each one and then write the correct ground state configuration.

(a) $1s^22s^22p^63s^24s^2$

(b) $1s^22s^22p^63s^23p^64s^23d^84p^2$

(c) $1s^22s^22p^63s^13p^4$

12. How many p-orbitals can be in the **n** = 3 principal shell? How many p-orbitals can be in the **n** = 4 principal shell?

13. How many unpaired electrons are in one atom of phosphorus?

14. In what group and period is this element: $1s^22s^22p^63s^23p^64s^23d^{10}4p^65s^2$?

15. Draw the Lewis symbols for Ar, Al, S, and K.

16. Arrange the following elements in order of decreasing atomic radius:

(a) Sr, Mg, Rb, Ca

(b) N, Si, Ga, P

17. What do the following pairs of ions have in common: S^{2-} and K^+; Li^+ and Be^{2+}?

18. Which element in each pair has the larger ionization energy: (a) Mg or Cl, (b) Ca or Ba?

19. What areas of the periodic table are called the s-block, p-block, and d-block?

20. Why is Mg^{2+} smaller than Mg?

Answers

1. **Order of increasing energy is order of decreasing wavelength: microwaves < yellow light < green light < violet light < ultraviolet light.** (page 223)

2. **Each line is of a specific wavelength and specific energy that equals the energy *differences* between energy levels (n) in the hydrogen atom.** Each transition of an electron from a higher to a lower energy level emits a specific energy of light. (page 224)

3. **They all have a spherical shape. They have different sizes, with the 3s-orbital > 2s-orbital > 1s-orbital.** (page 224)

4. **3p-subshell = 6 electrons; 3d-subshell = 10 electrons** (page 227)

5. **In order of increasing energy: 1s < 2s < 2p < 3s < 3d < 4p < 5p** (page 228)

6. **The three double-lobe p-orbitals are at right angles to one another.** (page 229)

7. **N = $1s^2 2s^2 2p^3$; Ti = $1s^2 2s^2 2p^6 3s^2 3p^6 4s^2 3d^2$; Cl = $1s^2 2s^2 2p^6 3s^2 3p^5$** (page 234)

8. **Mg^{2+} = $1s^2 2s^2 2p^6$; S^{2-} = $1s^2 2s^2 2p^6 3s^2 3p^6$; Br^- = $1s^2 2s^2 2p^6 3s^2 3p^6 4s^2 3d^{10} 4p^6$** (page 234)

9. **O = [He]$2s^2 2p^4$; Ti = [Ar]$4s^2 3d^2$; K = [Ar]$4s^1$; Cl = [Ne]$3s^2 3p^5$** (page 236)

10. **The group number equals the number of valence electrons for the main-group elements.** (page 240)

11. **(a) The last two electrons should be in the 3d-subshell: $1s^2 2s^2 2p^6 3s^2 3d^2$**

    (b) Fill the 3d-subshell before adding electrons to the 4p-subshell: $1s^2 2s^2 2p^6 3s^2 3p^6 4s^2 3d^{10}$

    (c) Completely fill the 3s-subshell before adding electrons to the 3p-subshell: $1s^2 2s^2 2p^6 3s^2 3p^3$

    (page 239)

12. **Only three p-orbitals (one p-subshell) can be in any principal shell with n = 2 or greater.** (page 239)

13. **P = [Ne]$3s^2 3p^3$, with three unpaired electrons in the 3p-subshell.** (page 239)

14. **With a configuration of [Kr]$5s^2$ the element is in the fifth period and Group IIA.** (page 240)

15. :Ä̤r: ·Ȧl· :Ṡ̤· K· (page 242)

16. **(a) Rb > Sr > Ca > Mg** **(b) Ga > Si> P > N** (page 247)

17. **S^{2-} and K^+ both have the electronic configuration of argon, [Ar]; Li^+ and Be^{2+} both have the electronic configuration of helium, [He].** (page 247)

18. **(a) Cl > Mg** **(b) Ca > Ba** (page 248)

19. **s-block: Groups IA and IIA**

    p-block: Groups IIIA through VIIIA

    d-block: The ten short columns of metals between Groups IIA and IIIA, the transition metals (page 240)

20. **In an Mg atom, the 12+ charge of the nucleus (12 protons) attracts 12 negative electrons. In the magnesium ion, Mg^{2+}, there are two fewer electrons, allowing the ten remaining electrons to be held more tightly and closer to the 12+ charge of the nucleus.** (page 240)

Chapter 10

Chemical Bonding—The Formation of Compounds

All chemical compounds can be classed as either *ionic compounds* or *covalent compounds,* which are categories based on the kind of bonding that holds them together. Ionic compounds are crystalline solids formed by neatly packed ions of opposite charge. Covalent compounds might be solid, liquid, or gas at room temperature, and they exist as molecules. For that reason, covalent compounds often are referred to as molecular compounds. Ionic compounds do not exist as molecules. The general properties of covalent and ionic compounds are discussed in Chapter 2.

What is it that holds atoms together in compounds? In ionic compounds, it is the natural attraction of positive ions to negative ions, and vice versa, but how do these ions form? What determines the size of the charges on the ions, a factor that governs the formulas of ionic compounds? In a similar vein, what holds the atoms together in covalent compounds? Although it is not as obvious as in the case of ionic compounds, it is again the natural attraction of negative to positive and positive to negative—negative electrons attracted to positive nuclei and positive nuclei attracted to negative electrons. In covalent molecules, atoms share negative electrons, and it is the mutual attraction of valence electrons for both nuclei that holds the atoms together.

The force of attraction between two species is called a **bond.** It might be a strong attraction or a weak one, but it is a bond nonetheless. Most ionic compounds are composed of metals and nonmetals, and most covalent compounds are composed of only nonmetals. Let's start the discussion by looking at the formation of ionic compounds.

Ionic Bonding—Ionic Compounds

Sodium metal, Na*(s)*, reacts vigorously with chlorine gas, $Cl_2(g)$, to form sodium chloride, NaCl*(s)*, an ionic compound that you know as common table salt.

$$2\ Na(s) + Cl_2(g) \rightarrow 2\ NaCl(s)$$

Sodium chloride is a crystalline solid composed of sodium ions, Na^+, and chloride ions, Cl^-, packed together in a highly organized way to maximize the attraction between ions of opposite charge. The **ionic bond** is the attraction between these positive and negative ions in a crystal, and compounds held together by ionic bonds are classed as **ionic compounds.** Ions, ionic compounds, their formulas, and names are discussed in Chapter 4. Examining the reaction between sodium and chlorine will show how each element forms an ion with a stable noble gas configuration by the loss or gain of a single electron. Let's go through the formation of NaCl in steps: first the formation of the sodium ion; then the formation of the chloride ion; and then combining these steps showing how the ions form simultaneously by the transfer of a single electron from sodium to chlorine. The final step arranges the ions forming the crystal of sodium chloride.

1. Sodium, a Group IA metal, has a single electron in its valence shell. Losing this electron forms a sodium ion, Na^+, that has the stable electronic configuration of neon. The ionization of sodium is shown three ways, first using symbols, then using symbols with electronic configurations, and finally using condensed electronic configurations. The single valence electron in sodium is underlined. See Chapter 9 if you need to review electronic configurations.

$$Na \rightarrow Na^+ + e^-$$
$$Na\ (1s^2 2s^2 2p^6 \underline{3s^1}) \rightarrow Na^+\ (1s^2 2s^2 2p^6) + e^-$$
$$Na\ [Ne]\underline{3s^1} \rightarrow Na^+\ [Ne] + e^-$$

2. Chlorine, a Group VIIA nonmetal, has a single vacancy in its valence shell. Adding one electron to chlorine fills this vacancy and forms a chloride ion, Cl^-, which has the stable argon configuration. The seven valence electrons in chlorine are underlined.

$$Cl + e^- \rightarrow Cl^-$$
$$Cl\ (1s^2 2s^2 2p^6 \underline{3s^2 3p^5}) + e^- \rightarrow Cl^-\ (1s^2 2s^2 2p^6 3s^2 3p^6)$$
$$Cl\ [Ne]\underline{3s^2 3p^5} + e^- \rightarrow Cl^-\ [Ar]$$

Combining both steps shows the simultaneous transfer of one electron from sodium to chlorine to form the sodium and chloride ions that make up the compound. The first equation uses Lewis symbols (Chapter 9) to show the electron transfer.

$$Na\cdot + \cdot\ddot{\underset{..}{Cl}}: \longrightarrow Na^+[Ne] + Cl^-[Ar] \longrightarrow NaCl$$

$$Na(1s^2 2s^2 2p^6 \underline{3s^1}) + Cl(1s^2 2s^2 2p^6 \underline{3s^2 3p^5}) \rightarrow Na^+(1s^2 2s^2 2p^6) + Cl^-(1s^2 2s^2 2p^6 3s^2 3p^6)$$
$$Na\ [Ne]3s^1 \qquad Cl\ [Ne]3s^2 3p^5 \qquad Na^+\ [Ne] \qquad Cl^-\ [Ar]$$

3. As sodium and chloride ions form, they arrange themselves in a highly structured array to maximize the attraction between ions of opposite charge. For sodium chloride, this arrangement is a cubic-shaped crystal, a small piece of which is shown in the following figure. In the actual crystal, the ions touch one another, although they are separated in the figure to emphasize the cubic packing of the ions. The arrangement of ions in an ionic crystal is called a **crystal lattice.** The following figure shows the crystal lattice of sodium chloride.

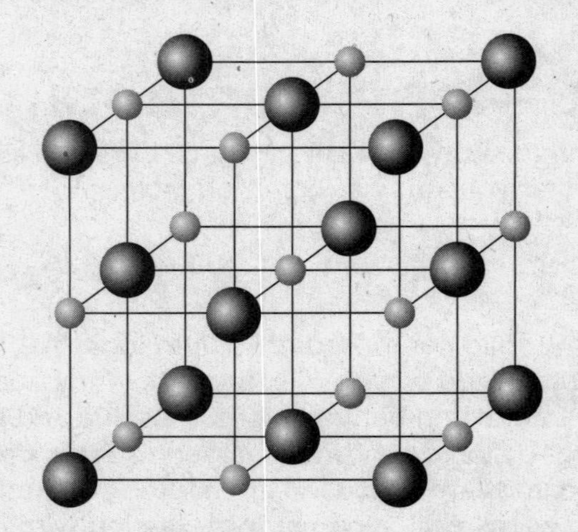

The sodium chloride lattice. The smaller spheres are Na^+; the larger are Cl^-.

Guidelines for the formation of ionic compounds:

- Ionic compounds form in reactions between metals and nonmetals. There are some exceptions to this, but it is usually true for the main-group elements.
- Metals tend to lose electrons to form positive ions (cations). The main-group metals, those in the A groups, *lose the necessary number of electrons to form ions with the electronic configuration of the next lower noble gas, lower in atomic number.* The ions will have eight electrons (an octet of electrons) in their valence shells. For main-group metals, the number of electrons lost equals the group number of the metal. The valence electrons are underlined in the following examples. You should work out the electronic configurations of the ions to ensure that they are the same as the next lower rare gas, with an octet of electrons in their valence shells. Of course, there is no octet for the lithium ion, Li^+. It has the helium configuration with two electrons in its valence shell. Notice that the valence shell of the cation, the outermost occupied principal shell of the cation, is the next lower shell compared to the neutral atom. The outermost shell of the atom is vacated as it becomes a cation.

  Group IA metals lose one electron to form cations with noble gas configurations.

  $$\text{Na } (1s^22s^22p^6\underline{3s^1}) \rightarrow \text{Na}^+ \text{ [Ne]} + 1\ e^-$$

  Group IIA metals lose two electrons to form cations with noble gas configurations.

  $$\text{Ca } (1s^22s^22p^63s^23p^6\underline{4s^2}) \rightarrow \text{Ca}^{2+} \text{ [Ar]} + 2\ e^-$$

  Group IIIA metals lose three electrons to form cations with noble gas configurations.

  $$\text{Al } (1s^22s^22p^6\underline{3s^2}\underline{3p^1}) \rightarrow \text{Al}^{3+} \text{ [Ne]} + 3\ e^-$$

- Nonmetals tend to gain electrons to form negative ions (anions). All nonmetals are main-group elements and will *gain the necessary number of electrons to form anions that have the electronic configuration of the next higher noble gas,* with eight electrons (an octet of electrons) in their valence shells. The number of electrons that must be gained equals eight minus the group number of the element. This is shown here for nonmetals from Groups VIIA, VIA, and VA.

  Group VIIA nonmetals must gain one electron to form anions with noble gas configurations.

  $$\text{F } (1s^2\underline{2s^2}\underline{2p^5}) + 1\ e^- \rightarrow \text{F}^- \text{ [Ne]}$$

  Group VIA nonmetals must gain two electrons to form anions with noble gas configurations.

  $$\text{S } (1s^22s^22p^6\underline{3s^2}\underline{3p^4}) + 2\ e^- \rightarrow \text{S}^{2-} \text{ [Ar]}$$

  Group VA nonmetals must gain three electrons to form anions with noble gas configurations.

  $$\text{N } (1s^2\underline{2s^2}\underline{2p^3}) + 3\ e^- \rightarrow \text{N}^{3-} \text{ [Ne]}$$

The observation that the main-group elements tend to form ions with eight electrons in their valence shells leads to a statement called the **octet rule:**

> *The octet rule:* In reactions involving main-group elements, atoms tend to gain, lose, or share the necessary number of electrons needed to achieve an octet of electrons in their valence shells.

Hydrogen cannot form an octet; it requires only two electrons to achieve the configuration of the noble gas, helium. Except for helium, all the noble gases have an octet of electrons in their valence shells, and this is why the phrases "noble gas configuration" and "octet of electrons" are often used interchangeably. The octet rule isn't perfect, but it is a useful guide.

The following three examples show the formation of ionic compounds by the complete transfer of one or more electrons from the metal to the nonmetal. In each case, the total number of electrons lost equals the total number gained. Note that each ion has the electronic configuration of a noble gas. As you inspect each example, notice how the number of electrons lost by the metal and gained by the nonmetal determines the formula of the ionic product. *The formula of the product can be predicted from the electronic configurations of the reactants.*

Example 1: Magnesium and fluorine react to form magnesium fluoride.

Both elements can form ions with the electronic configuration of neon.

$Mg\ (1s^2 2s^2 2p^6 \underline{3s^2}) \rightarrow Mg^{2+}\ (1s^2 2s^2 2p^6) + 2\ e^-$ Mg loses 2 e^-'s to form Mg^{2+} [Ne].

$F\ (1s^2 \underline{2s^2 2p^5}) + 1\ e^- \rightarrow F^-\ (1s^2 2s^2 2p^6)$ F gains 1 e^- to form F^- [Ne].

Clearly, it will require *two fluorine* atoms to accept the *two electrons* lost by *one magnesium* atom. Now you see why the formula of magnesium fluoride has to be MgF_2. It is convenient to show the reaction using Lewis symbols to show the transfer of electrons from magnesium to fluorine. The resulting ions are held together by ionic bonds in the MgF_2 solid.

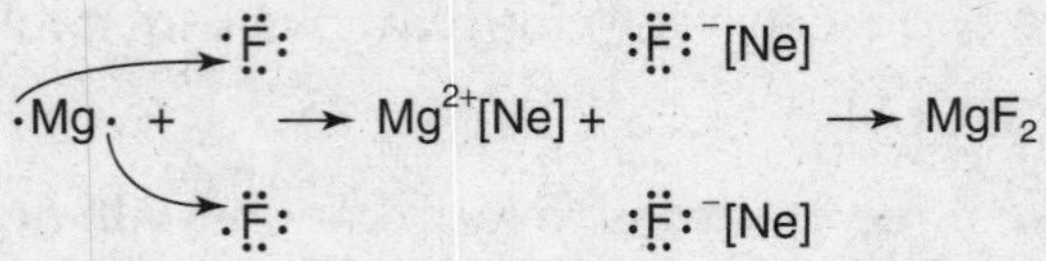

$$\cdot Mg \cdot + \begin{matrix} \cdot\ddot{\underset{\cdot\cdot}{F}}: \\ \cdot\ddot{\underset{\cdot\cdot}{F}}: \end{matrix} \rightarrow Mg^{2+}[Ne] + \begin{matrix} :\ddot{\underset{\cdot\cdot}{F}}:^-[Ne] \\ :\ddot{\underset{\cdot\cdot}{F}}:^-[Ne] \end{matrix} \rightarrow MgF_2$$

Example 2: Magnesium and sulfur react to form magnesium sulfide.

Magnesium can form a cation with the neon configuration and sulfur and an anion with the argon configuration.

$Mg\ (1s^2 2s^2 2p^6 \underline{3s^2}) \rightarrow Mg^{2+}\ (1s^2 2s^2 2p^6) + 2\ e^-$ Mg loses 2 e^-'s to form Mg^{2+} [Ne].

$S\ (1s^2 2s^2 2p^6 \underline{3s^2 3p^4}) + 2\ e^- \rightarrow S^{2-}\ (1s^2 2s^2 2p^6 3s^2 3p^6)$ S gains 2 e^-'s to form S^{2-} [Ar].

It will require *one sulfur* atom to accept the *two electrons* lost by *one magnesium* atom. The formula of magnesium sulfide must be MgS. Again, Lewis symbols will be used to show the electron transfer from magnesium to sulfur. As with all ionic compounds, the ions in the crystal of MgS are held together by ionic bonds, the natural attraction between ions of opposite charge.

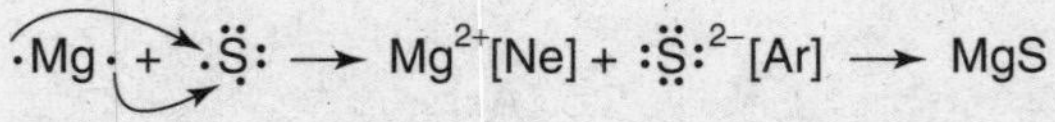

$$\cdot Mg \cdot + \cdot\ddot{\underset{\cdot\cdot}{S}}: \rightarrow Mg^{2+}[Ne] + :\ddot{\underset{\cdot\cdot}{S}}:^{2-}[Ar] \rightarrow MgS$$

Example 3: Aluminum and oxygen react to form aluminum oxide.

$Al\ (1s^2 2s^2 2p^6 \underline{3s^2 3p^1}) \rightarrow Al^{3+}\ (1s^2 2s^2 2p^6) + 3\ e^-$ Al loses 3 e^-'s to form Al^{3+} [Ne].

$O\ (1s^2 \underline{2s^2 2p^4}) + 2\ e^- \rightarrow O^{2-}\ (1s^2 2s^2 2p^6)$ O gains 2 e^-'s to form O^{2-} [Ne].

It will require *three oxygen* atoms to accept the *six electrons* lost by *two aluminum* atoms. Only by using two aluminum atoms will an even number of electrons be lost, which then can be gained, two at a time, by three oxygen atoms. The formula of aluminum oxide, Al_2O_3, shows this 2 to 3 ratio. The equation for the formation of Al_2O_3 using Lewis symbols to show the transfer of electrons from aluminum to oxygen follows. The crystals of aluminum oxide are exceedingly hard, the result of strong forces of attraction between cations with a 3+ charge and anions with a 2− charge.

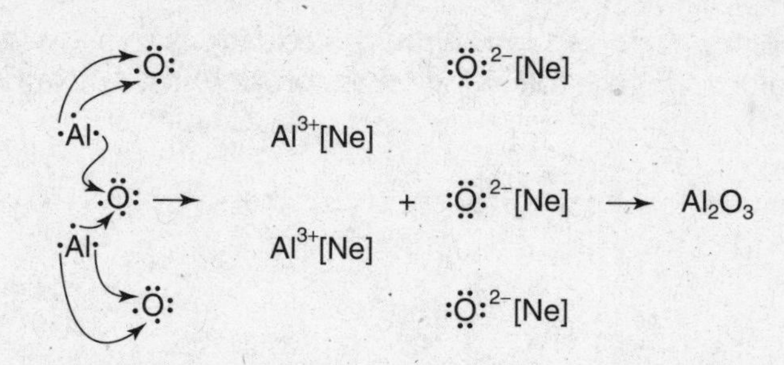

The previous examples have involved only main-group elements since they form monatomic ions with noble gas configurations in agreement with the octet rule. This predictable behavior lets you figure out formulas of ionic products if you know the electronic configurations of the reactants.

Many transition metals form ionic compounds, too, but the octet rule does not apply to transition metals. These metals lose electrons to form positive ions that do not have noble gas structures. The ionic compounds they form, such as $CrCl_2$, $CrCl_3$, and $NiBr_2$, are crystalline solids composed of ions held together by ionic bonds. When it comes to predicting the formulas of ionic compounds of transition metals, you should consult the table of common cations in Chapter 4 to see what ions each transition metal forms. We'll stick with the main-group metals here. In addition, many ionic compounds have polyatomic anions like sulfate, SO_4^{2-}, nitrate, NO_3^-, or phosphate, PO_4^{3-}. These ionic compounds are usually made in reactions other than the direct combination of the elements, as shown in Chapter 7. Yet, they are still ionic compounds with positive and negative ions in the correct ratio to form crystalline solids.

Example Problems

These problems have both answers and solutions given.

1. What is the ionic bond? Describe the physical state of ionic compounds.

   Answer: The ionic bond is the attraction between positive and negative ions in a crystal. Ionic compounds are crystalline solids.

2. State the octet rule. To which class of elements does it not apply?

   Answer: In reactions involving main-group elements, atoms tend to gain, lose, or share the necessary number of electrons needed to achieve an octet of electrons in their valence shells. It does not apply to transition metals, nor does it apply to hydrogen or lithium. Realize, an "octet of electrons" implies a "noble gas configuration."

3. Sodium loses one electron to form an ion with an octet of electrons in its valence shell. Write out the electronic configuration of the sodium ion, Na^+, and underline what are now the valence electrons for the ion.

   Answer: Na^+ ($1s^2\underline{2s^2}\underline{2p^6}$). The second principal shell is the valence shell of the sodium ion, and it is filled with an octet of electrons.

4. Using Lewis symbols for each element, write the equations to show the transfer of electrons and the formation of ions for the following ionic compounds: (a) Na_2O and (b) $AlCl_3$.

Answer:

(a)

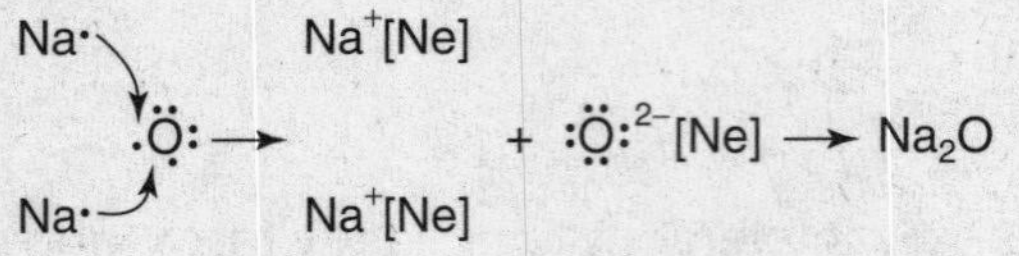

$$2\,Na\cdot + \cdot\ddot{O}: \rightarrow 2\,Na^+[Ne] + :\ddot{O}:^{2-}[Ne] \rightarrow Na_2O$$

(b)

$$\cdot\dot{Al}\cdot + 3\,\cdot\ddot{Cl}: \rightarrow Al^{3+}[Ne] + 3\,:\ddot{Cl}:^-[Ar] \rightarrow AlCl_3$$

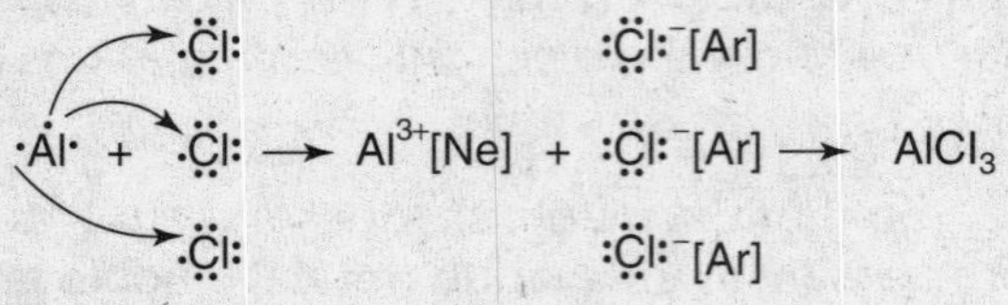

Work Problems

Use these problems for additional practice.

1. Which pairs of elements would react to form ionic compounds?

   (a) barium — iron; (b) lithium — oxygen; (c) sulfur — bromine; (d) iron — fluorine

2. Using complete electronic configurations for calcium and oxygen, show the transfer of electrons and the configurations of the resulting ions that lead to the formation of calcium oxide, CaO.

3. What would be the formula of the ionic compound formed in the reaction of sodium with nitrogen? The Lewis symbols for sodium and nitrogen are

   $Na\cdot \quad \cdot\dot{\underset{\cdot}{N}}\cdot$

   Use these symbols to write out the equation to show the formation of the ions.

Worked Solutions

1. **If two elements are to form an ionic compound, one must be a metal and the other a nonmetal. Only (b) and (d) meet that requirement.**

2. **Ca ($1s^22s^22p^63s^23p^64s^2$) + O ($1s^22s^22p^4$) → Ca^{2+} ($1s^22s^22p^63s^23p^6$), O^{2-} ($1s^22s^22p^6$) → CaO**

   Calcium has lost both its 4s electrons and oxygen gained them in its 2p-subshell.

3. **Na_3N, sodium nitride.**

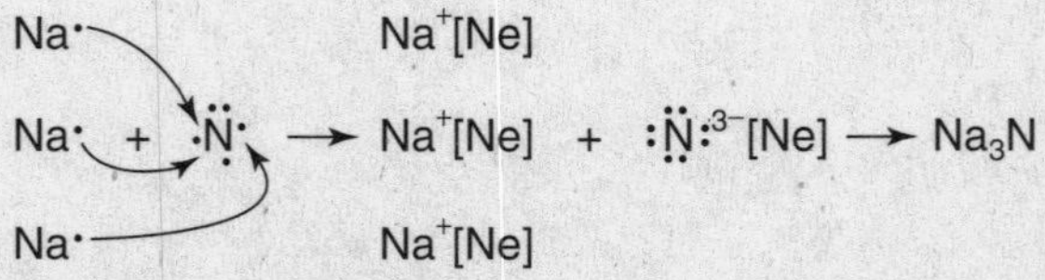

$$3\,Na\cdot + \cdot\dot{\underset{\cdot}{N}}\cdot \rightarrow 3\,Na^+[Ne] + :\ddot{\underset{\cdot\cdot}{N}}:^{3-}[Ne] \rightarrow Na_3N$$

Covalent Bonding—Covalent Compounds

In the previous section, the ionic bond and the reaction of metals with nonmetals to form ionic compounds were discussed. In this section, we examine the nature of the bond between atoms of two nonmetals, those elements to the right of the stair step on the periodic table. You will recall that in reactions between metals and nonmetals, metals lose electrons to form cations and nonmetals acquire them to from anions. But what if both elements are nonmetals? Nonmetals, like O, Br, or N, do not lose electrons easily; if anything, they prefer to acquire them. As a result, when two nonmetals bond to one another, they do so by *sharing* electrons forming bonds described as covalent. A **covalent bond** is a pair of electrons shared by two atoms. Compounds that are held together by covalent bonds are called **covalent compounds,** and they exist as individual molecules. Covalent bonding provides a second way for atoms to acquire an octet of electrons in their valence shells. The pair of electrons shared by two atoms becomes part of the valence shell of both atoms.

The simplest example of a covalent species is the hydrogen molecule, H_2. Each hydrogen atom contributes its one electron to form the pair of electrons shared by the two atoms. Since neither atom has a greater tendency to gain or lose electrons than the other (they are identical in all respects), each can acquire the stable noble gas configuration of helium by sharing its single electron with the other. The shared pair of electrons bonds the two atoms together. In the following figure, the formation of the hydrogen molecule is shown two ways. The first uses Lewis symbols showing the creation of the two-electron covalent bond. The second shows how the 1s-orbitals on both hydrogen atoms come together and *overlap* or merge to form a larger orbital containing both electrons and the two nuclei that share them. The spins of the shared electrons are paired (spinning in opposite directions). In the quantum mechanical description of covalent bonding, bonds are formed by the overlap of orbitals that allow electrons to be shared. Although bonds form through the overlap of orbitals, it is often more convenient to use Lewis symbols to construct pictures of molecules to show how all the valence electrons are involved in the covalent bonds that hold it together. Lewis symbols make the bookkeeping easier.

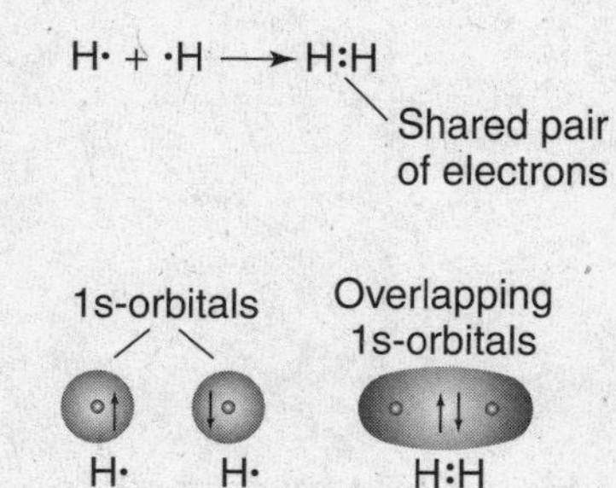

Two hydrogen atoms bonded together in a molecule represent a more stable arrangement (a lower energy arrangement) than two separate atoms. A detailed explanation of why this is so gets complicated, but it comes down to this: In the hydrogen molecule, each negative electron is attracted to *two* positive nuclei, and each positive nucleus is attracted to *two* electrons. This is a total of *four* forces of attraction in the molecule. In contrast, two separate hydrogen atoms experience a total of *two* attractive forces, one in each atom. The attractive forces *double* when the molecule is formed, as shown in the following figure. Covalent bonding is the force of attraction of positive for negative and negative for positive, not unlike the attractive forces that define ionic bonds; but remember, in covalent bonds, the attractions arise because electrons are shared, not transferred from one atom to another.

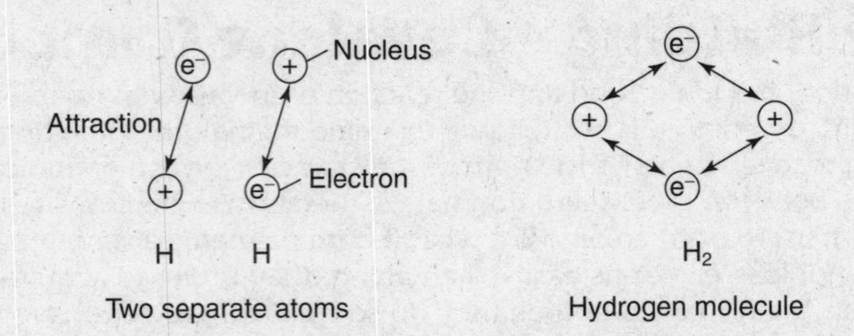

Now let's look at some other elements to see how they form covalent molecules. When two fluorine atoms combine to form the fluorine molecule, each atom shares one valence electron with the other, forming a shared pair of electrons. This is possible because each fluorine atom (Group VIIA) has seven valence electrons and one "vacancy" in its valence shell. Filling this vacancy gives fluorine the stable noble gas configuration of neon. *Two* fluorine atoms, each with one vacancy and one unpaired electron, can both achieve an octet by using their vacancy to accommodate the unpaired electron of the other, forming a shared pair of bonding electrons in the process. By sharing, each atom achieves a stable noble gas configuration. A single shared pair of electrons is termed a **single bond.** For clarity, a shared pair of electrons is usually represented in formulas with a dash (–) drawn between two symbols.

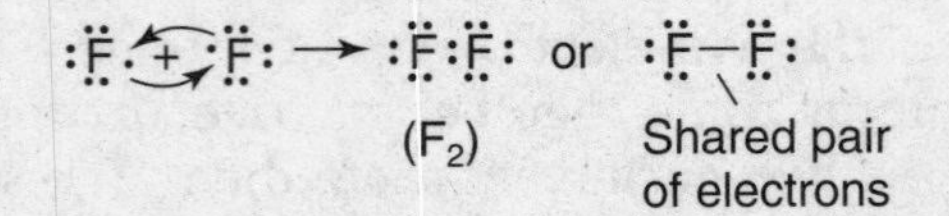

The way two fluorine atoms share two electrons is through the overlap of two p-orbitals, one from each atom, not unlike the way hydrogen atoms bond by overlapping their s-orbitals. The electronic configuration of each fluorine atom shows that one p-orbital in the valance shell holds only a single electron. This orbital has room for one more electron (one vacancy). Overlapping two singly occupied p-orbitals forms the larger bonding orbital that binds the two atoms.

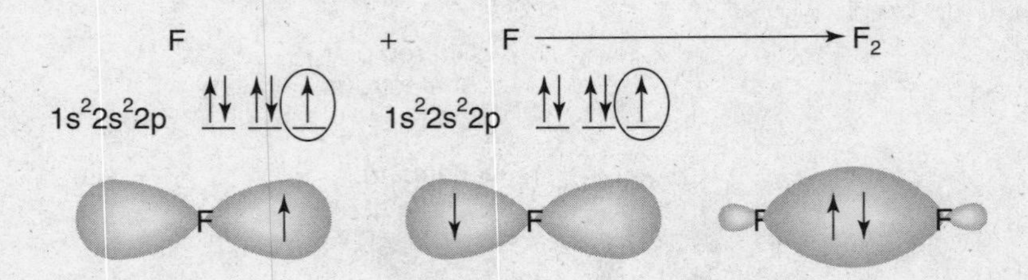

The elements chlorine, bromine, and iodine also exist as diatomic molecules with the two atoms bonded together with a single covalent bond.

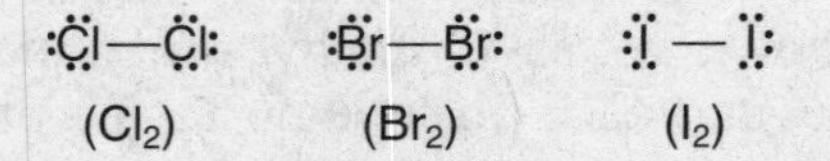

Oxygen and nitrogen are both found in nature as diatomic molecules, O_2 and N_2. The atoms in these molecules can achieve an octet of electrons in their valence shells by sharing electrons. Oxygen has six valence electrons (Group VIA) and *two* vacancies in its valence shell. (Notice how the number of valence electrons plus the number of vacancies always equal eight.) Filling those vacancies completes an octet. This is accomplished by sharing four electrons (two pair) with a second oxygen atom, as shown in the following figure. Each oxygen atom donates two electrons to the bond.

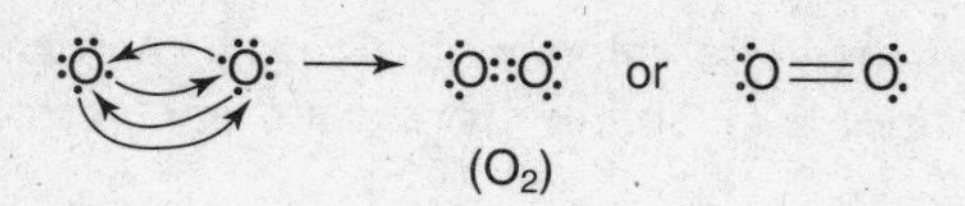

Two pairs of shared electrons represents a **double bond** between the two atoms.

Nitrogen (Group VA) has five valence electrons and *three* vacancies in its valence shell. Each nitrogen must share three pairs of electrons with a second nitrogen to achieve a noble gas configuration. The three shared pairs constitute a **triple bond** between the two atoms. Notice how three unpaired electrons on one nitrogen atom are connected with three vacancies on the other, and vice versa.

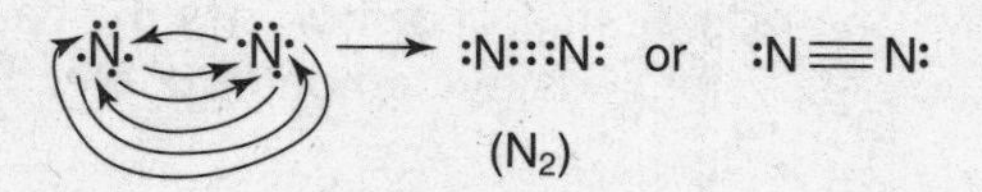

In the preceding examples, you have seen how two, four, and six electrons can be shared between two atoms in covalent bonding to fill each valence shell with an octet of electrons. It may be easier to see the octet about each atom by using overlapping circles to encompass the shared and unshared electrons about each atom. Shared electrons are counted in the octets of *both* atoms that share them.

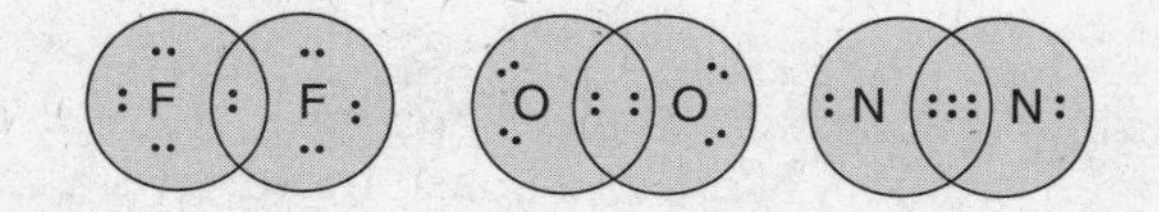

Triple bonds are stronger than double bonds, and double bonds are stronger than single bonds. The greater the number of electrons shared by two atoms, the stronger the bond between them. The strength of a bond is stated as the energy needed to break one mole of that bond. A comparison of the strengths of single, double, and triple bonds between two carbon atoms shows the increasing strength with increasing number of shared pairs of electrons.

C–C (347 kJ/moles) C=C (620 kJ/moles) C≡C (812 kJ/moles)

By far, most known compounds are covalent compounds. Let's look at some familiar ones. The formula of water is H_2O, and that of ammonia is NH_3, a result of the number of electrons (and vacancies) in the valence shell of oxygen and nitrogen. Oxygen has six valence electrons and *two* vacancies in its valence shell. *Two* of those six electrons are unpaired and available for bonding. Nitrogen has five valence electrons and *three* vacancies in its valence shell. *Three* of the five electrons are unpaired electrons and can engage in bonding. Each "vacancy-unpaired electron" combination in the valence shell can be used to form a two-electron bond. Oxygen has the capacity to bond to two hydrogen atoms, and nitrogen has the capacity to bond to three.

H· ·Ö· ·H → H:Ö:H or H–Ö–H
(H_2O)

H· ·N̈· ·H (·H) → H:N̈:H (H) or H–N̈–H (H)
(NH_3)

It should not be surprising to learn that sulfur and selenium, elements below oxygen in Group VIA, also have the capacity to bond to two hydrogen atoms, forming H_2S and H_2Se. Phosphorus, beneath nitrogen in Group VA, like nitrogen, can combine with three hydrogen atoms to form PH_3.

Carbon has four unpaired electrons in its valence shell and four vacancies. It can bond to four hydrogen atoms to form methane, CH_4, sharing four pairs of electrons in four single bonds.

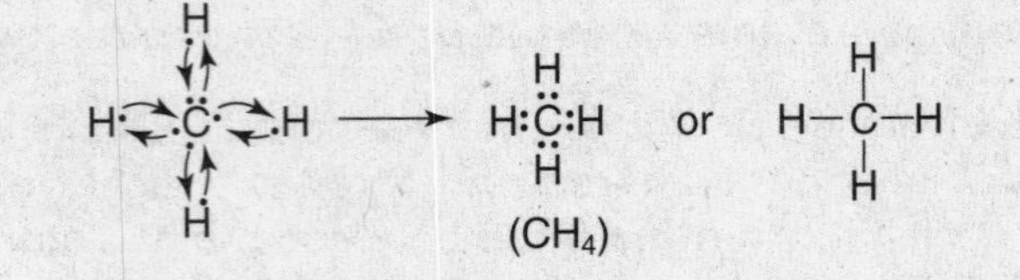

Carbon can also acquire an octet of electrons, and so can oxygen, in carbon dioxide, CO_2. Carbon is bonded to each oxygen by sharing two pairs of electrons in a double bond. Each oxygen has the capacity to form two bonds (two shared pairs), and carbon has the capacity to form four bonds (four shared pairs). Carbon and oxygen both achieve an octet of electrons by forming two double bonds.

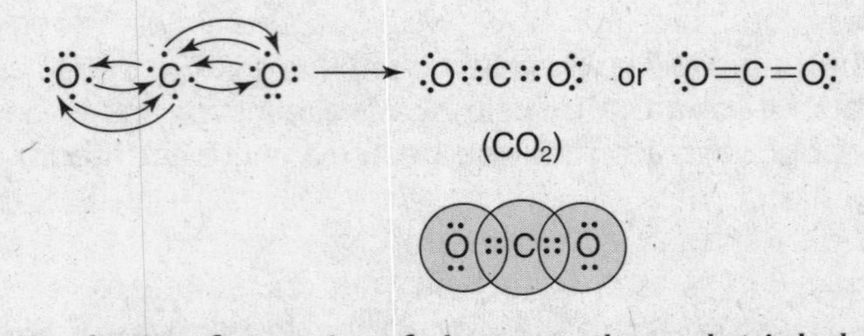

Carbon can also engage in sharing four pairs of electrons through triple bonding, as seen in acetylene, C_2H_2, and hydrogen cyanide, HCN. Each dash represents a two-electron bond, a shared pair of electrons.

H–C≡C–H H–C≡N:

If you go back and study each of the molecules discussed so far, you will note that each element forms a definite number of two-electron covalent bonds. Hydrogen forms one bond; oxygen forms two bonds; nitrogen forms three bonds; and carbon forms four. The two bonds formed by oxygen could be two single bonds or one double bond. In the following table, the *bonding patterns* used by four important elements are summarized. Each bonding pattern gives the element a filled valence shell of a noble gas.

Common Bonding Patterns of Four Elements

Element	*Number of Two-Electron Bonds*	*Bonding Pattern (example)*
H	1	1 single bond (H_2)
O	2	2 single bonds (H_2O)
		1 double bond (O=O)
N	3	3 single bonds (NH_3)
		1 single bond, 1 double bond (HO–N=O)
		1 triple bond (N≡N)
C	4	4 single bonds (CH_4)
		2 single bonds, 1 double bond (H_2C=O)
		2 double bonds (O=C=O)
		1 single bond, 1 triple bond (H–C≡N)

In the covalent bonds studied up to this point, half of the shared electrons came from each of the atoms joined by the bond. But in some cases *both* of the shared electrons come from only *one* of the two atoms. Covalent bonds in which both of the shared electrons are provided by only one atom are called **coordinate covalent bonds.** The only difference between an ordinary covalent bond and a coordinate covalent bond is the way they are formed. They are both covalent (electron sharing) bonds. An example of coordinate covalent bonding is the formation of the ammonium ion, NH_4^+, from an ammonia molecule and a hydrogen ion. The pair of nonbonding electrons on nitrogen is used to form the covalent bond with the hydrogen ion. The hydrogen ion can readily accept a pair of electrons, because the 1s-orbital is completely empty in the hydrogen ion. After the bond is formed, it is the same as the other nitrogen-hydrogen bonds.

```
      H                       H      +
      |                       |
  H—N:    +  H+   ——>     [ H—N—H ]
      |                       |
      H                       H
```

The structures of covalent molecules are usually drawn out using a dash (–) for a shared pair (bonding pair) of electrons and two dots for an unshared pair (nonbonding pair) of electrons. In the molecule that follows are eight nonbonding pairs of electrons (16 electrons) and four bonding pairs (8 electrons). The total number of bonding and nonbonding electrons equal the total number of valence electrons in the atoms that make up the molecule, 24 electrons. Count them to see whether this is so. Structures of molecules or polyatomic ions that show all the valence electrons as bonding or nonbonding pairs of electrons are called Lewis structures. You will learn more about them later in this chapter.

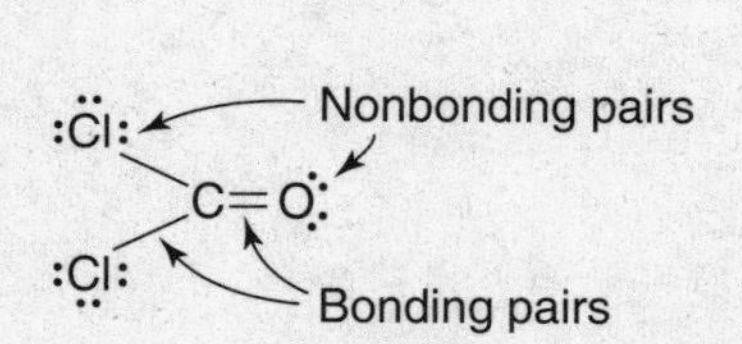

Guidelines for the formation of covalent compounds:

- Covalent compounds are composed of nonmetals. Polyatomic anions, sulfate, nitrate, and so on, are covalent species that bear a charge: covalent ions in ionic compounds. (NH_4Cl is also considered ionic, NH_4^+, Cl^-, although the ammonium ion is covalent.)
- The atoms in covalent molecules and polyatomic ions are held together by shared pairs of electrons in covalent bonds. These can be single bonds (one shared pair), double bonds (two shared pairs), or triple bonds (three shared pairs).
- Atoms will share the necessary number of electrons needed to acquire an octet of electrons in their valence shells. Hydrogen only needs two electrons.

Example Problems

These problems have both answers and solutions given.

1. (a) What is a covalent bond? (b) How is it different from an ionic bond? (c) What do the two bonds have in common?

   Answer: (a) A covalent bond is the attraction between two atoms due to the sharing of one or more pairs of electrons between them. (b) Covalent bonds involve sharing electron pairs; ionic bonds are the attractions between ions of opposite charge. (c) Both covalent and ionic bonds ultimately come down to the attraction between positive and negative charges, negative electrons for positive nuclei or negative ions for positive ions.

2. How many bonding pairs of electrons are in one molecule of acrylic acid that is used in plastics and adhesives? How many nonbonding pairs? Is the total number of electrons shown in the bonding and nonbonding pairs the same as the total number of valence electrons in the atoms?

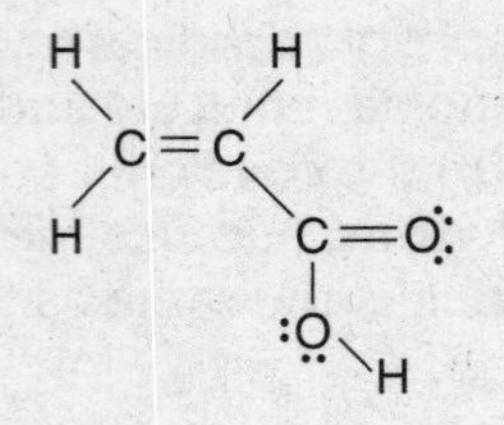

Answer: Each dash represents a shared pair of bonding electrons. There are ten bonding pairs and four nonbonding pairs, a total of 28 electrons. The total number of valence electrons is also 28: 3-C = 12, 2-O = 12 electrons, and 4-H = 4 electrons.

3. What is the formula of the covalent compound formed between one nitrogen atom and as much fluorine as needed to form the compound?

Answer: NF_3 Fluorine can form only one single bond; nitrogen can form three.

4. Draw circles about each atom in sulfur dioxide, SO_2, to show how each has acquired an octet of electrons.

$:\ddot{O}::\ddot{S}:\ddot{O}:$

Answer:

$:\ddot{O}::\ddot{S}:\ddot{O}:$

Work Problems

Use these problems for additional practice.

1. Which pairs of elements would link together with covalent bonds?

(a) H and F, (b) Li and O, (c) F and Cl, (d) P and O, (e) As and Cl, (f) Mg and I

2. What are the formulas of the covalent compounds involving: (a) one carbon atom and as much chlorine as needed, (b) one silicon atom and chlorine, and (c) one sulfur atom and chlorine?

3. How many bonding and how many nonbonding pairs of electrons are in one molecule of ethylene glycol, the compound used as antifreeze in automobile radiators? Does every atom in the molecule have a rare gas configuration?

```
      H  H
      |  |
H-Ö-C-C-Ö-H
      |  |
      H  H
```

Worked Solutions

1. **Nonmetals bond to nonmetals with covalent bonds.** Pairs of nonmetals are in (a), (c), (d), and (e).

2. **(a) CCl_4, carbon tetrachloride. (b) $SiCl_4$, silicon tetrachloride. Both carbon and silicon can form four single bonds; chlorine can form only one. (c) SCl_2, sulfur dichloride. Sulfur can form two single bonds.**

3. **Nine bonding pairs and four nonbonding pairs of electrons.** Every atom has acquired the necessary number of electrons to have the configuration of a rare gas.

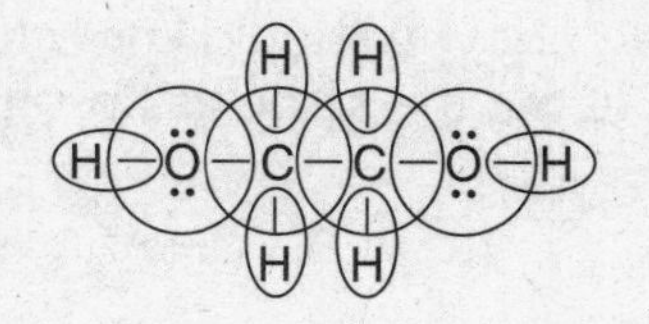

Polar and Nonpolar Covalent Bonds

Covalent bonds are pairs of electrons shared between two atoms, and you might wonder whether the electrons are shared equally by the atoms or unequally. The answer depends on the electron-giving and electron-attracting nature of the two atoms. Electrons shared by two identical atoms, as in H_2, O_2, or N_2, are shared equally. Neither atom has a greater tendency to attract the bonding pair to itself than the other; the atoms are the same in all respects. Covalent bonds in which electrons are shared equally are called **nonpolar covalent bonds.** If the two atoms sharing a pair of electrons are *not* alike, as in HF, the electron pair will *not* be shared equally, and the bond will be a **polar covalent bond.** Because the bond is between two *different* elements, one will inevitably have a greater attraction for the shared pair than the other.

In hydrogen fluoride, HF, fluorine draws the shared pair of electrons away from hydrogen and toward itself. The electrons are still shared, but shared unequally. This causes the entire HF molecule to act like an electric *dipole,* a body with two poles—one weakly negative and the other weakly positive. The dipole is the result of unequal electron sharing in the covalent bond. Because the negative bonding electrons are, on average, closer to fluorine and partially withdrawn from hydrogen, the fluorine end of the molecule is somewhat richer in negative charge, indicated with $\delta-$, at the expense of the hydrogen end, which is poorer in negative charge and left relatively positive, $\delta+$. To show polarity in a bond, the $\delta-$ is written above the atom that draws electrons to itself and $\delta+$ is written above the atom that has electrons drawn from it. The formulas of HF, HCl, and HBr are given here, showing the polarity in the bonds.

$$\overset{\delta^+}{H}-\overset{\delta^-}{F} \qquad \overset{\delta^+}{H}-\overset{\delta^-}{Cl} \qquad \overset{\delta^+}{H}-\overset{\delta^-}{Br}$$

In general, any covalent bond between two unalike atoms will be a polar covalent bond. The effect that polar bonds have on properties of compounds will be discussed in Chapter 12. They are not insignificant. The ability of an atom to attract electrons to itself is a measure of its electronegativity (elec-trow-neg-a-tiv-i-tee). Electronegativity is discussed in the next section.

Electronegativity of Atoms

The **electronegativity** of an element is a measure of the ability of that element to attract a shared pair of electrons to itself. The more strongly an element attracts electrons to itself, the greater its electronegativity. The electronegativities of elements are stated as numbers. The most electronegative element, fluorine, has the highest number, 4.0. Cesium, Cs, one of the least electronegative elements, is assigned a value of 0.7. All other elements have electronegativities between these two extremes. Hydrogen is 2.1. In general, nonmetals have higher electronegativities; metals have lower electronegativities, consistent with the fact that nonmetals tend to gain electrons and metals tend to lose them. In the following figure, the numbers beneath the symbols for several elements are their electronegativities. Notice the two important trends: Electronegativity increases left to right across a period and decreases top to bottom down a group. You need not memorize these values, but you should know the trends, the extremes, and that the electronegativity of hydrogen is about the same as boron, H (2.1), B (2.0).

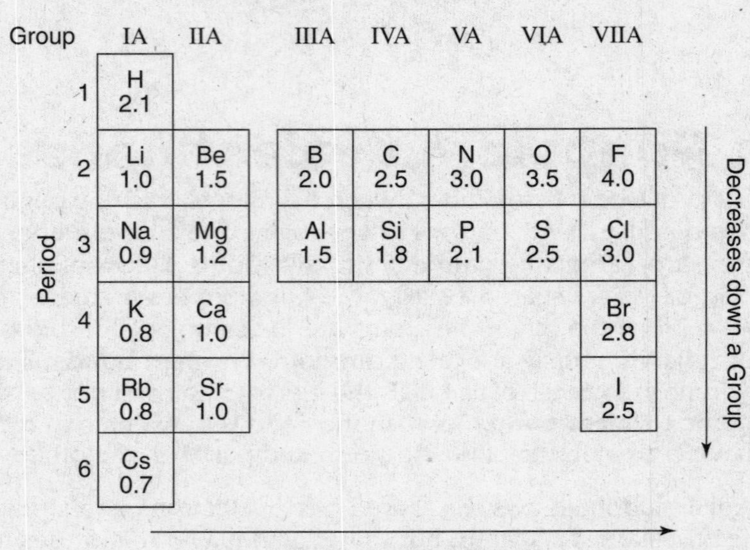

The electronegativities of several main-group elements.
Electronegativities increase across a period and decrease down a group.

The greater the difference in the electronegativities of two bonded atoms, the more polar will be the bond between them. The hydrogen-fluorine bond in HF is very polar. The difference between the electronegativities of H and F is 1.9 (H = 2.1, F = 4.0). The covalent bond in HCl is less polar than that in HF, since the electronegativity difference between H and Cl is only 0.9 (H = 2.1, Cl = 3.0). The covalent bond between two identical atoms is nonpolar because both atoms have the same electronegativity. The difference is zero. Generally, if the difference in electronegativity between two elements is greater than about 2.1, the bonding between them in a compound is likely to be ionic. In a sense, the ionic bond is the extreme case of a polar covalent bond. The following figure relates the polarity of a bond to the difference in electronegativity between the bonded atoms.

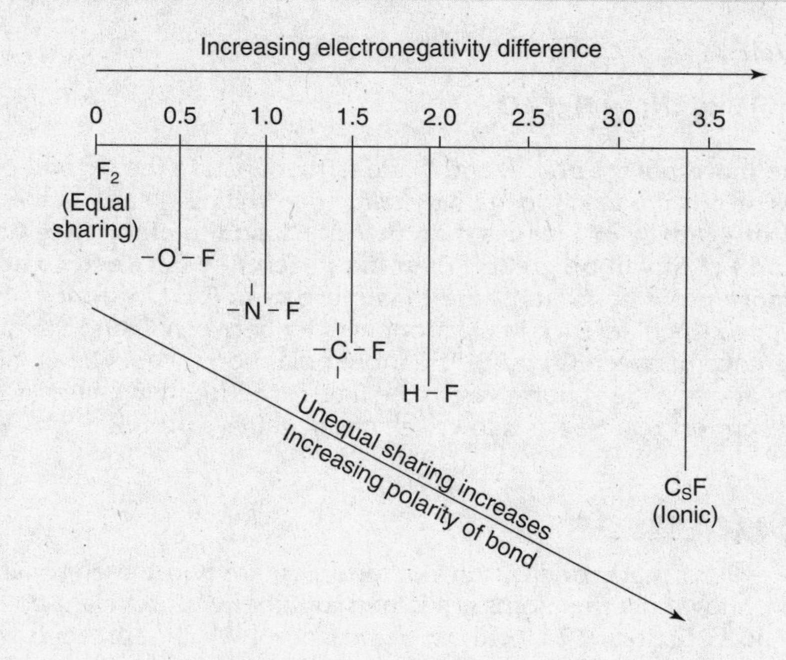

Example Problems

These problems have both answers and solutions given.

1. What is measured by the electronegativity of an element? Which element is the most electronegative?

   Answer: The electronegativity of an element is a measure of the ability of that element to attract a shared pair of electrons to itself. The most electronegative element is fluorine with a value of 4.0.

2. Using the trends in the periodic table, arrange the following elements in order of *increasing* electronegativity: (a) S, O, Na; (b) Cl, I, Br; (c) P, Al, S.

   Answer: Electronegativity increases across a period and decreases down a group. (a) Na<S<O; (b) I<Br<Cl; (c) Al<P<S

3. Which bond in each pair would be more polar, based on electronegativity difference? (a) H–F or H–Br; (b) N–H or O–H; (c) S–Cl or P–Cl.

   Answer: (a) H–F has a greater electronegativity difference (1.9) than HBr (0.7). (b) O–H is more polar. It has a difference of 1.4, greater than that of 0.9 for N–O. (c) P–Cl is more polar with an electronegativity difference of 0.9 compared to S–Cl, with a difference of 0.5.

Work Problems

Use these problems for additional practice.

1. Using the trends in the periodic table, arrange the following elements in order of *increasing* electronegativity: (a) Si, K, O; (b) N, B, C; (c) H, C, O.

2. For each pair of bonds, indicate the one that would be more polar based on difference in electronegativity: (a) B–O or N–O; (b) F–Cl or F–I; (c) O–S or O–Se.

Worked Solutions

1. **(a) $K<Si<O$; (b) $B<C<N$; (c) $H<C<O$**

2. **(a) B–O is the more polar bond.** B and O are farther apart in the second period than are N and O. This means the electronegativity *difference* will be greater in the case of B–O. (b) **The electronegativity of I is less than that of Cl, so the electronegativity difference between F and I (1.5) will be greater than the difference between F and Cl (1.0). F–I will be the more polar bond.** (c) **Same reasoning as in (b).** Electronegativity decreases down a group, so the difference in electronegativity between O and Se must be greater than the difference between O and S. The more polar bond would be O–Se. Even though the electronegativity of Se is not given in the figure, the trend in values going down a group would suggest that Se has a lower electronegativity than S.

Lewis Structures

Lewis structures (Lewis formulas or electron-dot formulas) are two-dimensional pictures of covalent species that show how the atoms are joined together with covalent bonds. A bond is shown as a pair of "dots" (2 dots = 2 electrons) or a dash (–), which represents a bonding or shared pair of electrons. A single dash represents two shared electrons; two dashes (=) represent four shared electrons; and so forth. In addition, Lewis structures also show the location of electron pairs *not* used in bonds, the nonbonding or *un*shared pairs of electrons. In a correct Lewis structure, all the valence electrons from every atom in the molecule or polyatomic ion must be accounted for, either in bonds or as nonbonding pairs (nb-pairs).

The Lewis structures of water and carbon dioxide are shown in the following illustration. In the molecule of water is one bonding pair (a single bond) of electrons between each hydrogen atom and the central oxygen atom, and two nonbonding or unshared pairs of electrons on the oxygen atom. In carbon dioxide are two shared pairs of electrons (a double bond—four electrons total) between each oxygen and the central carbon atom. In addition, two nonbonding pairs of electrons are located on each oxygen atom. There are eight valence electrons in the three atoms that make up the water molecule, and eight electrons are accounted for in the Lewis structure (2 single bonds = 4 e^-'s, 2 nb-pairs = 4 e^-'s). You should verify that the total of 16 valence electrons in the three atoms of carbon dioxide are all accounted for in its Lewis structure.

H – Ö – H   :Ö=C=Ö:

Lewis structures can tell a lot about the bonding in covalent species, and they can be derived for almost all small molecules and polyatomic ions by following a prescribed sequence of operations. Let's go through this sequence and derive the Lewis structures for both the nitrate ion, NO_3^-, and the formaldehyde molecule, H_2CO, at the same time.

1. Draw the **skeleton structure** of the ion or molecule, joining the bonded atoms with a single dash representing a shared pair of bonding electrons.

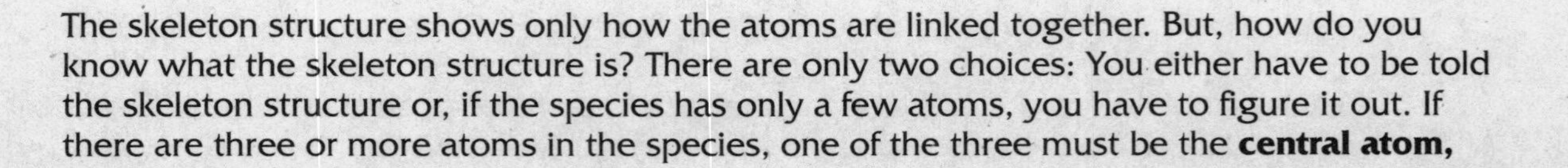

The skeleton structure shows only how the atoms are linked together. But, how do you know what the skeleton structure is? There are only two choices: You either have to be told the skeleton structure or, if the species has only a few atoms, you have to figure it out. If there are three or more atoms in the species, one of the three must be the **central atom,**

the atom to which all other atoms are joined. Oxygen is the central atom in H_2O, and carbon is the central atom in CO_2. After you identify the central atom, you can draw the skeleton structure. Here are a few guidelines for choosing a central atom:

(a) Neither hydrogen nor fluorine is ever a central atom.

(b) The central atom usually appears only once in the formula. The central atom in each of the following is underlined: $\underline{P}F_3$, $\underline{S}O_4^{2-}$, $\underline{C}Cl_4$, $\underline{N}H_4^+$.

(c) If there are two atoms that could be central atoms, most often the element of *lower electronegativity* will be the central atom. Notice that carbon, not oxygen, is the central atom in formaldehyde. Carbon has the lower electronegativity (C = 2.5, O = 3.5).

2. Total the valence electrons on all the atoms in the species and

   if it is an anion, *add* one electron to the total for each negative charge.

   if it is a cation, *subtract* one electron from the total for each positive charge.

   Consult a periodic table if you need help with this. For the A-groups, the group numbers equal the number of valence electrons for the elements in that group. Don't try to keep track of which electrons came from what atom, it is the total number of electrons that is important.

For NO_3^-:
N = 5 valence e^-'s
O = 6 valence e^-'s
O = 6 valence e^-'s
O = 6 valence e^-'s
−1 charge = 1 e^-
Total = 24 e^-'s

For H_2CO:
C = 4 valence e^-'s
O = 6 valence e^-'s
H = 1 valence e^-
H = 1 valence e^-
Total = 12 e^-'s

3. From the total number of electrons obtained in step 2, subtract two electrons for each single bond drawn in the skeleton structure. Both skeleton structures have three bonds, so a total of six electrons is subtracted.

   For NO_3^-: (24 e^-'s) − (6 e^-'s) = 18 e^-'s

   For H_2CO: (12 e^-'s) − (6 e^-'s) = 6 e^-'s

4. Distribute the remaining electrons as *nonbonding electron pairs* (unshared pairs) about each atom bonded to the central atom until each has eight electrons (an octet of electrons) about it, except for hydrogen, which can accommodate only the two electrons that bond it to another atom. *Any extra electrons should go on the central atom.*

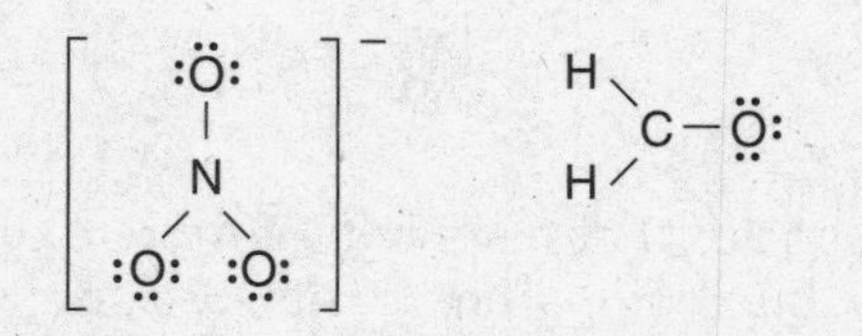

At this point, if the central atom does not have an octet of electrons, convert one or more nonbonding pairs to bonding pairs. This will change a single bond to a double bond or a triple bond. That pair of electrons, being a bonding pair, is now counted in the valence shell of both atoms.

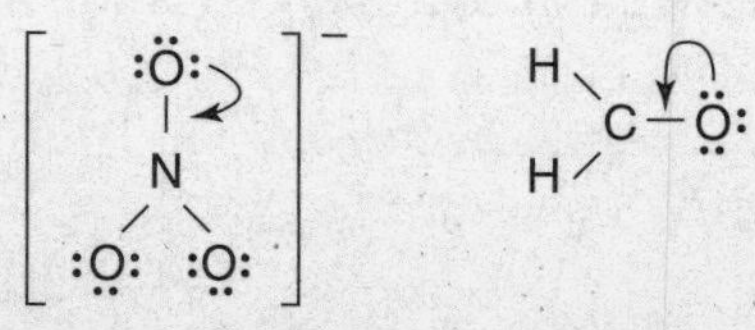

Bonding electrons are simultaneously counted for *both atoms* that share them. Satisfactory Lewis structures for the nitrate ion and formaldehyde are shown in the following illustration. Each nonhydrogen atom is associated with eight electrons and hydrogen with two. You should do an electron count to ensure that you agree with this. Lewis structures of ions are frequently placed in square brackets with the charge written outside the right bracket in the upper-right corner.

The nitrate ion displays an additional peculiarity that formaldehyde does not. Notice, in the case of the nitrate ion, how a nonbonding pair of electrons could have been taken from any one of the three oxygen atoms to form the double bond with nitrogen. The three Lewis structures for nitrate shown in the following illustration are energetically identical but differ only in the location of the double bond. Only electrons are moving in these pictures; the atoms do not move. The double-headed arrow indicates the ability of the three electronic arrangements to interchange with one another.

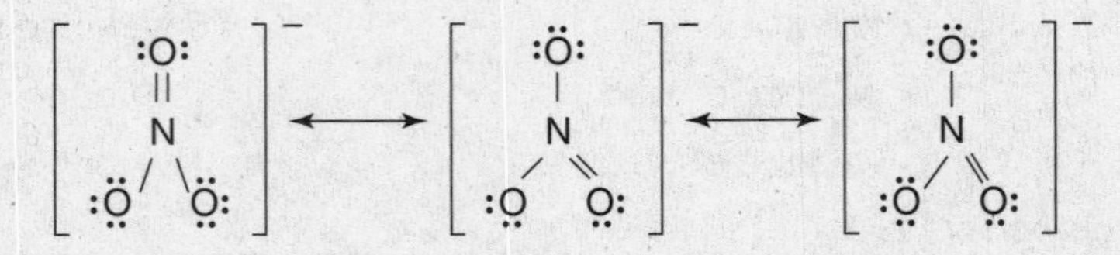

Whenever it is possible to draw two or more Lewis structures for a species that differ *only* in the location of a double bond between the same two kinds of atoms (N and O in this case), the true structure of the species is the *average* of the individual structures. In a sense, two of the four electrons in the double bond move from one N–O bond to the next so quickly that *each* bond is more than a single but not quite a double bond. That pair of electrons is not localized between two atoms; rather it is *delocalized* over the entire species. Delocalization gives rise to a condition known as **resonance,** which adds stability to the species. One way delocalization can be shown in a Lewis structure is to use dotted lines to represent the pair of delocalized electrons, as shown in the following figure. This figure is a composite of the three *resonance forms* of the nitrate ion.

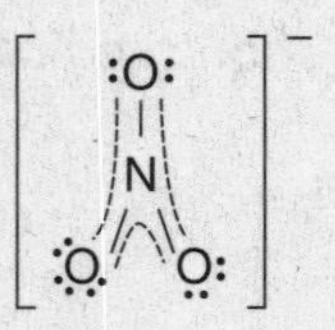

Always check for resonance in species that have two or more oxygen atoms attached to the central atom with single and double bonds. The possibility of delocalization does not exist in formaldehyde. Trioxygen, O_3, (also known as ozone), exhibits resonance. As its Lewis structure is developed, you come to a point where one nb-pair of electrons must be moved from one of the outside oxygen atoms to form a double bond with the central oxygen. Which oxygen do you choose? Both outside (terminal) oxygen atoms are identical. When you have two or more identical atoms from which to take a nonbonding pair, the Lewis structure you are developing *will display resonance.* The best description of trioxygen is the average of the two resonance forms with the delocalized pair of electrons spanning all three atoms. The bond between either terminal oxygen and the central oxygen is essentially a half single bond and half double bond.

One important goal when deriving Lewis structures is to associate each atom with an "octet" of electrons, the same number of electrons found in the valence shells of the noble gases. In reality, only a few elements *consistently* achieve an exact octet of electrons in covalent compounds, but those that do are the important elements found in the first and second periods of the periodic table, most notably H, C, N, O, and F. Elements in the third and higher periods have more empty orbitals (d-orbitals) in their valence shells and can expand their capacity to accommodate as many as 10, 12, or even 14 electrons. Elements like P, S, I, and several others can form compounds like PCl_5, SF_6, and IF_7. Yet, these same elements form many compounds and ions with an octet of electrons in their valence shells. Other elements, like boron (Group IIIA), have only three valence electrons, and when all are used to form bonds, as in BF_3, boron ends up with only six electrons in its valence shell.

So, does this mean the octet rule is useless? The answer is no, because it works as well for many main-group elements, especially the critically important elements C, N, and O, which are central atoms in thousands of compounds. As you learn more chemistry, you will learn when the octet rule is exceeded.

Example Problems

These problems have both answers and solutions given.

1. Identify the central atom in each of the following: (a) CH_4, (b) HCN, (c) ClO_2^-, (d) FNO_2.

   Answer: (a) C, (b) C (lower electronegativity than N), (c) Cl, (d) N (never F)

2. Total the valence electrons and construct the Lewis structures for: (a) NH_3 and (b) F_2CO.

   Answer:

   (a) NH_3: 8 valence e^-'s (b) F_2CO: 24 valence e^-'s; C is the central atom.

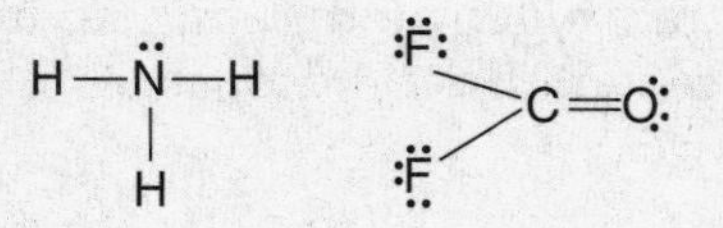

3. Total the valence electrons and construct the Lewis structures for: (a) H_3O^+ and (b) CN^-.

   Answer:

   (a) H_3O^+: 8 valence e^-'s (b) CN^-: 10 valence e^-'s

$[H—\ddot{O}—H \; (| \; H)]^+$

$[:C\equiv N:]^-$

4. Draw the Lewis structure of nitric acid, HNO_3, (H–O–NO_2). Does this molecule display resonance? The hydrogen is bonded to an oxygen atom, not nitrogen.

Answer: There are two ways to distribute the 24 valence electrons giving two Lewis structures that differ only in the location of the double bond. This predicts resonance.

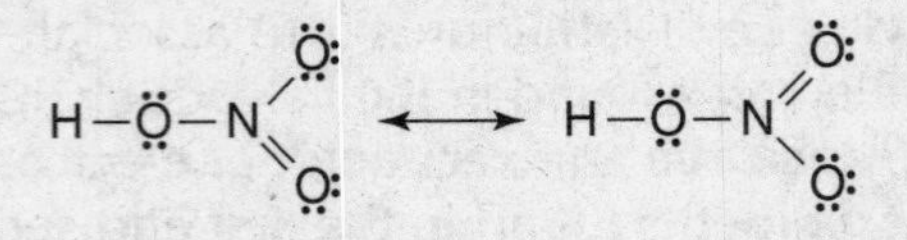

Work Problems

Use these problems for additional practice.

1. First, total the valence electrons and then construct the Lewis structures for: (a) PCl_3 and (b) F_2O.

2. First, total the valence electrons and then construct the Lewis structures for: (a) IBr_2^+ and (b) NO^-.

3. Which of these two will display resonance: FNO or FNO_2?

Worked Solutions

1. **(a) PCl_3: 26 valence electrons** **(b) F_2O: 20 valence electrons**

:Cl–P–Cl:
 |
:Cl:

:F–O–F:

2. **(a) IBr_2^+: 20 valence electrons** **(b) NO^-: 12 valence electrons**

$[:Br–I–Br:]^+$

$[N=O]^-$

3. There needs to be two or more oxygen atoms attached to the central atom (N) for resonance to be possible. Clearly, FNO cannot display resonance, but FNO_2 can, as shown in their Lewis structures.

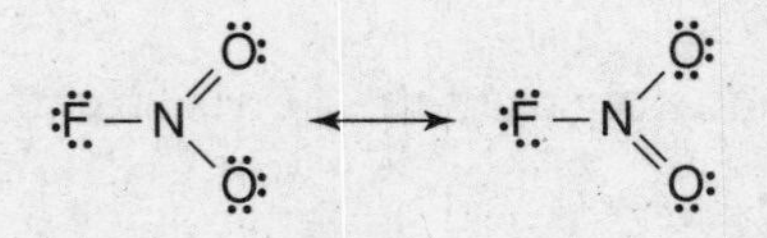

The Shapes of Molecules

Lewis structures describe, in a general way, the distribution of valence electrons among bonding pairs that link atoms together and nonbonding pairs that do not. Although Lewis structures are very useful, they do not show the three-dimensional shape of a molecule. A correct Lewis structure for water does not tell you whether the three atoms are arranged in a straight line, H–O–H, forming a "linear" molecule, or whether they form a bent, or V-shaped, molecule. You can draw it as bent or straight, but that doesn't mean it is. Covalent bonds are directional, and they extend from a central atom in definite directions with definite angles between them. As it turns out, the two bonds between oxygen and hydrogen in the water molecule do *not* point in opposite directions but are oriented about oxygen to form a bent molecule. You will learn why in the next section.

Two different kinds of measurements are used to describe the three-dimensional shapes of molecules: bond lengths and bond angles. A **bond length** is the average distance between the nuclei of two atoms that are bonded together. Bond lengths are often stated in picometers, pm (1 pm = 10^{-12} m). The **bond angle** is the angle of arc ∡ between any two bonds that have one atom in common. The H–O–H bond angle in water, ∡HOH, is 104.5°, and the H–O bond length, d_{H-O}, is 96 pm. Both H–O bond lengths are the same.

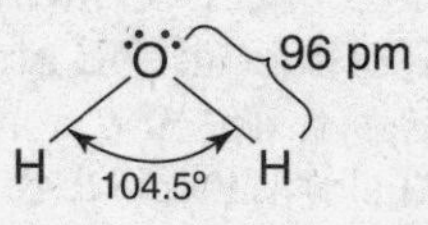

The molecular shape of the nitrate ion, NO_3^-, is shown here with its bond lengths and bond angles.

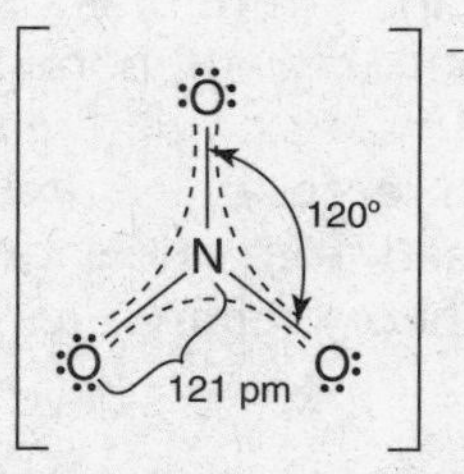

Both the water molecule and the nitrate ion are *planar;* all the atoms lie in the same plane. You could imagine these species lying flat on a tabletop, but most molecules are not planar. The shape of the methane molecule, CH_4, is *tetrahedral*. Connecting the four hydrogen atoms forms a pyramid with four identical sides, thus the name *tetra*hedral. All H–C–H bond angles are the same, 109.5°, the tetrahedral angle, and all C–H bond lengths are the same.

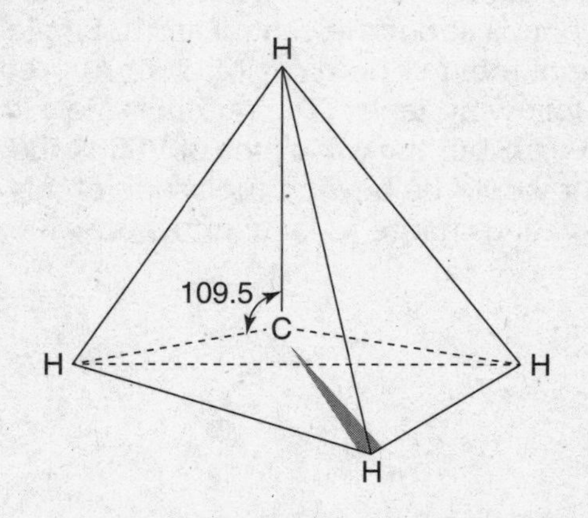

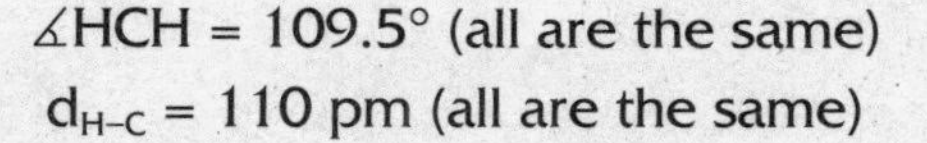

Of the two, bond angles and bond lengths, it is the bond angle that better indicates the three-dimensional shape of a species. Three of the most important bond angles are shown in the following three models. Note the name of each shape and the associated bond angle.

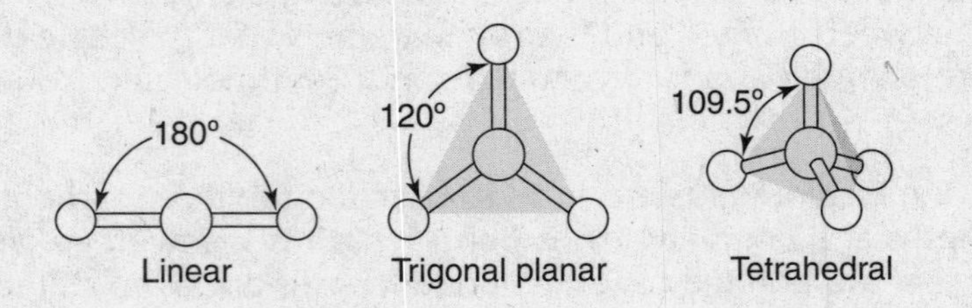

VSEPR Theory—Predicting Molecular Shape

The three-dimensional shapes of molecules and polyatomic ions are the result of the orientation of atoms about the *central atom*. Note that each bond angle in the three models shown here include the central atom. In the final analysis, it is the central atom that directs the shape of the species. **V**alence **S**hell **E**lectron **P**air **R**epulsion (**VSEPR**) theory focuses on the bonding and non-bonding electrons in the valence shell of the central atom and the role they play in determining molecular shape. The bonding and nonbonding pairs of electrons experience a natural electrostatic repulsion that pushes them as far apart from one another as possible. It is this repulsion between *domains* of negative charge (bonding and nonbonding electrons) that causes the molecule or ion to have a particular shape. It makes no difference if the electron pairs are bonding electrons or nonbonding electrons; each represents a region or domain of negative charge. Orienting the electron domains as far apart from one another as possible increases the stability of the molecule. You often hear VSEPR theory referred to "vesper theory," which is a bit easier to pronounce.

Here is how the VSEPR method is applied to predict the shape of a molecule or polyatomic ion:

1. Derive the Lewis structure of the species using the rules given previously. If the species displays resonance, choose one resonance form.
2. Identify the central atom. The central atom is the one to which all other atoms are joined.
3. Determine the number of electron domains about the central atom by totaling the number of nonbonding pairs and bonds about the central atom. Single, double, and triple bonds all count as *one* domain. The number of domains (2, 3, or 4) predicts a specific shape of the species, as shown in the following table. (The number of electron domains about a central atom can go as high as 6 or 7, but the principles of VSEPR theory can be demonstrated here using only 3. To go further would be beyond the scope of this book). Note that each shape places the domains of negative charge as far apart as possible.

Three Arrangements of Electron Domains About the Central Atom			
Number of Electron Domains	***Arrangement About Central Atom***	***Geometry of Domain Arrangement***	***Expected Bond Angle***
2	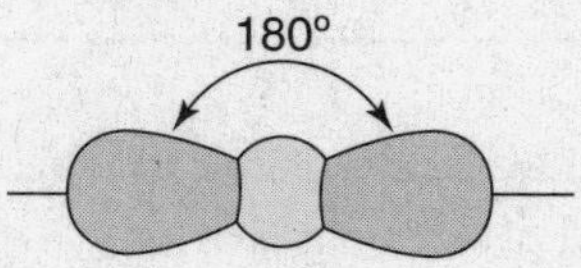 180°	linear	180°
3	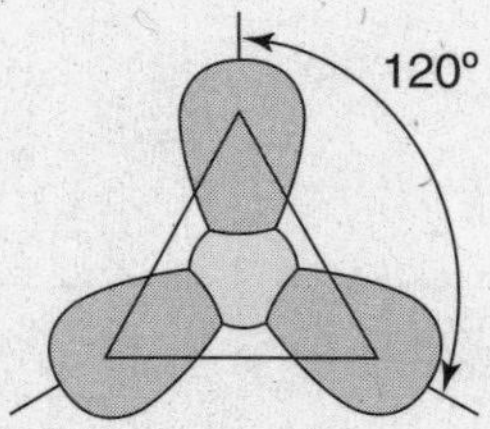 120°	trigonal planar	120°
4	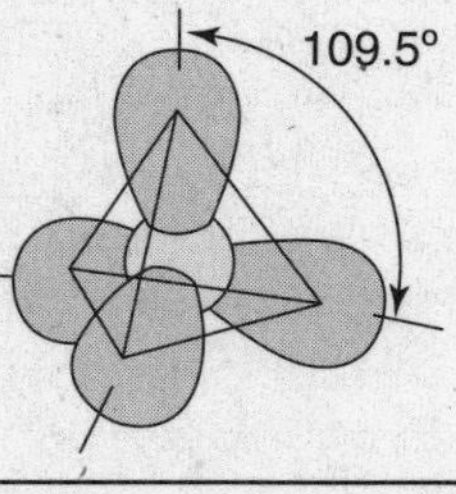 109.5°	tetrahedral	109.5°

Frequently, the three-dimensional shape of the molecule, the arrangement of just the atoms, is not described with the same label used for the arrangement of the electron domains about the central atom. Although nonbonding pairs of electrons occupy definite places, *it is the arrangement of the atoms that describes the shape of the molecule*. For example, the four electron domains around the oxygen atom in water would be oriented in a tetrahedral geometry, but considering *only* the three atoms, the shape of the molecule is bent (not tetrahedral). The HOH bond angle in water is predicted to be the tetrahedral angle, 109.5°. In reality it is 104.5°, just a little less than that predicted. Although it is not the perfect tetrahedral angle, 104.5° is a lot closer to tetrahedral than it is to the 120° or 180° angles of the trigonal planar and linear structures. In the following table, labels used to describe the geometry of the electron domains about the central atom are compared with the labels used to describe the arrangement of atoms in the molecule.

Comparing the Geometry of Electron Domains with Molecular Shapes

Electron Domains	*Bonding + Nonbonding Domains*	*Geometry of Domains*	*Molecular Shape*	*Example*
2	2-bonds	linear	linear	CO_2 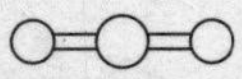 O=C=O
3	3-bonds	trigonal planar	trigonal planar	H_2CO H₂C=O
	2-bonds + 1-nonbonding pair	trigonal planar	bent	SO_2 
4	4-bonds	tetrahedral	tetrahedral	CH_4 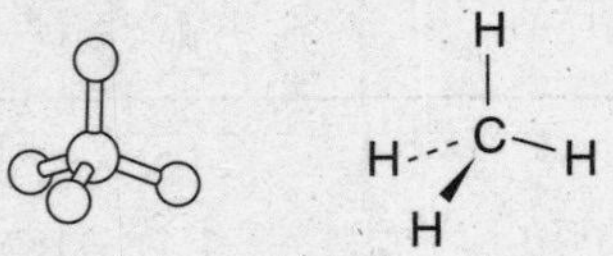
	3-bonds + 1-nonbonding pair	tetrahedral	trigonal pyramidal	NH_3 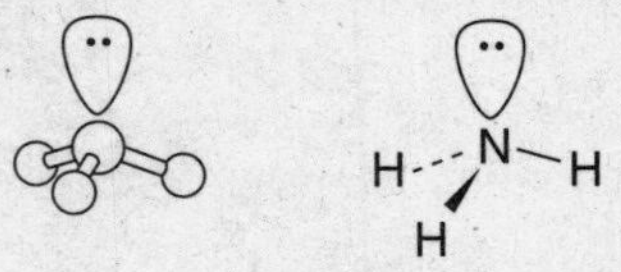
	2-bonds + 2-nonbonding pairs	tetrahedral	bent	H_2O 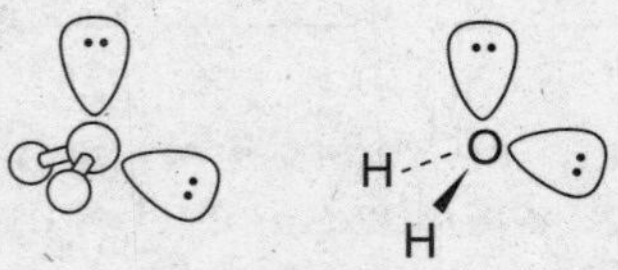

After the Lewis structure for a molecule or polyatomic ion is developed, the shape of the species can be determined using the information in this table as a guide.

The following table presents a step-by-step development of the Lewis structures and three-dimensional shapes of three chemical species. Only the critical facts are listed. Follow the analysis of each starting with the identification of the central atom and ending at the bottom with the three-dimensional shape.

Deriving Lewis Structures and Applying VSEPR Theory

	PO_4^{3-}	SOF_2	NO_2^-
Central Atom	P	S	N
Total valence e⁻'s + charge	32	26	18
Lewis Structure	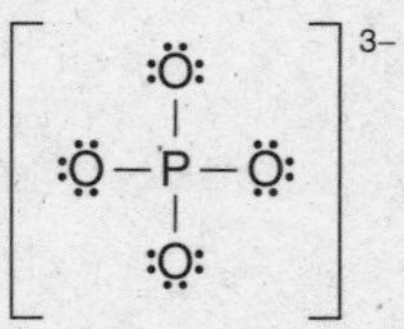 $[\ddot{O}{-}P(-\ddot{O})(-\ddot{O}){-}\ddot{O}]^{3-}$	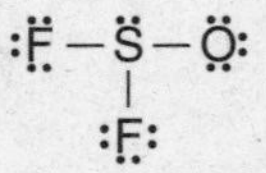 $:\ddot{F}{-}\ddot{S}(-\ddot{F}:){-}\ddot{O}:$	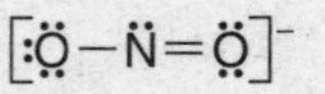 $[:\ddot{O}{-}\ddot{N}{=}\ddot{O}]^-$

	PO_4^{3-}	SOF_2	NO_2^-
Resonance (Y/N)	N	N	Y
Electron Domains about Central Atom	4	4	3
VSEPR Geometry of Domains	tetrahedral	tetrahedral	trigonal planar
Shape of Species	tetrahedral 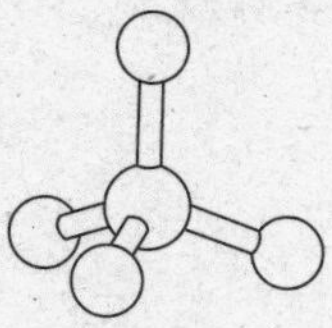	trigonal pyramid 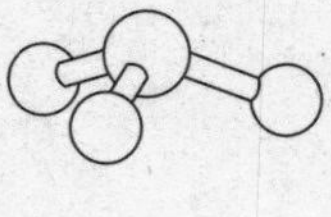	bent 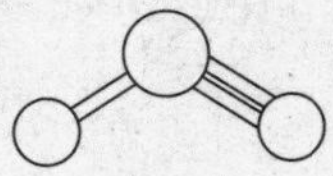

In summary, this section has focused on three arrangements of electron domains about the central atom: linear, trigonal planar, and tetrahedral. The critical bond angle for each is linear = 180°, trigonal planar = 120°, and tetrahedral = 109.5°. Five molecular shapes arise from these electron domain geometries: linear, trigonal planar, bent, tetrahedral, and trigonal pyramid. You should know and be able to recognize each of these. Models of each geometric shape appear in the previous two tables.

Example Problems

These problems have both answers and solutions given.

1. What property of the bonding and nonbonding electrons in the valence shell of a central atom is responsible for the shape of a molecule?

   Answer: Each bond and nonbonding pair on the central atom represents a domain of negative charge. It is the mutual repulsion of these negative domains that directs the bonds and determines the shape of the molecule.

2. Draw the Lewis structure of each of the following and give the term that describes the arrangement of the electron domains about the central atom *and* the term that describes the shape of the species. Estimate the bond angle(s) about the central atom.

   (a) HCN

   (b) SO_2

   (c) SO_4^{2-}

   Answer: (a) Two electron domains are about C, a single bond and a triple bond. The arrangement of the two domains is linear. The molecule is linear with an HCN bond angle of 180°. The Lewis structure of HCN is

   $$H{-}C{\equiv}N{:}$$

(b) SO_2 exhibits resonance. Examining one of the resonance forms shows there are three electron domains about S: a single bond, a double bond, and one nb-pair of electrons. The geometry of the three domains is trigonal planar, but the three atoms form a bent structure with an estimated OSO bond angle of 120°. The Lewis structure of SO_2 is

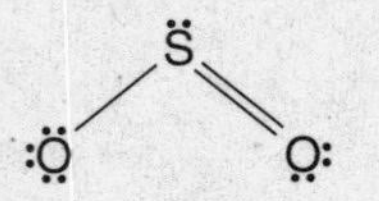

(c) Four electron domains are about S, each a bond to oxygen. The arrangement of the four domains is tetrahedral, and since there are no nb-pairs, the shape of the molecule is also tetrahedral with all OSO bond angles being 109.5°. The Lewis structure of SO_4^{2-} is

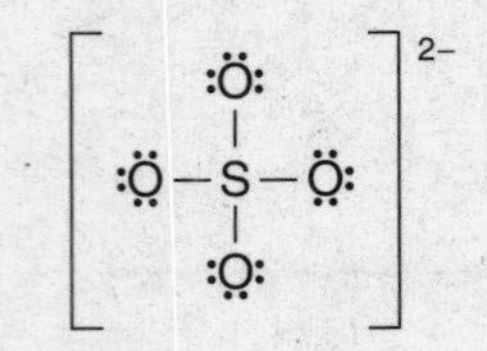

Work Problems

Use these problems for additional practice.

1. What bond angles are associated with a linear structure, trigonal planar structure, and a tetrahedral structure? Two bent structures can form, each with a different bond angle. What are those bond angles?

2. How is it possible for a molecule to have a tetrahedral arrangement of bonds and nb-pairs of electrons about a central atom but not be a tetrahedral molecule?

3. Draw the Lewis structures and predict the shapes of CF_4 and PF_3. What are the expected ∡FCF and ∡FPF values?

Worked Solutions

1. **Linear = 180°; trigonal planar = 120°; and tetrahedral = 109.5°.** One bent structure arises from the trigonal planar arrangement of electron domains and has a bond angle of 120°. The other arises from the tetrahedral arrangement of domains with a bond angle of 109.5°.

2. **The shape of the molecule is determined by the arrangement of its atoms.** If one or two of the electron domains about the central atom are nonbonding electron pairs, they are not atoms and are not considered when describing the shape of the molecule, though they play an important role in determining the shape of the molecule. Ammonia and water are typical examples of this.

3. **CF_4 is a tetrahedral molecule, and each ∡FCF is 109.5°. PF_3 is a trigonal pyramidal molecule by virtue of a tetrahedral arrangement of four electron domains. Each ∡FPF is 109.5°.**

```
     :F:
      |
  :F—C—F:        :F—P—F:
      |              |
     :F:            :F:
```

Chapter Problems and Answers

Problems

1. What is the ionic bond?

2. Which pairs of elements when reacted with each other would form ionic compounds?

   (a) Mg – I; (b) C – O; (c) Ni – Cl; (d) P – Br; (e) Al – F; (f) Ca – Mn; (g) N – Cl

3. What do the phrases "noble gas configuration" and "octet of electrons" have in common? What is the exception?

4. State the octet rule.

5. Following the octet rule, predict the charge on the ions that each of the following elements would form in reactions that produce ionic compounds:

   (a) P; (b) Sr; (c) F; (d) Se; (e) K; (f) Ga; (g) I

6. Using Lewis symbols for each element, write the equations to show the transfer of electrons and the formation of ions for the following ionic compounds: (a) K_2Se and (b) Mg_3P_2.

7. Predict the formulas of the ionic compounds that would be produced in the reaction of the following pairs of elements: (a) Ba and Cl; (b) Al and I; (c) K and O.

8. What is meant by the term "crystal lattice" as it relates to ionic compounds?

9. What is a covalent bond?

10. How does the "octet rule" apply in the formation of covalent compounds?

11. Derive the formula of the covalent compound that each pair of elements would form starting with one atom of the first element and as many atoms of the second as necessary: (a) one atom of Br and H; (b) one atom of P and F; (c) one atom of C and Cl; (d) one atom of Si and H; (e) one atom of S and Cl.

12. What is the difference between a coordinate covalent bond and a normal covalent bond? In what way are they similar? Show how the bond between a water molecule, H_2O, and a hydrogen ion, H^+, to form H_3O^+, would be considered a coordinate covalent bond.

13. In which molecules would the covalent bond(s) be described as "polar"? What determines the degree of polarity of a covalent bond?

    (a) HI; (b) O_2; (c) N_2; (d) H_2O; (e) SO_2

14. Illustrate each of the following with the appropriate diagram using pairs of dots for electron pairs: (a) a double bond between two oxygen atoms; (b) nonbonding pairs of electrons in the F_2 molecule; (c) resonance in the trioxygen (ozone) molecule, O_3.

15. What property of an atom is described by its "electronegativity"? What is the most electronegative element found in compounds? In terms of groups and periods, what are the trends in electronegativity values in the periodic table?

16. Using a periodic table, determine which element in each pair is more electronegative:

(a) P, Mg; (b) S, Si; (c) Cl, Br; (d) O, P

17. What is wrong with this statement? "One molecule of oxalic acid, $H_2C_2O_4$, has eight bonding pairs and eight nonbonding pairs of electrons."

18. For each bond, write the symbols δ+ and δ– over the appropriate atoms to indicate properly the polarity of the bond: (a) C – O; (b) H – F; (c) P – O.

19. How many pairs of electrons are shared between two atoms in a: (a) single bond; (b) double bond; (c) triple bond?

20. What information about a molecule or polyatomic ion can be gained by creating its Lewis structure?

21. Draw the Lewis structures of difluorine, F_2, dioxygen, O_2, and dinitrogen, N_2. How many bonding pairs and how many nonbonding pairs of electrons are present in each structure?

22. Which atom would be the central atom in: (a) NH_4^+, (b) SOF_2, (c) ClO_2^-, (d) F_3PO?

23. Determine the total number of electrons that must be distributed as bonding pairs and nonbonding pairs when developing the Lewis structures of the following molecules and ions: (a) AsF_3, (b) COS, (c) $(HO)_2SiCl_2$, (d) $AlCl_4^-$, (e) IF_2^+, (f) PCl_4^+, (g) AsO_4^{3-}

24. Determine the Lewis structures for nitrous acid, (HO)NO, and the nitrite ion, NO_2^-. Does either exhibit resonance?

25. Draw all the resonance forms for the carbonate ion, CO_3^{2-}. Does resonance make the ion more stable or less stable?

26. Develop the Lewis structures for each of the following. Indicate whether a species will display resonance.

(a) H_2S (f) ICN

(b) NCl_3 (g) SO_3^{2-}

(c) SiH_4 (h) SO_2

(d) SCl_3^+ (i) O_3

(e) HCO_2^- (j) BrO_3^-

27. What do the letters VSEPR stand for? There are two different kinds of "EP" in molecules. What are they? Can they both be referred to as "electron domains"?

28. (a) How would you describe the shape of the I_3^- ion if you know it has a bond angle of 180°?

    (b) How would you describe the shape of the BF_4^- ion if you know it has a bond angle of 109.5°?

    (c) How would you describe the shape of the H_3O^+ ion if you know it has a bond angle of 109.5°?

    (d) There are two bond angles observed in bent molecules. What are they?

29. Apply the VSEPR method to determine the three-dimensional shape of each of the species listed in Question 26. As you do so: (a) state the number of bonding and nonbonding electron domains about the central atom; (b) state the predicted geometric arrangement of those domains; and (c) describe the shape of the species with the correct term and predicted bond angle(s).

30. What is wrong with each of these Lewis structures?

    (a) SO_3 :Ö–S̈–Ö:
    |
    :Ö:

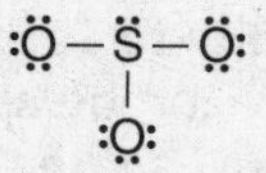

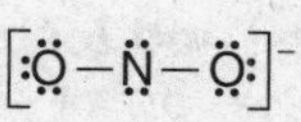

    (b) CH_4O
    H
    |
    H–C=Ö:
    |
    H

    (c) NO_2^- [:Ö–N̈–Ö:]$^-$

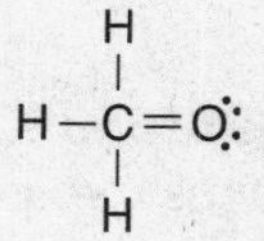

    (d) SiF_4
    :F̈:
    |
    :F̈–Si–F̈–F̈:

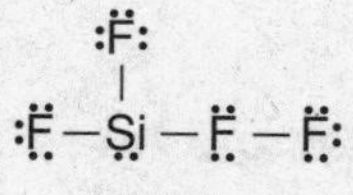

Answers

1. **The ionic bond is the attraction between positive and negative ions in a crystal.**

2. **Metals and nonmetals react to form ionic compounds.** Only (a) Mg – I; (c) Ni – Cl; and (e) Al – F would qualify.

3. **With the exception of helium, all noble gases have an octet of electrons in their valence shells, so any species with an octet of electrons in its valence shell has a noble gas configuration.**

4. **The octet rule: In reactions involving main-group elements, atoms tend to gain, lose, or share the necessary number of electrons needed to achieve an octet of electrons in their valence shells.**

5. **(a) P^{3-}; (b) Sr^{2+}; (c) F^-; (d) Se^{2-}; (e) K^+; (f) Ga^{3+}; (g) I^-.**

6.

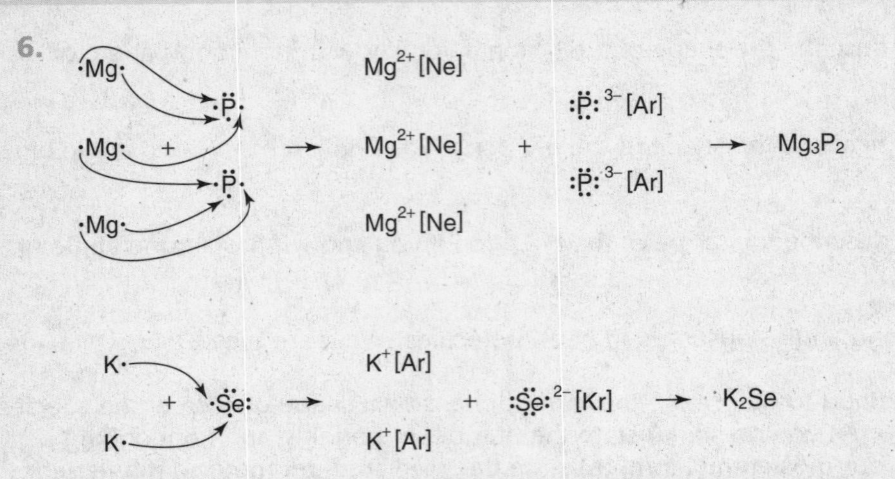

7. **(a) $BaCl_2$ Barium is a Group IIA metal and loses two electrons to form Ba^{2+}.** Chlorine is a nonmetal in group VIIA and gains one electron to form Cl^-. It requires two chlorine atoms to gain the two electrons lost by one barium atom, so the formula must show this 1 to 2 ratio of barium and chlorine: $BaCl_2$.

(b) AlI_3 Aluminum is a Group IIIA metal and loses three electrons to form Al^{3+}. Iodine is in Group VIIA and gains one electron to form I^-. It requires three iodine atoms to gain the three electrons lost by one aluminum atom, so the formula must show the 1 to 3 ratio of aluminum and iodine: AlI_3.

(c) K_2O Potassium is a Group IA metal and loses one electron to form K^+. Oxygen is a Group VIA nonmetal and gains two electrons to form O^{2-}. It requires two potassium atoms to provide the two electrons gained by a single oxygen atom, so the formula must be K_2O.

8. **The crystal lattice of an ionic compound describes the highly organized arrangement of ions in the crystal.**

9. **A covalent bond is the attraction between two atoms due to the sharing of one or more pairs of electrons between them.**

10. **Atoms engaged in covalent bonding will share the necessary number of electrons to acquire an octet of electrons in their valence shells.**

11. **(a) HBr; (b) PF_3; (c) CCl_4; (d) SiH_4; (e) SCl_2**

12. **In a coordinate covalent bond, both shared electrons are provided by only one of the bonded atoms unlike a normal covalent bond in which each bonded atom provides half the shared electrons.** Once formed, both bonds are similar in that they both behave like normal covalent bonds. The hydrogen ion and water molecule form a coordinate-covalent bond using a nonbonding pair of electrons on the oxygen atom of water to form the bond with the hydrogen ion.

H—Ö: + H^+ ⟶ [H—Ö—H]$^+$
(each O also bonded to a lower H)

13. **Any covalent bond between two different atoms is polar.** The polar bonds are those in: (a) HI; (d) H_2O; and (e) SO_2. The greater the difference in electronegativity between two bonded atoms, the greater the polarity of the bond.

14. **(a)** :O::O: **; (b)** :F:F: ← nonbonding pairs **; (c)** O:O:O ⟷ O:O:O

15. **The electronegativity of an atom is a measure of the ability of that atom to draw electrons to itself that it shares in a covalent bond.** Fluorine is the most electronegative element (4.0). In the periodic table, electronegativity increases left to right across a period and decreases down a group.

16. **Following the trends in periods and groups: (a) P is farther to the right in the second period and is more electronegative then Mg; (b) S is farther to the right in the third period and is more electronegative then Si; (c) Cl is above Br in Group VIIA and is more electronegative; (d) O is higher and to the right of P and, therefore, is more electronegative.**

17. **Eight bonding pairs and eight nonbonding pairs of electrons total 32 electrons.** The total number of valence electrons in oxalic acid, $H_2C_2O_4$, is 34 (2-H = 2, 2-C = 8, 4-O = 24). There should be 34 electrons accounted for in the bonding and nonbonding electron pairs. In fact, there are nine bonding pairs and eight nonbonding pairs in one molecule of oxalic acid.

18. **Of the two bonded atoms, the atom that is more electronegative will draw the shared pair away from the other atom and more to itself, increasing negative charge about that end of the bond.** That atom would be designated $\delta-$; the other $\delta+$: (a) $C^{\delta+} - O^{\delta-}$; (b) $H^{\delta+} - F^{\delta-}$; (c) $P^{\delta+} - O^{\delta-}$.

19. **(a) One pair of electrons is shared in a single bond; (b) two pairs in a double bond; and (c) three pairs in a triple bond.**

20. **Lewis structures show how the atoms in a molecule or polyatomic ion are joined together.** Beyond that, Lewis structures also reveal nonbonding pairs, whether covalent links are single, double, or triple bonds and whether resonance is a possibility.

21. **F_2 has one bonding pair, six nonbonding pairs.** O_2 has two bonding pairs, four nonbonding pairs. N_2 has three bonding pairs, two nonbonding pairs.

:F–F: :O=O: :N≡N:

22. **(a) N; (b) S (lower electronegativity than O); (c) Cl; (d) P (lower electronegativity than O).**

23. **(a) 26 e^-'s; (b) 16 e^-'s; (c) 32 e^-'s; (d) 32 e^-'s; (e) 20 e^-'s; (f) 32 e^-'s; (g) 32 e^-'s.**

24. **Nitrous acid, (HO)NO.** Nitrous acid will not exhibit resonance.

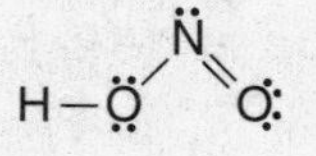

Nitrite ion, NO_2^-. The nitrite ion will exhibit resonance.

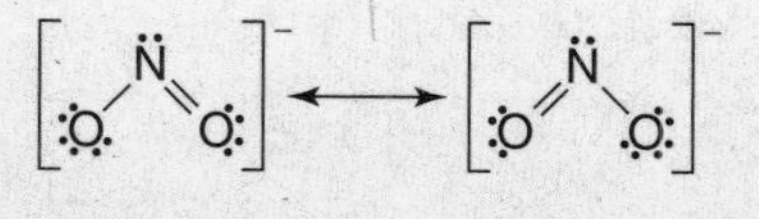

25. **Resonance makes the carbonate ion more stable.**

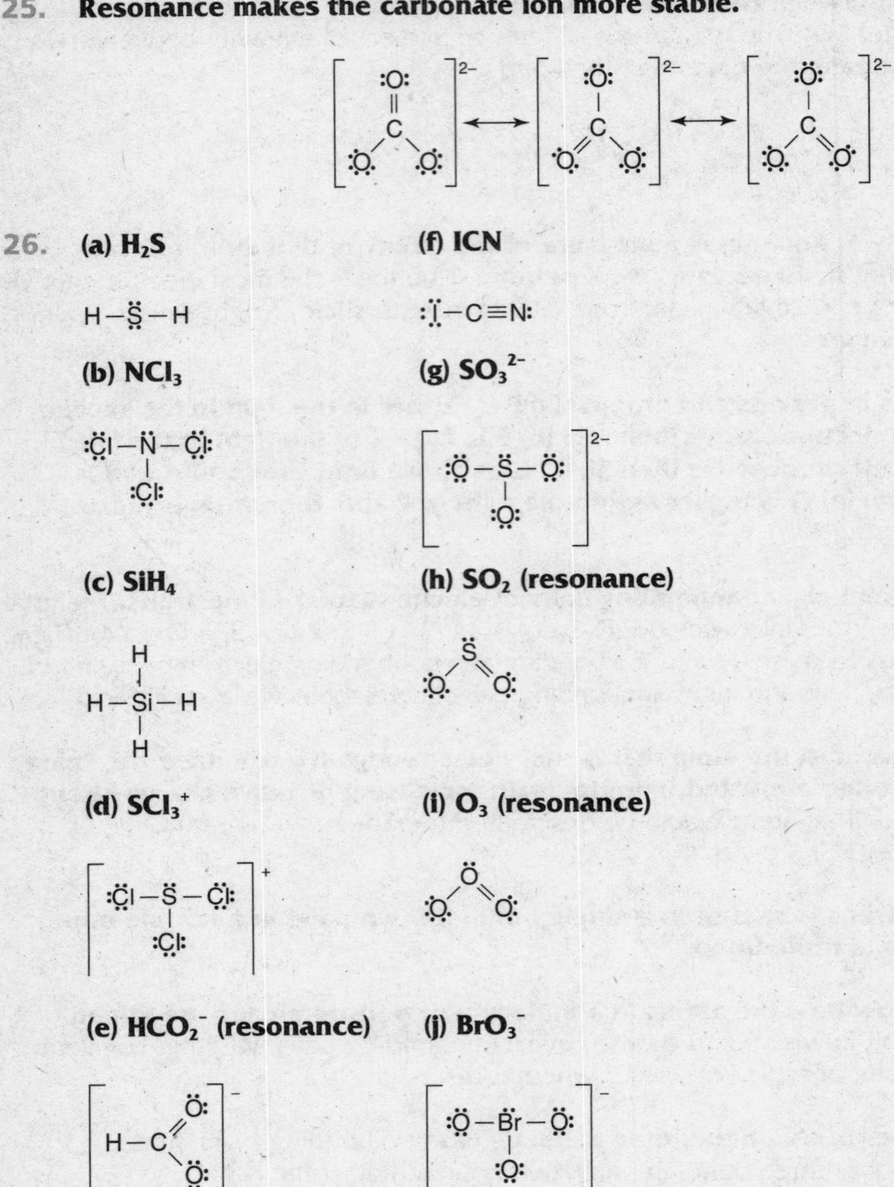

27. **Valence Shell Electron Pair Repulsion.** Both bonding pairs of electrons and nonbonding pairs are represented by "EP." Both act as negative electron domains.

28. **(a) I_3^- with three atoms must be linear.** 180° is the linear bond angle.

(b) BF_4^- must be tetrahedral. 109.5° is the tetrahedral bond angle.

(c) H_3O^+ is trigonal pyramidal. With three bonding pairs and one nonbonding pair, the 109.5° bond angle shows that the structure is based on the tetrahedral arrangement of the four electron domains.

(d) The two bond angles seen in bent molecules are 120° and 109.5° based on the trigonal planar arrangement of three electron domains and the tetrahedral arrangement of four electron domains.

29. **(a) H_2S – bent**

2-bonding and 2-nonbonding domains

4 domains total – tetrahedral geometry

bent shape – ∡HSH = 109.5°

(b) NCl_3 – trigonal pyramid

3-bonding and 1-nonbonding domain

4 domains total – tetrahedral geometry

trigonal pyramid shape – ∡ClNCl = 109.5°

(c) SiH_4 – tetrahedral

4-bonding domains

4 domains total – tetrahedral geometry

tetrahedral shape – ∡HSiH = 109.5°

(d) SCl_3^+ – trigonal pyramid

3-bonding and 1-nonbonding domain

4 domains total – tetrahedral geometry

trigonal pyramid shape – ∡ClSCl = 109.5°

(e) HCO_2^- – trigonal planar

3-bonding domains

3 domains total – trigonal planar geometry

trigonal planar shape – ∡HCO = 120°

(f) ICN – linear

2-bonding domains

2 domains total – linear geometry

linear shape – ∡ICN = 180°

(g) SO_3^{2-} – trigonal pyramid

3-bonding and 1-nonbonding domain

4 domains total – tetrahedral geometry

trigonal pyramid shape – ∡OSO = 109.5°

(h) SO_2 – bent

2-bonding and 1 nonbonding domain

3 domains total – trigonal planar geometry

bent shape – ∡OSO = 120°

(i) O_3 – bent

2-bonding and 1-nonbonding domain

3 domains total – trigonal planar geometry

bent shape – ∡OOO = 120°

(j) BrO_3^- – trigonal pyramid

3-bonding and 1-nonbonding domain

4 domains total – tetrahedral geometry

trigonal pyramid shape – ∡OBrO = 109.5°

30. **(a) SO_3 has too many electrons.** There are 26 electrons showing but only 24 valence electrons in the four atoms. **(b) CH_4O has too many electrons about C. (c) NO_2^- has too many electrons; 20 are shown but only 18 valence electrons are available. (d) SiF_4 has silicon as the central atom; fluorine cannot form two bonds.**

Supplemental Chapter Problems

Problems

1. For each of the following pairs of atoms, use Lewis symbols to show the transfer of electrons leading to the ions that would form an ionic compound: (a) Ba and Cl; (b) Al and P.

2. What type of compound, ionic or covalent, would each pair of elements form? (a) N and Cl; (b) Cs and O; (c) P and I; (d) Mn and F; (e) C and Se.

3. How many 2-electron covalent bonds can be formed by each of the following atoms? (a) N; (b) F; (c) O; (d) H; (e) C.

4. Write the Lewis structures of the three resonance forms of sulfur trioxide, SO_3, then combine them in a fourth composite structure showing the delocalization of the pair of electrons over all atoms in the molecule.

5. Following the octet rule, what would be the expected charge on the ions formed by each of the following elements: (a) S; (b) I; (c) P; (d) N; (e) O.

6. What bonding patterns can be used by carbon to acquire a noble gas configuration in covalent compounds?

7. Which atom is the central atom in: (a) CI_4; (b) NO_3^-; (c) NOF; (d) $POCl_3$; (e) $SOCl_2$; (f) ClF_3.

8. What is the HCF bond angle in CHF_3?

9. What is the expected HNH bond angle in ammonia, NH_3? What is the expected OSO bond angle in SO_3?

10. Draw the Lewis structures and predict the three-dimensional shape of the following: (a) ClO_2^-; (b) BF_4^-; (c) SO_3^{2-}.

Answers

1.

·Ba· + ·C̤̈l: + ·C̤̈l: → Ba^{2+} [Xe] + :C̤̈l:$^-$ [Ar] + :C̤̈l:$^-$ [Ar] → $BaCl_2$

·Al· + ·P̤· → Al^{3+} [Ne] + :P̤:$^{3-}$ [Ar] → AlP

(page 262)

2. **(a) covalent; (b) ionic; (c) covalent; (d) ionic; (e) covalent** (pages 259, 265)

3. **(a) 3; (b) 1; (c) 2; (d) 1; (e) 4** (page 265)

4. 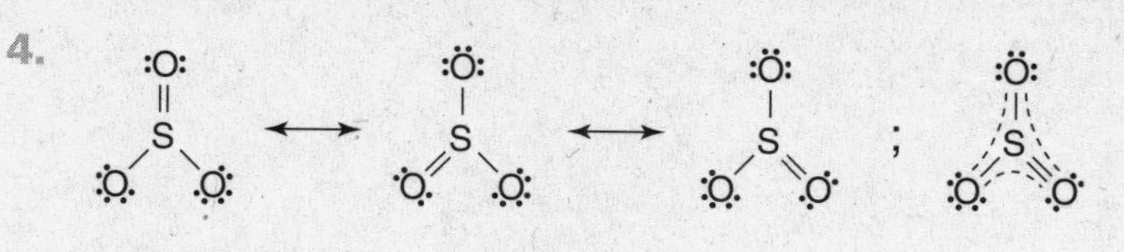

(page 274)

5. **(a) S, 2–; (b) I, 1–; (c) P, 3–; (d) N, 3–; (e) O, 2–** (page 261)

6. **Four single bonds, two double bonds, two single bonds and one double bond, or one single bond and one triple bond.** (page 265)

7. **The central atom is underlined: (a) $\underline{C}I_4$; (b) $\underline{N}O_3^-$; (c) $\underline{N}OF$; (d) $\underline{P}OCl_3$; (e) $\underline{S}OCl_2$; (f) $\underline{Cl}F_3$** (page 274)

8. **The tetrahedral angle, 109.5°.** (page 279)

9. **In NH_3, the expected HNH angle is 109.5°.** In SO_3, the expected OSO angle is 120°. (page 280)

10. **(a) ClO_2^- is a bent ion.** Four domains about Cl.

$$[:\ddot{O}-\ddot{Cl}-\ddot{O}:]^-$$

(b) BF_4^- is a tetrahedral ion. Four domains about B.

$$\left[\begin{array}{ccc} & :\ddot{F}: & \\ & | & \\ :\ddot{F}- & B & -\ddot{F}: \\ & | & \\ & :\ddot{F}: & \end{array}\right]^-$$

(c) SO_3^{2-} is a trigonal pyramidal ion. Four domains about S.

$$\left[\begin{array}{ccc} :\ddot{O}- & \ddot{S} & -\ddot{O}: \\ & | & \\ & :\ddot{O}: & \end{array}\right]^{2-}$$

(page 281)

Chapter 11

Gases and the Gas Laws

We live at the bottom of a sea of gas, the atmosphere, a mixture that is about 78 percent nitrogen gas, $N_2(g)$, and 21 percent oxygen gas, $O_2(g)$, with the remaining 1 percent being mostly argon gas, $Ar(g)$. The air is the most important gas in our lives since it is the source of life-sustaining oxygen. Of the three states of matter, solid, liquid, and gas, the gaseous state was the first to be described quantitatively through equations, which we call the gas laws. In the early days of science, the 1600s, the predictable behavior of gases made them easier to study and to be described with mathematical equations. In addition, the air, which is a mixture of gases, behaved in the same predictable way as the pure gases themselves. All gases behaved the same. Several properties of gases show them to be quite unlike solids and liquids.

- **Gases completely fill the container that holds them.** A sample of gas, unlike a solid or liquid, must be confined. The shape of a gas is the shape of its container. Every bit of volume is filled with gas, top to bottom.
- **Gases can be compressed to smaller volumes or expanded to larger volumes.** Unlike liquids and solids that have fixed volumes, the volume of a gas will contract or expand in response to a change in external pressure or a change in temperature. Gases are transported in heavy steel cylinders under high pressure. The volume of the cylinder may be 40 L, but the gas it contains when released into the much lower pressure of the atmosphere expands to about 7,000 L, a 175-fold volume expansion. The changes in volume of a gas can be calculated using the gas laws.
- **Gases spontaneously mix with one another to form homogeneous mixtures.** The particles that make up a gas (molecules and atoms) are in constant motion and are widely separated from one another. For these reasons, gases readily mix together. If two gases are in a 10 L container, each gas has the same volume, 10 L, because each gas merges into the other occupying the same volume.
- **Gases exert a constant and uniform pressure in all directions simultaneously.** This is because gases completely fill their containers and contact the entire inner surface.

Because gases can change their volume, a large part of this chapter will deal with the classic gas laws that describe the effects of pressure, temperature, and the amount of gas on its volume. The final topic will briefly present the postulates of the kinetic-molecular theory of gases that describe the behavior of gases on the molecular level. Let's begin with one of the most important properties of a gas, its pressure.

Pressure

A gas exerts pressure as its molecules collide with the surface of the container that holds it. The impact of each collision exerts a tiny force against the surface, and at any instant, there are trillions of impacts exerting force against each square millimeter of surface. The **pressure** of a gas is the force of these impacts divided by the unit area receiving that force, P = force/area. The pressure of the air inside a tire is given in pounds per square inch, the force exerted by the collisions of the molecules at any instant on each square inch of surface inside the tire. The atmosphere exerts pressure by virtue of the collisions of its molecules with every surface it contacts. As you move away from the surface of the earth, the pressure of the atmosphere decreases, as the air becomes thinner and molecular collisions decrease. A **barometer** is used to measure atmospheric pressure. A barometer is constructed by using a glass tube that is sealed at one end, completely filling it with mercury, and then, while covering the open end, inverting the tube and placing the open end beneath the surface of mercury in an open dish. When the cover is removed from the open end of the tube, the column of mercury will fall a little and then hold. There is no air (no pressure) in the tube above the mercury, but the atmosphere is exerting pressure on the surface of mercury in the dish. It is the pressure of the atmosphere that supports the column of mercury, as shown in the figure. *The length of the mercury column is a measure of pressure,* stated in millimeters of mercury, mmHg. At sea level, the average atmospheric pressure is 760 mmHg, which is used as a standard of atmospheric pressure. A unit of one atmosphere (atm) of pressure is equivalent to 760 mmHg. A pressure of exactly 1 atmosphere, 760 mmHg, is the **standard pressure,** a benchmark pressure used when comparing volumes of gases. The following figure shows a mercury barometer.

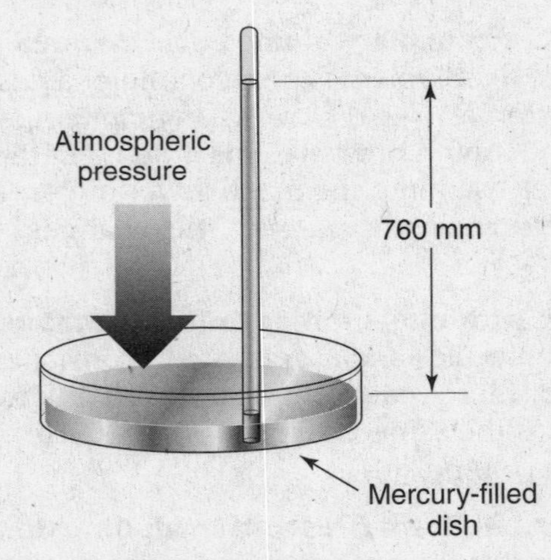

Although pressure is formally defined as force ÷ unit area, it is more commonly expressed in milli-meters of mercury, **mmHg,** or atmospheres, **atm.** The mmHg is also called the **torr** in honor of Torricelli, who built the first barometer in 1643. The SI unit of pressure is the Pascal, although it is not widely used in the United States. One atmosphere is 1.013×10^5 Pascal (Pa) or 101.3 kilopascal (kPa). The relationships between the units of pressure are

$$1 \text{ atm} = 760 \text{ mmHg} = 760 \text{ torr} = 1.013 \times 10^5 \text{ Pa} = 101.3 \text{ kPa}$$

Note that the relationships between 1 atm, 760 mmHg, and 760 torr are exact.

You should be comfortable converting from one unit of pressure to another.

Problem 1: 485 mmHg = _____ atm.

A conversion factor to replace mmHg with atm can be derived from: 1 atm = 760 mmHg (exactly).

$$\text{pressure in atm} = (485\ \cancel{\text{mmHg}}) \times \left(\frac{1\ \text{atm}}{760\ \cancel{\text{mmHg}}}\right) = 0.638\ \text{atm}$$

Answer 1: 485 mmHg = 0.638 atm.

Problem 2: 2.50 atm = _____ torr.

A conversion factor can be derived from: 1 atm = 760 torr (exactly).

$$\text{pressure in torr} = (2.50\ \cancel{\text{atm}}) \times \left(\frac{760\ \text{torr}}{1\ \cancel{\text{atm}}}\right) = 1.90 \times 10^3\ \text{torr}$$

Answer 2: 2.50 atm = 1.90×10^3 torr.

You need to be aware of the Pascal and the kilopascal, but because neither is used widely in the United States, pressures will be given in atmospheres, mmHg, and torr in the following discussions.

There are two ways of describing the pressure of a gas: the pressure exerted *by* the gas or the pressure exerted *on* the gas. Think of a balloon filled with helium. The helium exerts a pressure on the inner wall of the balloon that is identical to the pressure exerted by the atmosphere against the outer surface. If they were not the same, the balloon would either expand or contract until they were equal.

What is a vacuum? Simply stated, a vacuum is the absence of pressure—the absence of gaseous particles. Very high vacuums are stated as very low pressures, on the order of 10^{-6} mmHg.

Temperature

When working problems involving gases, temperature is *always* in Kelvin, not Celsius. To convert a Celsius temperature to Kelvin:

$$\text{temperature in Kelvin} = \text{temperature in Celsius} + 273$$
$$\text{K} = {}^\circ\text{C} + 273$$

If additional significant figures are needed, 273.15 can be used in place of 273. The **standard temperature** for gases is 0°C, which is 273 K, a benchmark temperature used when comparing volumes of gases. No degree symbol is used with Kelvin temperatures. More information about the Kelvin temperature scale appears in the discussion of Charles' law.

The Gas Laws

The volume occupied by a gas depends on three things: (a) the pressure of the gas, (b) the temperature of the gas, and (c) the number of moles of gas. The relationship between volume and each of these is shown in the three classic gas laws.

- **Boyle's law:** The effect of pressure on volume.
- **Charles' law:** The effect of temperature on volume.
- **Avogadro's law:** The effect of the quantity of gas on volume.

Boyle's Law

Robert Boyle (1627–1691), an English chemist, studied how pressure applied to air influenced its volume. His conclusions are stated in a law bearing his name.

> **Boyle's law:** The volume of a gas varies *inversely* with the pressure applied to it as long as the temperature and amount of gas remain constant.

There are two things not to miss in Boyle's law. First, only pressure and volume are allowed to change; temperature and moles of gas, both of which can affect volume, are held constant. Second, volume and pressure are *inversely* related. If the pressure applied to the gas doubles, its volume will be halved. Pressure increases by a factor of two; volume decreases by a factor of two. The two variables move in precise, opposite directions. Boyle's law applies to all gases and mixtures of gases. For a given quantity of gas at a constant temperature, Boyle's law can be stated mathematically in this way:

$$P_iV_i = P_fV_f$$

P_i = the initial pressure of the gas

V_i = the initial volume of the gas

P_f = the final pressure of the gas

V_f = the final volume of the gas

If any three terms of this equation are known, the fourth can be calculated. Any units of pressure and volume can be used, but do not mix different units of pressure or different units of volume in the same equation.

- **Calculating a new volume caused by a change in pressure.**

Example: A helium balloon has a volume of 45.2 L at a pressure of 745 mmHg. What would its volume be after floating high in the air where the pressure is 640 mmHg? No helium leaks from the balloon, and the temperature does not change.

First, identify each term: P_i = 745 mmHg; P_f = 640 mmHg; V_i = 45.2 L; V_f = ?

The pressure exerted on the gas is decreasing from 745 to 640 mmHg. This means the volume will expand. Boyle's equation needs to be rearranged to isolate the final volume on one side of the equation. This is done by dividing both sides of the equation by P_f; that cancels P_f on the right side of the equation and places it in the denominator on the left side.

$\frac{(V_iP_i)}{P_f} = \frac{(V_f\cancel{P_f})}{\cancel{P_f}}$, which becomes: $\frac{V_iP_i}{P_f} = V_f$ or, reversed: $V_f = \frac{V_iP_i}{P_f}$

Once rearranged, substitute the given values for V_i, P_i, and P_f and solve for the final volume,

$$V_f = \frac{V_iP_i}{P_f} = \frac{(45.2\ \text{L})(745\ \text{mmHg})}{(640.\ \text{mmHg})} = 52.6\ \text{L}$$

Answer: The volume of the helium balloon will expand from 45.2 L to 52.6 L as the pressure exerted on the gas falls from 745 to 640 mmHg.

- **Calculating a new pressure to cause a desired change in volume.**

Example: What pressure is needed to compress 10.0 L of air at 0.985 atm to 3.75 L? The quantity of air and temperature remain constant.

Identify each term: P_i = 0.985 atm; P_f = ?; V_i =10.0 L; V_f = 3.75 L

To compress the volume of air to a smaller volume will require exerting more pressure; P_f will be greater than P_i. Rearranging Boyle's law and substituting the known values:

$$P_f = \frac{V_i P_i}{V_f} = \frac{(10.0\ \text{L})(0.985\ \text{atm})}{(3.75\ \text{L})} = 2.63\ \text{atm}$$

Answer: It requires 2.63 atm of pressure to compress 10.0 L of air, initially at a pressure of 0.985 atm, to a volume of 3.75 L at a constant temperature.

Example Problems

These problems have both answers and solutions given.

1. Given a pressure of 650 mmHg, convert to atmospheres.

   Answer: 0.855 atm The conversion factor can be derived from: 1 atm = 760 mmHg (exactly).

   $$\text{pressure in atm} = (650.\ \cancel{\text{mmHg}}) \times \left(\frac{1\ \text{atm}}{760\ \cancel{\text{mmHg}}}\right) = 0.855\ \text{atm}$$

2. A high-pressure gas cylinder contains 95.0 L of oxygen gas at a pressure of 60.0 atm. The valve is opened releasing the oxygen into the atmosphere where the pressure is 0.982 atm. What volume will the oxygen occupy at this lower pressure? The temperature does not change.

   Answer: 5.80×10^3 L

   Identify each term: P_i = 60.0 atm; P_f = 0.982 atm; V_i = 95.0 L; V_f = ?

   $$V_f = \frac{V_i P_i}{P_f} = \frac{(95.0\ \text{L})(60.0\ \text{atm})}{(0.982\ \text{atm})} = 5.80 \times 10^3\ \text{L}$$

3. A medical syringe contains 10.0 mL of air at a pressure of 745 mmHg. The end of the syringe is sealed, and the plunger is pulled back until the volume is 32.0 mL. What is the pressure of the air in the syringe at this final volume?

   Answer: 233 mmHg

   Identify each term: P_i = 745 mmHg; P_f = ?; V_i = 10.0 mL; V_f = 32.0 mL.

   Rearranging Boyle's law to solve for P_f:

   $$P_f = \frac{V_i P_i}{V_f} = \frac{(10.0\ \text{mL})(745\ \text{mmHg})}{(32.0\ \text{mL})} = 233\ \text{mmHg}$$

Work Problems

Use these problems for additional practice.

1. Air in a scuba tank can have a pressure as high as 120. atm. What is this pressure in torr?

2. An 8.00 L scuba tank is filled with compressed air at a pressure of 120. atm. What volume would this air occupy when released into the atmosphere where the pressure is 0.980 atm? Both volumes are measured at the same temperature.

3. 1,500. L of nitrogen gas at 1.00 atm is compressed into a 16.0 L tank. What is the pressure of the nitrogen gas in the tank? Temperature remains constant.

Worked Solutions

1. **P = 9.12 × 10⁴ torr**

$$\text{pressure in torr} = (120.\ \cancel{\text{atm}}) \times \left(\frac{760\ \text{torr}}{1\ \cancel{\text{atm}}}\right) = 9.12 \times 10^4\ \text{torr}$$

2. **V = 980 L**

$$V_f = \frac{V_i P_i}{P_f} = \frac{(8.00\ \text{L})(120.\ \cancel{\text{atm}})}{(0.980\ \cancel{\text{atm}})} = 979.6\ \text{L} = 980\ \text{L}$$

3. **P = 93.8 atm**

$$P_f = \frac{V_i P_i}{V_f} = \frac{(1500.\ \cancel{\text{L}})(1.00\ \text{atm})}{(16.0\ \cancel{\text{L}})} = 93.8\ \text{atm}$$

Charles' Law

Jacques Charles (1746–1823) studied the effect of temperature on the volume of a gas. His careful observations showed that the volume of *any* gas at 0°C, and constant pressure, would increase by 1/273 for each 1°C increase in temperature. It would increase by 25/273 if the temperature increased 25°C. Not only did the volume increase with increasing temperature, it increased by a predictable amount. Lowering the temperature *reduced* the volume of gas. If the temperature of 100 L of gas at 0°C were lowered by 50°C, the volume would *contract* by 50/273 of the original volume. The new volume would be 100 L – (50/273) 100 L = 81.7 L. What would happen if the temperature were lowered by 273°C? Theoretically, a 273°C drop in temperature would reduce the volume of the gas to zero. This does not happen when the experiment is done because gases liquefy before reaching a temperature of –273°C. Yet the idea led William Thomson, a British physicist with the title of Lord Kelvin, in 1848 to propose an absolute scale of temperature. The lowest temperature on this scale was zero, **absolute zero** that corresponded to a temperature of –273°C (–273.15°C, to be exact). There were no temperatures below absolute zero, so all temperatures above absolute zero were positive. The absolute scale came to be known as the Kelvin scale, with temperatures given in Kelvins (K). A temperature change of 1 Kelvin degree is the same as a change of 1 Celsius degree. The principal difference between the Kelvin and Celsius scales is the location of the zero temperature. The Kelvin temperature scale is used in all equations concerning volumes of gases and is part of Charles' law.

Charles' law: The volume of a gas varies *directly* with its Kelvin temperature as long as the pressure and amount of gas remain constant.

Volume and Kelvin temperature vary directly. This means that if the Kelvin temperature of a gas doubles, its volume will double. If the Kelvin temperature is cut in half, its volume will be cut in half. Of course, this exact relationship requires that both the amount of gas and the pressure of the gas remain constant. The dependence of volume on temperature can be shown mathematically in this useful form of Charles' law.

$$\frac{V_i}{T_i} = \frac{V_f}{T_f}$$

V_i = the initial volume of the gas

T_i = the initial temperature of the gas in Kelvins

V_f = the final volume of the gas

T_f = the final temperature of the gas in Kelvins

It is important to remember that Charles' law holds only if the pressure and amount of gas remain constant. Charles' law applies to all gases and mixtures of gases. If three of the four terms are known, the fourth can be calculated.

- **Calculating a new volume caused by a change in temperature.**

  Example: A 750. mL volume of air is heated from 25°C to 100°C at constant pressure. What is the volume of air at the higher temperature?

  Both temperatures need to be in Kelvin:

  T_i = 25°C = (25 + 273)K = 298 K

  T_f = 100°C = (100 + 273)K = 373 K

  Identify each term: V_i = 750. mL; V_f = ?; T_i = 298 K; T_f = 373 K

  Multiply both sides of the equation by T_f to isolate the final volume, V_f, on one side of the equal sign,

  $$T_f\left(\frac{V_i}{T_i}\right) = \left(\frac{V_f}{\cancel{T_f}}\right)\cancel{T_f}, \text{ which becomes } \frac{T_f V_i}{T_i} = V_f \text{ or, reversed, } V_f = \frac{T_f V_i}{T_i}.$$

  Then substitute the known values for V_i, T_i, and T_f into the equation and solve for V_f.

  $$V_f = \frac{(373\ \cancel{K})(750.\ mL)}{298\ \cancel{K}} = 938.8\ mL = 939\ mL$$

  Answer: A 750. mL volume of air will expand to 939 mL as its temperature increases from 25.0° to 100.0°C, at constant pressure.

- **Calculating a new temperature to cause a desired change in volume.**

  Example: A gas thermometer contains exactly 226 mL of air at 21°C. At what Celsius temperature will the volume be to 240. mL? Pressure remains constant.

  T_i = 21°C = (21 + 273)K = 294 K

  Identify the terms: V_i = 226 mL; V_f = 240 mL; T_i = 294 K; T_f = ?

  Rearranging Charles' law to solve for T_f:

  $$T_f = \frac{V_f T_i}{V_i} = \frac{(240.\ \cancel{mL})(294\ K)}{226\ \cancel{mL}} = 312\ K$$

  Converting to Celsius: °C = (K − 273)°C = (312 − 273)°C = 39°C.

  Answer: The volume of gas in the thermometer will be 240. mL at 39°C.

Example Problems

These problems have both answers and solutions.

1. A 50.0 L volume of oxygen gas is heated from 23.0°C to 250.0°C at constant pressure. What is the final volume of the gas at the higher temperature?

   Answer: V_f = 88.3 L

   Converting the temperatures:

   T_i = 23°C = (23 + 273)K = 296 K

$T_f = 250°C = (250 + 273)K = 523\ K$

$$V_f = \frac{V_i T_f}{T_i} = \frac{(50.0\ L)(523\ \cancel{K})}{296\ \cancel{K}} = 88.3\ L$$

2. To what temperature, in Kelvins, must 1,500. mL of helium at 300 K be cooled to reduce its volume to 850. mL, at constant pressure?

Answer: $T_f = 170\ K$ (This corresponds to a temperature of −103°C.)

$$T_f = \frac{V_f T_i}{V_i} = \frac{(850.\ \cancel{mL})(300.\ K)}{1{,}500.\ \cancel{mL}} = 170.\ K$$

Work Problems

Use these problems for additional practice.

1. A 500. L volume of nitrogen gas is cooled from 20°C to −75°C. What is the volume of the nitrogen at the lower temperature if all volumes are measured at the same pressure?

2. To what temperature must 5.0 L of air at 300. K be cooled to reduce its volume to 2.5 L, while pressure remains constant?

Worked Solutions

1. **$V_f = 338\ L$.**

   Convert the temperatures:

   $20°C = (20 + 273)K = 293\ K = T_i$

   $-75°C = (-75 + 273)K = 198\ K = T_f$

   Rearranging Charles' law and substituting the appropriate values:

   $$V_f = \frac{V_i T_f}{T_i} = \frac{(500.\ L)(198.\ \cancel{K})}{293\ \cancel{K}} = 338\ L$$

2. **Charles' law says to reduce a volume by half, reduce the Kelvin temperature to half its initial value, in this case to a final temperature of 150 K.**

   $$T_f = \frac{V_f T_i}{V_i} = \frac{(2.5\ L)(300.\ K)}{5.0\ L} = 150\ K$$

Avogadro's Law

Amedeo Avogadro (1776–1856) proposed the relationship between volume and quantity of gas in 1811, which has stood the test of time: Equal volumes of two different gases at the same temperature and pressure contain the same number of molecules. The last phrase "same number of molecules" can also be stated as "same number of moles." An enlargement of this observation has given us the gas law that bears his name.

Avogadro's law: The volume of a gas varies *directly* with the number of moles of gas as long as the pressure and temperature remain constant.

Like Charles' law, Avogadro's law is a "direct" relationship. If the number of moles of gas doubles, the volume will double. If the number of moles of gas is cut in half, the volume will be cut in half. Of course, this precise relationship requires that the temperature and pressure of the gas do not vary. Avogadro's law applies to all gases. Mathematically, Avogadro's law can be stated this way:

$$\frac{V_i}{n_i} = \frac{V_f}{n_f}$$

V_i = the initial volume of the gas

n_i = the initial number of moles of gas

V_f = the final volume of the gas

n_f = the final number of moles of gas

- **Calculating a new volume caused by a change in the number of moles of gas.**

  Example: 0.15 mole of helium gas has a volume of 3.6 L. If an additional 0.25 mole of helium is added, what will be the new volume? Temperature and pressure do not change.

  Identifying the terms: V_i = 3.6 L; V_f = ?; n_i = 0.15 mole; n_f = (0.15 + 0.25) mole = 0.40 mole

  To solve Avogadro's law for V_f, multiply both sides by n_f, then divide identical terms to isolate the final volume, V_f, on one side of the equal sign.

  $$n_f\left(\frac{V_i}{n_i}\right) = \left(\frac{V_f}{\cancel{n_f}}\right)\cancel{n_f}\text{, which becomes } \frac{V_i n_f}{n_i} = V_f\text{, or reversed, } V_f = \frac{V_i n_f}{n_i}$$

  Substitute the known values for V_i, n_i, and n_f and solve for V_f.

  $$V_f = \frac{V_i n_f}{n_i} = \frac{(3.6\ \text{L})(0.40\ \cancel{\text{mole}})}{0.15\ \cancel{\text{mole}}} = 9.6\ \text{L}$$

  Answer: Increasing the number of moles of helium from 0.15 to 0.40 mole will increase the volume of the gas from 3.6 L to 9.6 L, if the pressure and temperature remain constant.

- **Calculating a new number of moles of gas from a new volume.**

  Example: 10.3 L of chlorine gas, 0.420 mole of Cl_2, leaked from a gas cylinder. Two hours later, a total of 16.0 L of chlorine had escaped. During this time, the pressure and temperature did not change. What is the total number of moles of $Cl_2(g)$ that escaped?

  Identifying the terms: V_i = 10.3 L; V_f = 16.0 L; n_i = 0.420 mole; n_f = ?

  Avogadro's law must be rearranged to solve for n_f. Because n_f is in the denominator, a two-step rearrangement is necessary. First, cross-multiply the terms to bring both denominators to the top of the opposite sides. Then divide both sides of the equation by V_i.

  Cross-multiplying rearranges the terms on one line. Then both sides of this equation are divided by V_i to bring n_f to one side of the equal sign.

  $$\frac{n_f \cancel{V_i}}{\cancel{V_i}} = \frac{V_f n_i}{V_i} \text{ becomes } n_f = \frac{V_f n_i}{V_i}$$

  Substituting the known values for V_i, n_i, and V_f gives the final number of moles.

  $$n_f = \frac{V_f n_i}{V_i} = \frac{(16.0\ \cancel{\text{L}})(0.420\ \text{mole})}{10.3\ \cancel{\text{L}}} = 0.652\ \text{mole}$$

  Answer: The 16.0 L volume of $Cl_2(g)$ represents 0.652 mole of $Cl_2(g)$.

Example Problems

These problems have both answers and solutions given.

1. If 0.250 mole of ammonia occupies 4.00 L, what volume will be occupied by 0.650 mole of ammonia if all volumes are measured at the same temperature and pressure?

   Answer: 10.4 L

$$V_f = \frac{V_i n_f}{n_i} = \frac{(4.00\text{ L})(0.650\ \cancel{\text{mole}})}{0.250\ \cancel{\text{mole}}} = 10.4\text{ L}$$

2. 0.35 mole of hydrogen occupies a volume of 8.50 L. A sample of helium has a volume of 20.0 L. How many moles of helium do you have? All volumes are measured at the same temperature and pressure.

   Answer: 0.824 mole of helium

   Avogadro's law applies equally well to all gases, so comparing moles of hydrogen with moles of helium is fine. If 0.350 mole of hydrogen has a volume of 8.50 L, the number of moles of helium that occupies 20.0 L can be determined using the direct relationship of Avogadro's law.

$$\frac{V_{hydrogen}}{n_{hydrogen}} = \frac{V_{helium}}{n_{helium}}$$

$$n_{helium} = \frac{V_{helium} n_{hydrogen}}{V_{hydrogen}} = \frac{(20.0\ \cancel{\text{L}})(0.350\text{ mole})}{8.50\ \cancel{\text{L}}} = 0.824\text{ mole}$$

Work Problems

Use these problems for additional practice.

1. If 6.00 moles of oxygen have a volume of 2,000. L, what volume would be occupied by 3.00 moles if both volumes are compared at the same temperature and pressure?

2. 1.0 mole of nitrogen gas has a volume of 25 L before 3.0 moles of oxygen gas are added to it. What is the volume of the nitrogen-oxygen mixture if all volumes are measured at the same temperature and pressure?

Worked Solutions

1. **3.00 moles have a volume of 1,000. L, exactly half of that is occupied by 6.00 moles.**

$$V_f = \frac{(2{,}000.\text{ L})(3.00\text{ moles})}{6.00\text{ moles}} = 1{,}000.\text{ L}$$

2. **The mixture would have a volume of 100 L.** Four moles of gas (1.0 mole nitrogen + 3.0 moles oxygen) would occupy four times the volume as one mole at the same temperature and pressure.

$$V_{mixture} = \frac{V_{nitrogen} n_{mixture}}{n_{nitrogen}} = \frac{(25\text{ L})(4.0\text{ moles})}{1.0\text{ mole}} = 1.0\times 10^2\text{ L}$$

Standard Temperature and Pressure

Boyle's law and Charles' law show how the volumes of gases depend on their pressures and temperatures. To ensure there is a level playing field when comparing volumes of two or more gases, all gases must be compared at the same temperature and pressure. For this purpose, a temperature of 0°C and a pressure of 1.00 atm have been universally accepted as the **standard temperature and pressure (STP)** for gases.

Standard Temperature and Pressure = **STP** = 0°C (273 K) and 1.00 atm (760 mmHg)

The volume of one mole of *any* gas at *STP* is 22.4 L. This is the **standard molar volume** of gases, and remember, this is true for all gases.

Standard molar volume = 22.4 L at STP

For example, 1.00 mole of oxygen gas, $O_2(g)$, occupies a volume of 22.4 L at STP (1.00 molar volume) and has a mass of 32.0 g (1.00 molar mass).

Example: What is the volume of 30.8 g of $CO_2(g)$ at STP? The molar mass of CO_2 is 44.01 g.

There are two approaches to the solution:

Approach 1: Convert 30.8 g of CO_2 to moles and then use Avogadro's law and the fact that 1.00 mole of $CO_2(g)$ occupies 22.4 L at STP.

$$\text{moles } CO_2 = (30.8 \text{ g } \cancel{CO_2}) \times \left(\frac{1 \text{ mole } CO_2}{44.01 \text{ g } \cancel{CO_2}}\right) = 0.700 \text{ mole}$$

$$\frac{V}{0.700 \text{ mole}} = \frac{22.4 \text{ L}}{1.00 \text{ mole}} \qquad \text{Volume} = \frac{(22.4 \text{ L})(0.700 \cancel{\text{mole}})}{1.00 \cancel{\text{mole}}} = 15.7 \text{ L at STP}$$

Approach 2: Convert 1.00 mole of CO_2 to mass, 44.01 g of CO_2, and then restate the molar volume of CO_2 like this: 44.01 g of CO_2 occupies 22.4 L at STP. Mass can be used in place of moles in Avogadro's law as long as both masses are of the *same* compound. Remember, the mass of a compound is directly proportional to the moles of that compound.

$$\frac{V}{30.8 \text{ g}} = \frac{22.4 \text{ L}}{44.01 \text{ g}} \qquad \text{Volume} = \frac{(22.4 \text{ L})(30.8 \cancel{\text{g}})}{44.01 \cancel{\text{g}}} = 15.7 \text{ L at STP}$$

Answer: Both approaches show that 30.8 g of $CO_2(g)$ have a volume of 15.7 L at STP.

The Combined Gas Law

The **combined gas law** unites Boyle's law and Charles' law in a single equation:

$$\frac{P_i V_i}{T_i} = \frac{P_f V_f}{T_f}$$

The pressure-volume relationship of Boyle's law and the volume-temperature relationship of Charles' law are both seen in the combined gas law. As with Charles' law, temperature must be in Kelvins. The combined gas law applies to all gases and mixtures of gases. If five of the six terms are known, the sixth can be calculated.

Problem: A weather balloon has a volume of 500. L at ground level, where the temperature is 21° (294 K) and pressure is 749 mmHg. What would be the volume of the balloon when it reaches an altitude where the temperature is −30°C (244 K) and the pressure is 547 mmHg?

The decrease in pressure would increase the volume, but the decrease in temperature would reduce the volume. The combined gas law accommodates both changes. First, identify the terms:

$V_i = 500.\ L$ $P_i = 749\ mmHg$ $T_i = 294\ K$

$V_f = ?$ $P_f = 547\ mmHg$ $T_f = 244\ K$

The combined gas law equation will need to be rearranged to solve for V_f. First cross-multiply the temperatures to put the equation on a single line:

$$P_i V_i T_f = P_f V_f T_i$$

Then divide both sides by P_f and T_i to isolate V_f on one side of the equal sign:

$$\frac{P_i V_i T_f}{P_f T_i} = \frac{\cancel{P_f} V_f \cancel{T_i}}{\cancel{P_f} \cancel{T_i}}, \text{ which gives } \frac{P_i V_i T_f}{P_f T_i} = V_f$$

Rewriting the equation groups together the effects of the change in pressure and the change in temperature on the initial volume of the balloon, V_i.

$$V_f = V_i\left(\frac{T_f}{P_f}\right)\left(\frac{P_i}{T_i}\right) = (500.\ L)\left(\frac{749\ \cancel{mmHg}}{547\ \cancel{mmHg}}\right)\left(\frac{244\ \cancel{K}}{294\ \cancel{K}}\right) = 568\ L$$

Answer: The volume expansion due to the pressure drop was greater than the volume decrease due to the temperature drop, and the overall change was an increase in volume to 568 L.

Example Problems

These problems have both answers and solutions given.

1. A sample of hydrogen gas, $H_2(g)$, had a volume of 455 L at 20°C (293 K) and 755 mmHg. What would its volume be at 100°C (373 K) and 565 mmHg?

   Answer: Volume = 774 L. The combined gas law is used.

   $$V_f = V_i\left(\frac{P_i}{P_f}\right)\left(\frac{T_f}{T_i}\right) = (455\ L)\left(\frac{755\ \cancel{mmHg}}{565\ \cancel{mmHg}}\right)\left(\frac{373\ \cancel{K}}{293\ \cancel{K}}\right) = 774\ L$$

2. Knowing that 1.00 mole of $Cl_2(g)$ has a volume of 22.4 L at STP, what will its volume be at 300.°C (573 K) and 1,000. mmHg?

   Answer: Volume = 35.7 L.

   The initial temperature and pressure are the values of STP. The initial volume, V_i, is 22.4 L.

   $$V_f = V_i\left(\frac{P_i}{P_f}\right)\left(\frac{T_f}{T_i}\right) = (22.4\ L)\left(\frac{760\ \cancel{mmHg}}{1{,}000.\ \cancel{mmHg}}\right)\left(\frac{573\ \cancel{K}}{273\ \cancel{K}}\right) = 35.7\ L$$

Work Problems

1. 2,000. g of helium are put in a weather balloon. What is the volume of the balloon at STP? The molar mass of He(*g*) is 4.00 g.

2. What is the volume of 0.670 mole of oxygen gas, $O_2(g)$, at 125°C (398 K) and 1.50 atm? *Hint:* First calculate the volume of 0.670 mole of oxygen gas, $O_2(g)$, at STP and then correct for the change in pressure and temperature using the combined gas law.

Worked Solutions

1. 2,000. g of helium at STP has a volume of 1.12×10^4 L.

   2,000. g of He(*g*) are 500 moles of helium.

   $$\text{mole He} = (2{,}000.\ \text{g}\ \cancel{\text{He}}) \times \left(\frac{1\ \text{mole He}}{4.00\ \text{g}\ \cancel{\text{He}}}\right) = 500.\ \text{moles He}$$

   If 1.00 mole of He(*g*) has a volume of 22.4 L at STP, according to Avogadro's law, 500 moles will have a volume 500 times larger.

   $$\frac{22.4\ \text{L}}{1.00\ \text{mole}} = \frac{\text{Volume}}{500.\ \text{moles}} \qquad \text{Volume} = \frac{(22.4\ \text{L})(500.\ \cancel{\text{moles}})}{1.00\ \cancel{\text{mole}}} = 1.12 \times 10^4\ \text{L He}$$

2. 0.65 mole of oxygen gas has a volume of 14.6 L at 125°C and 1.50 atm.

   1.00 mole of oxygen gas at STP has a volume of 22.4 L. Application of Avogadro's law shows that 0.670 mole of oxygen gas would have a volume of 15.0 L at STP.

   $$\frac{22.4\ \text{L}}{1.00\ \text{mole}} = \frac{\text{Volume}}{0.670\ \text{mole}} \qquad \text{Volume} = \frac{(22.4\ \text{L})(0.670\ \cancel{\text{mole}})}{1.00\ \cancel{\text{mole}}} = 15.0\ \text{L}\ O_2$$

   The volume is affected by two opposing changes: An increase in pressure from 1.00 atm (P_1) to 1.50 atm (P_2) would reduce the volume, but an increase in temperature from 273 K (T_1) to 398 K (T_2) would increase it. V_i = 15.0 L.

   $$V_f = V_i\left(\frac{P_i}{P_f}\right)\left(\frac{T_f}{T_i}\right) = (15.0\ \text{L})\left(\frac{1.00\ \cancel{\text{atm}}}{1.50\ \cancel{\text{atm}}}\right)\left(\frac{398\ \cancel{\text{K}}}{273\ \cancel{\text{K}}}\right) = 14.6\ \text{L}$$

The Ideal Gas Law

The equations for the three classic gas laws, Boyle's law, Charles' law, and Avogadro's law, can be combined into a single equation called the ideal gas law.

The Ideal Gas Law: PV = nRT

All the terms in this equation should be familiar to you, except for R. The "R" symbolizes the ideal gas law constant that has a value of 0.0821 L-atm/mole-K.

Ideal gas law constant = R = 0.0821 L-atm/mole-K

Boyle's law, Charles' law, and Avogadro's law always deal with a change in one term brought about by a change in another, a change in V with a change in P, T, or n. The ideal gas law allows direct calculation of pressure, volume, temperature, or moles of gas. If you know three of the variables, V, P, n, or T, you can calculate the fourth. Notice the units of R: liter-atmosphere per mole-Kelvin. This requires when using the ideal gas law that:

- V must be in liters.
- P must be in atmospheres.
- T must be in Kelvins.
- n, the quantity of gas, must be in moles.

The ideal gas law applies to all gases. The word "ideal" in its name may seem a bit unusual, but it comes from that fact that the equation can be developed from the properties of a hypothetical, ideal gas, a gas that behaves perfectly (ideally) at all temperatures and pressures. The ideal gas will be part of the final discussion in this chapter, but suffice it to say, the ideal gas law, and all the gas laws for that matter, work well with real gases (as opposed to an ideal gas) as long as temperatures are not exceedingly low or pressures exceedingly high.

In many ways, the ideal gas law is the most versatile of the gas laws. All the gas laws studied up to now can be derived from it. The ideal gas equation can be rearranged to solve for each of the four variables:

$$V = \frac{nRT}{P} \qquad P = \frac{nRT}{V} \qquad n = \frac{PV}{RT} \qquad T = \frac{PV}{nR}$$

Let's use the ideal gas law to solve several problems:

Problem 1: What is the volume of 110. g of $CO_2(g)$ at STP, 273 K, 1.00 atm? The molar mass of CO_2 is 44.01 g.

The mass of CO_2 must be converted to moles before used in the ideal gas law.

$$\text{moles } CO_2 = (110.\ \text{g } \cancel{CO_2}) \times \left(\frac{1 \text{ mole } CO_2}{44.01\ \text{g } \cancel{CO_2}}\right) = 2.50 \text{ moles } CO_2$$

Identifying terms: n = 2.50 moles; T = 273 K; P = 1.00 atm; R = 0.0821 L-atm/mole-K.

Rearrange the ideal gas equation and solve for volume:

$$V = \frac{nRT}{P} = \frac{(2.50 \text{ moles})(0.0821 \text{ L–atm/mole–K})(273 \text{ K})}{1.00 \text{ atm}} = 56.0 \text{ L}$$

The units of moles, K, and atm cancel leaving only L, the unit of volume, for the answer. Only the units appear in the following analysis. Note how the units of moles and K in the gas constant divide out leaving L-atm over atm. Then, dividing out atm leaves the desired unit for the volume. You should analyze the units in each of the following problems to ensure they cancel as needed.

$$\frac{(\cancel{\text{mole}})\left(\frac{\text{L–atm}}{\cancel{\text{mole}}\text{–}\cancel{\text{K}}}\right)(\cancel{\text{K}})}{\text{atm}} = \frac{(\text{L–}\cancel{\text{atm}})}{\cancel{\text{atm}}} = \text{L}$$

There is an alternative way to do this problem without using the ideal gas law. By definition, one mole of any gas occupies 22.4 L at STP, the standard molar volume of gases. So, if 1.00 mole of CO_2 occupies 22.4 L, 2.50 moles would occupy 2.50 moles × 22.4 L/mole = 56.0 L, the identical volume calculated with the ideal gas equation.

Answer 1: 110. g of $CO_2(g)$ have a volume of 56.0 L at STP.

The ideal gas law makes calculating the number of moles of gas a one-step calculation.

Problem 2: How many moles of helium gas, He*(g)*, are in an 8.50 L metal cylinder at a pressure and temperature of 20.0 atm and 22°C (295 K)?

Identifying terms: V = 8.50 L; T = 295 K; P = 20.0 atm, R = 0.0821 L-atm/mole-K.

Rearrange the ideal gas equation and solve for the number of moles:

$$n = \frac{PV}{RT} = \frac{(20.0\ \text{atm})(8.50\ \text{L})}{(0.0821\ \text{L–atm/mole–K})(295\ \text{K})}$$

Answer 2: A 8.50 L cylinder at 22°C and 20.0 atm contains 7.02 moles He*(g)*.

The temperature of a gas can be calculated with the ideal gas law.

Problem 3: 2.80 g of $O_2(g)$ have a volume of 3.05 L at 742 mmHg. What is the Celsius temperature of the gas? The molar mass of O_2 is 32.00 g.

First, the amount of oxygen and its pressure must be converted to the units of moles and atmosphere, units required in the ideal gas law.

$$\text{moles } O_2 = (2.80\ \cancel{g}) \times \left(\frac{1\ \text{mole } O_2}{32.00\ \cancel{g\,O_2}}\right) = 0.0875\ \text{mole } O_2$$

$$\text{P in atm} = (742\ \cancel{\text{mmHg}}) \times \left(\frac{1.00\ \text{atm}}{760\ \cancel{\text{mmHg}}}\right) = 0.976\ \text{atm}$$

Identifying terms: V = 3.05 L; n = 0.0875 mole; P = 0.976 atm; R = 0.0821 L-atm/mol-K.

Rearrange the ideal gas equation and solve for temperature:

$$T = \frac{PV}{nR} = \frac{(0.976\ \text{atm})(3.05\ \text{L})}{(0.0875\ \text{mole})(0.0821\ \text{L–atm/mole–K})} = 414\ \text{K}$$

The temperature needs to be converted from Kelvins to Celsius for the final answer:

°C = (K – 273)°C = (414 – 273)°C = 141°C

Answer 3: 2.80 g of $O_2(g)$ occupying 3.05 L at 742 mmHg have a temperature of 141°C.

Calculating the pressure of a confined sample of gas.

Problem 4: How much pressure is exerted by 10.0 moles of hydrogen gas, $H_2(g)$, confined in a 3.00 L steel gas-bottle at 25°C (298 K)?

Identifying terms: *n* = 10.0 mole; T = 298 K; V = 3.00 L; R = 0.0821 L-atm/mole-K.

Rearrange the ideal gas equation and solve for pressure:

$$P = \frac{nRT}{V} = \frac{(10.0\ \text{moles})(0.0821\ \text{L–atm/mole–K})(298\ \text{K})}{3.00\ \text{L}} = 81.6\ \text{atm}$$

Answer 4: 10.0 moles of $H_2(g)$ in a 3.00 L volume at 298 K exert a pressure of 81.6 atm.

The numerical value of the ideal gas law constant, R, can be derived from the standard molar volume of gases.

Problem 5: 1.00 mole of any gas at 1.00 atm and 273 K has a volume of 22.4 L. From this, calculate the value of R.

Identifying terms: n = 1.00 mole; T = 273 K; P = 1.00 atm; V = 22.4 L.

$$R=\frac{PV}{nT}=\frac{(1.00\text{ atm})(22.4\text{ L})}{(1.00\text{ mole})(273\text{ K})}=0.0821\frac{\text{L-atm}}{\text{mole-K}}$$

Example Problems

These problems have both answers and solutions given.

1. How many kilograms of oxygen, $O_2(g)$, are in an 85.0 L steel gas-cylinder that at 20°C (293 K) shows a pressure of 135 atm? The molar mass of O_2 is 32.00 g.

   Answer: 15.3 kg of $O_2(g)$

   Identifying terms: V = 85.0 L; P = 135 atm; T = 293 K; R = 0.0821 L-atm/mole-K.

   First, calculate the number of moles of oxygen and then convert moles O_2 to mass.

$$n=\frac{PV}{RT}=\frac{(135\text{ atm})(85.0\text{ L})}{(0.0821\text{ L-atm/mole-K})(293\text{ K})}=477\text{ moles } O_2$$

$$\text{mass of } O_2=(477\ \cancel{\text{moles } O_2})\times\left(\frac{32.00\text{ g}\ \cancel{O_2}}{1\ \cancel{\text{mole } O_2}}\right)\left(\frac{1\text{ kg}}{1{,}000\ \cancel{\text{g}}}\right)=15.3\text{ kg } O_2$$

2. What is the pressure exerted by 0.250 mole of $N_2(g)$ in a 10.0 L cylinder at 400°C?

   Answer: P = 1.38 atm

   T = 400°C = (400 + 273)K = 673 K

   Identifying terms: n = 0.250 mole; T = 673 K; V = 10.0 L; R = 0.0821 L-atm/mole-K.

$$P=\frac{nRT}{V}=\frac{(0.250\text{ mole})(0.0821\text{ L-atm/mole-K})(673\text{ K})}{10.0\text{ L}}=1.38\text{ atm}$$

3. A weather balloon has a maximum volume of 10,000. L. 1,400. g of helium, He(*g*), were put in the balloon. Will the volume of the helium exceed the volume of the balloon at an elevation where the pressure is 0.685 atm and the temperature is −50°C (223 K)? The molar mass of He is 4.00 g.

   Answer: No, the helium will only expand to a volume of about 9,350 L.

   First, find the moles of helium and then calculate the volume it would occupy at this T and P.

$$\text{mole He}=(1{,}400.\text{ g}\ \cancel{\text{He}})\times\left(\frac{1\text{ mole He}}{4.00\text{ g}\ \cancel{\text{He}}}\right)=350\text{ moles He}$$

Identifying terms: n = 350 mole; P = 0.685 atm; T = 223 K; R = 0.0821 L-atm/mole-K.

$$V = \frac{nRT}{P} = \frac{(350.\text{ moles})(0.0821\text{ L-atm/mole-K})(223\text{ K})}{0.685\text{ atm}} = 9,354\text{ L} = 9.35 \times 10^3\text{ L}$$

Work Problems

Use these problems for additional practice.

1. It requires 2.1 moles of ammonia in a particular chemical reaction. What volume of ammonia would this be if measured at 760 mmHg and 0°C?

2. A safety seal on a gas cylinder will begin to leak at a pressure of 180. atm. 150.0 moles of oxygen gas are in a 65.0 L gas cylinder. At what temperature will the contents of the cylinder reach the critical pressure of 180. atm?

3. A laboratory experiment produced 4.50 L of oxygen gas, $O_2(g)$, at 745 mmHg and 22°C. How many grams of O_2 were produced? The molar mass of O_2 is 32.0 g.

Worked Solutions

1. **V = 47 L.** Notice that the temperature and pressure given in the question are the standard T and P for gases. Therefore, if 1 mole of ammonia occupied 22.4 L at STP, 2.1 moles would occupy 2.1 moles × 22.4 L/mole = 47 L. Alternatively, using the ideal gas equation:

$$V = \frac{nRT}{P} = \frac{(2.1\text{ moles})(0.0821\text{ L-atm/mole-K})(273\text{ K})}{1.00\text{ atm}} = 47\text{ L}$$

2. **The safety seal should begin to leak at a temperature of 950 K (677°C).**

   Calculate the temperature at which the pressure in the tank equals 180 atm.

   Identifying terms: P = 180 atm; *n* = 150 moles; V = 65.0 L; R = 0.0821 L-atm/mole-K.

$$T = \frac{PV}{nR} = \frac{(180\text{ atm})(65.0\text{ L})}{(150\text{ moles})(0.0821\text{ L-atm/mole-K})} = 950\text{ K}$$

3. **5.83 g of O_2 were produced.** Pressure and temperature must be in atmospheres and Kelvins.

   P = 745 mmHg = 0.980 atm and T = 22°C = (22 + 273)K = 295 K

   Calculate the moles of O_2 produced and then convert moles of O_2 to mass of O_2.

   Identifying terms: P = 0.980 atm; T = 295 K; V = 4.50 L; R = 0.0821 L-atm/mole-K.

$$n = \frac{PV}{RT} = \frac{(0.980\text{ atm})(4.50\text{ L})}{(0.0821\text{ L-atm/mole-K})(295\text{ K})} = 0.182\text{ mole } O_2$$

   Converting to mass:

$$\text{mass of } O_2 = (0.182\ \cancel{\text{mole}}\ O_2) \times \left(\frac{32.0\text{ g } O_2}{1.00\ \cancel{\text{mole}}\ O_2}\right) = 5.83\text{ g } O_2$$

Dalton's Law of Partial Pressures

Up to this point, nearly all discussions and problems dealt with pure samples of gases, not mixtures of gases. But, one gas law specifically addresses mixtures; this law was put forth by John Dalton in 1801 during his studies of the composition of the atmosphere. See Chapter 3 for more information on John Dalton's work.

Dalton's law: The pressure exerted by a mixture of gases equals the sum of the individual pressures exerted by each gas in the mixture.

A **partial pressure** is the pressure exerted by an individual gas in a mixture of gases; it is part of the total pressure. If you have a mixture of three gases, a, b and c, the total pressure of the mixture, P_{total}, is the sum of the individual pressures, that is, the partial pressures of the three gases, p_a, p_b, p_c, and so on.

$$P_{total} = p_a + p_b + p_c$$

Dalton's law clearly implies that each gas in the mixture acts independently of the others, and this is so. The total pressure exerted by the gases in a mixture is directly related to the sum of the moles of gas in the mixture, n_{total}. The number of moles of each gas is symbolized by n_a, n_b, n_c, and so on.

$$n_{total} = n_a + n_b + n_c + \cdots$$

$$P_{total} = \frac{n_{total}RT}{V}$$

Example 1: The total pressure of a mixture of hydrogen gas, $H_2(g)$, argon gas, $Ar(g)$, and nitrogen gas, $N_2(g)$, equals 930 mmHg. The partial pressure of hydrogen is 40 mmHg. The partial pressure of argon is 230 mmHg. What is the partial pressure of nitrogen gas?

$$P_{total} = p_{H_2} + p_{Ar} + p_{N_2} = 930 \text{ mmHg}$$

Rearranging and substituting partial pressures of H_2 and Ar allows the calculation of the partial pressure of nitrogen gas.

$$p_{N_2} = P_{total} - p_{H_2} - p_{Ar} = 930 \text{ mmHg} - 40 \text{ mmHg} - 230 \text{ mmHg} = 660 \text{ mmHg}$$

Answer 1: The partial pressure of $N_2(g)$ is 660 mmHg.

Example 2: A mixture of 0.25 mole of carbon dioxide gas, $CO_2(g)$, and 0.30 mole of oxygen gas, $O_2(g)$, is in a 10.0 L container at 25°C (298 K).

(2a) What is the total pressure of the mixture? (2b) What is the partial pressure of CO_2? (2c) What is the partial pressure of O_2?

The total pressure, P_{total}, depends on the total number of moles of gas, n_{total}, that is 0.55 mole (n_{total} = 0.25 mole + 0.30 mole). P_{total} is found using the ideal gas equation.

Identifying terms: n_{total} = 0.55 mole; V = 10.0 L; T = 298 K; R = 0.0821 L-atm/mole-K.

$$P_{total} = \frac{n_{total}RT}{V} = \frac{(0.55 \text{ moles})(0.0821 \text{ L-atm/mole-K})(298 \text{ K})}{10.0 \text{ L}} = 1.35 \text{ atm}$$

Answer 2a: The total pressure exerted by the mixture is 1.35 atm.

(2b) The partial pressure exerted by just the CO_2 is proportional to the number of moles of CO_2 in the mixture, 0.25 mole.

$$P_{CO_2} = \frac{n_{CO_2}RT}{V} = \frac{(0.25\text{ mole})(0.0821\text{ L–atm/mole–K})(298\text{ K})}{10.0\text{ L}} = 0.61\text{ atm}$$

Answer 2b: The partial pressure of CO_2 is 0.61 atm.

(2c) The partial pressure of O_2 can be determined by difference. Knowing that the sum of the partial pressures equals the total pressure, it follows that:

$$p_{O_2} = P_{total} - p_{CO_2} = 1.35\text{ atm} - 0.61\text{ atm} = 0.74\text{ atm}$$

Answer 2c: The partial pressure of O_2 is 0.74 atm.

When gases are generated in small quantities in the laboratory, they are often collected by displacing water from inverted bottles filled with water, as shown in the following figure.

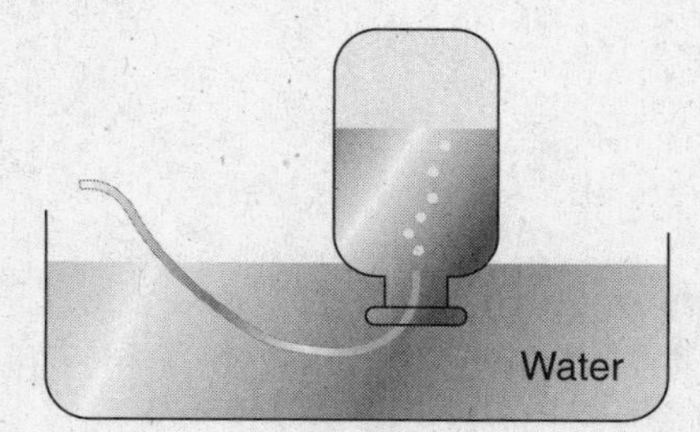

Oxygen gas, which is produced by heating potassium chlorate, $KClO_3$, can be collected in this manner.

$$2\ KClO_3(s) \xrightarrow[\Delta]{MnO_2} 2\ KCl(s) + 3\ O_2(g)$$

Collecting the oxygen by water displacement quickly saturates it with water vapor (gaseous water) producing a mixture of $O_2(g)$ and $H_2O(g)$. The total pressure of the mixture is the sum of the partial pressures of $O_2(g)$ and $H_2O(g)$.

$$P_{total} = p_{oxygen} + p_{water\ vapor}$$

The pressure of water vapor changes with the temperature of water and is listed in tables for quick reference. The vapor pressure of water at four temperatures is given in the following table.

The Vapor Pressure of Water at Four Temperatures

Temperature (°C)	*Vapor Pressure (mmHg)*
15	12.8
20	17.5
25	23.8
30	31.8

Example 3: 220 mL of oxygen was collected over water. The total pressure of oxygen plus water vapor was 745.8 mmHg at 25°C (298 K). How many moles of oxygen gas were collected?

The number of moles of oxygen can be calculated using the ideal gas equation and the partial pressure of oxygen. The pressure exerted by only the oxygen is calculated using Dalton's law. From the table, the vapor pressure of water at 25°C is 23.8 mmHg.

$$p_{oxygen} = P_{total} - p_{water\ vapor} = 745.8 \text{ mmHg} - 23.8 \text{ mmHg} = 722.0 \text{ mmHg}$$

In atmospheres, the pressure of dry O_2 = (722 mmHg) × (1.00 atm/760 mmHg) = 0.950 atm.

Identifying terms: T = 298 K; V = 0.220 L; P = 0.0950 atm; R = 0.0821 L-atm/mole-K.

$$n_{O_2} = \frac{P_{O_2} V_{O_2}}{RT_{O_2}} = \frac{(0.950 \text{ atm})(0.220 \text{ L})}{(0.0821 \text{ L-atm/mole-K})(298 \text{ K})} = 8.54 \times 10^{-3} \text{ mole } O_2$$

Answer 3: 8.54×10^{-3} mole of $O_2(g)$ has been collected.

Example Questions

These problems have both answers and solutions given.

1. A gaseous mixture containing 0.75 mole of $N_2(g)$ and 0.75 mole of $O_2(g)$ has a volume of 30.0 L at 300. K. What is the pressure exerted by the mixture? What fraction of that total pressure is contributed by $N_2(g)$; what fraction by $O_2(g)$?

   Answer: The total pressure is 1.23 atm. Each gas contributes exactly half of the total pressure.

   The total number of moles of gas in the mixture, n_{total}, equals 1.50 moles. The pressure of the mixture depends of the *total* number of moles of gas.

   $$P_{total} = \frac{n_{total} RT}{V} = \frac{(1.50 \text{ moles})(0.0821 \text{ L-atm/mole-K})(300. \text{ K})}{30.0 \text{ L}} = 1.23 \text{ atm}$$

   Since the number of moles of each gas is identical, each gas contributes an identical pressure (both are at the same volume and temperature). Half the total pressure is contributed by each gas.

2. Hydrogen gas, $H_2(g)$, is collected over water at 20°C. The total pressure of hydrogen gas and water vapor is 753.0 mmHg. What is the pressure exerted by only the hydrogen gas?

   Answer: 735.5 mmHg

   $P_{total} = p_{hydrogen} + p_{water\ vapor}$. From the table, the vapor pressure of water at 20°C is 17.5 mmHg.

   $$p_{hydrogen} = P_{total} - p_{water\ vapor} = 753.0 \text{ mmHg} - 17.5 \text{ mmHg} = 735.5 \text{ mmHg}$$

Work Problems

Use these problems for additional practice.

1. A mixture containing 8.00 g of $O_2(g)$ and 16.0 g of $N_2(g)$ has a volume of 51.3 L at 300 K. What is the pressure of the mixture? The molar mass of O_2 is 32.00 g, and that for N_2 is 28.02 g.

2. Three noble gases are mixed together in a 10.0 L container. The pressure of neon is 285 torr, that of helium is 488 torr, and that of argon is 50 torr. What is the total pressure of the mixture?

Worked Solutions

1. **The pressure of the $O_2(g)$–$N_2(g)$ mixture is 0.394 atm.**

   The pressure of the mixture can be calculated using the ideal gas equation. The quantity of each gas must be in moles.

$$n_{\text{oxygen}} = \left(8.00\ \cancel{g\,O_2}\right)\left(\frac{1\text{ mole }O_2}{32.00\ \cancel{g\,O_2}}\right) = 0.250\text{ mole }O_2$$

$$n_{\text{nitrogen}} = \left(46.0\ \cancel{g\,N_2}\right)\left(\frac{1\text{ mole }N_2}{28.02\ \cancel{g\,N_2}}\right) = 0.571\text{ mole }N_2$$

   The total number of moles of gas equals 0.821 mole.

$$P_{\text{total}} = \frac{n_{\text{total}}RT}{V} = \frac{(0.821\text{ mole})(0.0821\text{ L–atm/mole–K})(300\text{ K})}{51.3\text{ L}} = 0.394\text{ atm}$$

2. **The total pressure is 823 torr.**

$$P_{\text{total}} = p_{\text{neon}} + p_{\text{helium}} + p_{\text{argon}} = 285\text{ torr} + 488\text{ torr} + 50\text{ torr} = 823\text{ torr}$$

Stoichiometry of Reactions Involving Gases

Stoichiometry deals with calculations based on balanced chemical equations (see Chapter 8). Recall how the coefficients in a balanced equation correspond to the number of moles of each species in the equation. For gases in a balanced equation, the ideal gas law conveniently relates moles of a gas to its volume, pressure, and temperature.

Moles and Volumes of Gases

If, in a chemical reaction, you know the number of moles of a gaseous reactant or product, you can calculate its volume or pressure using the ideal gas law.

Example 1: 0.800 mole of $KClO_3(s)$ is decomposed with heating to produce oxygen gas. What volume of $O_2(g)$ will be produced if measured at STP? The balanced equation is:

$$2\,KClO_3(s) \xrightarrow[\Delta]{MnO_2} 2\,KCl(s) + 3\,O_2(g)$$

The coefficients in the balanced equation relate the moles of $KClO_3(s)$ to the moles of $O_2(g)$. You should remember the following equation from Chapter 8 that uses these coefficients:

$$\text{moles of sought species} = \left(\frac{\text{coefficient of sought species}}{\text{coefficient of known species}}\right) \times (\text{moles of known species})$$

The problem is asking about oxygen, so the sought species is $O_2(g)$. The known species is $KClO_3(s)$; you know you have 0.800 mole of $KClO_3$. The coefficient of $O_2(g)$ is 3 and that of $KClO_3(s)$ is 2.

First calculate the number of moles of $O_2(g)$ produced from 0.800 mole of $KClO_3$:

$$\text{moles of } O_2 = \left(\frac{3 \text{ moles } O_2}{2 \text{ } \cancel{\text{moles } KClO_2}}\right) \times (0.800 \text{ } \cancel{\text{mole } KClO_2}) = 1.20 \text{ moles } O_2$$

Knowing that 1.00 mole of any gas at STP occupies 22.4 L, you can use Avogadro's law to calculate the volume occupied by 1.20 moles of $O_2(g)$ at STP.

$$\text{volume } O_2 = (1.20 \text{ } \cancel{\text{moles } O_2}) \times \left(\frac{22.4 \text{ L } O_2}{1.00 \text{ } \cancel{\text{mole } O_2}}\right) = 26.9 \text{ L } O_2$$

Answer 1: Decomposition of 0.800 mole of $KClO_3$ will produce 26.9 L of $O_2(g)$ measured at standard temperature and pressure.

Aluminum metal reacts vigorously with hydrochloric acid to produce hydrogen gas, $H_2(g)$.

Example 2: What mass of aluminum must be used to produce 2.50 L of hydrogen gas, $H_2(g)$, at 0.975 atm and 24°C (297 K)? The molar mass of aluminum is 26.98 g.

$$2 \text{ Al}(s) + 6 \text{ HCl}(aq) \rightarrow 2 \text{ AlCl}_3(aq) + 3 \text{ H}_2(g)$$

First, use the ideal gas law to calculate the *moles* of hydrogen to be produced:

$$n = \frac{PV}{RT} = \frac{(0.975 \text{ atm})(2.50 \text{ L})}{(0.0821 \text{ L–atm/mole–K})(297 \text{ K})} = 0.100 \text{ moles } H_2$$

Then, using the balanced equation, determine the number of moles of aluminum required:

$$\text{moles of Al} = \left(\frac{2 \text{ moles Al}}{3 \text{ } \cancel{\text{moles } H_2}}\right) \times (0.100 \text{ } \cancel{\text{moles } H_2}) = 0.0667 \text{ mole Al}$$

The mass of aluminum required to produce the desired volume of hydrogen is

$$\text{mass of Al} = (0.0667 \text{ } \cancel{\text{mole Al}}) \times \left(\frac{26.98 \text{ g Al}}{1 \text{ } \cancel{\text{mole Al}}}\right) = 1.80 \text{ g Al}$$

Answer 2: It requires 1.80 g of aluminum to prepare 2.50 L of $H_2(g)$ measured at 0.975 atm and 24°C.

Volumes of Gases in Balanced Equations

The coefficients in a balanced equation involving gases can be read in terms of "moles" of each gas, *and* they can be read in terms of "volumes" of each gas (at the same T and P). Knowing that one mole of any gas at STP occupies 22.4 L, notice how the following balanced equation can be interpreted in terms of "moles" or "volumes" of the gases:

2 H_2(g)	+	***O_2(g)***	→	***2 H_2O(g)***
2 moles H_2*(g)*		**1** mole O_2*(g)*		**2** moles H_2O*(g)*
2(22.4 L) H_2*(g)*		**1**(22.4 L) O_2*(g)*		**2**(22.4 L) H_2O*(g)* all at STP
2.0 L H_2*(g)*		1.0 L O_2*(g)*		2.0 L H_2O*(g)* all at the same T and P
500 mL H_2*(g)*		250 mL O_2*(g)*		500 mL H_2O*(g)* all at the same T and P

Comparison of gas volumes in balanced equations *requires* that *all* volumes be at the same temperature and pressure; notice in the preceding equation how the volumes are in the exact same ratios as the coefficients in the balanced equation. This observation was first recognized by Gay-Lussac (1778–1823) and stated in his *law of combining volumes* in 1808: "At a given temperature and pressure, gases combine in a ratio of small whole numbers." This observation was later interpreted by Avogadro who related the volumes of gas to moles of gas. Avogadro's work showed that the small whole numbers observed by Gay-Lussac were tied to the coefficients of the balanced equation.

Example: How many liters of O_2*(g)* are required to consume just 2.0 L of NH_3*(g)*? How many liters of water vapor, H_2O*(g)*, will be formed? All volumes are compared at the same temperature and pressure.

$$4\ NH_3(g) + 5\ O_2(g) \rightarrow 4\ NO(g) + 6\ H_2O(g)$$

The coefficients in the balanced equation can be used in a *volume ratio*. Seeking O_2*(g)*:

$$\text{volume of } O_2 = \left(\frac{5 \text{ volumes } O_2}{4 \text{ volumes } \cancel{NH_3}}\right) \times \left(2.0 \text{ L } \cancel{NH_3}\right) = 2.5 \text{ L } O_2$$

$$\text{volume of } H_2O = \left(\frac{6 \text{ volumes } H_2O}{4 \text{ volumes } \cancel{NH_3}}\right) \times \left(2.0 \text{ L } \cancel{NH_3}\right) = 3.0 \text{ L } H_2O$$

Answer: 2.0 L of NH_3*(g)* require 2.5 L of O_2*(g)* for complete reaction, and 3.0 L of H_2O*(g)* will be produced when measured at the same T and F.

Example Problems

These problems have both answers and solutions given.

1. What volume of CO_2*(g)*, measured at STP, would be produced by the complete decomposition of 65.0 g of $CaCO_3$*(s)*? The molar mass of $CaCO_3$ is 100.1 g.

   $$CaCO_3(s) \xrightarrow{\Delta} CaO(s) + CO_2(g)$$

   Answer: 14.5 L of CO_2*(g)*.

   The 1-to-1 coefficient ratio of the balanced equation tells that the number of moles of CO_2*(g)* produced will be identical to the number of moles of $CaCO_3$*(s)* decomposed.

   $$\text{moles } CaCO_3 = \left(65.0 \text{ g } \cancel{CaCO_3}\right) \times \left(\frac{1 \text{ mole } CaCO_3}{100.1 \text{ g } \cancel{CaCO_3}}\right) = 0.650 \text{ mole } CaCO_3$$

0.649 mole of $CaCO_3(s)$ will produce 0.649 mole $CO_2(g)$. Knowing that 1.00 mole of CO_2 would have a volume of 22.4 L at STP, the volume of 0.649 mole of $CO_2(g)$ is

$$\text{volume } CO_2 = (0.649 \text{ }\cancel{\text{mole } CO_2}) \times \left(\frac{22.4 \text{ L } CO_2}{1.00 \text{ }\cancel{\text{mole } CO_2}}\right) = 14.5 \text{ L } CO_2$$

2. How many grams of oxygen gas, $O_2(g)$, are consumed as 100.0 L of methane, $CH_4(g)$, are burned? The volumes of methane and oxygen are measured at 25°C (298 K) and 0.925 atm. The molar mass of $O_2(g)$ is 32.00 g. The balanced combustion equation is

$$CH_4(g) + 2\ O_2(g) \rightarrow CO_2(g) + 2\ H_2O(g)$$

Answer: 242 g $O_2(g)$

First determine the number of moles of methane being consumed:

$$n = \frac{PV}{RT} = \frac{(0.925 \text{ atm})(100.0 \text{ L})}{(0.0821 \text{ L–atm/mole–K})(298 \text{ K})} = 3.78 \text{ moles } CH_4$$

Use the coefficients of the balanced equation to calculate the moles of $O_2(g)$ consumed.

$$\text{moles of } O_2 = \left(\frac{2 \text{ moles } O_2}{1 \text{ }\cancel{\text{mole } CH_4}}\right) \times (3.78 \text{ }\cancel{\text{moles } CH_4}) = 7.56 \text{ moles } O_2$$

This represents $(7.56 \text{ }\cancel{\text{moles } O_2}) \times (32.0 \text{ g } O_2/\cancel{\text{mole } O_2}) = 242 \text{ g } O_2(g)$.

This problem could also be solved by realizing that 200 L of $O_2(g)$ would be consumed in burning 100 L of methane and then using the ideal gas law to calculate the moles of oxygen gas. Try it.

3. Consider the equation for the synthesis of ammonia from hydrogen gas and nitrogen gas. Interpret the balanced equation in terms of volumes of each gas and complete the table:

$3\ H_2(g)$ +	$N_2(g)$ →	$2\ NH_3(g)$
(a) 3 volumes	_____	_____
(b) 600 mL	200 mL	_____
(c) _____	22.4 L	_____
(d) _____	80.0 mL	_____

Answer:

(a) 3 volumes	1 volume	2 volumes
(b) 600 mL	200 mL	400 mL
(c) 67.2 L	22.4 L	44.8 L
(d) 240 mL	80.0 mL	160 mL

Work Problems

Use these problems for additional practice.

1. What volume of oxygen gas, $O_2(g)$, measured at 32°C at 748 torr, will be produced in the complete reaction of 12.0 g of sodium peroxide, $Na_2O_2(s)$ with water? The molar mass of $Na_2O_2(s)$ is 77.98 g. The balanced equation for the reaction is

   $$2\ Na_2O_2(s) + 2\ H_2O(l) \rightarrow 4\ NaOH(aq) + O_2(g)$$

2. What volume of $SO_2(g)$ is formed if 16.0 L of $O_2(g)$ are consumed, and all volumes are compared at the same temperature and pressure?

   $$4\ FeS(s) + 7\ O_2(g) \rightarrow 2\ Fe_2O_3(s) + 4\ SO_2(g)$$

Worked solutions

1. **1.96 L of oxygen gas will be produced.** The number of moles of Na_2O_2 consumed is

   $$\text{moles } Na_2O_2 = (12.0\text{ g } Na_2O_2) \times \left(\frac{1\text{ mole } Na_2O_2}{77.98\text{ g } Na_2O_2}\right) = 0.154\text{ mole}$$

   Using the coefficients of the balanced equation:

   $$\text{moles } O_2 = \left(\frac{1\text{ mole } O_2}{2\ \cancel{\text{moles } Na_2O_2}}\right) \times (0.154\ \cancel{\text{mole } Na_2O_2}) = 0.0770\text{ mole}$$

   The volume of oxygen is determined using the ideal gas law. Converting the temperature and pressure: 32°C = (32 + 273)K = 305 K and 748 torr equals 748 torr × (1 atm/760 torr) = 0.984 atm.

   $$V_{O_2} = \frac{nRT}{P} = \frac{(0.0770\text{ mole})(0.0821\text{ L-atm/mole-K})(305\text{ K})}{0.984\text{ atm}} = 1.96\text{ L}$$

2. **9.14 L $SO_2(g)$ are produced.**

   $$\text{volume of } SO_2 = \left(\frac{4\text{ volumes } SO_2}{7\text{ volumes } \cancel{O_2}}\right) \times (16.0\text{ L } \cancel{O_2}) = 9.14\text{ L } SO_2$$

The Kinetic-Molecular Theory of Gases

The purpose of the **kinetic molecular theory** (KMT) is to explain the behavior of gases using the properties of an "ideal" gas, one that behaves exactly as predicted by the gas laws. Real gases and the ideal gas have similar properties as long as the temperature is not too low or the pressure too high. It is difficult to say what temperature or pressure because they are not the same for all real gases, but generally, pressure less than a few atmospheres and temperatures several tens of degrees above the point at which a gas liquefies give a reasonable working area for the gas laws.

The kinetic-molecular theory is primarily a mathematical model of a gas based on the following postulates:

- Gases are composed of particles (atoms or molecules) that are, on average, widely separated from one another, which is why gases are mostly empty space.

Comment: 1.00 mole of helium occupies 22.4 L at STP, a volume about 20 percent larger than a 5-gallon gas can. 1.00 mole of helium atoms (6.022×10^{23} atoms) has a volume less than 0.03 percent of the volume of the gas. At STP, 99.97 percent of the volume of the gas is empty space. The fraction of its volume taken up by atoms decreases still further as its volume increases. Gases *are* mostly empty space!

- The particles of a gas are in *constant, random, straight-line motion* colliding with each other and the walls of the container. No energy is lost in the collisions, although energy can be transferred from one particle to another.

Comment: The straight-line motion and collision behavior of molecules is similar to the way billiard balls move and collide with one another on a billiard table, with an important exception: Billiard balls eventually will slow to a stop because of friction; the particles in a gas do not; they continue to move.

Pressure is caused by molecular collisions with the container walls. At constant temperature, a gas exerts greater pressure as its volume is compressed, which increases the concentration of particles and the number of collisions for each instant. The relationship between the pressure applied to a gas and its volume is related in Boyle's law.

- The atoms or molecules that compose a gas have no attraction for one another or repulsion from one another. Each particle is completely independent of all others.

Comment: The particles of real gases *do* attract one another when they are squeezed close together by very high pressures. Low temperatures enhance this attraction. If this were not so, it would not be possible to liquefy gases. The fact that attractive forces are exhibited in real gases at high pressures and low temperatures causes the gas laws to be less accurate under these conditions.

- The average kinetic energy, the energy of motion, is directly proportional to the absolute temperature (Kelvin temperature) of the gas.

Comment: Kinetic energy is the energy of motion. All the particles of a gas have different kinetic energies, but it is the average value that determines the temperature of the sample. If the absolute temperature doubles, the average kinetic energy of the gaseous particles doubles. As the average kinetic energy increases, the particles move faster, and the number and force of collisions increase. This is why gases expand when heated and contract when cooled, a fact stated in Charles' law.

The kinetic-molecular theory produces a model of a gas that explains the properties of real gases quite well. As temperatures get lower and pressures higher, the deviation of real gases from ideal behavior increases. Yet, the KMT describes gas behavior at the molecular level very well, and for that reason, it contributes to our understanding of gases.

Chapter Problems and Answers

Problems

1. Convert the following pressure values to the unit requested:

   (a) 745 mmHg = _____ torr

   (b) 2.50 atm = _____ mmHg

   (c) 325 torr = _____atm

   (d) 1.50 atm = _____ torr

2. What is standard temperature and pressure in °C – mmHg and K – atm?

3. Each of the following gas laws requires one or more of the following terms (P, V, T, or *n*) to be held constant. Give these terms for each of the following:

   (a) Boyle's law

   (b) Charles' law

   (c) Avogadro's law

   (d) Combined gas law

4. Which gas law (Boyle's law, Charles' law, Avogadro's law, or the combined gas law) would apply in each of the following scenarios? Only the variables that change are given. You need to state any terms (P, V, T, or *n*) that must remain constant for the gas law you choose to be correct.

   (a) The volume of a gas changes as more gas is added.

   (b) The volume of a gas changes as temperature decreases.

   (c) The volume of a gas changes as both pressure and temperature change.

   (d) The pressure of a gas changes as it is compressed.

   (e) The volume of a gas changes as it is compressed.

   (f) The pressure exerted by a gas changes as its temperature is increased and volume is reduced.

5. State Boyle's law.

6. A fixed quantity of oxygen gas, $O_2(g)$, is held at a constant temperature and occupies 545 mL at 1,250 mmHg. What is the volume of oxygen at:

   (a) 1,650 mmHg

   (b) 2.550 atm

7. A sample of helium occupies 1.65 L at 750. torr and 60°C. If the temperature does not change, what is the pressure of helium if the volume:

   (a) increases to 4.00 L

   (b) decreases to 1.00 L

8. Oxygen gas is shipped in 60.0 L steel cylinders at a pressure of 150. atm. What volume would this oxygen occupy at standard pressure, both volumes measured at the same temperature?

9. State Charles' law.

10. A sample of nitrogen gas, $N_2(g)$, at 100°C occupies a volume of 245 mL. What will the volume be if the temperature is lowered to 0°C? Both volumes are measured at the same pressure.

11. A 1,000. L volume of oxygen gas at 125°C is cooled. If pressure does not change, at what temperature will this sample of oxygen gas occupy 600. L?

12. According to Charles' law, if the volume of a fixed mass of hydrogen gas is cut in half, what must happen to its temperature? Pressure does not change.

13. State Avogadro's law.

14. What will be the new volume of argon gas, $Ar(g)$, if 0.25 additional mole of argon is added to a 0.40 mole sample of argon that occupies a volume of 20.0 L? Both volumes are measured at the same temperature and pressure.

15. What number of moles of nitrogen gas, $N_2(g)$, occupies 250 L at STP? Answer the same question for helium gas, $He(g)$, and sulfur dioxide gas, $SO_2(g)$.

16. Do a rough estimate to judge which of the following samples in each pair contains the larger number of molecules? Remember, 1.000 mole = 6.022×10^{23} molecules.

(a) 2.0 g of $H_2(g)$ or 15 L $H_2(g)$ at STP (molar mass H_2 = 2.016 g)

(b) 2.0×10^{23} molecules of $SO_2(g)$ or 5.2 L $SO_2(g)$ at STP

(c) 32 L of $He(g)$ or 30 L of $CO_2(g)$, both at STP

17. What volume would 10.0 L of air at STP occupy at −25°C and 0.40 atm?

18. Knowing that 1.00 mole of argon gas occupies 22.4 L at STP, what will the volume be at 25°C and 0.798 atm?

19. What is the ideal gas law constant, R, including its units?

20. What is the volume in liters of 1.45 moles of helium gas at 100°C (373 K) and 1.50 atm?

21. What is the pressure of 50.2 g of $CO_2(g)$ stored in a 4.50 L cylinder at 300 K? The molar mass of CO_2 is 44.0 g.

22. How many moles of methane gas, $CH_4(g)$, are in a 350. L cylinder at a pressure of 3.65 atm and 295 K?

23. What pressure is exerted by 7.50 g of $N_2(g)$ contained in a 2,500. mL tank at 298 K? The molar mass of N_2 is 28.0 g.

24. An extremely fine vacuum system can create a pressure as low as 4.0×10^{-17} atm at 305 K. What number of moles and number of molecules are in each 1.0 L of volume at this pressure?

25. What is the Kelvin temperature of 0.355 mole of nitrogen gas that occupies 30.2 L at 735 mmHg (0.967 atm)?

26. The total pressure of a mixture of oxygen gas and nitrogen gas is 755 torr. The pressure exerted by just the nitrogen gas is 245 torr. What is the pressure exerted by the oxygen gas?

27. 0.450 mole of helium gas and 0.600 mole of argon gas are in a 8.65 L glass vessel at a temperature of 300 K. What is the total pressure of the mixture? What is the partial pressure of each gas?

28. A 1.06 L volume of hydrogen gas, $H_2(g)$, is collected over water at 30°C (303 K). The total pressure of the hydrogen gas plus water vapor is 745 mmHg. The vapor pressure of water at 20°C is 31.8 mmHg. How many moles of hydrogen gas are collected?

29. What volume of oxygen gas, $O_2(g)$, measured at STP, is required to consume just 0.250 mole of ethane, $C_2H_6(g)$? What volume of $CO_2(g)$ is produced, also measured at STP? The equation for the combustion reaction is:

$$2\ C_2H_6(g) + 7\ O_2(g) \rightarrow 4\ CO_2(g) + 6\ H_2O(g)$$

30. What volume of hydrogen gas is produced if 2.16 g of zinc metal are consumed with sulfuric acid? The hydrogen gas is measured at 748 mmHg and 22°C. The molar mass of zinc is 65.4 g. The balanced equation is

$$Zn(s) + H_2SO_4(aq) \rightarrow ZnSO_4(aq) + H_2(g)$$

31. What property of gases can you describe that indicates gases are mostly empty space?

32. One container holds oxygen gas at 50°C, another holds nitrogen gas at 50°C. What can you say about the average kinetic energies of the molecules of the two gases?

33. How many liters of oxygen gas, $O_2(g)$, would be consumed in the combustion of 800. L of ethane gas, $C_2H_6(g)$? The balanced equation is

$$2\ C_2H_6(g) + 7\ O_2(g) \rightarrow 4\ CO_2(g) + 6\ H_2O(g)$$

Answers

1. **(a) 745 torr**

$$\text{P in torr} = (745\ \cancel{\text{mmHg}}) \times \left(\frac{1\ \text{torr}}{1\ \cancel{\text{mmHg}}}\right) = 745\ \text{torr}$$

(b) 1.90×10^3 mmHg

$$\text{P in mmHg} = (2.50\ \cancel{\text{atm}}) \times \left(\frac{760\ \text{mmHg}}{1\ \cancel{\text{atm}}}\right) = 1.90 \times 10^3\ \text{mmHg}$$

(c) 0.428 atm

$$\text{P in atm} = (325\ \cancel{\text{torr}}) \times \left(\frac{1\ \text{atm}}{760\ \cancel{\text{torr}}}\right) = 0.4276\ \text{atm} = 0.428\ \text{atm}$$

(d) 1.14×10^3 torr

$$\text{P in torr} = (1.50\ \cancel{\text{atm}}) \times \left(\frac{760\ \text{torr}}{1\ \cancel{\text{atm}}}\right) = 1.14 \times 10^3\ \text{torr}$$

2. **STP = 0°C – 760 mmHg = 273 K – 1.00 atm**

3. **(a) Boyle's law: constant T and *n***

   (b) Charles' law: constant P and *n*

   (c) Avogadro's law: constant P and T

   (d) Combined Gas law: constant *n*

4. **(a) Avogadro's law applies if P and T are constant.**

   (b) Charles' law applies if P and *n* are constant.

   (c) The combined gas law applies as long as *n* is constant.

   (d) Boyle's law applies as long as T and *n* are constant.

   (e) Boyle's law applies as long as T and *n* are constant.

   (f) The combined gas law applies as long as *n* is constant.

5. **Boyle's law: The volume of a gas varies *inversely* with the pressure applied to it as long as the temperature and amount of gas remain constant.**

6. **(a) V = 413 mL**

$$V_f = \frac{V_i P_i}{P_f} = \frac{(545\text{ mL})(1{,}250\ \cancel{\text{mmHg}})}{1{,}650\ \cancel{\text{mmHg}}} = 413\text{ mL}$$

   (b) V = 352 mL

2.55 atm = 1,940 mmHg

$$V_f = \frac{V_i P_i}{P_f} = \frac{(545\text{ mL})(1{,}250\ \cancel{\text{mmHg}})}{1{,}938\ \cancel{\text{mmHg}}} = 351.5\text{ mL} = 352\text{ mL}$$

7. **(a) P = 309 torr**

$$P_f = \frac{P_i V_i}{V_f} = \frac{(750.\text{ torr})(1.65\ \cancel{\text{L}})}{4.00\ \cancel{\text{L}}} = 309\text{ torr}$$

   (b) P = 1.24 × 10^3 torr

$$P_f = \frac{P_i V_i}{V_f} = \frac{(750.\text{ torr})(1.65\ \cancel{\text{L}})}{1.00\ \cancel{\text{L}}} = 1.24 \times 10^3\text{ torr}$$

8. **V = 9.00 × 10^3 L Standard pressure = 1.00 atm**

$$V_f = \frac{V_i P_i}{P_f} = \frac{(60.0\text{ L})(150.\ \cancel{\text{atm}})}{1.00\ \cancel{\text{atm}}} = 9.00 \times 10^3\text{ L}$$

9. **Charles' law: The volume of a gas varies *directly* with its Kelvin temperature as long as the pressure and amount of gas remain constant.**

10. **V = 179 mL** 100°C = (100 + 273)K = 373 K and 0°C = (0 + 273)K = 273 K

$$V_f = \frac{V_i T_f}{T_i} = \frac{(245\ \text{mL})(273\ \cancel{K})}{373\ \cancel{K}} = 179\ \text{mL}$$

11. **T = 239 K = −34°C**

$$T_f = \frac{V_i n_f}{n_i} = \frac{(600.\ \cancel{L})(398\ \text{K})}{1{,}000.\ \cancel{L}} = 239\ \text{K}$$

12. **According to Charles' law, if the volume is cut in half, the absolute temperature of the gas must be half of what it was (at constant P and *n*).**

13. **Avogadro's law: The volume of a gas varies *directly* with the number of moles of gas as long as the pressure and temperature remain constant.**

14. **V = 32.5 L**

$$V_f = \frac{V_i n_f}{n_i} = \frac{(20.0\ \text{L})(0.65\ \cancel{\text{mole}})}{0.40\ \cancel{\text{mole}}} = 32.5\ \text{L}$$

15. ***n* = 11.2 moles of each of the three gases.** It is the same for each gas because for all gases, one mole occupies a volume of 22.4 L at STP.

$$n_f = \frac{V_f n_i}{V_i} = \frac{(250.\ \cancel{L})(1.00\ \text{mole})}{22.4\ \cancel{L}} = 11.2\ \text{moles}$$

16. **(a) There are more molecules in 2.0 g of H_2.** One mole of H_2 is about 2.0 g of H_2, (6.0×10^{23} molecules). At STP, 15 L of $H_2(g)$ is less than 22.4 L, the volume occupied by *one* mole of H_2 at STP.

(b) The larger number of molecules is 2.0×10^{23} molecules, which is about one-third of a mole of SO_2. At STP, 5 L of SO_2 is 5/22.4 mole, about 0.22 mole of SO_2.

(c) 32 L of He(*g*). Both are at the same temperature and pressure, so the larger volume has the larger number of moles of gas, and the larger number of molecules.

17. **V = 22.7 L** −25°C = (−25 + 273)K = 248 K

$$V_f = V_i\left(\frac{P_i}{P_f}\right)\left(\frac{T_f}{T_i}\right) = (10.0\ \text{L})\left(\frac{1.00\ \cancel{\text{atm}}}{0.40\ \cancel{\text{atm}}}\right)\left(\frac{248\ \cancel{K}}{273\ \cancel{K}}\right) = 22.7\ \text{L}$$

18. **V = 30.6 L** 25°C = 298 K

$$V_f = V_i\left(\frac{P_i}{P_f}\right)\left(\frac{T_f}{T_i}\right) = (22.4\ \text{L})\left(\frac{1.00\ \cancel{\text{atm}}}{0.798\ \cancel{\text{atm}}}\right)\left(\frac{298\ \cancel{K}}{273\ \cancel{K}}\right) = 30.6\ \text{L}$$

19. **R = 0.0821 L-atm/mole-K**

20. **V = 29.6 L**

$$V = \frac{nRT}{P} = \frac{(1.45\ \text{moles})(0.0821\ \text{L–atm/mole–K})(373\ \text{K})}{1.50\ \text{atm}} = 29.6\ \text{L}$$

21. **P = 6.24 atm**

$$\text{moles } CO_2 = (50.2\ \text{g}\ \cancel{CO_2}) \times \left(\frac{1.00\ \text{mole } CO_2}{44.0\ \text{g}\ \cancel{CO_2}}\right) = 1.14\ \text{moles } CO_2$$

$$P = \frac{nRT}{V} = \frac{(1.14\ \text{moles})(0.0821\ \text{L–atm/mole–K})(300\ \text{K})}{4.50\ \text{L}} = 6.24\ \text{atm}$$

22. ***n* = 52.7 moles methane gas.**

$$n = \frac{PV}{RT} = \frac{(3.65\ \text{atm})(350.\ \text{L})}{(0.0821\ \text{L–atm/mole–K})(295\ \text{K})} = 52.7\ \text{moles}$$

23. **P = 2.62 atm**

$$\text{moles } N_2 = (7.50\ \text{g}\ \cancel{N_2}) \times \left(\frac{1.00\ \text{mole } N_2}{28.0\ \text{g}\ \cancel{N_2}}\right) = 0.268\ \text{mole } N_2$$

$$P = \frac{nRT}{V} = \frac{(0.268\ \text{mole})(0.0821\ \text{L–atm/mole–K})(298\ \text{K})}{2.50\ \text{L}} = 2.62\ \text{atm}$$

24. ***n* = 1.6×10^{-18} moles = 9.6×10^{5} molecules in each liter.**

$$n = \frac{PV}{RT} = \frac{(4.0 \times 10^{-17}\ \text{atm})(1.0\ \text{L})}{(0.0821\ \text{L–atm/mole–K})(305\ \text{K})} = 1.6 \times 10^{-18}\ \text{moles}$$

$$\text{molecules} = (1.6 \times 10^{-18}\ \cancel{\text{moles}}) \times \left(\frac{6.02 \times 10^{23}\ \text{molecules}}{1.00\ \cancel{\text{mole}}}\right) = 9.6 \times 10^{5}\ \text{molecules}$$

25. **T = 1.00×10^{3} K**

$$T = \frac{PV}{nR} = \frac{(0.967\ \text{atm})(30.2\ \text{L})}{(0.355\ \text{mole})(0.0821\ \text{L–atm/mole–K})} = 1.00 \times 10^{3}\ \text{K}$$

26. **p_{oxygen} = 510 torr**

$P_{total} = p_{nitrogen} + p_{oxygen}$ $p_{oxygen} = P_{total} - p_{nitrogen}$ = 755 torr − 245 torr = 510 torr

27. **P_{total} = 2.99 atm; p_{He} = 1.28 atm; p_{Ar} = 1.71 atm**

n_{total} = 0.450 mole He + 0.600 mole Ar = 1.05 moles

$$P_{total} = \frac{n_{total}RT}{V} = \frac{(1.05\ \text{moles})(0.0821\ \text{L–atm/mole–K})(300\ \text{K})}{8.65\ \text{L}} = 2.99\ \text{atm}$$

$$P_{He} = \frac{n_{He}RT}{V} = \frac{(0.450\ \text{mole})(0.0821\ \text{L–atm/mole–K})(300.\ \text{K})}{8.65\ \text{L}} = 1.28\ \text{atm}$$

$p_{Ar} = P_{total} - p_{He}$ = 2.99 atm − 1.28 atm = 1.71 atm

28. **0.0400 mole H_2*(g)* is collected.**

$P_{hydrogen} = P_{total} - p_{water}$ = (745 mmHg − 32 mmHg) = 713 mmHg = 0.938 atm

$$n = \frac{PV}{RT} = \frac{(0.938\ \text{atm})(1.06\ \text{L})}{(0.0821\ \text{L–atm/mole–K})(303\ \text{K})} = 0.0400\ \text{mole}$$

29. **V_{oxygen} = 19.6 L; $V_{carbon\ dioxide}$ = 11.2 L**

From the balanced equation: mole $O_2(g)$ = (7/2)(0.250 mole) = 0.875 mole $O_2(g)$

Volume $O_2(g)$ at STP = (0.875 mole) × (22.4 L/1.00 mole) = 19.6 L $O_2(g)$

From the balanced equation: mole $CO_2(g)$ = (4/2)(0.250 mole) = 0.500 mole $CO_2(g)$

Volume $CO_2(g)$ at STP = (0.500 mole) × (22.4 L/1.00 mole) = 11.2 L $CO_2(g)$

30. **$V_{hydrogen}$ = 0.812 L**

$$\text{moles Zn} = (2.16\ \text{g Zn}) \times \left(\frac{1\ \text{mole Zn}}{65.4\ \text{g Zn}}\right) = 0.0330\ \text{mole Zn}$$

$$\text{moles } H_2 = \left(\frac{1\ \text{mole } H_2}{1\ \text{mole Zn}}\right)(0.0330\ \text{mole Zn}) = 0.0330\ \text{mole } H_2$$

22°C = (22 + 273)K = 295 K and 748 mmHg (1.00 atm/760 mmHg) = 0.984 atm

$$V_{hydrogen} = \frac{nRT}{P} = \frac{(0.0330\ \text{mole})(0.0821\ \text{L-atm/mole-K})(295\ \text{K})}{0.984\ \text{atm}} = 0.812\ \text{L}$$

31. **The fact that gases are able to be compressed to smaller volumes supports the idea that gases are mostly empty space.**

32. **Both gases are at the same temperature, so the molecules of both have the same average kinetic energy.**

33. **2.80 × 10^3 L of $O_2(g)$.**

$$\text{volume } O_2 = \left(\frac{7\ \text{moles } O_2}{2\ \text{moles } C_2H_6}\right) \times (800.\ \text{L } C_2H_6) = 2.80 \times 10^3\ \text{L}$$

Supplemental Chapter Problems

Problems

1. (a) 1.08 atm = _____ mmHg; 421 mmHg = _____ atm

   (b) 550 torr = _____ mmHg; 6.25 atm = _____ torr

2. A steel cylinder contained helium gas at a pressure of 95.5 atm. When released to the atmosphere where the pressure is 0.978 atm, it occupied a volume of 168 L. The temperature did not change. What is the volume of the steel cylinder?

3. A quantity of hydrogen had a volume of 50.0 L at 35°C. What is its volume at 70°C? Both volumes are measured at the same pressure.

4. A weather balloon has a volume of 4,500 L at STP. What volume will it have at 18,000 feet where the temperature is −45°C and the pressure is 0.485 atm?

5. There are two identical cylinders of propane gas, $C_3H_8(g)$. One cylinder contains 100 grams of propane, the other 120. grams of propane. What pressure would be exerted by the propane if both cylinders are vented into a 1,000. L container at 300 K? The molar mass of propane is 44.09 g.

6. A 4.00 L flask contains 0.00250 mole of oxygen gas and 0.00185 mole of nitrogen gas at 300 K. What is the pressure of the gaseous mixture? What is the pressure of just the oxygen gas?

7. What volume of hydrogen gas, measured at 22°C and 745 mmHg, is produced when 3.40 g of aluminum metal are consumed with hydrochloric acid? The molar mass of aluminum is 26.98 g. The balanced equation is

   $2\ Al(s) + 6\ HCl(aq) \rightarrow 2\ AlCl_3(aq) + 3\ H_2(g)$

8. A chemical reaction produced 230 mL of oxygen gas at 27°C and 745 mmHg. How many moles of oxygen gas are produced?

9. When a sample of calcium carbonate, $CaCO_3$, is strongly heated, it decomposes forming calcium oxide and carbon dioxide, $CO_2(g)$.

   $CaCO_3(s) \rightarrow CaO(s) + CO_2(g)$

   If 3.45 L of carbon dioxide are formed at 700°C and 0.980 atm, what number of moles and what mass of calcium carbonate has decomposed? The molar mass of $CaCO_3$ is 100.1 g.

10. What is the volume of 3.41 moles of sulfur dioxide, $SO_2(g)$, at 645 mmHg and 100°C?

11. What is the pressure exerted by 12.5 moles of methane gas, $CH_4(g)$, when confined to a 30.0 L container at −30°C?

12. How many liters of fluorine gas, $F_2(g)$, measured at STP, are required to just completely consume 0.65 mole of sulfur? How many liters of sulfur hexafluoride form?

    $S(s) + 3\ F_2(g) \rightarrow SF_6(g)$

13. What is the pressure of hydrogen gas collected over water at 15°C? The total pressure of the collected mixture, hydrogen gas plus water vapor, is 752 mmHg.

14. The temperature of a 50.0 L volume of argon gas is 140°C. What will be the volume of the argon if the temperature falls to −140°C, the pressure remaining constant.

15. 8.20 moles of chlorine gas, $Cl_2(g)$, are pumped into a heavy-walled steel 150.0 L tank at 300 K. What is the pressure of the chlorine gas?

16. 1.00 L of air at 28°C and 750 mmHg are heated to 400°C while the pressure is lowered to 500 mmHg. What is the volume of air at these new temperature and pressure conditions?

17. A 23.1 g mass of fluorine gas, $F_2(g)$, has a volume of 15.0 L at 748 mmHg and 23°C. The molar mass of F_2 is 38.0 g.

    (a) Use Boyle's law to calculate the volume of $F_2(g)$ if the pressure increases to 2.50 atm as temperature remains constant.

(b) Use Charles' law to calculate the volume of $F_2(g)$ at 165°C as pressure remains constant.

(c) Use Avogadro's law to calculate the new volume of $F_2(g)$ after an additional 14.9 g of F_2 is added, while both pressure and temperature remain constant.

(d) Use the combined gas law to calculate the volume after the pressure increases to 4.00 atm and temperature increases to 400°C.

18. Calculate the absolute temperature of a gas if 1.50×10^{-2} moles occupy a volume of 650 mL at 730 torr.

19. Calcium hydride, $CaH_2(s)$, is used as a portable source of hydrogen gas. How many grams of $CaH_2(s)$ must be used to fill a 120 L weather balloon with hydrogen at 750 mmHg and 30°C? The molar mass of CaH_2 is 42.1 g. The equation for the generation of hydrogen is

$CaH_2(s) + 2\ H_2O(l) \rightarrow Ca(OH)_2(aq) + 2\ H_2(g)$

20. What volume of hydrogen gas, $H_2(g)$, will be consumed in the synthesis of 100 L of ammonia, $NH_3(g)$, both volumes measured at the same temperature and pressure?

$N_2(g) + 3\ H_2(g) \rightarrow 2\ NH_3(g)$

Answers

1. **(a) 821 mmHg; 0.554 atm** **(b) 550 mmHg; 4.75×10^3 torr** (page 296)
2. **1.72 L** (pages 297–298)
3. **55.7 L** (page 300)
4. **7.75×10^3 L** (pages 297–300)
5. **0.123 atm** (page 302)
6. **0.0268 atm total; 0.0154 atm oxygen gas** (pages 302–303)
7. **4.67 L** (pages 315–316)
8. **9.15×10^{-3} mole** (pages 315–316)
9. **0.0423 moles $CaCO_3$; 4.24 g $CaCO_3$** (page 300)
10. **123 L** (pages 296–297)
11. **8.31 atm** (page 298)
12. **43.7 L $F_2(g)$; 14.6 L $SF_6(g)$** (pages 305–306)
13. **$p_{hydrogen}$ = 739 mmHg** (pages 305–306)
14. **16.1 L** (pages 305–306)

15. **1.35 atm** (page 300)

16. **3.35 L** (page 305)

17. **(a) 5.90 L** **(b) 22.2 L** **(c) 24.7 L** **(d) 8.39 L** (pages 297–303)

18. **507 K** (page 300)

19. **100 g** (page 302)

20. **150 L of $H_2(g)$** (page 305)

Chapter 12

The Forces between Molecules—Solids and Liquids

Water, H_2O, methane, CH_4, and glucose, $C_6H_{12}O_6$, are all molecular compounds, yet water is a liquid at room temperature, methane is a gas, and glucose is a solid. Why aren't they all gases, or all liquids? The answer to that question lies in the strength of the attractive forces that exist between the molecules when they are close together. These forces of attraction are called **intermolecular forces** and the kind and size of these forces determine the physical state of molecular substances at room temperature as well as their boiling and freezing points. In Chapter 10, you learned about the forces of attraction that hold the atoms or ions together in compounds. They were the covalent and ionic bonds, which, for comparison, can be regarded as *intra*molecular forces. In this chapter, we are going to look at the three principal kinds of attractive forces that can exist *between* molecules, and the effect they have on the properties of molecular compounds. Just as the explanation of chemical bonds came down to positive attracting negative and negative attracting positive, the explanation of intermolecular forces will come down to the same thing, except that in molecules the negative and positive poles arise from polarity in covalent bonds and unsymmetrical motions of electrons about the nuclei.

Polar Bonds—Polar Molecules

An atom's ability to attract *shared* electrons to itself is indicated by its **electronegativity.** Electronegativity is introduced in Chapter 10, and the numerical values of electronegativity for several elements are given. In general, nonmetals have larger electronegativity values, ranging from about 2.0 to 4.0, and metals have smaller values, starting as low as 0.8, up to about 1.8. The larger the electronegativity number, the greater ability that atom has to attract electrons it shares with other atoms in a covalent bond. Fluorine has the highest electronegativity of any element commonly found in compounds, 4.0. In the following figure, several elements are placed along a line of electronegativity values. This arrangement will make it easier to get a measure of the "difference in electronegativity" between two elements that share electrons in a covalent bond—the farther the elements are apart, the greater the difference in their electronegativities.

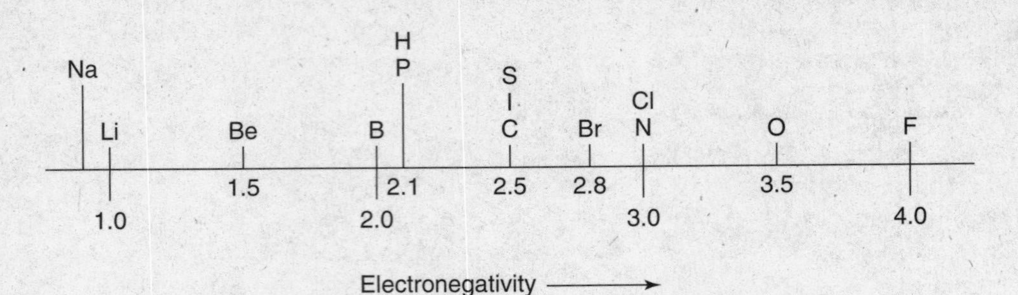

A pair of electrons shared between two atoms is a covalent bond. If the two atoms are identical, as would be the case in H–H or F–F, the electron pair is shared equally between the two atoms. Both atoms have the same electronegativity, and neither has a greater attraction for the shared pair, so they are shared equally. But if a pair of electrons is shared between two different atoms, as in H–F, the bonding pair is shared unequally because the electronegativities of the two atoms are not the same. Fluorine has an electronegativity of 4.0, and hydrogen has an electronegativity of 2.1, causing the electron pair to be drawn closer to fluorine and away from hydrogen. The fluorine end of the HF molecule becomes richer in negative charge, δ–, at the expense of the hydrogen end, which becomes partially positive, δ+. The small Greek delta, δ, indicates "partial charge"—one that is not as large as the full 1– or 1+ charge of an electron or proton. The unequal sharing of the bonding pair makes the bond *polar,* which, in turn, causes the entire HF molecule to be polar—that is, acting as an electric dipole. An *electric dipole* is a body with two electric poles, one negative, δ–, and the other positive, δ+. In the following figure, an arrow with a cross at one end is used to symbolize the dipole, with the head of the arrow pointing toward the δ– end of the bond, with the crossed end placed at the δ+ end. This kind of arrow is also used to indicate polarity in individual molecules. The polar covalent bond in the HF molecule causes the entire molecule to act as a dipole.

$$\delta^{+}\,H—F\,\delta^{-}$$

The greater the difference in electronegativity of two bonded atoms, the more polar is the bond between them. The polarity of the bond in HF is relatively large because F and H differ in electronegativity by a relatively large 1.9 units: $\Delta EN = 4.0_{(F)} - 2.1_{(H)} = 1.9$. The bond in HCl is less polar than the bond in HF because the electronegativity difference between Cl and H is smaller: $\Delta EN = 3.0_{(Cl)} - 2.1_{(H)} = 0.9$. H and Cl are not as widely separated on the electronegativity line compared to H and F.

The polar bonds in HF and HCl cause both molecules to be described as polar molecules. CO, HBr, and ICl are also polar molecules. But some molecules that have polar bonds do not end up being polar molecules. If the shape of a molecule is symmetric (linear, trigonal planar, or tetrahedral), it is possible for the polarity of the two, three, or four bonds to cancel the polarity of one another, resulting in a molecule with no overall polarity, a nonpolar molecule. Carbon dioxide, CO_2, boron trifluoride, BF_3, and carbon tetrachloride, CCl_4, are nonpolar molecules, even though every bond in them *is* polar. (You might want to review the VSEPR shapes of molecules in Chapter 10.) The following figure shows three molecules that are nonpolar yet have polar bonds. The polarities of the linear, trigonal planar, and tetrahedral arrangements of identically polar bonds cancel, and the result in each case is a nonpolar molecule.

O = C = O

Carbon dioxide
Linear

F, B, F, F

Boron trifluoride
Trigonal planar

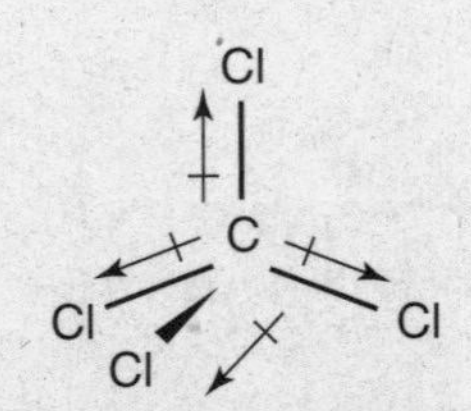

Carbon tetrachloride
Tetrahedral

Not *every* linear, trigonal planar, or tetrahedral molecule is nonpolar. If all the bonds are identical and have the same polarity, then they cancel to form a nonpolar species; like a tug-of-war with 2, 3, or 4 identically strong teams, it's a standoff, and the strength of each team is nullified. But if the bonds are not all the same and have different degrees of polarity, they will not perfectly cancel, and the molecule will have an overall polarity. The following figure shows linear, trigonal planar, and tetrahedral molecules that are polar.

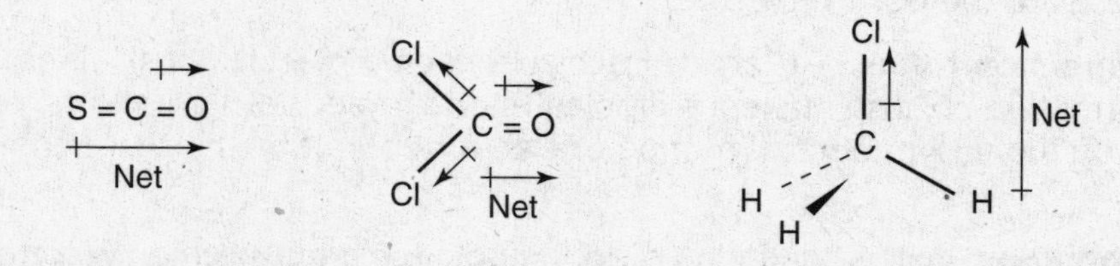

Water, H_2O, is a bent molecule, and ammonia, NH_3, has the trigonal pyramid structure. Both have polar bonds that, because of their shapes, do not cancel the effect of each other. They combine to make the molecules polar. In the following figure, you can see that the bent shape of water and the trigonal pyramidal shape of ammonia cause both to be polar molecules.

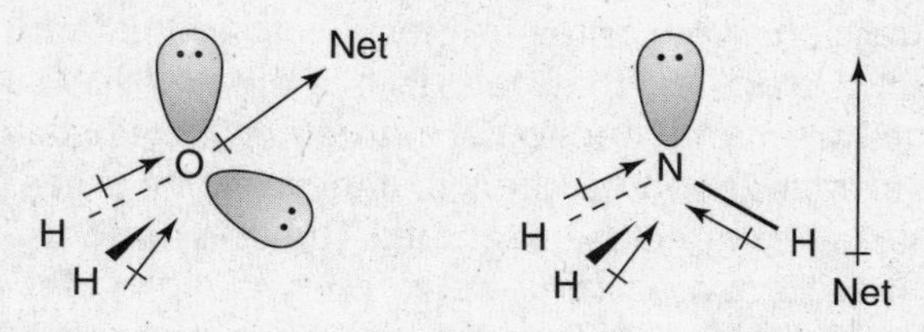

A physical property that measures the polarity of a molecule is its dipole moment. Without going through the details, the more polar the molecule, the larger its dipole moment. Several molecules are compared in the following table. The unit of the dipole moment is the Debye, (D).

The Dipole Moment of Several Molecules

Molecule	***Molecular Shape***	***Dipole Moment (D)***
HF		1.82
HCl		1.08
H_2O	bent	1.85
NH_3	trigonal pyramid	1.47
CO_2	linear	0 (nonpolar)
SCO	linear	0.71
BF_3	trigonal planar	0 (nonpolar)
$Cl_2C=O$	trigonal planar	1.17
CCl_4	tetrahedral	0 (nonpolar)
$CHCl_3$	tetrahedral	1.01

Polar covalent bonds can cause a molecule to act like an electric dipole, possessing a permanent dipole, partially negative on one end, partially positive on the other. When molecules with permanent dipoles get close together, the natural attraction of δ+ for δ– and δ– for δ+ draw them together, and this is the basis of the intermolecular forces of attraction.

The Intermolecular Forces of Attraction

Intermolecular forces are sometimes referred to as **van der Waals forces,** after the Dutch scientist Johannes van der Waals (1837–1923). The three most important intermolecular forces that can exist between neutral molecules are:

- **Dipole-dipole forces:** Attractive forces between molecules that possess permanent dipoles; these are the polar molecules.
- **London dispersion forces:** Attractive forces between molecules of all kinds, which result from the formation of "instantaneous dipoles" in the molecules. The short-lived dipoles are caused by the unsymmetrical motion of electrons about the nuclear framework of the molecule.
- **Hydrogen bonding:** A special kind of dipole-dipole force between a hydrogen atom, which is bonded to either F, O, or N, and the nonbonding electrons on a second, very electronegative F, O, or N atom. The small hydrogen atom bridges the two very electronegative atoms, drawing them together.

One or more of these forces hold neutral molecules together in liquids and solids. It is the balance between these forces of attraction, which draw molecules together, and the kinetic energy of motion, which tends to separate molecules, that determines the physical state of a substance at any particular temperature. If the forces of attraction are greater, the substance is liquid or solid. If the kinetic energy exceeds the attractive forces, the substance will be a gas. Intermolecular forces are only effective if the molecules are very close together. The only time these forces can be effective in a gas is during that brief instant of a collision between two molecules. If the attractive forces are great enough, the molecules will stick together, the first step in condensing a gas to form a liquid. This will happen in all gases if the temperature (kinetic energy) is low enough. This is why many substances we think of as gases are liquids only at very low temperatures.

Dipole-Dipole Forces

As the name implies, dipole-dipole forces are those between dipoles when they are very close together or touching one another. The $\delta-$ end of one polar molecule is attracted to the $\delta+$ end of another, drawing them together. Depending on the orientation of the dipoles, the forces can be either attractive or repulsive, but the *more stable* and more likely arrangement will maximize the attractive interactions. Only molecules that possess permanent dipoles, like H_2O or HCl, can engage in dipole-dipole interactions; nonpolar molecules like CCl_4, CO_2, or I_2 cannot. The following figure shows how dipoles align to maximize attractions between them.

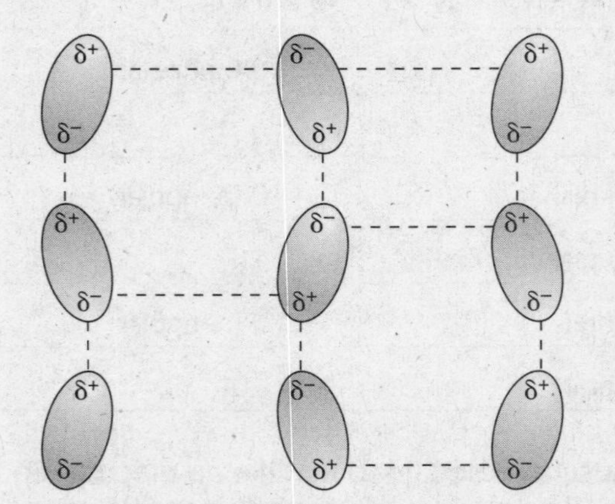

As you might expect, the strength of dipole-dipole attractive forces increases as the polarity of molecules increases. As with all the intermolecular forces, dipole-dipole forces are not as strong as covalent bonds between atoms, yet they play an important role as a force between molecules.

London Dispersion Forces

It is not difficult to imagine attractive forces between molecules that are dipoles, but what kind of intermolecular forces exist between nonpolar species, like CCl_4? Carbon tetrachloride would not be a liquid at room temperature if there were no attractive forces keeping the molecules together. It turns out that *all* molecules can form very short-lived dipoles, "instantaneous dipoles," as their electron clouds briefly shift off-center from the nuclear arrangement (electrons move much more quickly than the heavier nuclei). It is as if the electron cloud sways a little to one side of the molecule, momentarily making one end of the molecule a bit more negative, δ–, than the opposite end, δ+. The next instant, the electron cloud sways in another direction, again creating a weak but effective dipole. Adjacent molecules will respond to the neighboring dipole and be induced to couple movement of their electron clouds to generate a dynamic network of attractive forces. The short-lived δ+ end of one molecule will induce a δ– charge in an adjacent molecule, forming another instantaneous dipole, which induces the next molecule, and the next throughout the liquid. Then in the next instant, the dipoles reform in another arrangement. The overall effect is a significant intermolecular attractive force.

The following figure shows that London dispersion forces are attractive forces between "instantaneous dipoles" that result from the flexing of the negative electron cloud about the positive nuclear framework of an atom or molecule. The short-lived dipoles induce formation of more dipoles in adjacent molecules, which attracts them to one another. After an instant, the whole attractive structure of "instantaneous dipoles" will form again in a different direction.

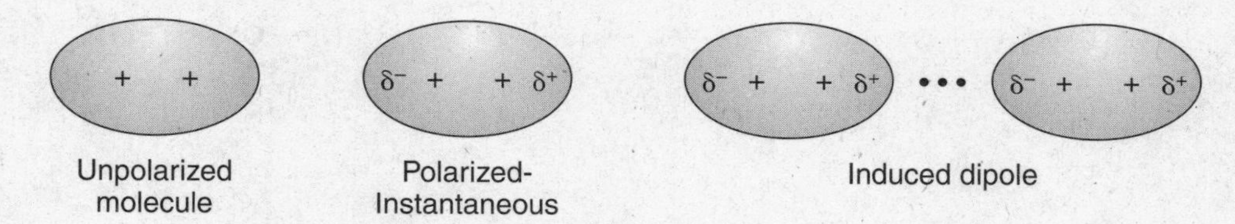

These are the London dispersion forces of attraction, or London forces, proposed in 1928 by Fritz London. Every atom, ion, or molecule can engage in London forces, as long as it has at least one electron. London forces are the *only* intermolecular forces possible in nonpolar substances. London forces are significant *only* when molecules are very close together, essentially touching. They are the forces that are responsible for carbon tetrachloride being a liquid at room temperature. The magnitude of London forces increases as the molecular masses of molecules increase.

In the diatomic halogen molecules, London forces get larger from Cl_2 (71 amu) to Br_2 (160 amu) to I_2 (254 amu), which helps us understand why I_2 is a solid, Br_2 a liquid, and Cl_2 a gas at room temperature. The London forces are strong enough in I_2 to hold the molecules tightly together in a solid, but in Br_2 they are only strong enough to keep Br_2 a liquid at room temperature. The kinetic energy of Cl_2 molecules at room temperature is large enough to overcome the London forces between them, keeping the rapidly moving molecules in the gas phase.

For polar molecules with large molecular masses, like $HCCl_3$, ICl, or SCl_2, the London forces can be a more significant attractive force between molecules than the dipole-dipole forces, although with smaller polar molecules, the dipole-dipole forces are usually stronger than the London forces.

Hydrogen Bonding

Hydrogen bonding is the strongest of the three intermolecular forces, and it exists only in those molecules that have a hydrogen atom bonded to fluorine (H–F), oxygen (H–O), or nitrogen (H–N). These H–X covalent bonds are highly polar with the small hydrogen atom bearing a sizable δ+ charge that enhances its attraction for the nonbonding electrons on a second, small, δ–, highly electronegative F, O, or N atom in a neighboring molecule. The small size of the hydrogen atom allows it to be a δ+ bridge between two δ– atoms, holding them together.

Although it is called a hydrogen "bond," it is not a bond in the same sense that a covalent bond is a bond. The following figure shows a hydrogen bond between two water molecules

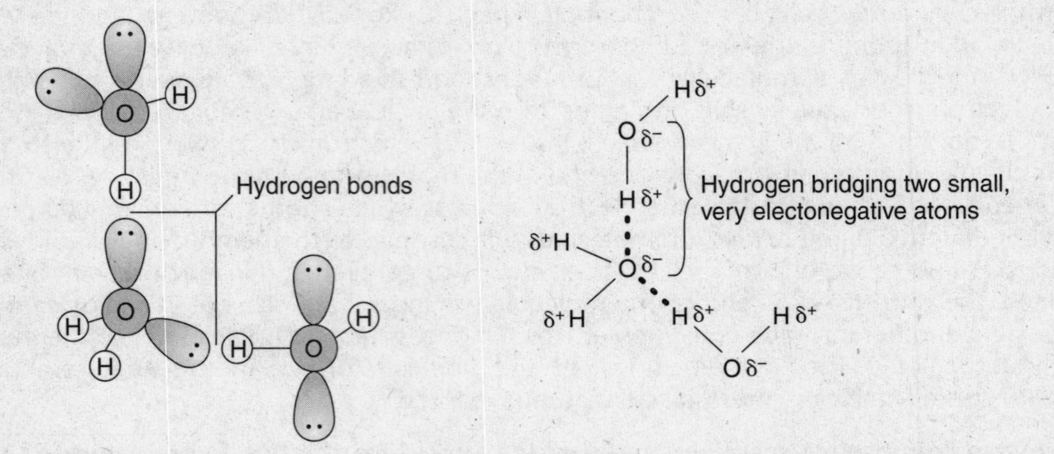

The next figure shows examples of hydrogen bonding. The hydrogen atom bridges two, small, highly electronegative atoms.

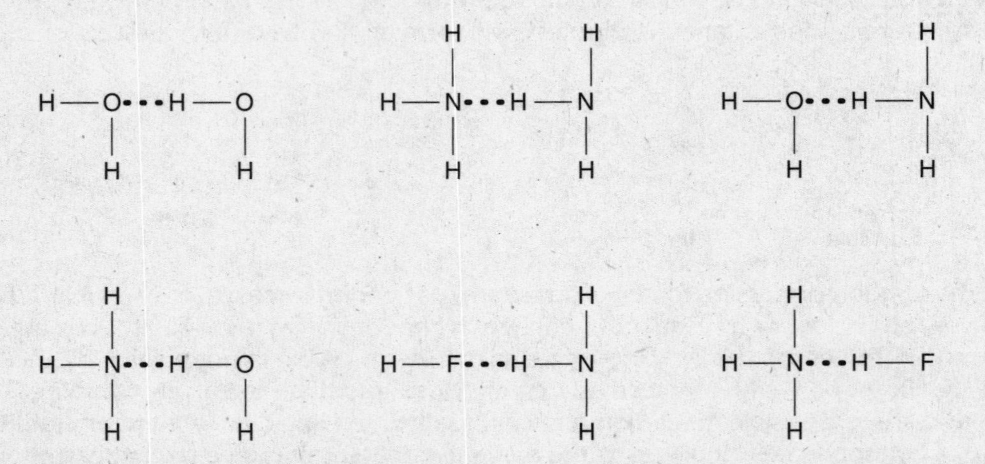

Unlike London dispersion forces that exist in *all* species, or dipole-dipole forces that exist in polar molecules, hydrogen bonding is restricted to only those molecules that have an H–F, H–O, or H–N bond. There is only one molecule that has the H–F bond, and that is hydrogen fluoride, HF, but the other two bonds appear in thousands of molecules. Hydrogen bonding is the strongest of the three intermolecular forces, although still far weaker than normal covalent bonds between atoms. It is the dominant force holding strands of nucleic acids together to form the double helix. It is responsible for water being a liquid at room temperature and for ice being less dense than liquid water and floating on its surface. When other liquids freeze, the solid crystals settle to the bottom.

Comparing the Intermolecular Forces

If intermolecular forces of attraction did not exist, most molecular substances would be gases at room temperature and pressure. There would be no attractive force between molecules causing them to cling together to form liquids or solids. You can determine the intermolecular forces in a substance using the following flowchart.

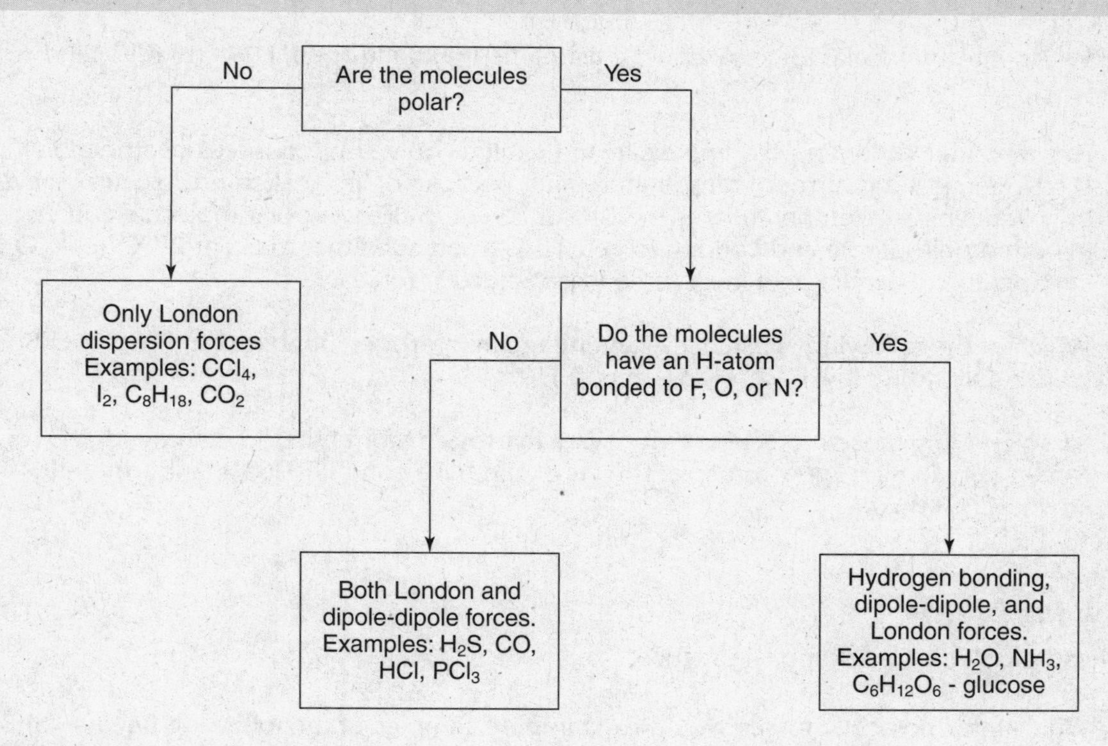

In summary,

- In terms of the relative strengths of the intermolecular forces:

  covalent bonds >>> hydrogen bonds > dipole-dipole forces > London dispersion forces

  There will be instances in which London forces may be the stronger force compared to dipole-dipole forces, especially with larger molecules. For small molecules, the general trend is as shown.

- Intermolecular forces operate only when molecules are very close together or touching one another. They are short-range forces.

- The strength of dipole-dipole forces increases with the polarity of the molecules. The more polar the molecule, the stronger the dipole-dipole attractive force. Only polar molecules can engage in dipole-dipole forces.

- The strength of London dispersion forces increases with the molecular masses of the molecules. The larger the molecular mass, the stronger the London attractive force. London forces exist in all substances.

- Hydrogen bonding occurs only in molecules that have an H-atom bonded to F, O, or N. The $\delta+$ hydrogen atom is attracted to nonbonding electron pairs on F, O, or N atoms in an adjacent molecule, bridging between the two very electronegative atoms.

Example Problems

These problems have both answers and solutions given.

1. List the following nonpolar molecules in order of increasing strength of the London forces of attraction: CH_4, CCl_4, CI_4, and CF_4.

   Answer: $CH_4 < CF_4 < CCl_4 < CI_4$ The strength of the London forces increases with increasing molecular mass (amu): CH_4 (16), CF_4 (88), CCl_4 (154), and CI_4 (520).

2. Which intermolecular force(s) would exist in: (a) liquid nitrogen, $N_2(l)$; (b) $ICl(l)$; and (c) $H_2O(l)$?

Answer: (a) N_2 is a nonpolar molecule and exhibits only London forces of attraction. Dinitrogen is a gas at room temperature and, because of its weak intermolecular forces, only liquefies at a temperature of −196°C or lower. (b) ICl is a polar molecule and engages in both dipole-dipole and London forces. ICl is a soft solid that melts at 27°C. (c) H_2O is very polar and can engage in all three intermolecular forces.

3. Which of the following substances can engage in hydrogen bonding: (a) PH_3; (b) H_2S; (c) C_2H_5OH; (d) CH_3NH_2; (e) H_2CF_2; and (f) H_2?

Answer: Only those molecules that have a hydrogen atom bonded directly to F, O, or N can engage in hydrogen bonding. They are (c) ethyl alcohol, $C_2H_5\mathbf{OH}$, and (d) methyl amine, $CH_3\mathbf{NH_2}$.

Work Problems

Use these problems for additional practice.

1. Which intermolecular forces would exist in pure samples of the following liquids: (a) $Br_2(l)$; (b) $CS_2(l)$; (c) $PCl_3(l)$; (d) $H_5C_3(OH)_3$ – glycerin?

2. The balance of intermolecular forces between molecules and the kinetic energy of those molecules determines whether the substance will be a molecular solid, liquid, or gas at a given temperature. What is it about these two properties that makes this so?

3. Explain the following observations: (a) CH_4 is a gas at room temperature, but CCl_4 is a liquid. (b) H_2O is a liquid at room temperature, but H_2S is a gas.

Worked Solutions

1. **(a) Only London forces are possible in nonpolar Br_2. (b) CS_2 is a linear molecule like CO_2 and is nonpolar.** Only London forces are possible in CS_2. **(c) PCl_3 is a trigonal pyramidal molecule with polar P–Cl bonds.** Both London and dipole-dipole forces are expected in PCl_3. **(d) The presence of –O–H groups in glycerin indicates that it can engage in hydrogen bonding.** It also can engage in dipole-dipole and London forces. (If you need to review molecular shapes, see Chapter 10.)

2. **Intermolecular forces are forces of attraction that draw and hold molecules together.** Kinetic energy is the energy of motion that can be great enough to overcome these attractive forces. One property seeks to bring molecules together; the other tends to keep them apart. Whichever is greater will determine whether the substance is solid, liquid, or gas at a given temperature. Temperature is important because the kinetic energy of molecules increases with increasing temperature; it doubles when absolute temperature doubles.

3. **(a) Both CH_4 and CCl_4 are tetrahedral, nonpolar molecules and can engage only in London forces, which are larger in CCl_4, sufficient to cause it to be a liquid at room temperature. (b) Both H_2O and H_2S are bent, polar molecules, but water can engage in hydrogen bonding, a much stronger force of attraction that causes it to be a liquid at room temperature.**

Phase Changes

When we speak of the **phase** of a substance, we are referring to its physical state: solid, liquid, or gas. Water has three common phases: solid (ice), liquid, and gas (steam). The change from one phase to another is called a phase change. **Phase changes** are physical changes of state; there is no change in the identity of the substance—that is, there is no decomposition. There are six kinds of phase changes, and each is accompanied by either the absorption of heat (an *endo*thermic process) or the release of heat (an *exo*thermic process). Each phase change involves increasing or decreasing (making or breaking) intermolecular forces. It requires energy to break or decrease the number of intermolecular forces, and energy is released with an increase in attractive forces.

- **Phase Changes for Solids:**

  Melting or **fusion** is the change of a solid to a liquid. Heat is *absorbed* as intermolecular forces are partially overcome, allowing the solid to melt to a liquid.

  Example: Heat + $H_2O(s) \rightarrow H_2O(l)$ (melting ice to cool a glass of tea)

  Sublimation is the change of a solid directly to a gas without going through the liquid state. Heat is *absorbed* as a solid sublimes.

  Example: Heat + $H_2O(s) \rightarrow H_2O(g)$ and Heat + $CO_2(s) \rightarrow CO_2(g)$ (the sublimation of dry ice).

  Although we are not aware of it happening, ice and snow sublime slowly on winter days. Water molecules escape directly from the surface of ice and enter the atmosphere as a gas. Dry ice gets its name from the fact that as a solid at −78°C, it can disappear (sublime) at normal atmospheric pressure without becoming a liquid. Freeze-dried foods are made by freezing them and removing the water by sublimation at low pressures.

- **Phase Changes for Liquids:**

  Freezing is the change of a liquid to a solid. Heat energy is *released* as a liquid freezes.

  Example: $H_2O(l) \rightarrow H_2O(s)$ + Heat (freezing water to make ice cubes)

  Vaporization, or **evaporation,** is the change of a liquid to a gas. Heat energy is *absorbed* as a liquid vaporizes. We have all sensed the cooling effect as water evaporates from our skin. Heat is removed from the body and absorbed by the liquid as it evaporates to a gas.

  Example: Heat + $H_2O(l) \rightarrow H_2O(g)$ (vaporizing water in a steam iron)

- **Phase Changes for Gases:**

  Condensation is the change of a gas to a liquid or directly to a solid. Both of these changes *release* heat energy as they take place. The change of a gas directly to a solid is also called **deposition.**

  Examples: $H_2O(g) \rightarrow H_2O(l)$ + Heat (gaseous water in the air condensing to form a fog)

  $H_2O(g) \rightarrow H_2O(s)$ + Heat (the formation of frost on cold windows in the winter)

Phase changes are constant temperature processes. The energy absorbed or released is not used to raise or lower the temperature of the substance. It is absorbed to overcome attractive forces or released as attractive forces increase in number. The temperature does not change. After a solid has melted to a liquid, the addition of more heat will raise the temperature of the liquid until it reaches its boiling point; then the temperature remains constant as the liquid turns to a gas. If more heat is added, the temperature of the gas will rise. The following figure shows the heating curve for water. The temperature does not change during the phase changes.

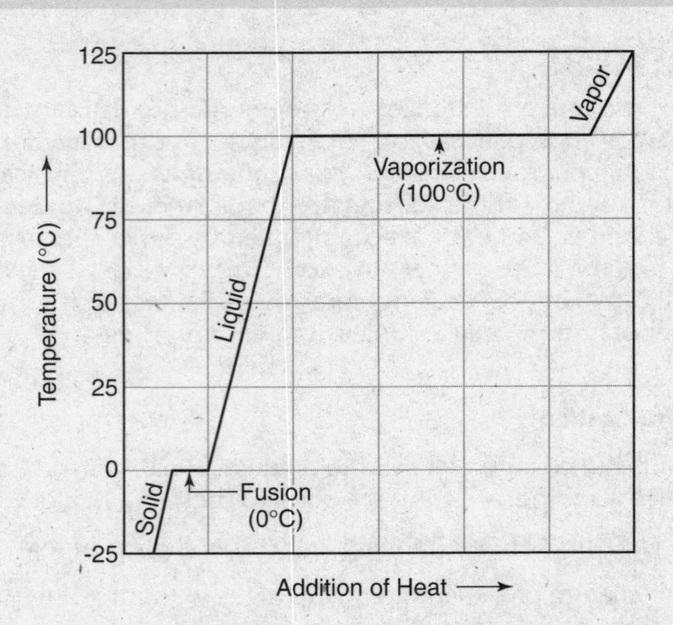

Each phase change can be paired with an opposing change.

Melting (fusion) and freezing: *(s)* →*(l)* and *(l)* → *(s)* — Heat + $H_2O(s)$ ⟷ $H_2O(l)$

Sublimation and deposition: *(s)* → *(g)* and *(g)* →*(s)* — Heat + $H_2O(s)$ ⟷ $H_2O(g)$

Vaporization and condensation: *(l)* → *(g)* and *(g)* → *(l)* — Heat + $H_2O(l)$ ⟷ $H_2O(g)$

Notice how each equation on the right uses an arrow with two points, ⟷, to indicate the two opposing changes. In each equation, the change from left to right (→) *consumes* heat energy; the change from right to left (←) *produces* heat energy. It requires the *same* amount of heat energy to melt one mole of water at 0°C as is released when one mole of liquid water freezes at 0°C.

The Molar Heat of Fusion and Vaporization

It requires heat energy to melt (fuse) a solid or vaporize a liquid. The amount of heat energy required to melt *one mole* of a solid is called the **molar heat of fusion.** The amount of heat energy required to vaporize *one mole* of a liquid is called the **molar heat of vaporization.** These phase changes are constant temperature processes, and the amount of heat is that involved in the phase change itself. Notice in the following table that the vaporization values are larger than those for fusion.

Molar Heats of Fusion and Vaporization

Compound	*Molar Heat of Fusion*	*Molar Heat of Vaporization*
Water, H_2O	6.01 kJ/mole	40.7 kJ/mole
Ethyl alcohol, C_2H_5OH	5.01 kJ/mole	38.7 kJ/mole
Carbon tetrachloride, CCl_4	2.51 kJ/mole	30.0 kJ/mole

The opposite of melting (fusion) is freezing. If 6.01 kJ of heat is *absorbed* to melt one mole of water, then 6.01 kJ of heat will be *evolved* if one mole of liquid water freezes.

$6.01\text{ kJ} + H_2O(s) \rightarrow H_2O(l)$ $\Delta H_{fusion} = +6.01\text{ kJ/mole}$

$H_2O(l) \rightarrow H_2O(s) + 6.01\text{ kJ}$ $\Delta H_{freezing} = -6.01\text{ kJ/mole}$

The same kind of thing holds for the vaporization and condensation of water.

$40.7\text{ kJ} + H_2O(l) \rightarrow H_2O(g)$ $\Delta H_{vaporization} = +40.7\text{ kJ/mole}$

$H_2O(g) \rightarrow H_2O(l) + 40.7\text{ kJ}$ $\Delta H_{condensation} = -40.7\text{ kJ/mole}$

The heat associated with chemical and physical changes, ΔH, and the sign of ΔH are discussed in Chapter 7. Calculations involving molar heats of fusion and vaporization are done in the following examples.

Example Problems

These problems have both answers and solutions given.

1. Identify each phase change with the proper term: vaporization, condensation (deposition), sublimation, melting (fusion), or freezing.

   (a) $CO_2(s) \rightarrow CO_2(g)$ (c) $C_6H_6(l) \rightarrow C_6H_6(g)$

   (b) $ICl(s) \rightarrow ICl(l)$ (d) $C_6H_4Cl_2(g) \rightarrow C_6H_4Cl_2(s)$

   Answer: (a) sublimation, (b) melting or fusion, (c) vaporization, and (d) condensation or deposition.

2. Identify each phase change in Question 1 as exothermic or endothermic and explain your choice.

   Answer: (a) Endothermic. It requires heat energy to separate molecules in a solid to form a gas. (b) Endothermic. It requires heat energy to overcome some of the intermolecular forces in a solid to form a liquid. (c) Endothermic, for the same reasons given in (a). (d) Exothermic. Energy is released as intermolecular forces form when the gaseous molecules pack together in the solid.

3. How much heat energy is required to melt 100. g of ice to liquid water at 0°C? The molar heat of fusion for water is 6.01 kJ/mole. The molar mass of water is 18.01 g. Also, state the answer in terms of a change in heat energy: ΔH = ______?

   Answer: 33.4 kJ of heat energy are required. ΔH = +33.4 kJ.

   First calculate the number of moles of solid water that are melted:

   $$\text{moles } H_2O = 100.\text{ g } \cancel{H_2O} \times \left(\frac{1\text{ mole } H_2O}{18.01\text{ g } \cancel{H_2O}}\right) = 555\text{ moles } H_2O$$

   Then calculate the amount of heat required to melt 5.55 moles of solid water. Use the molar heat of fusion as a conversion factor to convert moles of water to kJ of heat energy.

   $$\text{amount of heat absorbed} = 5.55\ \cancel{\text{moles } H_2O} \times \left(\frac{6.01\text{ kJ}}{1.00\ \cancel{\text{mole } H_2O}}\right) = 33.36\text{ kJ} = 33.4\text{ kJ}$$

   Since heat is absorbed, the sign of ΔH is +. ΔH = + 33.4 kJ.

4. How many kilojoules of heat are evolved if 2.76 g (0.0599 mole) of ethyl alcohol vapor condense to liquid on a cool surface? Choose the appropriate heat of vaporization from the previous table. Also, state the answer as a change in heat energy: $\Delta H =$ ______?

Answer: 2.32 kJ of heat energy are released. $\Delta H = -2.32$ kJ.

Condensation is the opposite of vaporization. If 38.7 kJ of heat are required to vaporize one mole of ethyl alcohol, 33.7 kJ of heat are evolved when one mole of alcohol condenses.

$$\text{amount of heat evolved} = 0.0599\ \cancel{\text{mole alcohol}} \times \left(\frac{38.7\ \text{kJ}}{1.00\ \cancel{\text{mole alcohol}}}\right) = 2.32\ \text{kJ}$$

Since heat energy is evolved, the sign of ΔH is $-$. $\Delta H = -2.32$ kJ.

5. 100. kJ of heat were absorbed by a block of ice at 0°C. How many grams of solid ice melted to liquid water? The molar heat of fusion equals 6.01 kJ/mole.

Answer: 299 g of ice melted.

First, calculate the number of moles of ice that 100. kJ of heat would melt.

$$\text{moles } H_2O = 100.\ \cancel{\text{kJ}} \times \left(\frac{1.00\ \text{mole } H_2O}{6.01\ \cancel{\text{kJ}}}\right) = 16.6\ \text{moles } H_2O$$

Then convert the moles of water to mass of water.

$$\text{mass } H_2O = 16.6\ \cancel{\text{moles } H_2O} \times \left(\frac{18.01\ \text{g } H_2O}{1.00\ \cancel{\text{mole } H_2O}}\right) = 299\ \text{g } H_2O$$

Work Problems

Use these problems for additional practice.

1. How many kilojoules of heat are required to vaporize 20.5 moles of water at the boiling point of water? The molar heat of vaporization for water is 40.7 kJ/mole. State the answer as a change in heat energy, $\Delta H =$ ______?

2. How much heat is evolved if 225 g of liquid water freeze at 0°C? The molar heat of fusion of water is 6.01 kJ. State the answer as a change in heat energy, $\Delta H =$ ______?

3. 1,000. kJ of heat are absorbed by a beaker of CCl_4 at its boiling point. How many grams of CCl_4 will be vaporized by 1,000. kJ of heat? The molar heat of vaporization of CCl_4 is 30.0 kJ, and the molar mass of CCl_4 is 153.8 g.

Worked Solutions

1. **834 kJ of heat will be absorbed; therefore, $\Delta H = +834$ kJ.**

$$\text{amount of heat absorbed} = 20.5\ \cancel{\text{moles } H_2O} \times \left(\frac{40.7\ \text{kJ}}{1.00\ \cancel{\text{moles } H_2O}}\right) = 834\ \text{kJ}$$

2. **75.1 kJ of heat energy are released, so $\Delta H = -75.1$ kJ.**

225 g of water are 12.5 moles water. You should calculate this yourself to verify.

Realizing that fusion and freezing are opposite phase changes, the amount of heat energy evolved as 12.5 moles of liquid water freeze to ice is

$$\text{amount of heat evolved} = 12.5 \text{ moles } H_2O \times \left(\frac{6.01 \text{ kJ}}{1.00 \text{ mole } H_2O}\right) = 75.1 \text{ kJ}$$

3. **5.12×10^3 g of CCl_4**

First, calculate the number of moles of CCl_4 that can be vaporized with 1,000. kJ of heat energy.

$$\text{moles } CCl_4 = 1{,}000. \text{ kJ} \times \left(\frac{1.00 \text{ mole } CCl_4}{30.0 \text{ kJ}}\right) = 33.3 \text{ moles } CCl_4$$

Then convert moles of CCl_4 to mass.

$$\text{mass } CCl_4 = 33.3 \text{ moles } CCl_4 \times \left(\frac{153.8 \text{ g } CCl_4}{1.00 \text{ mole } CCl_4}\right) = 5.12 \times 10^3 \text{ g } CCl_4$$

Liquids

A liquid, like a gas, is a fluid and can flow from one container to another. But unlike a gas, the molecules in a liquid are very close together and touching. Because the molecules are in contact with each other, liquids cannot be compressed to smaller volumes, they are *incompressible*. Because liquids can flow, they take the shape of their container and have a horizontal surface.

The intermolecular forces described in the previous section can help us understand some of the properties of liquids: Viscosity, surface tension, vapor pressure, and boiling points.

Viscosity

Water pours out of a glass easily, but honey does not; it pours as slowly as molasses. Some liquids flow with ease; others do not. Those that don't are described as *viscous liquids*. **Viscosity** is a measure of a fluid's resistance to flow; the greater the resistance to flow, the greater the viscosity. As you might suspect, the viscosity of a liquid is affected by the strength of the intermolecular forces in the liquid. The stronger the intermolecular forces, the greater the resistance to flow.

The shape of larger, longer molecules also affects viscosity if they can intertwine about each other, making it more difficult to flow by each other. Viscosity is an important property of motor oils that relates to the ability of the oil to lubricate at high or low temperatures. Oils with lower viscosity flow more easily and lubricate better at lower temperatures; those with higher viscosity are used at higher temperatures. One way viscosity is measured is by the time it takes for a measured volume of liquid to drain through a small hole in the bottom of a cup under the influence of gravity alone.

Surface Tension

Small water insects can move on the surface of water because the surface acts like a thin membrane that supports them. The water molecules on the surface experience forces of attraction that bind them across the surface and draw them into the body of the liquid, unlike the molecules within the body of the liquid that experience these forces equally in *all* directions, as shown in the following figure. The result is a tendency for the surface to act like a thin elastic membrane that attempts to contract to as small a surface as possible. This is why droplets of water suspended in air on the space shuttle are spherical. A sphere has the smallest surface area per unit volume of any geometrical shape. The following figure shows surface tension as the result of an imbalance in forces of attraction. Molecules within the liquid are attracted equally in all directions, while those on the surface are drawn across the surface and into the liquid.

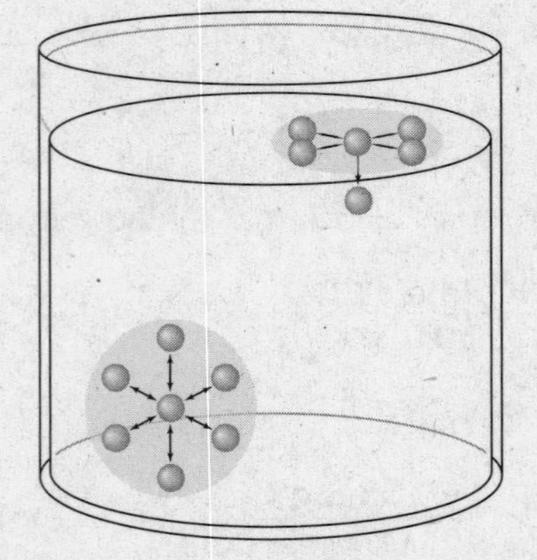

As you might expect, the magnitude of the surface tension of a liquid is affected by the strength of its intermolecular forces. The greater the intermolecular forces, the greater the surface tension. Because water has strong intermolecular forces, it has a very high surface tension. Soaps and detergents can disrupt the surface of liquids, and in doing so, greatly reduce the surface tension.

The Vapor Pressure of Liquids

Liquids are constantly evaporating; molecules at the surface break away from the liquid and enter the gas phase. If a glass of water is left out in a room, it's just a matter of time before all the water evaporates. But, what would happen if the water was in a closed container, so that molecules that evaporated were not allowed to escape into the atmosphere? The liquid would continue to evaporate, increasing the concentration of molecules in the gas phase, $(l) \rightarrow (g)$. Then, as the concentration of gaseous molecules increases (they cannot escape), the likelihood of some of them crashing into the surface of the liquid and returning to the liquid increases, $(g) \rightarrow (l)$. Both vaporization and condensation take place, and in time, a condition of **equilibrium** is established—each time a molecule leaves the liquid and enters the gas phase (vaporization), another molecule leaves the gas phase and returns to liquid (condensation).

At equilibrium, the rates of evaporation and condensation are the same. The concentration of molecules in the gas phase no longer changes, because each time a molecule leaves the gas phase, another enters; the rate in and out is the same. The state of dynamic equilibrium can be symbolized in an equation using two arrows that point in opposite directions, $\leftrightarrows$, to indicate the two opposing changes, vaporization and condensation, taking place at the same time and at the same rate.

$$H_2O(l) \leftrightarrows H_2O(g)$$

The **vapor pressure** of a liquid is the pressure exerted by its vapor (gas) when the liquid and vapor are at equilibrium.

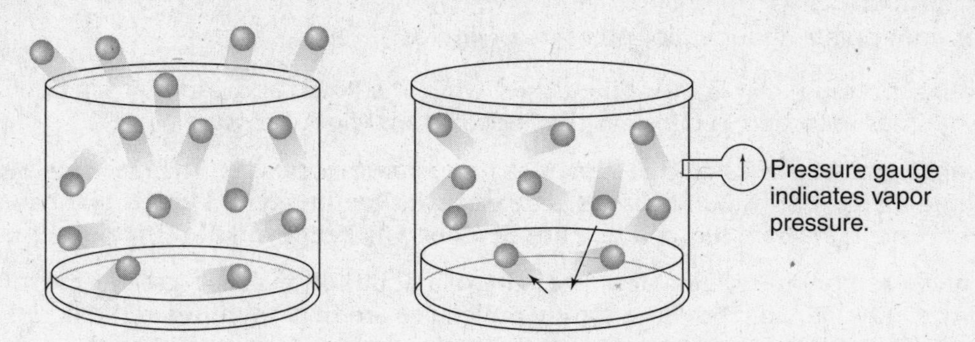

If evaporating molecules are not allowed to escape into the atmosphere, an equilibrium between evaporation and condensation will be established. The vapor pressure of the liquid is the pressure exerted by its vapor at this equilibrium.

The vapor pressures of all liquids increase with temperature. At higher temperatures, molecules have greater kinetic energies, and more molecules are able to escape the liquid surface and enter the gas phase. The vapor pressures of three liquids from 0°C to 100°C are shown in the following figure. All three liquids are polar molecules, but unlike water and ethyl alcohol, diethyl ether cannot engage in hydrogen bonding. The sequence of the strength of intermolecular forces is: diethyl ether (least) < ethyl alcohol < water (greatest). The **normal boiling point** of a liquid is the temperature at which its vapor pressure equals 760 mmHg, as shown in the figure. The effect of atmospheric pressure on boiling points of liquids can also be seen in the figure. Tracing the curve for water, as pressure falls from 760 mmHg, the boiling temperature becomes less than 100°C. A liquid will boil when its vapor pressure equals the external pressure of the atmosphere. The average atmospheric pressure on the top of Mt. Everest is about 275 mmHg. At that pressure, water boils at 73°C. Verify this using the figure.

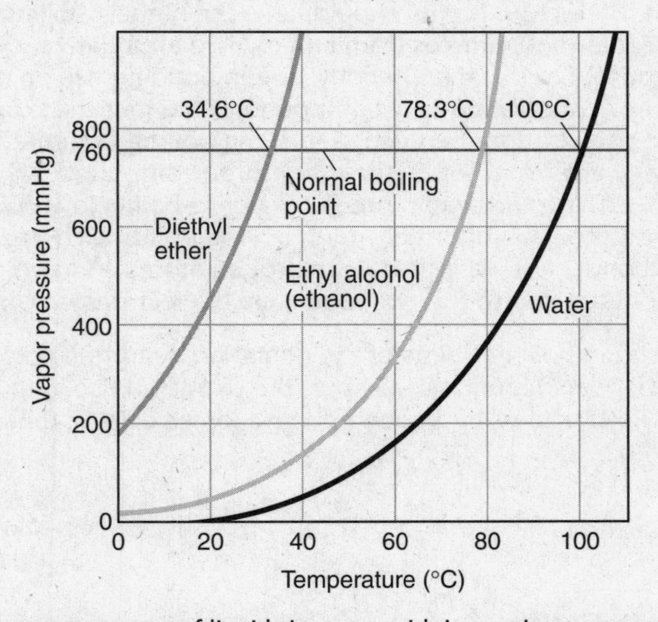

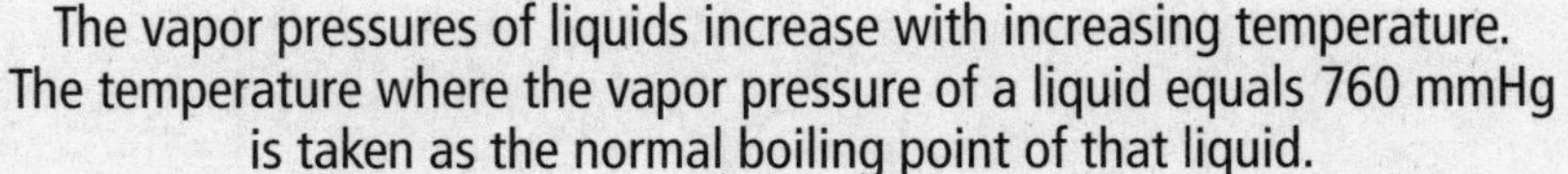
The vapor pressures of liquids increase with increasing temperature. The temperature where the vapor pressure of a liquid equals 760 mmHg is taken as the normal boiling point of that liquid.

Freezing points (melting points) of solids are not as severely affected by pressure as boiling points, still the **normal freezing point** of a liquid is that measured at an external pressure of 760 mmHg.

A few important points about vapor pressure of liquids:

- The vapor pressure of a liquid is measured while the liquid and vapor are in equilibrium. It is sometimes recorded in tables as the "equilibrium vapor pressure."
- The vapor pressure of a liquid *increases* as temperature *increases*. The rate of increase is not the same for all liquids, but it always increases with temperature. Liquids that have a high vapor pressure at room temperature are described as being **volatile.** They evaporate quickly.
- At a given temperature, the vapor pressure of a liquid is lower the greater the intermolecular forces in the liquid. The more tightly molecules are held together in the liquid phase, the more difficult it is for them escape into the gas phase.
- The temperature at which the vapor pressure of a liquid equals 760 mmHg is taken as the **normal boiling point** of that liquid.

The Boiling Point of Liquids

Intermolecular forces have a marked effect on the normal boiling points of liquids. The stronger the intermolecular forces, the higher the temperature needed to bring the liquid to a boil. As temperature increases, the kinetic energy of the molecules increases until they are moving with sufficient vigor to overcome the forces holding them together throughout the entire volume of the liquid. At the boiling temperature, rapid vaporization occurs throughout the entire liquid, not just at the surface. This is why liquids bubble as they boil.

Perhaps the most remarkable example of the effect of intermolecular forces on boiling points is the trend seen in the hydrogen compounds of the Group VIA elements, H_2O, H_2S, H_2Se, and H_2Te. All are polar molecules with increasing London forces from H_2O to H_2Te. If only London forces were considered, H_2Te (129.6 amu) should have the highest boiling point and water (18.0 amu) the lowest. The dipole-dipole forces from H_2S to H_2Te are relatively weak compared to water. But, here is the surprise. Look at the effect hydrogen bonding has on the boiling point of water in the following figure. Water is the *only* compound of the four than can engage in hydrogen bonding, a strong force of attraction. Because of this, the boiling temperature of water deviates markedly from the downward trend set by the other three compounds: H_2Te (–2°C), H_2Se (–42°C), and H_2S (–60°C). The trend would predict water to boil below –60°C, but in reality it has what would be considered an abnormally high boiling point. Each water molecule can engage in four hydrogen bonds, two with its own hydrogen atoms and two with hydrogen atoms from adjacent molecules attracted to the two lone pairs of electrons on oxygen.

For comparison, the trend in boiling points of the nonpolar, hydrogen compounds of the Group IVA elements—CH_4, SiH_4, GeH_4, and SnH_4—shows the expected rise with increasing strength of London forces. No dipole-dipole or hydrogen bonding forces exist in these compounds.

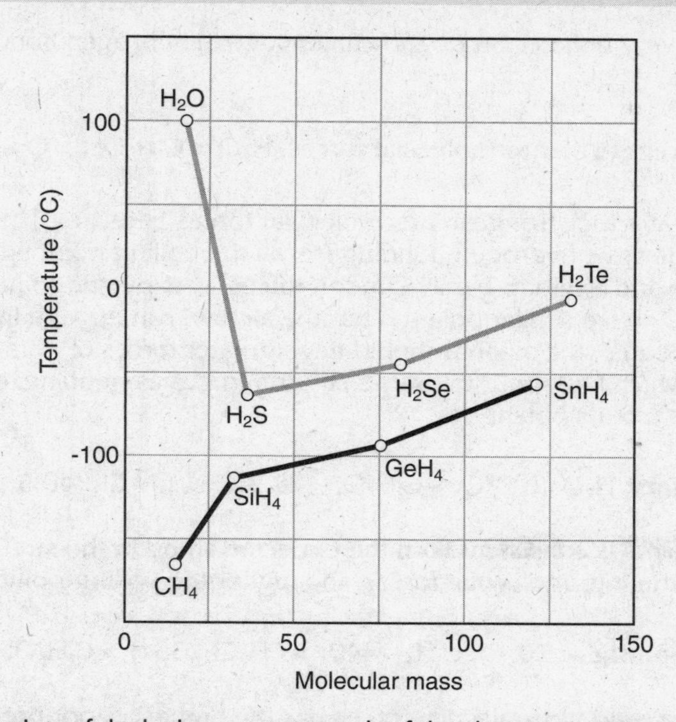

The boiling points of the hydrogen compounds of the elements in Groups VIA and IVA.

Comparing Intermolecular Forces and Properties of Liquids

In the preceding sections, some properties of liquids were related to the nature and strength of intermolecular forces within them.

- **Viscosity:** The viscosity of liquids increases as the strength of the intermolecular forces increases.
- **Surface tension:** The surface tension of liquids increases as the strength of intermolecular forces increases.
- **Vapor pressure:** At a given temperature, the greater the strength of the intermolecular forces in the liquid, the lower the vapor pressure of the liquids.
- **Normal boiling point:** The greater the strength of the intermolecular forces in the liquid, the lower the normal boiling points of the liquids.

Let's compare the trends of two of these properties, boiling points and vapor pressures, using four liquids that span a range of intermolecular forces. The molecular mass, in parentheses, is in amu.

- Pentane, C_5H_{12}: (72) A nonpolar liquid. Its only intermolecular force is the London force.
- Dichloromethane, CH_2Cl_2: (85) A polar liquid that can engage in both London and dipole-dipole forces. Its molecular mass is similar to that of pentane; expect similar London forces.
- Ethyl alcohol, C_2H_5OH: (46) A polar liquid that can engage in all three intermolecular forces. The polarity of ethyl alcohol (1.69 D) is nearly the same as that of dichloromethane (1.60 D); expect similar dipole-dipole forces.

- Water, H_2O: (18) Very polar (1.85 D) and can engage in hydrogen bonding more extensively than ethyl alcohol.

Expected strength of intermolecular forces: $H_2O > C_2H_5OH > CH_2Cl_2 > C_5H_{12}$

Water is expected to have the strongest intermolecular forces because of its greater polarity and ability to engage in extensive hydrogen bonding (recall the boiling point figure). Pentane, not being a large molecule and exhibiting only London forces, is expected to have the weakest. Ethyl alcohol and CH_2Cl_2 have similar polarity, but the alcohol can engage in hydrogen bonding. All else being roughly equal, the alcohol should have greater forces of attraction compared to CH_2Cl_2. And, CH_2Cl_2, which has about the same London forces as pentane, exceeds pentane in attractive forces because of its polarity.

Normal boiling points: H_2O (100°C) > C_2H_5OH (78.3°C) > CH_2Cl_2 (40.5°C) > C_5H_{12} (36.1°C)

The trend in boiling points is consistent with the expected trend in the strength of intermolecular forces; the greater the intermolecular forces, the higher the boiling points.

Vapor pressure (mmHg at 20°C): C_5H_{12} (440) > CH_2Cl_2 (351) > C_2H_5OH (45) > H_2O (18)

The liquid with the weakest intermolecular forces has the highest vapor pressure at 20°C, and the trend is consistent with the trend in boiling temperatures; the liquid with the highest vapor pressure at 20°C has the lowest normal boiling point.

Example Problems

These problems have both answers and solutions given.

1. Examining the figure of vapor pressures at various temperatures, estimate the boiling point of ethyl alcohol at 600 mmHg.

   Answer: A vapor pressure of 600 mmHg corresponds to approximately 73°C.

2. Why is it necessary to state boiling points of liquids as normal boiling points?

   Answer: The external pressure, most commonly the pressure of the atmosphere, affects the boiling temperature of liquids. So that boiling points of liquids are compared under the same conditions, the temperature at which a liquid boils under standard atmospheric pressure, 760 mmHg, is chosen as the benchmark. It levels the playing field when comparing boiling temperatures.

3. Water is the lowest molecular mass compound known that is a liquid at room temperature. Why is water, H_2O (18.02 amu), a liquid at 25°C while methane, CH_4, with a similar molecular mass (16.04 amu), is a gas?

   Answer: Water, unlike CH_4, can engage in extensive hydrogen bonding, the strongest intermolecular force. Because of this, water molecules are much more strongly attracted to each other than are the molecules in CH_4, which only experience London forces. For this reason, CH_4 is a gas at a temperature at which water is liquid.

Work Problems

Use these problems for additional practice.

1. The vapor pressure of a liquid increases with increasing temperature. (a) What is special about the temperature of the liquid when its vapor pressure is 760 mmHg? (b) Tabulated values of vapor pressure are measured under the conditions of equilibrium. In terms of the two processes, vaporization and condensation, describe that equilibrium.

2. What is the relationship between the strength of intermolecular forces and the vapor pressure of a liquid at a particular temperature?

3. Arrange these three compounds in order of decreasing vapor pressure at 20°C. They are all nonpolar molecules: Pentane, C_5H_{12}, octane, C_8H_{18}, and hexane, C_6H_{14}.

Worked Solutions

1. **(a) The temperature of a liquid when its vapor pressure is 760 mmHg is the normal boiling point of the liquid. (b) At equilibrium, the vaporization of the liquid and the condensation of its vapor occur at the same rate, which maintains a constant concentration of molecules in the gas phase.**

2. **The stronger the intermolecular forces between the molecules in a liquid, the lower is its vapor pressure at any particular temperature.**

3. **These nonpolar molecules will experience only London dispersion forces of attraction that increase with the molecular mass of the compounds.** The sequence of decreasing vapor pressure will be in the order of increasing molecular mass: C_5H_{12} (72 amu) > C_6H_{14} (86 amu) > C_8H_{18} (114 amu).

Solids

Solids have a definite shape and volume and, unlike liquids, do not flow. Like liquids, they are incompressible. Most solids are **crystalline solids,** ones in which the particles that make them up, be they atoms, molecules, or ions, are arranged in highly organized structures and held rigidly in place by strong attractive forces. Other solids are not crystalline; they are amorphous (without form). **Amorphous solids** do not have a highly organized arrangement of particles; they are often composed of long-chain molecules, intertwined, and held together in random arrangements, as is seen in rubber, glass, and many plastics. Some solids are pure substances, such as copper, ice, sucrose (table sugar), and sodium chloride (table salt), and others are solid mixtures, like stone, hard candy, wood, and brass. The focus in this section will be on crystalline solids that are pure substances.

Crystalline solids exist as crystals that might be microscopically small or quite large. One feature all crystalline solids share is a highly organized, three-dimensional arrangement of particles: ions, molecules, or atoms. Depending on the nature of the attractive forces holding the particles together, the crystals might be hard and brittle or soft and crushable, with very high or low melting points. Metallic solids can conduct electricity; ionic and molecular solids cannot. Many metallic solids are *ductile* (capable of being drawn into wires), *malleable* (can be hammered into thin sheets), and flexible. Crystalline solids can be broken down into four classes based on the kinds of forces that hold the structural units together: molecular, metallic, covalent-network, and ionic. Let's take a look at each class.

Molecular Solids

Molecular solids are composed of molecules or atoms. If the molecules are nonpolar, they are held together with London forces forming crystals that are usually soft and melt at lower temperatures. Examples are solid carbon tetrachloride, $CCl_4(s)$ (–23°C), and the solid benzene, $C_6H_6(s)$ (5.5°C). Other examples are the low-temperature solids of the noble gases: Ar*(s)* (–189°C) and Xe*(s)* (–112°C). These solids are composed of atoms bound by London forces.

Molecular solids made up of polar molecules form crystals that are a little harder and melt at temperatures a little higher because of the additional dipole-dipole or hydrogen bonding attractive forces. Examples are ice, sucrose (table sugar), and naphthalene (moth balls). The crystals of these solids are not very rigid and can be crushed rather easily. You can hear the crystals of sugar spilled on the kitchen floor crushing underfoot. Molecular solids do not conduct electricity.

Metallic Solids

All solid metallic elements are metallic solids in which metal atoms are held in a highly organized structure. The atoms are held together by a unique bonding arrangement that is neither purely ionic nor purely covalent. It is described as *metallic bonding* that, while holding the atoms together, allows most metals to be electrically conductive, that is, permitting a flow of electrons through the metal. The bonding between atoms can be relatively weak or quite strong, causing some metals to be soft (Na, K) and others very hard (Ti, Cr) with low or high melting temperatures, respectively. The unique bonding in metals allows many to be malleable and/or ductile. Some of our most important metals are aluminum, iron, copper, and gold.

Covalent-Network Solids

Covalent-network solids are three-dimensional networks of atoms held together by covalent bonds. Because covalent bonds are much stronger than intermolecular forces, and are highly directional, covalent-network solids are much harder and have much higher melting temperatures than molecular solids. Many of the hardest known substances are of this class: diamond (C), boron nitride (BN), and silicon carbide (SiC). Covalent-network solids are not electrical conductors with one important exception, graphite. Graphite, like diamond, is made up of carbon atoms covalently bonded together. In diamond, *every* carbon atom is covalently bonded to a tetrahedron of four other carbon atoms, a structure that repeats throughout the solid. But in graphite, the atoms are arranged in layers that are stacked on top of one another like pancakes. The layers are held together by London forces. The unique bonding *within* the layers allows them to conduct electricity and, because of the weak bonding *between* the layers, the layers slide easily over one another, which accounts for its use as a lubricant.

Ionic Solids

Ionic compounds are crystalline solids at room temperature, a consequence of the very strong ionic bonds that hold the positive and negative ions together. The arrangement of ions in the solid maximize the attraction between ions of opposite charge while minimizing the repulsion between ions of the same charge. Ionic crystals are usually hard and brittle with high melting temperatures. The higher the charge on the ions, the greater the ionic bonds holding them together, which is reflected in their melting points. The ions in sodium chloride each bear a single charge, Na^+ and Cl^-, and the compound has a melting point of 801°C. The ions in calcium oxide each bear a double charge, Ca^{2+} and O^{2-}, causing its melting point to be much higher, 2,707°C. Ionic solids do not conduct electricity, but when melted, the ions are free to flow in the liquid, and the ionic liquids do conduct electricity. Ionic compounds were discussed in Chapters 2 and 4.

Several solids of each class are listed in the following table with their melting points.

Examples and Properties of Solid Compounds

Species	*Type of Solid*	*Forces within Solid*	*Melting Point*
Pentane, C_5H_{12}	molecular	London	−130°C
Chloroform, $CHCl_3$	molecular	London/dipole	−64°C
Water, H_2O	molecular	London/dipole/H-bond	0°C
Sodium, Na	metallic	metallic bond	98°C
Chromium, Cr	metallic	metallic bond	1,857°C
Diamond, C	covalent-network	covalent bonds	3,500°C
Silicon carbide, SiC	covalent-network	covalent bonds	2,700°C (sublimes)
Sodium chloride, NaCl	ionic	ionic bonds	801°C
Calcium fluoride, CaF_2	ionic	ionic bonds	1,418°C
Magnesium oxide, MgO	ionic	ionic bonds	2,825°C

Chapter Problems and Answers

Problems

1. What is the difference between intramolecular forces and intermolecular forces?
2. What must be present in a molecule if it is a polar molecule?
3. Why is the bond between carbon and oxygen polar in carbon dioxide? If the carbon-oxygen bond is polar, why is the CO_2 molecule not polar?
4. What is meant by a change of state? Why is the decomposition of $H_2O(l)$ to form two gases, $H_2(g)$ and $O_2(g)$, not a change of state?
5. What is the term used to describe each of the following phase changes?

   (a) $Br_2(l) \rightarrow Br_2(s)$ (d) $Hg(s) \rightarrow Hg(l)$

   (b) $CO_2(s) \rightarrow CO_2(g)$ (e) $C_6H_6(l) \rightarrow C_6H_6(g)$

   (c) $H_2O(g) \rightarrow H_2O(s)$ (f) $C_5H_{12}(g) \rightarrow C_5H_{12}(l)$

6. List the type(s) of intermolecular forces that would be present in the following substances:

   (a) $Cl_2(l)$ (d) $Cl\text{-}NH_2(l)$

   (b) $Cl_2CH_2(l)$ (e) $CO_2(s)$

   (c) $HCF_3(l)$ (f) $Xe(l)$

7. Why is the boiling point of a polar liquid generally higher than the boiling point of a nonpolar liquid of similar molecular mass?

8. Under what condition is the vapor pressure of liquids measured? Does the vapor pressure of liquids increase or decrease as temperature increases?

9. What is the general difference between a gas and a vapor, between vaporization and boiling, and between a crystalline solid and an amorphous solid?

10. List the following compounds in order of increasing London dispersion force: CCl_4, SiF_4, CH_4, and CBr_4.

11. The boiling point of liquid chlorine, $Cl_2(l)$, is −34°C. Which would be a reasonable boiling point for liquid bromine, $Br_2(l)$: 59°C, −170°C, or −50°C?

12. What is the general relationship between the vapor pressure of a liquid and the strength of the intermolecular forces in the liquid?

13. Arrange these molecules in order of increasing strength of the dipole-dipole force: HI, HBr, HF, HCl. Justify your sequence with an explanation.

14. The molar heat of fusion for water is 6.01 kJ/mole. How much heat energy is needed to melt (fuse) 1,802 g (100. moles) of water at 0°C?

15. How many grams of solid benzene, C_6H_6, can be melted if 1,500 kJ of heat energy are absorbed at its melting temperature of 5.5°C? The molar heat of fusion for benzene is 9.87 kJ/mole. The molar mass of benzene is 78.1 g.

16. Thinking in terms of the kinetic energy of molecules, why does the vapor pressure of a liquid increase with increasing temperature?

17. Why is the vapor pressure of ethyl alcohol, $C_2H_5OH(l)$, higher than that of water, $H_2O(l)$, at the same temperature?

18. Both dimethyl ether and ethyl alcohol have the same formula, C_2H_6O. The molar heats of vaporization are 21.5 kJ/mole and 38.7 kJ/mole, respectively. Only one of the two can engage in hydrogen bonding. Which one is it? Justify your choice.

19. Explain why the normal boiling point of HF (19.4°C) is higher than the normal boiling point of HBr (–66.7°C), while the boiling point of Br_2 (58.8°C) is higher than that of F_2 (–188°C).

20. What mass of carbon disulfide, CS_2, can be vaporized by absorbing 1.35×10^3 kJ of heat energy at its boiling point? The molar heat of vaporization of CS_2 is 27.4 kJ/mole. The molar mass of CS_2 is 76.15 g.

Answers

1. **Intramolecular forces are those within the molecule, the bonds between atoms, whereas intermolecular forces are those between molecules.**

2. **If a molecule is to be polar, it must have a polar-covalent bond, a bond between two atoms of differing electronegativity.**

3. **The electrons shared between carbon and oxygen are drawn closer to oxygen because it has the greater electronegativity (O = 3.5, C = 2.5).** Both carbon-oxygen bonds are polar, but CO_2 is a linear molecule orienting the two bonds to be opposite each other. This cancels the polarity of each bond, making the molecule nonpolar overall.

4. **A change of state is a physical change from one state, solid, liquid, or gas, to another.** The decomposition of liquid water to form $H_2(g)$ and $O_2(g)$ is a chemical change, a decomposition.

5. **(a) freezing**

   (b) sublimation

   (c) condensation or deposition

   (d) melting or fusion

   (e) vaporization or evaporation

   (f) condensation

6. **(a) London**

   (b) London, dipole-dipole

   (c) London, dipole-dipole

   (d) London, dipole-dipole, and hydrogen bonding

   (e) London

   (f) London

7. **If the molecular masses of the molecules are similar, then the London forces in each should be similar.** But if one molecule can also engage in dipole-dipole attractive forces, it will have the greater total intermolecular attractive forces that will cause it to have the higher boiling point.

8. **The vapor pressure of liquids is measured at the point of equilibrium between the liquid and gaseous phases.** The vapor pressure of liquids increases with increasing temperature.

9. **A vapor is the gaseous phase of a substance that is normally a liquid at room temperature ($H_2O(g)$, $CCl_4(g)$), while a substance regarded as a gas is normally a gas at room temperature (N_2, O_2).** Both vapors and gases behave as gases. Vaporization is the transfer of molecules on the surface of a liquid to the gas phase; boiling occurs when vaporization occurs throughout the liquid as it converts to a gas. In a crystalline solid, the particles are packed in a highly organized way, but in an amorphous solid, they are packed together randomly.

10. **They are all tetrahedral, nonpolar molecules, and the strength of the London dispersion force increases with molecular mass.** The order of increasing London force is CH_4 (16.0 amu) < SiF_4 (104.1 amu) < CCl_4 (153.8 amu) < CBr_4 (331.6 amu).

11. **59°C is the best choice.** The London forces increase with molecular mass, so the boiling point for $Br_2(l)$ should be greater than that for $Cl_2(l)$. The other temperatures are lower and not expected.

12. **The stronger the intermolecular forces in a liquid, the lower its vapor pressure at a given temperature.** The stronger the intermolecular forces, the more energy it requires to escape the liquid phase and enter the gas phase.

13. **These are all two-atom molecules, so the strength of dipole-dipole force will depend only on the polarity of the bond.** The polarity increases with the increasing electronegativity difference between the two bonded atoms. Knowing that the electronegativity of H is 2.1

(Chapter 10), you can determine that the differences in electronegativity are HI (2.5 – 2.1 = 0.4), HBr (2.8 – 2.1 = 0.7), HCl (3.0 – 2.1 = 0.9), and HF (4.0 – 2.1 = 1.9). The order of increasing dipole-dipole force is HI < HBr < HCl < HF.

14. **601 kJ of heat energy are required.**

$$\text{amount of heat absorbed} = 100.\ \cancel{\text{moles } H_2O} \times \left(\frac{6.01\ \text{kJ}}{1.00\ \cancel{\text{mole } H_2O}}\right) = 601\ \text{kJ}$$

15. **1.19×10^4 g of benzene will be melted.** First, calculate the number of moles of benzene that can be melted with 1,500 kJ of heat energy, and then convert moles to mass of benzene.

$$\text{amount of heat absorbed} = 1{,}500\ \text{kJ} \times \left(\frac{1.00\ \text{mole}}{9.87\ \text{kJ/mole}}\right) = 152\ \text{moles}$$

$$\text{mass } C_6H_6 = 152\ \text{moles } C_6H_6 \times \left(\frac{78.1\ \text{g } C_6H_6}{1.00\ \text{mole } C_6H_6}\right) = 1.19 \times 10^4\ \text{g } C_6H_6$$

16. **If a molecule is to escape from a liquid and enter the gas phase, it must overcome the intermolecular forces holding it in the liquid phase.** As temperature increases, the average kinetic energy of the molecules in the liquid increases, giving more molecules the energy to overcome the forces holding them in the liquid phase and allowing them to evaporate, increasing the concentration of molecules in the vapor phase. Thus, vapor pressure increases.

17. **Both liquids, water and ethyl alcohol, can engage in all three intermolecular forces, but water can hydrogen bond more extensively than can the alcohol (water has two –O–H bonds; the alcohol has one), giving water stronger intermolecular forces to overcome in vaporization.** At a given temperature with lower intermolecular forces, it will be easier for the alcohol to escape the liquid phase than water, so it will always have a greater vapor pressure.

18. **The greater the forces of attraction in the liquid, the greater the heat of vaporization.** The greater heat of vaporization for the **ethyl alcohol** indicates that it has the greater forces of attraction, which would be so if it engaged in hydrogen bonding, which it does.

19. **The boiling point of HF is higher than that of HBr because HF can engage in strong hydrogen bonding, something HBr cannot do.** It's a different story with the diatomic elements. The only intermolecular force in these nonpolar molecules is the London force, which is greater in Br_2 (159.8 amu) than in F_2 (38.0 amu), causing Br_2 to have the higher boiling point.

20. **3.75×10^3 g of CS_2 can be vaporized.** First calculate the number of moles of CS_2 that can be vaporized with 1.35×10^3 kJ of heat energy, and then convert moles of CS_2 to mass.

$$\text{moles } CS_2 = 1.35 \times 10^3\ \text{kJ} \times \left(\frac{1.00\ \text{mole } CS_2}{27.4\ \text{kJ}}\right) = 49.3\ \text{moles } CS_2$$

$$\text{mass } CS_2 = 49.3\ \text{moles} \times \left(\frac{76.15\ \text{g } CS_2}{1.00\ \text{mole}}\right) = 3{,}754\ \text{g} = 3.75 \times 10^3\ \text{g } CS_2$$

Supplemental Chapter Problems

Problems

1. Which of the following molecules have polar bonds and are polar molecules?

   (a) O_3 – bent; (b) HCN – linear; (c) SO_3 – trigonal planar; (d) CH_3Br – tetrahedral

2. Why does water have such a high boiling point?

3. What does it mean when fusion and vaporization are described as constant temperature processes?

4. The boiling point of $NH_3(l)$ is −33°C, and that of $H_2S(l)$ is −60°C. Which would have the greater vapor pressure at −70°C?

5. In which of the following is hydrogen bonding an important intermolecular force? (a) FCH_3; (b) H_2S; (c) $C_6H_{12}O_6$ – glucose; (d) H_2O_2; (e) C_2H_5OH

6. How much heat energy is required to vaporize 0.850 kg of octane, C_8H_{18}? The molar heat of octane is 41.5 kJ/mole. The molar mass of C_8H_{18} is 114.2 g.

7. What kind of species exhibit London dispersion forces?

8. Why are the melting points of covalent-network solids so very high?

9. Arrange the following in order of increasing London forces: He, OCS, H_2O, HCl, and SO_3.

10. How much heat energy is evolved as 250 g of liquid ammonia freeze to form solid ammonia at its normal freezing point? The molar heat of fusion of ammonia is 5.65 kJ/mole. The molar mass of ammonia is 17.0 g. State the answer in terms of a change in heat energy, ΔH.

11. In terms of molecular arrangement, what must occur if intermolecular forces are to have any appreciable effect in a substance?

12. Compare the crystal hardness and melting temperatures of molecular solids and ionic solids.

13. The molar heat of vaporization of ethane, C_2H_6, a nonpolar molecule, is 14.7 kJ/mole. Which of the following values would you expect to be the molar heat of vaporization of propane, C_3H_8, also a nonpolar molecule: 8.2 kJ/mole, 18.8 kJ/mole, or 12.7 kJ/mole? Justify your selection.

14. If you spill fingernail polish remover on your skin, you sense a cooling sensation as it quickly evaporates. Why?

15. How many kilojoules of heat energy are evolved as 5.00 kg of water are frozen to ice at 0°C? The molar heat of fusion of water is 6.01 kJ/mole, and the molar mass of water is 18.02 g. State the answer in terms of a change in heat energy, ΔH.

Answers

1. **(a) no polar bonds, nonpolar molecule; (b) polar bonds and polar molecule; (c) polar bonds but nonpolar molecule; (d) polar bonds and polar molecule.** (page 331)

2. **The high boiling point of water is because of the extensive hydrogen bonding in liquid water.** Each water molecule can engage in a maximum of four hydrogen bonds, two with its own hydrogen atoms and two between its oxygen atom and hydrogen atoms from two other water molecules. (page 346)

3. **In fusion and vaporization, all the heat energy goes into breaking down intermolecular forces not changing the temperature of the substance.** (page 339)

4. **The one with the lower boiling point will have the higher vapor pressure at −70°C, H_2S.** (page 348)

5. **Hydrogen bonding is important in (c) glucose, a sugar, (d) H_2O_2, and (e) C_2H_5OH, ethyl alcohol.** (page 335)

6. **309 kJ.** (page 340)

7. **As long as one electron is in the species, it will exhibit the London dispersion force, and that includes everything except the hydrogen ion, H^+.** (pages 335–336)

8. **If a covalent-network solid is to melt to a liquid, many covalent bonds must be simultaneously broken, and that requires a great deal of energy, thus the high temperature.** (page 350)

9. **Order of increasing London force: He < H_2O < HCl < OCS < SO_3.** (page 346)

10. **$\Delta H = -83.1$ kJ.** (page 340)

11. **The molecules must be very close together or touching.** (page 347)

12. **Molecular solids have softer crystals with lower melting temperatures than ionic solids.** (page 350)

13. **Propane is a larger molecule, so expect a higher value, 18.8 kJ/mole.** (page 340)

14. **As the liquid evaporates, intermolecular forces between molecules in the liquid are broken, and this requires energy.** That energy can be drawn from your body, which you sense as cooling. Evaporation is a cooling process. (page 339)

15. **277.5 moles of water evolve 1,668 kJ of heat energy. $\Delta H = -1,668$ kJ.** (page 340)

Chapter 13
Solutions and Solution Concentrations

A **solution** is a homogeneous mixture of two or more substances (Chapter 2). A solution is considered a *mixture*, such as a solution of sugar in water, because the ratio of sugar to water can vary; a solution is considered *homogeneous*, because it has the same composition throughout. Every part of the solution has the same ratio of sugar to water, and the molecules of sugar stay in solution and do not settle out in time. When we think of solutions, it is common to think of one substance being dissolved in another; sugar being dissolved in water. The substance being dissolved is called the **solute.** The substance that dissolves the solute is the **solvent.** Many solutions have two or more solutes, but they can have only one solvent.

Most solutions you encounter are liquids, but solutions can also be gaseous or solid. The atmosphere is a gaseous solution, a mixture of several gases, with nitrogen gas, $N_2(g)$, acting as the solvent because it is most abundant. A 14-Karat gold ring is a solution of silver dissolved in gold; brass is a solution of zinc in copper. Both are solid solutions with gold and copper acting as solvents, respectively. Although gaseous and solid solutions are important, the emphasis in this chapter will be on those that are liquids.

A few specifics about liquid solutions:

- In liquid solutions, the solvent is always a molecular species, like water, alcohols, or organic liquids (hexane, benzene, and so on).
- In a solution, the solute exists in its smallest state of subdivision.

  (a) If the solute is a molecular substance, like I_2, sugars or alcohols, the substance is dispersed throughout the solvent as individual molecules separated from one another by molecules of the solvent. Molecular solutes are called **nonelectrolytes.** Their solutions do not conduct electricity.

  (b) If the solute is an ionic compound, like NaCl or KNO_3, the solute will *ionize* as it goes into solution. The compound enters the solvent as positive and negative ions as the crystalline solid comes apart. The ions disperse throughout the solvent being kept apart and insulated from one another by the solvent. Solutes that exist as ions in solution are called **electrolytes.** Solutions of electrolytes conduct electricity.

- If both the solute and solvent are liquid, and both liquids are soluble in each other in all proportions, they are said to be **miscible.** If two liquids are not soluble in one another, they are **immiscible.**

- The **solubility** of a substance describes the *maximum* mass of that substance that can dissolve in a given mass or volume of solvent at a specific temperature to form a stable solution. The solution is *saturated* with solute. The solubility of sodium chloride, NaCl, at 20°C is 35.9 g in 100 g (100 mL) of water. For potassium iodide, KI, it is 144 g/100 g of water at 20°C. Yes, this liquid solution has a mass of solute greater than the mass of solvent. Water is classed as the solvent because both it and the solution are liquid.

Factors Affecting Solubility

Many factors govern the solubility of a substance in a solvent. Is the potential solute a gas, liquid, or solid? If it is a solid, is it an ionic compound, or is it molecular? If molecular, are the molecules large or small, and are they polar or nonpolar? If ionic, how tightly are the ions held together in the crystals of the compound?

The nature of the solvent must also be considered. How well does the solvent accommodate the solute? If a substance is to be soluble, it needs to interact with the solvent in a manner similar to the way the solvent molecules interact with themselves. The intermolecular forces between molecules of the solvent that are disrupted (this requires energy) should be replaced, to some degree, by the new attractive forces between the solute and the solvent (this produces energy). Clearly, the characteristics of both the solute and the solvent contribute to the solubility of a substance, and one of the most important is the polarity or lack of polarity of the solute and solvent.

Polar versus Nonpolar

There is an old saying in science that says, "like dissolves like." It means that substances that are similar should form a solution. It also implies that substances that are not similar should not form a solution. As a general rule of thumb, "like dissolves like" works pretty well. The term "like" refers to the overall polarity of the solvent molecule (whether polar or nonpolar) and the overall polarity of the solute (whether polar, nonpolar, or an ionic species that is accommodated by polar solvents).

- **Polar species:** Molecules with a permanent dipole and/or the ability to engage in hydrogen bonding (Chapter 12). Water is a perfect example of a polar species. It is a bent, polar molecule with a substantial ability to hydrogen bond. Alcohols are a little less polar, but they and sugars can engage in strong hydrogen bonding. Water, because of its polarity, is uniquely suited to accommodate positive and negative ions in solution, making it an excellent solvent for many ionic compounds.
- **Nonpolar species:** Molecules that do not have a permanent dipole and do not have the ability to engage in hydrogen bonding. Many covalent, organic liquids fall into this category: oils, solvents derived from petroleum, carbon tetrachloride (CCl_4), and so on.

Polar and ionic compounds are more likely to dissolve in a polar solvent, like H_2O, than in a nonpolar solvent like CCl_4 or oil. Polar water molecules are not soluble in oil or carbon tetrachloride, both of which are nonpolar liquids. However, nonpolar compounds, like oils, are soluble in nonpolar solvents but are not soluble in polar solvents like water. The likelihood of forming a solution using solvents and solutes of differing polarity is summarized in the following table. Keep in mind that polar/ionic-nonpolar comparisons have some limitations. Not all ionic compounds are soluble in water, but those that are do not dissolve in nonpolar solvents.

Solutions: Polar versus Nonpolar		
Solvent	*Solute*	*Probable Solution?*
Polar liquid (H_2O)	Polar molecules or ionic compound	Yes
Polar liquid (H_2O)	Nonpolar molecules	No
Nonpolar liquid (CCl_4)	Polar molecules or ionic compound	No
Nonpolar liquid (CCl_4)	Nonpolar molecules	Yes

At the molecular level, the process of solution formation with nonpolar compounds can be viewed as one of simple mixing since neither compound would have meaningful forces of repulsion or attraction for one another. They easily blend together to form a homogeneous mixture. Such is not the case with polar molecules. If two polar compounds are to form a solution, the intermolecular forces of attraction that exist in each compound (dipole-dipole forces and/or hydrogen bonding) must exist between the two compounds when they are mixed together in a solution. There is no reason for a polar compound, whose molecules are held together by strong intermolecular forces, to dissolve in a solvent that cannot engage in those same kinds of intermolecular attractions. Oil doesn't dissolve in water because it cannot compete with the strong hydrogen bonding that binds water molecules together. Even if oil is vigorously shaken with water to form a milky-white mixture, in time the oil is simply squeezed out as water molecules reform their hydrogen bonds with one another. The oil being less dense rises to the top and floats on the water. Large sugar molecules are soluble in water because each molecule of sugar can engage in multiple hydrogen bonds with water, interacting with water in a way similar to the way water interacts with itself. Again, like dissolves like.

Sodium chloride, NaCl, dissolves in water because the attractive forces between the ions and the water molecules overcome the force of attraction between the ions in the crystal of NaCl. As each ion enters the solution, it is immediately surrounded by molecules of water. The $\delta-$ ends of the water molecules are oriented toward the positive sodium ions, and the $\delta+$ ends of the molecules are oriented toward the negative chloride ions, as shown in the following figure. The cloak of water about each ion keeps them apart in solution, insulating the positive ions from the negative ions. This kind of interaction between the solvent and the solute is called **solvation.** Water solvates the ions, keeping them in solution. The following figure shows how polar water molecules solvate Na^+ and Cl^- in solution.

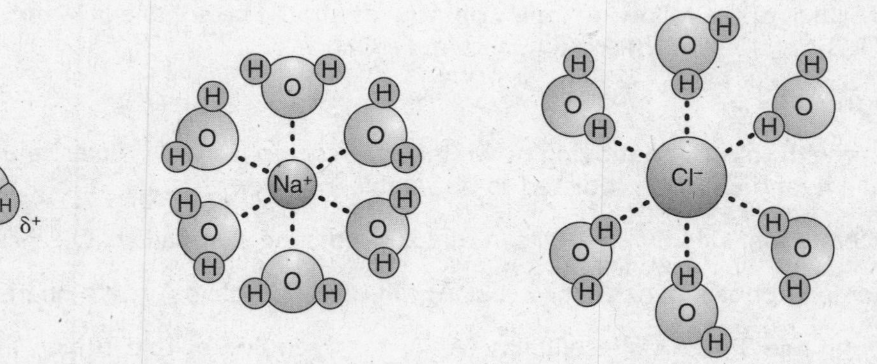

The solubility of most solids in water increases as the temperature of the solution increases. Just the opposite is true with gases; the solubility of gases decreases as temperature increases. Increasing the pressure of a gas in contact with the solvent will increase its solubility. The increased pressure forces more gas into the solvent. But, when the pressure is reduced, the gas begins to bubble out of solution, which is what happens when a can of soda is opened.

Predicting Solubility of Ionic Compounds

Not every ionic compound is soluble in water, even at elevated temperatures. As the size of the charges on the positive and negative ions gets larger (2+, 3+, 2−, 3−), the forces holding the ions together in the crystal become so great they cannot be easily overcome. Potassium nitrate, KNO_3, is very soluble in water; both the K^+ and NO_3^- ions bear a single charge, and water is capable of separating and keeping them in solution. Barium carbonate, $BaCO_3$, is not soluble in water. Both ions are doubly charged, Ba^{2+} and SO_4^{2-}, and strongly bound together in the barium carbonate crystal. Water has great difficulty separating them and keeping them in solution. Yet, the other combinations of these ions, K_2CO_3 in potassium carbonate and $Ba(NO_3)_2$ in barium nitrate, are both very soluble in water. In each compound one ion bears a single charge and the other a double charge, making it easier for water to separate the ions and keep them in solution.

The factors that determine whether or not an ionic compound will be soluble in water are complex, making predictions difficult. As a result, a series of statements or rules have come into being that guide predictions. These rules are called the **solubility rules.** As with most general rules, exceptions exist, but they are correct most of the time. The following rules are organized in a hierarchal structure—that is, the first rule takes precedence over the second, the second over the third, and so forth.

1. All sodium (Na^+), potassium (K^+), and ammonium (NH_4^+) compounds are *soluble* in water.
2. All nitrate (NO_3^-) and acetate ($C_2H_3O_2^-$) compounds are *soluble* in water.
3. All silver (Ag^+), lead(II) (Pb^{2+}), and mercury(I) (Hg_2^{2+}) compounds are *insoluble* in water. (Because rule 2 precedes this rule, the nitrates and acetates of these cations are soluble in water.)
4. All chloride (Cl^-), bromide (Br^-), and iodide (I^-) compounds are *soluble* in water, except those of Ag^+, Pb^{2+}, and Hg_2^{2+}, as stated in the previous rule.
5. All carbonates (CO_3^{2-}), phosphates (PO_4^{3-}), sulfides (S^{2-}), and oxalates ($C_2O_4^{2-}$) are *insoluble* in water, unless paired with one of the cations in rule 1, Na^+, K^+, or NH_4^+.
6. All sulfates (SO_4^{2-}) are *soluble* in water, except those of Ca^{2+}, Ba^{2+}, and the cations in rule 3.
7. All oxides (O^{2-}) and hydroxides (OH^-) are *insoluble* in water, except those of Na^+, K^+, and to a lesser degree, Ca^{2+} and Ba^{2+}.

Let's apply these rules to predict the solubility of a few ionic compounds.

Problem: Which of the following ionic compounds should be soluble in water: (a) $CaCl_2$; (b) $Ba(C_2H_3O_2)_2$; (c) AgBr; (d) $AgNO_3$; and (e) $Fe(OH)_3$?

Answer:

(a) Starting with rule 1 and moving down the sequence of solubility rules, the first ion to appear is Cl^- in rule 4, and $CaCl_2$ is predicted to be soluble in water.

(b) The acetate ion, $C_2H_3O_2^-$, appears in rule 2, predicting that $Ba(C_2H_3O_2)_2$ is soluble in water.

(c) Neither ion appears in rule 1 or 2, but Ag^+ in rule 3 predicts AgBr to be insoluble in water.

(d) Although rule 3 predicts insolubility (Ag^+), it is overruled by rule 2 (NO_3^-), and $AgNO_3$ would be predicted to be soluble in water.

(e) Fe^{3+} appears in none of the rules, but rule 7 predicts that $Fe(OH)_3$ should be insoluble in water.

Only $CaCl_2$, $Ba(C_2H_3O_2)_2$, and $AgNO_3$ of the five compounds are predicted to be soluble in water.

Example Problems

These problems have both answers and solutions given.

1. Use these four terms in a single sentence: solution, solvent, homogenous, and solute.

   Answer: Something like this: A solution is a homogeneous mixture in which a solute is dissolved in a solvent.

2. A vinegar and oil salad dressing (vinegar is an aqueous solution) is vigorously shaken, but in time the two liquids separate. Does it need to be shaken more vigorously or longer?

   Answer: More shaking isn't the answer. The aqueous solution of vinegar is mostly water, a polar compound that is not naturally soluble in oil, a nonpolar liquid. They will always separate in time, but you can shake them more if you want to.

3. Which of these ionic compounds would be soluble in water: (a) $Pb(NO_3)_2$; (b) $Ba_3(PO_4)_2$; and (c) $(NH_4)_2CO_3$?

   Answer: Rules 2 and 1, respectively, predict that $Pb(NO_3)_2$ and $(NH_4)_2CO_3$ would be soluble in water. Rule 5 indicates that $Ba_3(PO_4)_2$ would not be soluble in water.

Work Problems

Use these problems for additional practice.

1. What is the significance of the saying "like dissolves like" as applied to the formation of solutions?

2. Give an example of a solution that is a gas, a liquid, and a solid.

3. Which of these ionic compounds would be soluble in water: (a) Al_2O_3; (b) PbS; and (c) $CuSO_4$?

Worked Solutions

1. **Polar molecules are more likely to dissolve in a polar solvent than in a nonpolar solvent. Nonpolar molecules are more likely to dissolve in a nonpolar solvent than in a polar solvent.**

2. **Gaseous solutions: the atmosphere, anesthetic gas mixtures, helium and air in a balloon**

   Liquid solutions: salt water, iced tea, club soda

   Solid solutions: brass, 14 K gold, bronze, stainless steel

3. **Only $CuSO_4$ is predicted to be soluble in water, rule 6.** Al_2O_3 is judged insoluble by rule 7, and PbS by rule 3 with reinforcement from rule 5.

Solution Concentrations

The **concentration** of a solution states the amount of solute (in mass or moles) dissolved in a given amount of solution or solvent. Because many chemical reactions take place in solution, it is important that you understand the most common concentration schemes.

The concentration of solutions can be expressed in several ways. Some are qualitative; others are quantitative. Let's look at some of the qualitative terms used to relate the concentration of solutions: A **dilute solution** contains a relatively *small* amount of solute dissolved in a given amount of solution. A **concentrated solution** contains a relatively *large* amount of solute dissolved in a given amount of solution. Acids and bases are purchased from suppliers as concentrated solutions that are then diluted with water to form dilute solutions. In some cases, "dilute" and "concentrated" refer to definite concentrations, but you always need to check. Another qualitative term is unsaturated. A solution is **unsaturated** if it is capable of dissolving more solute. When it can dissolve no more solute, it is described as being **saturated.** Many substances become more soluble in water at higher temperatures. When some saturated solutions that were prepared at higher temperatures cool down, the excess solute might remain in solution, forming a supersaturated solution. **Supersaturated** solutions are very unstable, and the slightest vibration will destroy them, causing a rapid precipitation of the excess solute as it becomes a saturated solution at the lower temperature. Qualitative descriptions of concentration can be useful, but the quantitative methods are the ones you should master.

Three ways of quantitatively expressing the concentration of a solution will be presented here: Mass/mass percent, %(m/m), mass/volume percent, %(m/v), and molarity, M. A fourth, molality, will appear later in this chapter. You should know an interesting fact about concentrations. No matter what size sample of a solution you have, be it a teaspoonful or a bucketful, the concentration is the same for both. This is because concentrations are stated in terms of the amount of solute in a fixed amount of solvent: 100 g, 100 mL, or 1.00 L. It's like density. The density of mercury is 13.6 g/mL. If I have 100 mL or three drops of mercury, the density of mercury is still 13.6 g/mL. Neither density nor concentration depends on the size of the sample.

Mass/Mass Percent

The **mass/mass percent,** %(m/m), is a concentration term that states the mass of solute, as a percent, in a given mass of solution. This concentration scheme is also referred to as "mass percent" and sometimes as weight/weight percent. The (m/m) following the percent sign indicates that both solute and solution are given in units of mass, and to avoid confusion, (m/m) should always accompany the percent symbol. If the concentration of sugar in a solution is stated as 5%(m/m), there are 5 grams of sugar in 100 grams of solution, (5 g sugar/100 g solution). Percent values are always in terms of "parts per hundred," 100 grams of solution in this case. The defining equation for mass/mass percent is

$$\%(\text{m/m}) = \frac{\text{mass of solute}}{\text{mass of solution}} \times 100\%$$

Commonly, the unit of mass for solute and solution is gram, but any unit can be used as long as the same unit is used for both. After all, the unit of mass divides out in the calculation. Let's work a few problems involving mass/mass percent.

Problem 1: 55 g of sodium chloride, NaCl, are dissolved in 875 g of water. What is the %(m/m) concentration of NaCl in the final solution? How many grams of NaCl are in 200 g of this solution?

The mass of the final solution equals the mass of the NaCl plus the mass of water:

Mass of solution = 55 g NaCl + 875 g H_2O = 930. g.

Substituting into the defining equation, the %(m/m) concentration of NaCl is

$$\%(\text{m/m})\,\text{NaCl} = \left(\frac{55\ \cancel{g}\ \text{NaCl}}{930.\ \cancel{g}\ \text{solution}}\right) \times 100\% = 5.9\%(\text{m/m})\,\text{NaCl}.$$

The mass percent indicates that 5.9 g NaCl are in each 100 g of solution. So, if 5.9 g of NaCl are in 100 g of solution, 2 × 5.9 g NaCl, or 11.8 g NaCl will be in 200 grams of solution.

Answer 1: Dissolving 55 g of NaCl in 875 g H_2O produces a solution that has a mass percent concentration of 5.9%(m/m), and 200 g of that solution contain 11.8 g NaCl.

Problem 2: How many grams of sugar are in 1,500 g of a 1.50%(m/m) solution of sugar in water?

The mass percent gives us the mass of sugar in 100 g of solution. This information can be used to make a conversion factor to solve the problem. Two conversion factors can be written:

$$\left(\frac{1.50\ \text{g sugar}}{100.\ \text{g solution}}\right) \text{ and inverted } \left(\frac{100.\ \text{g solution}}{1.50\ \text{g sugar}}\right)$$

Multiplying the given mass of solution by the first conversion factor provides the answer:

$$\text{g sugar} = 1{,}500.\ \text{g}\ \cancel{\text{solution}} \times \left(\frac{1.50\ \text{g sugar}}{100.\ \text{g}\ \cancel{\text{solution}}}\right) = 22.5\ \text{g sugar}$$

Answer 2: There are 22.5 g sugar in 1,500 g of a 1.50%(m/m) solution.

Problem 3: You want to prepare an aqueous solution of calcium chloride, $CaCl_2$. What mass of a 5.0%(m/m) solution of calcium chloride can be prepared using 100 g of calcium chloride?

The interpretation of the mass/mass percent number tells us that 5.0 g of $CaCl_2$ are used to prepare 100. g of solution, which provides the conversion factor to find the mass of solution that can be prepared using 100 g of $CaCl_2$.

$$\text{g solution} = 100.\ \text{g}\ \cancel{\text{CaCl}_2} \times \left(\frac{100.\ \text{g solution}}{5.0\ \text{g}\ \cancel{\text{CaCl}_2}}\right) = 2.0 \times 10^3\ \text{g solution}$$

Answer 3: 2.0×10^3 g of a 5%(m/m) solution can be prepared using 100. g $CaCl_2$.

Mass/Volume Percent

The **mass/volume percent,** %(m/v), states the mass of solute dissolved in a given volume of solution. *Almost always, mass is in grams and volume in milliliters.* Mass/volume percent is widely used in medicine for injectable medications and IV solutions. If a solution has a concentration of 2.5%(m/v), it contains 2.5 g of solute in 100 mL of solution. The units of "g" and "mL" do not appear in the concentration term, you must remember that with %(m/v), mass is in grams and volume in milliliters.

$$\%(\text{m/v}) = \frac{\text{mass of solute (g)}}{\text{volume of solution (mL)}} \times 100\%$$

Problem 1: What mass of sugar is in 2,500. mL of a 0.50%(m/v) solution?

The mass/volume percent figure tells that 0.50 g of sugar is in 100 mL of solution. This allows two conversion factors to be written:

$$\left(\frac{0.50\text{ g sugar}}{100.\text{ mL solution}}\right)\text{ and inverted }\left(\frac{100.\text{ mL solution}}{0.50\text{ g sugar}}\right)$$

The first conversion factor eliminates the unit of volume and gives the answer in mass:

$$\text{g sugar} = 2{,}500.\ \cancel{\text{mL solution}} \times \left(\frac{0.50\text{ g sugar}}{100.\ \cancel{\text{mL solution}}}\right) = 12.5\text{ g sugar}$$

Answer 1: 2,500. mL of a 0.50%(m/v) sugar solution contains 12.5 g of sugar.

Problem 2: What volume of a 2.5%(m/v) solution of sodium chloride, NaCl, in water can be prepared using exactly 1.000 kg of NaCl?

First, convert kilogram to gram: 1.000 kg of NaCl equals 1,000 g of NaCl. A 2.5%(m/v) solution of sodium chloride contains 2.5 g of NaCl in 100 mL of solution. This relationship provides the conversion factor needed to solve the problem. The answer has two significant figures:

$$\text{volume of solution} = 1{,}000.\ \cancel{\text{g NaCl}} \times \left(\frac{100.\text{ mL}}{2.5\ \cancel{\text{g NaCl}}}\right) = 4.0 \times 10^4\text{ mL}$$

Answer 2: 4.0×10^4 mL (40 L) of a 2.5%(m/v) sodium chloride solution can be prepared from 1,000 g of NaCl.

Problem 3: An injectable medication solution contains 0.0750 g of medication in 1.00 mL of solution. (a) What is the %(m/v) of this solution? (b) How many milliliters of this solution must be used to give a patient 0.0300 g of medication?

(a) From the fact that 0.0750 g of medication is in 1.00 mL of solution; the %(m/v) is

$$\%(\text{m/v}) = \left(\frac{0.0750\text{ g medication}}{1.00\text{ mL solution}}\right) \times 100\% = 7.50\%(\text{m/v})$$

(b) Either the mass/volume percent or the fact that 0.0750 g of drug is in 1.00 mL of solution can provide the necessary conversion factor. Let's use the percent value calculated in part (a):

$$\text{volume in mL} = 0.0300\ \cancel{\text{g medication}} \times \left(\frac{100.\text{ mL solution}}{7.50\ \cancel{\text{g medication}}}\right) = 0.400\text{ mL}$$

Answer 3: (a) 0.0750 g of medication in 1.00 mL of solution is a 7.5%(m/v) concentration. (b) Administering 0.400 mL of this solution will provide a patient with 0.0300 g of medication.

Example Problems

These problems have both answers and solutions given.

1. What is the mass/mass percent concentration of a solution prepared by dissolving 72.5 g of sodium hydroxide, NaOH, in 850.0 g of water?

Answer: 7.86%(m/m)

First, calculate the mass of the final solution: 72.5 g NaOH + 850.0 g H_2O = 922.5 g.

$$\%(m/m) = \frac{72.5 \not{g} \text{ NaOH}}{922.5 \not{g} \text{ solution}} \times 100\% = 7.86\%(m/m)$$

2. What is the mass/volume percent concentration of a solution prepared by dissolving 72.5 g of sodium hydroxide, NaOH, in enough water to produce 850.0 mL of solution?

Answer: 8.53%(m/v)

$$\%(m/v) = \frac{72.5 \text{ g}}{850.0 \text{ mL}} \times 100\% = 8.53\%(m/v)$$

3. How many grams of urea are in 385 g of a 0.750%(m/m) solution of urea in water?

Answer: 2.89 g urea

The mass/mass percent value indicates that 0.750 g of urea is in 100. g of solution. This fact provides the necessary information for the conversion factor.

$$\text{g urea} = 385 \text{ g solution} \times \left(\frac{0.750 \text{ g urea}}{100. \text{ g solution}}\right) = 2.89 \text{ g urea}$$

4. How many liters of a 3.00%(m/v) aqueous sugar solution can be prepared using exactly 1.500 kg of sugar?

Answer: 50.0 L

The mass/volume percent indicates the solution is to contain 3.00 g of sugar in each 100. mL of solution. This provides the conversion factor to find the volume of solution that contains 1,500 g of sugar (1.500 kg × 1,000g/kg = 1,500 g).

$$\text{volume of solution} = 1{,}500. \text{ g sugar} \times \left(\frac{100. \text{ mL solution}}{3.00 \text{ g sugar}}\right) = 5.00 \times 10^4 \text{ mL}$$

Converting milliliters to liters:

$$\text{volume in L} = 5.00 \times 10^4 \text{ mL} \times \left(\frac{1.00 \text{ L}}{1{,}000 \text{ mL}}\right) = 50.0 \text{ L}$$

Work Problems

Use these problems for additional practice.

1. What mass of nickel(II) sulfate, $NiSO_4$, must be used to prepare 500. g of a 7.50%(m/m) solution in water?

2. A solution of copper(II) chloride, $CuCl_2$, is 8.75%(m/m). How many grams of this solution must be used to obtain 0.525 g of $CuCl_2$ for an experiment?

3. 30.0 g of methyl alcohol is used to form 250. mL of an aqueous solution. What is the mass/volume percent of alcohol in the solution?

4. What volume of a 1.35%(m/v) sodium hydroxide solution contains 10.0 g of NaOH?

Worked Solutions

1. **37.5 g of $NiSO_4$ are required.**

$$\text{g } NiSO_4 \text{ required} = 500.\ \text{g solution} \times \left(\frac{7.50\ \text{g } NiSO_4}{100.\ \text{g solution}}\right) = 37.5\ \text{g}$$

2. **6.00 g of solution are needed.**

$$\text{g solution required} = 0.525\ \text{g } CuCl_2 \times \left(\frac{100.\ \text{g solution}}{8.75\ \text{g } CuCl_2}\right) = 6.00\ \text{g}$$

3. **The concentration of the solution is 12.0%(m/v).**

$$\%(\text{m/v}) = \frac{30.0\ \text{g alcohol}}{250.\ \text{mL solution}} \times 100\% = 12.0\%(\text{m/v})$$

4. **741 mL of 1.35%(m/v) NaOH contain 10.0 g NaOH.**

$$\text{volume of solution} = 10.0\ \text{g NaOH} \times \left(\frac{100.\ \text{mL solution}}{1.35\ \text{g NaOH}}\right) = 741\ \text{mL}$$

Molarity

Molarity is the most widely used concentration scheme in chemistry, largely because it measures the amount of solute in moles as opposed to grams. The equation that defines molarity, M, states the amount of solute in moles and the volume of solution in liters, V_L:

$$\text{molarity} = \frac{\text{number of moles of solute}}{\text{volume of solution in liters}} \text{ or } M = \frac{\text{moles of solute}}{V_L}$$

This equation that defines molarity can be rearranged to solve for any of the three terms:

$$M = \frac{\text{moles of solute}}{V_L} \qquad V_L = \frac{\text{moles of solute}}{M} \qquad \text{moles of solute} = M \times V_L$$

The units of molarity are mole/liter (of solution), but they are commonly replaced with a capital M, which symbolizes molarity. Yet, there will be times when you will need to replace M with "mole/liter" when analyzing units and solving problems. If a sodium hydroxide solution is labeled 2 M (read as two-molar), it means that 2 moles of NaOH are dissolved in 1 L of solution, 2 moles/liter. If you need to brush up on mass-mole conversions, review the pertinent material in Chapter 5. In all the problems dealing with molar solutions, molarity will be written as a conversion factor to emphasize the canceling and retention of units, just as was done with the percent concentrations. The molarity term for a solution that is 0.55 M in NaOH could be written in four ways to make the required conversion factor:

$$\left(\frac{0.55\ \text{moles NaOH}}{1.00\ \text{L solution}}\right) \quad \left(\frac{0.55\ \text{moles NaOH}}{1{,}000.\ \text{mL solution}}\right) \quad \left(\frac{1.00\ \text{L solution}}{0.55\ \text{moles NaOH}}\right) \quad \left(\frac{1{,}000.\ \text{mL solution}}{0.55\ \text{moles NaOH}}\right)$$

Problem 1: What is the molarity of a solution that contains 50.0 g of sodium hydroxide, NaOH, in 850 mL of solution? The molar mass of NaOH is 40.0 g.

To calculate molarity, the amount of NaOH must be in moles, and the volume of solution must be in liters. 850 mL is 0.850 L. Knowing that 1.00 mole of NaOH has a mass of 40.0 g, the number of moles of NaOH in the solution is

$$\text{moles of NaOH} = 50.0\ \text{g NaOH} \times \left(\frac{1.00\ \text{mole NaOH}}{40.0\ \text{g NaOH}}\right) = 1.25\ \text{moles}$$

The molarity of the solution is

$$M_{NaOH} = \frac{1.25 \text{ moles NaOH}}{0.850 \text{ L}} = 1.47 \text{ M}$$

Answer 1: 850 mL of a solution that contains 50.0 g of NaOH has a molarity of 1.47 M.

Problem 2: What volume of 1.50 M $CuSO_4$ contains 35.0 g $CuSO_4$? The molar mass of copper(II) sulfate is 159.6 g.

First, the amount of solute must be in moles. From the molar mass, the mass of 1.00 mole of $CuSO_4$ is 159.6 g.

$$\text{mole } CuSO_4 = 35.0 \text{ } \cancel{\text{g } CuSO_4} \times \left(\frac{1.00 \text{ mole } CuSO_4}{159.6 \text{ } \cancel{\text{g } CuSO_4}}\right) = 0.219 \text{ mole } CuSO_4$$

A 1.50 M solution of $CuSO_4$ contains 1.50 moles of $CuSO_4$ in 1.00 L of solution. This fact provides the factor to convert moles of $CuSO_4$ to volume of solution in liters, V_L.

$$V_L = 0.219 \text{ } \cancel{\text{mole } CuSO_4} \times \left(\frac{1.00 \text{ L solution}}{1.50 \text{ } \cancel{\text{moles } CuSO_4}}\right) = 0.146 \text{ L}$$

Answer 2: 35.0 g $CuSO_4$ are in 0.146 L (146 mL) of 1.50 M $CuSO_4$ solution.

Problem 3: How many grams of NaCl are in 2.53 L of 0.750 M NaCl? The molar mass of NaCl is 58.4 g.

Starting with the fact that 0.750 mole of NaCl is in 1.00 L of solution, determine the number of moles of NaCl in 2.53 L of solution. Then, convert moles of NaCl to mass of NaCl.

$$\text{moles NaCl} = 2.53 \text{ } \cancel{\text{L}} \times \left(\frac{0.750 \text{ mole NaCl}}{1.00 \text{ } \cancel{\text{L}} \text{ solution}}\right) = 1.898 \text{ moles} = 1.90 \text{ moles NaCl}$$

$$\text{mass NaCl} = 1.90 \text{ } \cancel{\text{moles NaCl}} \times \left(\frac{58.4 \text{ g NaCl}}{1.00 \text{ } \cancel{\text{mole NaCl}}}\right) = 111 \text{ g NaCl}$$

Answer 3: There are 111 g of NaCl in 2.53 L of 0.750 M NaCl solution.

Dilution of Molar Solutions

Many chemical reagents are supplied in solutions of much higher concentration than their use requires. The concentrated solutions need to be diluted with solvent to prepare solutions of lower concentration. Adding solvent increases the volume of the solution without changing the number of moles of solute, and the concentration of the solution decreases. For molar solutions, the volume of the solution in liters times the molarity of the solution equals the number of moles of solute. Because the number of moles of solute remains constant with dilution, a relationship in terms of volumes and molarities can be derived that is very useful for dilution problems. This is called the **dilution equation.**

$$V_{conc} \times M_{conc} = V_{dil} \times M_{dil}$$

V_{conc} and M_{conc} are the volume and molarity of the concentrated solution (the one of greater molarity); V_{dil} and M_{dil} are the volume and molarity of the diluted solution. If any three of the four terms are known, the fourth can be calculated. When using the dilution equation, volumes can be in liters or milliliters. If two volumes are used in the equation, the usual case, they both must be in the same unit.

Problem: Hydrochloric acid is obtained commercially at a concentration of 12.1 M. How many milliliters of 12.1 M HCl(*aq*) must be used to prepare 2,000. mL of 0.500 M HCl(*aq*)?

Gathering terms: M_{conc} = 12.1 M; M_{dil} = 0.500 M; V_{conc} = ?; V_{dil} = 2,000 mL

The dilution equation is rearranged to solve for the volume of 12.1 M HCl(*aq*), V_{conc}:

$$V_{conc} = \frac{V_{dil}M_{dil}}{M_{conc}} = \frac{(2{,}000.\ \text{mL})(0.500\ \cancel{M})}{(12.1\ \cancel{M})} = 82.6\ \text{mL}$$

Answer: To prepare the dilute solution of hydrochloric acid, 82.6 mL of 12.1 M HCl(*aq*) are diluted with water to a final volume of 2,000. mL. The final solution would be 0.500 M HCl(*aq*).

After you know the volume of 12.1 M HCl(*aq*) needed for the final solution, how is the solution prepared? You want to make sure you end up with exactly 2,000 mL of 0.500 M HCl(*aq*). Here's how it is done:

1. Obtain a large graduated cylinder or volumetric flask that has the 2,000 mL volume clearly marked.
2. Fill the cylinder or flask about half full of pure water and then slowly add 82.6 mL of concentrated HCl(*aq*) while stirring the solution.
3. While stirring, continue to add pure water until the volume is close to the 2,000 mL mark. Remove the stirring rod and slowly add water until the meniscus (the curved surface of the solution) just rests on the 2,000 mL mark. Thoroughly mix the final solution. You now have a *final volume* of 2,000 mL and a concentration of 0.500 M HCl(*aq*).

Following this sequence ensures that you end up with the correct *final volume*. If you simply added 82.6 mL of acid to 2,000 mL of water, the final volume would exceed 2,000 mL, and the concentration of the acid would be something less than 0.500 M.

Example Problems

These problems have both answers and solutions given.

1. A 50.0 mL volume of a calcium bromide solution contains 3.50 g of solute. What is the molarity of calcium bromide? The molar mass of $CaBr_2$ is 199.9 g.

   Answer: 0.350 M $CaBr_2$

   To calculate molarity, volume must be in liters, and quantity of solute must be in moles. 50.0 mL is 0.0500 L. The number of moles of $CaBr_2$ is

   $$\text{moles } CaBr_2 = 3.50\ \cancel{g} \times \left(\frac{1\ \text{mole}}{199.9\ \cancel{g}}\right) = 0.0175\ \text{mole}$$

   The molarity is

   $$M = \frac{0.0175\ \text{mole}}{0.0500\ \text{L}} = 0.350\ \text{M}$$

2. What volume of 2.50 M NaOH, in milliliters, contains 1.00 mole of NaOH?

   Answer: 400 mL

In a 2.50 M solution, each 1.00 L (1,000 mL) of solution contains 2.50 moles of NaOH. This fact provides the factor to convert moles of NaOH to volume of solution.

$$\text{volume of solution} = 1.00\ \cancel{\text{mole NaOH}} \times \left(\frac{1{,}000.\ \text{mL}}{2.50\ \cancel{\text{moles NaOH}}}\right) = 400\ \text{mL}$$

3. How many grams of potassium hydroxide, KCl, are in 800 mL of 0.150 M KCl solution. The molar mass of KCl is 56.1 g.

Answer: 6.73 g KCl

The number of moles of KCl in 800 mL (0.800 L) of 0.150 M KCl is

$$\text{moles KCl} = 0.800\ \cancel{\text{L}} \times \left(\frac{0.150\ \text{mole KCl}}{1.00\ \cancel{\text{L}}}\right) = 0.120\ \text{mole KCl}$$

Converting mole KCl to mass of KCl:

$$\text{mass KCl} = 0.120\ \cancel{\text{mole}} \times \left(\frac{56.1\ \text{g KCl}}{1.00\ \cancel{\text{mole}}}\right) = 6.73\ \text{g KCl}$$

4. How many milliliters of 18.0 M sulfuric acid, $H_2SO_4(aq)$, must be diluted with water to prepare 5.00×10^3 mL of 0.250 M sulfuric acid? Briefly, how would you prepare this solution?

Answer: 69.4 mL of 18.0 M H_2SO_4

$$V_{conc} = \frac{V_{dil}M_{dil}}{M_{conc}} = \frac{(5.00 \times 10^3\ \text{mL})(0.250\ \cancel{M})}{(18.0\ \cancel{M})} = 69.4\ \text{mL}$$

To prepare this solution, obtain a large volumetric flask or cylinder that marks the point where it contains exactly 5,000 mL. Fill it half full of pure water. Then slowly add, while stirring, 69.4 mL of 18 M H_2SO_4. Continue to add water with stirring until the surface of the solution is 20–30 mL below the 5,000 mL mark. Remove the stirring rod and slowly add water until the meniscus just rests on the 5,000 mL mark. Stir.

Work Problems

Use these problems for additional practice.

1. What is the molarity of 945 mL of a solution that contains 55.0 g of $Fe(NO_3)_3$? The molar mass of $Fe(NO_3)_3$ is 241.9 g.

2. It requires 0.650 mole of NaOH to neutralize a sample of hydrochloric acid. How many milliliters of 2.00 M NaOH are needed to supply that amount of NaOH?

3. What mass of copper(II) sulfate, $CuSO_4$, is needed to prepare 2.50 L of 0.500 M $CuSO_4$? The molar mass of $CuSO_4$ is 159.6 g.

4. You need 1,500. mL of 0.100 M aqueous ammonia, $NH_3(aq)$, for a project, but the only available solution of this reagent is 2.00 M. How many milliliters of the 2.00 M solution must you use to prepare 1,500. mL of 0.100 M aqueous ammonia?

Worked Solutions

1. **The solution is 0.241 M in $Fe(NO_3)_3$.**

$$\text{mole } Fe(NO_3)_3 = 55.0 \text{ g} \times \left(\frac{1.00 \text{ mole } Fe(NO_3)_3}{241.9 \text{ g}}\right) = 0.227 \text{ mole}$$

$$M = \frac{0.227 \text{ mole } Fe(NO_3)_3}{0.945 \text{ L}} = 0.241 \frac{\text{mole}}{\text{L}} = 0.241 \text{ M}$$

2. **325 mL of 2.00 M NaOH are required.**

$$\text{volume NaOH} = 0.650 \text{ mole NaOH} \times \left(\frac{1.00 \text{ L}}{2.00 \text{ moles NaOH}}\right) = 0.325 \text{ L or } 325 \text{ mL}$$

3. **200 g of $CuSO_4$ are required.**

$$\text{mole } CuSO_4 \text{ required} = 2.50 \text{ L} \times \left(\frac{0.500 \text{ mole}}{1.00 \text{ L}}\right) = 1.25 \text{ moles}$$

$$\text{mass } CuSO_4 = 1.25 \text{ moles} \times \left(\frac{159.6 \text{ g}}{1.00 \text{ mole}}\right) = 199.5 \text{ g} = 200 \text{ g}$$

4. **75.0 mL of 2.0 M NH_3*(aq)* would be needed.**

$$V_{conc} = \frac{V_{dil} M_{dil}}{M_{conc}} = \frac{(1,500. \text{ mL})(0.100 \text{ M})}{(2.00 \text{ M})} = 75.0 \text{ mL}$$

Solution Stoichiometry

Stoichiometry concerns calculations based on balanced chemical equations, a topic that was presented in Chapter 8. Remember that the coefficients in the balanced equations indicate the number of moles of each reactant and product. Because many reactions take place in solution, and because the molarity of solutions relates to moles of solute and volumes, it is possible to extend stoichiometric calculations to reactions involving solutions of reactants and products. The calculations involving balanced equations are the same as those done in Chapter 8, but with the additional need to do some molarity calculations. Let's get our feet wet by working a couple of problems involving solutions in chemical reactions.

Problem 1: How many milliliters of 0.500 M sulfuric acid, H_2SO_4*(aq)*, are needed to just completely react with 12.0 g of sodium hydroxide, NaOH? The molar mass of NaOH is 40.0 g, and the balanced equation for the reaction is

$$2\ NaOH(s) + H_2SO_4(aq) \rightarrow Na_2SO_4(aq) + 2\ H_2O(l)$$

After the number of moles of NaOH is calculated, the balanced equation will be used to determine the required number of moles of acid. The pathway to the answer is

mass NaOH → mole NaOH → mole H_2SO_4 → volume 0.500 M H_2SO_4

Number of mole of NaOH = 12.0 g × (1 mole NaOH/40.0 g) = 0.300 mole NaOH.

The balanced equation shows that 2 moles of NaOH react with 1 mole of sulfuric acid. We are seeking the mole of acid that will just consume 0.300 mole NaOH. The coefficient ratio of sought (1 mole H_2SO_4) over known (2 mole NaOH) equals (1/2).

$$\text{moles } H_2SO_4 \text{ required} = \left(\frac{1}{2}\right)(0.300 \text{ mole}) = 0.150 \text{ mole}$$

1.00 L (1,000 mL) of sulfuric acid contains 0.500 mole of H_2SO_4. The volume of 0.500 M acid that contains 0.150 mole of H_2SO_4 is

$$\text{volume } H_2SO_4 = 0.150 \cancel{\text{mole}} \times \left(\frac{1{,}000. \text{ mL}}{0.500 \cancel{\text{mole}}}\right) = 300 \text{ mL}$$

Answer 1: 300 mL of 0.50 M H_2SO_4*(aq)* will just exactly consume 12.0 g of NaOH.

Problem 2: What volume of 0.250 M HCl*(aq)* is necessary to just consume the solute in 0.150 L of 0.100 M $Pb(NO_3)_2$? The balanced equation is

$Pb(NO_3)_2(aq) + 2\ HCl(aq) \rightarrow PbCl_2(s) + 2\ HNO_3(aq)$

The pathway to the answer is

solution of $Pb(NO_3)_2(aq) \rightarrow$ mole $Pb(NO_3)_2 \rightarrow$ mole HCl $\rightarrow$ volume 0.250 M HCl

We first need to calculate the number of moles of $Pb(NO_3)_2$ in 150 mL of the 0.100 M solution.

$$\text{moles } Pb(NO_3)_2 = 0.150 \cancel{\text{L}} \times \left(\frac{0.100 \text{ mole}}{1.00 \cancel{\text{L}}}\right) = 0.0150 \text{ mole } Pb(NO_3)_2$$

The balanced equation shows that *one* mole of $Pb(NO_3)_2$ requires *two* moles of HCl*(aq)*. Therefore, 0.0150 mole $Pb(NO_3)_2$ will require 0.0300 mole of HCl*(aq)*.

The volume of 0.250 M HCl*(aq)* that contains 0.0300 mole is

$$\text{volume HCl} = 0.0300 \cancel{\text{mole}} \times \left(\frac{1.00 \text{ L}}{0.250 \cancel{\text{mole}}}\right) = 0.120 \text{ L HCl}$$

Answer 2: 0.120 L (120 mL) of 0.250 M HCl*(aq)* is required to just exactly consume the lead(II) nitrate in 0.150 L of a 0.100 M solution of $Pb(NO_3)_2$.

Example Problems

These problems have both answers and solutions given.

1. How many mole of carbon dioxide gas, $CO_2(g)$, will form if 300 mL of 0.500 M HCl*(aq)* are added to an excess of sodium carbonate? The balanced equation is

   $Na_2CO_3(s) + 2\ HCl(aq) \rightarrow 2\ NaCl(aq) + H_2O(l) + CO_2(g)$

   Answer: 0.0750 mole $CO_2(g)$

Here's the pathway to the answer: solution $HCl(aq) \rightarrow$ mole HCl $\rightarrow$ mole $CO_2(g)$.

In 300 mL (0.300 L) of 0.500 M HCl, there is $(0.300\ \cancel{L}) \times (0.500\ \text{mole}/1.00\ \cancel{L}) = 0.150$ mole $HCl(aq)$.

The balanced equation shows that 1 mole of $CO_2(g)$ forms as 2 moles of $HCl(aq)$ are consumed. The coefficient ratio of (sought/known) is (1/2). The number of moles of $CO_2(g)$ produced by 0.150 mole $HCl(aq)$ is

$$\text{moles } CO_2 = \left(\frac{1}{2}\right) \times 0.150 \text{ mole} = 0.0750 \text{ mole}$$

2. What volume of 0.250 M $AgNO_3(aq)$ is needed to just precipitate all the chromate ion, CrO_4^{2-}, in 0.500 L of 0.800 M K_2CrO_4? The balanced equation is

$$K_2CrO_4(aq) + 2\ AgNO_3(aq) \rightarrow Ag_2CrO_4(s) + 2\ KNO_3(aq)$$

Answer: 3.20 L

Starting with the solution of $K_2CrO_4 \rightarrow$ mole $K_2CrO_4 \rightarrow$ mole $AgNO_3 \rightarrow$ volume $AgNO_3$.

In 0.500 L of 0.800 M $K_2CrO_4(aq)$, there is $(0.500\ \cancel{L}) \times (0.800\ \text{mole}/1.00\ \cancel{L}) = 0.400$ mole K_2CrO_4.

From the balanced equation, each mole of K_2CrO_4 reacts with 2 moles of $AgNO_3(aq)$, so 0.400 mole of K_2CrO_4 would consume: 2×0.400 mole = 0.800 mole of $AgNO_3(aq)$.

The volume of 0.250 M $AgNO_3(aq)$ that contains 0.800 mole of $AgNO_3$ is

$$\text{volume } AgNO_3 = 0.800\ \cancel{\text{mole}} \times \left(\frac{1.00\text{ L}}{0.250\ \cancel{\text{mole}}}\right) = 3.20\text{ L}$$

Work Problems

Use these problems for additional practice.

1. How many milliliters of 0.450 M $HCl(aq)$ are needed to consume 0.0250 mole of iron metal? The balanced equation is

   $$6\ HCl(aq) + 2\ Fe(s) \rightarrow 2\ FeCl_3(aq) + 3\ H_2(g)$$

2. 50.0 mL of 0.130 M nitric acid just exactly neutralized all the $Ca(OH)_2$ in 3.60 L of a calcium hydroxide solution. What was the molarity of the $Ca(OH)_2$ solution? The balanced equation is

   $$2\ HNO_3(aq) + Ca(OH)_2(aq) \rightarrow Ca(NO_3)_2(aq) + 2\ H_2O(l)$$

Worked Solutions

1. **167 mL of acid are needed.**

The coefficients of the balanced equation (sought/known) indicate that 0.0250 mole of Fe*(s)* will require: (6/2) × (0.0250 mole) = 0.0750 mole of HCl*(aq)*. The volume of 0.450 M HCl*(aq)* that provides this amount of HCl*(aq)* is

$$\text{volume HCl} = 0.0750\ \cancel{\text{mole}} \times \left(\frac{1.00\ \text{L}}{0.450\ \cancel{\text{mole}}}\right) = 0.167\ \text{L} = 167\ \text{mL}$$

2. **The calcium hydroxide solution is 9.03×10^{-4} M.**

The number of moles of HNO_3*(aq)* consumed = (0.0500 $\cancel{L}$) × (0.130 mole HNO_3/1.00 $\cancel{L}$) = 0.00650 mole HNO_3*(aq)*.

From the balanced equation, the number of moles of $Ca(OH)_2$ consumed is half the number of mole of HNO_3. The coefficient ratio (sought/known) is (1/2).

$$\text{mole Ca(OH)}_2 = \left(\frac{1}{2}\right)(0.00650\ \text{mole}) = 0.00325\ \text{mole}$$

The molarity of the $Ca(OH)_2$ solution is

$$\text{M Ca(OH)}_2 = \frac{\text{moles Ca(OH)}_2}{\text{volume in L}} = \frac{0.00325\ \text{mole}}{3.60\ \text{L}} = 0.000903\ \text{M} = 9.03 \times 10^{-4}\ \text{M}$$

Properties of Solutions

If you were asked the temperature at which water freezes to ice, you would probably answer 0°C and you would be right. But, how would you answer if asked the temperature at which a solution of sugar in water freezes? You might say 0°C, just like water, but this time you would not be right. The solution of sugar in water would freeze at a *lower* temperature than pure water. How about the boiling point of water? You would say 100°C, and you would be right. But what about the sugar solution? As it turns out, the sugar solution boils at a temperature *higher* than that of pure water. The lowering of the freezing point and the elevation of the boiling point are two of the four colligative properties of solutions. The **colligative properties** of solutions are those that depend only on the concentration of solute *particles* (molecules or ions) in the solution and not on the identity of those particles. There are four colligative properties:

1. **The depression of the freezing point.** Solutions freeze at *lower* temperatures than the solvent used to prepare the solutions.
2. **The elevation of the boiling point.** Solutions boil at *higher* temperatures than the solvent used to prepare the solutions.
3. **The depression of the vapor pressure.** Solutions have a *lower* vapor pressure than the pure solvent, both compared at the same temperature.
4. **Osmotic pressure.** The ability of solvent molecules in a solution to pass through a semi-permeable membrane is described by the osmotic pressure of the solution. Solvent passes through the membrane from a solution of lower osmotic pressure into one of higher osmotic pressure.

Three of the four colligative properties will be examined in this book.

The phrase "concentration of solute particles" in the definition of colligative property needs clarification. There are two kinds of solutes, those that exist in solution as neutral molecules and those that ionize when dissolved and exist in solution as ions. Compounds that ionize in water are called **electrolytes.** Those that do not ionize are **nonelectrolytes.** Glucose is a nonelectrolyte and exists as neutral molecules in solution. A 1.0 M solution of glucose is 1.0 M in "solute particles." Sodium chloride, NaCl, is an electrolyte and exists in solution as separated sodium and chloride ions, $NaCl(s) \rightarrow Na^+(aq) + Cl^-(aq)$. A 1.0 M solution of NaCl is 2.0 M in "solute particles," 1.0 M $Na^+(aq)$ plus 1.0 M $Cl^-(aq)$. The concentration of solute particles for compounds that ionize in solution will be some whole number multiple of the concentration of the compound itself. For those solutes that do not ionize when dissolved in water, the concentration of the compound and the concentration of the solute particles (molecules) will be the same. A listing of common nonelectrolytes and electrolytes in water appears in the following table.

Common Nonelectrolytes and Electrolytes in Water

Nonelectrolytes (molecular)	*Electrolytes (ions)*
Glucose – $C_6H_{12}O_6$	$NaCl(s) \rightarrow Na^+(aq) + Cl^-(aq)$
Sucrose – $C_{12}H_{22}O_{11}$	$CaCl_2(s) \rightarrow Ca^{2+}(aq) + 2\ Cl^-(aq)$
Urea – CH_4N_2O	$NaOH(s) \rightarrow Na^+(aq) + OH^-(aq)$
Ethylene glycol – $C_2H_6O_2$	$KOH(s) \rightarrow K^+(aq) + OH^-(aq)$
Methyl alcohol – CH_4O	$HCl(aq) \rightarrow H^+(aq) + Cl^-(aq)$
Ethyl alcohol – C_2H_6O	$HNO_3(aq) \rightarrow H^+(aq) + NO_3^-(aq)$

Freezing-Point Depression

Solutions freeze at lower temperatures than the solvents used to prepare them. When antifreeze (ethylene glycol, $C_2H_6O_2$) is added to the water in the radiator of a car, the water-glycol solution will remain liquid at temperatures far below the normal freezing point of water itself. Salt and calcium chloride spread on snow-covered roads in the winter dissolve in the snow, forming solutions that are liquid at temperatures at which pure water is solid. The depression of the freezing point is symbolized ΔT_{FP} and is used as a positive number.

$$\Delta T_{FP} = \text{freezing point of the solvent} - \text{freezing point of the solution}$$

Quantitatively, the number of degrees the freezing point is lowered is directly proportional to the concentration of solute particles when that concentration is expressed in **molality.** The molality (m) of a solution is *not* the same as its molarity (M) because molality is based on the *mass of solvent* (kg) and not the *volume of solution* (L) as is used in molarity. The defining equation for molality is

$$\text{molality} = \text{m} = \frac{\text{moles of solute}}{\text{mass of solvent in kg}}$$

The units of molality are mole per kilogram, but molality is usually symbolized with a small "m." Molality is not as widely used as molarity by most chemists, although it is the preferred concentration unit for two of the colligative properties.

The equation used to calculate the size of the freezing point depression, ΔT_{FP}, is

$$\Delta T_{FP} = \text{m}\ K_f$$

where K_f is a property of the solvent itself and is called the *freezing point depression constant*. For water, K_f equals 1.86°C/m. The value of K_f tells us that a 1.00 m solution of glucose (1.00 m in glucose molecules) would freeze 1.86°C *lower* than pure water. The freezing point of the solution, $FP_{solution}$, would be −1.86°C. Every solvent has its own unique value of K_f. The equation given previously for calculating ΔT_{FP} is for solutes like sugars and alcohols, nonelectrolytes that do not break apart into ions when dissolved in water. If the solute is an electrolyte, the molality of the *ions* will be a whole number multiple of the molality of the compound.

Problem 1: At what temperature would a 0.50 m methyl alcohol solution in water freeze? K_f for water is 1.86°C/m. Methyl alcohol is a nonelectrolyte.

Solve for ΔT_{FP} and then calculate the freezing temperature of the solution. Notice how K_f is written and used as a conversion factor in the calculation.

$$\Delta T_{FP} = mK_f = 0.50\ \cancel{m} \times \left(\frac{1.86^\circ C}{1.00\ \cancel{m}}\right) = 0.93^\circ C$$

The freezing point of water is 0.00°C. The freezing point of the solution is then:

$$FP_{solution} = FP_{water} - \Delta T_{FP} = 0.00^\circ C - 0.93^\circ C = -0.93^\circ C$$

Answer 1: A 0.50 m solution of alcohol in water will freeze at −0.93°C.

Problem 2: What is the freezing temperature of a 0.50 m NaOH solution in water? K_f for water is 1.86°C/m. NaOH is listed in the previous table as an electrolyte.

The molality of solute particles is 2 × the molality of NaOH since each NaOH unit forms 2 ions in solution. The molality of solute particles is 1.0 m. Solving for ΔT_{FP}:

$$\Delta T_{FP} = mK_f = 1.00\ \cancel{m} \times \left(\frac{1.86^\circ C}{1.00\ \cancel{m}}\right) = 1.86^\circ C$$

The freezing point of water is 0.00°C. The freezing point of the solution is then:

$$FP_{solution} = FP_{water} - \Delta T_{FP} = 0.00^\circ C - 1.86^\circ C = -1.86^\circ C$$

Answer 2: A 0.50 m solution of NaOH in water will freeze at −1.86°C.

Problem 3: A solution is prepared by dissolving 35.0 g of glucose, $C_6H_{12}O_6$, in 650 g (0.650 kg) of water. At what temperature will this solution freeze? K_f for water is 1.86°C/m. The molar mass of glucose is 180.2 g.

First calculate the molality of glucose. The number of moles of glucose equals (35.0 g) × (1.00 mole/180.2 g) = 0.194 mole. The mass of solvent is 0.650 kg.

$$m = \frac{0.194\ \text{mole}}{0.650\ \text{kg}} = 0.299\frac{\text{mole}}{\text{kg}} = 0.299\ \text{m}$$

$$\Delta T_{FP} = mK_f = 0.299\ \cancel{m} \times \left(\frac{1.86^\circ C}{1.00\ \cancel{m}}\right) = 0.556^\circ C$$

$$FP_{solution} = FP_{water} - \Delta T_{FP} = 0.00^\circ C - 0.56^\circ C = -0.56^\circ C$$

Answer 3: A solution of 35.0 g of glucose in 650 grams of water will freeze at −0.56°C.

Boiling-Point Elevation

Solutions boil at higher temperatures than the solvents used to prepare them, and the amount of elevation is directly proportional to the molality of solute particles. The elevation of the boiling point is symbolized ΔT_{BP} and is used as a positive number.

$$\Delta T_{BP} = \text{boiling point of the solution} - \text{boiling point of the pure solvent}$$

The amount of elevation of the boiling temperature is calculated using an equation similar to that used for freezing point depressions. As before, the concentration of the solute is in molality, but the constant, K_b, specifically applies to boiling points.

$$\Delta T_{BP} = m\ K_b$$

K_b is the *boiling point elevation constant,* and for water equals 0.52°C/m. Each solvent has its own unique value for K_b, and the value of K_b for water indicates that a 1.0 m solution of glucose, a nonelectrolyte, would boil 0.52°C higher than that of pure water, 100.52°C. As with the equation used to calculate freezing point depressions, if the solute is an electrolyte, the molality of the ions will be a whole number multiple of the molality of the compound.

Problem 1: What is the boiling point of a 0.75 m solution of urea in water? K_b for water is 0.52°C/m. Urea is a nonelectrolyte.

$$\Delta T_{BP} = mK_b = 0.75\ \cancel{m} \times \left(\frac{0.52^\circ C}{1.00\ \cancel{m}}\right) = 0.39^\circ C$$

The boiling point of the solution is elevated above that of pure water, so ΔT_{BP} is added to the normal boiling point of water, 100.00°C.

$$BP_{solution} = BP_{water} + \Delta T_{BP} = 100.00^\circ C + 0.39^\circ C = 100.39^\circ C$$

Answer 1: The boiling point of a 0.75 M urea solution is elevated to 100.39°C.

Problem 2: What is the boiling point elevation, ΔT_{BP}, of a 1.50 m solution of $CaCl_2$ in water? K_b for water is 0.52°C/m. Calcium chloride, $CaCl_2$, is an electrolyte. Each formula unit forms three ions in solution. The molality of solute particles (ions) is 3 × 1.50 m = 4.50 m.

$$\Delta T_{BP} = mK_b = 4.50\ \cancel{m} \times \left(\frac{0.52^\circ C}{1.00\ \cancel{m}}\right) = 2.34^\circ C$$

Answer 2: The boiling point of a 1.50 m $CaCl_2$ solution is elevated 2.34°C above the boiling point of pure water.

Problem 3: A solution of sucrose in water has a freezing point of −2.6°C. What would be the boiling point of this solution? K_f for water is 1.86°C/m; K_b is 0.52°C/m. Sucrose is a nonelectrolyte.

Pure water normally freezes at 0.00°C, so if the solution freezes 2.6°C lower than that, ΔT_{FP} is 2.6°C. Knowing both ΔT_{FP} and K_f for water, the molality of this sucrose solution is

$$m_{sucrose} = \frac{\Delta T_{FP}}{K_f} = 2.6^\circ \cancel{C} \times \left(\frac{1.00\ m}{1.86^\circ \cancel{C}}\right) = 1.4\ m$$

When the molality of sucrose is known, the elevation of the boiling point can be calculated:

$$\Delta T_{BP} = mK_b = 1.4\ \cancel{m} \times \left(\frac{0.52^\circ C}{1.0\ \cancel{m}}\right) = 0.73^\circ C$$

The boiling point of the solution is then 100.00°C + 0.73°C = 100.73°C.

Answer 3: An aqueous solution that freezes at −2.6°C will boil at 100.73°C.

Example Problems

These problems have both answers and solutions given.

1. What is the molality of a solution of ethyl alcohol in water that freezes at −14.5°C? K_f for water is 1.86°C/m.

   Answer: 7.80 m

   Knowing the normal freezing point of water is 0.00°C, ΔT_{FP} = 14.5°C. Solving for molality:

   $$m = \frac{\Delta T_{FP}}{K_f} = 14.5^\circ \cancel{C} \times \left(\frac{1.00\ m}{1.86^\circ \cancel{C}}\right) = 7.80\ m$$

2. What is the boiling point of a 2.0 m solution of sodium hydroxide, NaOH? K_b for water is 0.52°C/m.

   Answer: 102.1°C

   NaOH is an electrolyte ionizing to Na^+(aq) and OH^-*(aq)*. The concentration of solute particles (ions) is 2 × 2.0 m = 4.0 m.

   $$\Delta T_{BP} = mK_b = 4.0\ \cancel{m} \times \left(\frac{0.52^\circ C}{1.0\ \cancel{m}}\right) = 2.08^\circ C = 2.1^\circ C$$

   The boiling point of the solution is 100.0°C + ΔT_{BP} = 100.0°C + 2.1°C = 102.1°C.

Work Problems

Use these problems for additional practice.

1. What is the freezing point and boiling point of a solution prepared by dissolving 0.450 mole of glucose in 1,200 g of water? K_f for water is 1.86°C/m; K_b is 0.52°C/m.
2. What is the molality of an aqueous urea solution that freezes at −0.85°C? K_f for water is 1.86°C/m.

Worked Solutions

1. **The freezing point is −0.70°C, and the boiling point is 100.20°C.**

   The molality of the glucose solution is (0.450 m) ÷ (1.200 kg) = 0.375 m. Glucose is a nonelectrolyte.

$$\Delta T_{FP} = mK_f = 0.375\ \cancel{m} \times \left(\frac{1.86^\circ C}{1.00\ \cancel{m}}\right) = 0.698^\circ C = 0.70^\circ C$$

$$FP_{solution} = FP_{water} - \Delta T_{FP} = 0.00^\circ C - 0.70^\circ C = -0.70^\circ C$$

$$\Delta T_{BP} = mK_b = 0.375\ \cancel{m} \times \left(\frac{0.52^\circ C}{1.0\ \cancel{m}}\right) = 0.195^\circ C = 0.20^\circ C$$

$$BP_{solution} = BP_{water} + \Delta T_{BP} = 100.00^\circ C + 0.20^\circ C = 100.20^\circ C$$

2. **The urea solution is 0.46 m.**

$$\Delta T_{FP} = 0.85^\circ C$$

$$m = \frac{\Delta T_{FP}}{K_f} = 0.85^\circ \cancel{C} \times \left(\frac{1.00\ m}{1.86^\circ \cancel{C}}\right) = 0.457\ m = 0.46\ m$$

Osmosis and Osmotic Pressure

Assuming that water is the solvent, **osmosis** is the net movement of solvent (water) through a semipermeable membrane from a dilute solution (or from pure water) into a concentrated solution. Water moves through the membrane in the direction that would make the concentrations of the two solutions the same. As water leaves the dilute solution, that solution becomes more concentrated, and as that water enters the concentrated solution, it becomes more dilute. A **semipermeable membrane** is one that allows passage of certain species through it while restricting passage of others. The membrane (cellophane or gut tissue) acts like a screen or a sieve with holes just large enough to let water molecules through but not large enough for big species, like glucose molecules or ions, to pass through. Water can pass through the membrane in both directions, but it travels faster in the direction *from* the dilute *into* the concentrated solution. That is why osmosis is described as the "net movement" of water. It moves faster in one direction than the other.

All solutions exhibit **osmotic pressure,** which is symbolized with the Greek Pi, π. The osmotic pressure of a solution is directly proportional to the concentration of solute particles in that solution. The equation used to calculate the osmotic pressure of a solution also shows the importance of temperature on π:

Osmotic pressure = $\pi = M_{solute}RT$

M_{solute} = Molarity of solute particles

R = The ideal gas law constant, 0.0821 L-atm/mole-K

T = The Kelvin temperature

Be aware that concentration is in molarity (M), not molality. Because osmotic pressure involves the motion of molecules, both absolute temperature and the ideal gas law constant are involved in its calculation. In fact, the equation used to calculate osmotic pressure can be derived from the ideal gas law. In the following problem, note how the units cancel leaving only the unit of pressure.

Problem: What is the osmotic pressure of a 0.5 M solution of glucose in water at 22°C?

Converting the Celsius temperature to Kelvin: T = 22°C = (273 + 22)K = 295 K.

$$\Pi = MRT = \left(0.50\frac{\cancel{mole}}{\cancel{L}}\right)\left(0.0821\frac{\cancel{L}-atm}{\cancel{mole}-\cancel{K}}\right)(295\ \cancel{K}) = 12\ atm$$

Answer: A 0.50 M glucose solution at 22°C has an osmotic pressure of 12.1 atm.

Notice that osmotic pressure has the unit of atmosphere (1 atm = 760 mmHg). But, what does it mean when a solution has an osmotic pressure of 12.1 atm? The explanation requires comparing this solution with pure water. If the 0.50 M glucose solution was separated from *pure water* by a semipermeable membrane, it would require a pressure of 12.1 atmospheres pushing on the glucose solution to just bring the net flow of water into the glucose solution to a halt. The apparatus in the following figure shows how the osmotic pressure of the glucose solution could be measured. The osmotic pressures of solutions calculated using $\pi = MRT$ are the pressures versus pure solvent, water in this case.

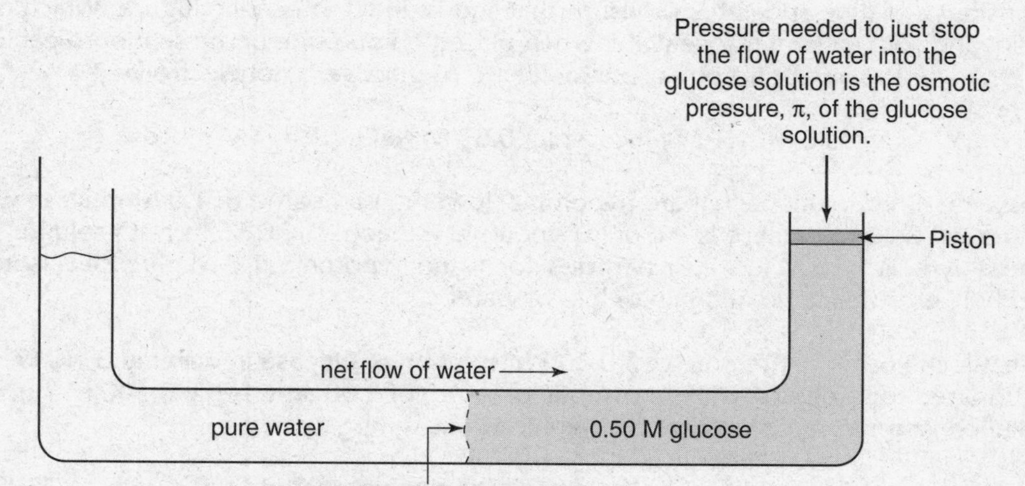

The osmotic pressure of the glucose solution is that needed to just bring the net flow of water into the glucose solution to a halt.

Here are some important facts about osmosis and osmotic pressure:

- The osmotic pressure of pure solvent (pure water) is *zero*. Solutions have osmotic pressures; pure solvents do not.
- The net flow of solvent (water) through the semipermeable membrane is *from the solution of lower osmotic pressure into the solution of higher osmotic pressure.*
- If two solutions have the same osmotic pressure, they are said to be *isotonic*. A solution with a greater osmotic pressure than another is *hypertonic*. A solution with a lower osmotic pressure than another is *hypotonic*. Compared to pure water, all solutions are hypertonic.

Osmosis and osmotic pressure are of critical importance in the chemistry of living things. Cell walls, the lining of the digestive tract, and the walls of blood vessels are all semipermeable. The osmotic pressure of blood serum and the solution within red blood cells is 7.65 atm at 37°C, normal body temperature. The solutions are isotonic. The walls of red blood cells being semipermeable allow water to pass freely through them. Intravenous solutions, such as physiological saline, 0.91%(m/v) NaCl in water or 5.5%(m/v) glucose in water, are isotonic with blood serum. As these solutions enter the blood, they do not disturb the delicate balance of osmotic pressures. But imagine if pure water were pushed into a vein diluting the serum and lowering its osmotic pressure. Immediately, water would flow from the diluted serum (lower π) *into* the red blood cells (higher π) causing them to expand and burst like small balloons. Not a healthy thing to do. On the other hand, administering an IV solution of greater osmotic pressure than normal blood serum would have the opposite effect. The osmotic pressure of blood serum would rise above that of the red blood cells, causing water to flow *out* of the cells, collapsing them and destroying their oxygen-carrying ability. The body continually responds to the stresses put on it by diet and exercise to keep osmotic pressures in balance.

Example Problems

These problems have both answers and solutions given.

1. What is osmosis?

   Answer: In terms of water as the solvent, osmosis is the net movement of water through a semipermeable membrane from a dilute solution or pure water into a concentrated solution.

2. Consider a U-tube apparatus similar to that shown in the preceding figure. Which of the following solution(s) will *lose* water when placed on one side of the semipermeable membrane separating it from a solution of 1.0 M glucose, a nonelectrolyte?

   (a) 0.25 M glucose; (b) 1.1 M glucose; (c) 0.55 M NaCl; (d) 1.0 M glucose

   Answer: Only solutions that are hypotonic (lower π) compared to 1.0 M glucose will lose water. (a) 0.25 M glucose is the only hypotonic solution. The 0.55 M NaCl solution, an electrolyte, is 1.10 M in solute particles (ions) and hypotonic. 1.0 M glucose is isotonic and will experience no net gain or loss of water.

3. What is the osmotic pressure of a 0.300 M solution of glucose in water at 37°C (310 K)? Compared to a solution with an osmotic pressure of 5.00 atm, is the 0.300 M glucose solution isotonic, hypotonic, or hypertonic at this temperature?

   Answer: π = 7.64 atm. This is a hypertonic solution compared to one with π = 5.00 atm.

$$\Pi = MRT = \left(0.30\frac{\cancel{\text{mole}}}{\cancel{\text{L}}}\right)\left(0.0821\frac{\cancel{\text{L}}\text{-atm}}{\cancel{\text{mole}}\text{-K}}\right)(310.\ \cancel{\text{K}}) = 7.64\ \text{atm}$$

Work Problems

Use these problems for additional practice.

1. When a small cucumber is soaked in brine (concentrated NaCl in water), it shrivels up to form a pickle. Explain what is happening here in terms of osmosis.

2. What is the osmotic pressure of the hot water in a mineral bath that is 1.35 M in various ions and has a temperature of 40°C? If you soak in this hot mineral bath for an hour, will you lose or gain water, or will there be no effect? Remember, the osmotic pressure of the fluids in body tissue is around 7.7 atm.

Worked Solutions

1. **The brine solution has a much greater osmotic pressure than the aqueous solutions in the tissues of the cucumber.** Water will naturally flow out of the cucumber into the higher osmotic pressure environment of the brine solution, causing the cucumber to shrivel and become a pickle.

2. **The mineral bath has an osmotic pressure of 34.7 atmospheres.**

$$\Pi = MRT = \left(1.35\frac{\cancel{\text{moles}}}{\cancel{\text{L}}}\right)\left(0.0821\frac{\cancel{\text{L}}\text{-atm}}{\cancel{\text{mole}}\text{-}\cancel{\text{K}}}\right)(313\ \cancel{\text{K}}) = 34.7\ \text{atm}$$

Since the osmotic pressure of the solution in your body cells (about 7.7 atm) is substantially less than 34.7 atm, your body will lose water as you soak in this mineral bath; you will temporarily be a bit thinner, and you will have something in common with the pickle.

Chapter Problems and Answers

Problems

1. Define the terms: solution, solute, solvent, miscible, and immiscible.

2. What is the state of subdivision of the solute in: (a) a solution of potassium iodide, KI, in water; (b) a solution of glucose, $C_6H_{12}O_6$, in water?

3. What characteristic(s) are possessed by the molecules of a solvent that make it a polar solvent?

4. Sodium chloride is described as an electrolyte, and glucose is a nonelectrolyte. What do these terms tell about these compounds in solution?

5. Which of these ionic compounds should be soluble in water: (a) $(NH_4)_2SO_4$; (b) CaS; (c) $Ca(OH)_2$?

6. Which of these ionic compounds should be soluble in water: (a) $Hg_2(NO_3)_2$; (b) K_2S; (c) CaC_2O_4?

7. Calculate the %(m/m) of the following solutions.

   (a) 62.0 g of NH_4Cl in 750 g of solution.

   (b) 8.30 g NaI in 100.0 g of water.

   (c) 45.5 g K_2SO_4 in 500 g of water.

   (d) 0.685 mole of NaCl in 350 g of water. The molar mass of NaCl is 58.4 g.

8. What mass of a 3.50%(m/m) solution of $NiCl_2$ would contain:

   (a) 7.00 g $NiCl_2$

   (b) 100.0 g $NiCl_2$

   (c) 0.50 mole $NiCl_2$. The molar mass of $NiCl_2$ is 129.6 g.

9. What mass of a 5.00%(m/m) aqueous sugar solution can be prepared from 20.0 g of sugar?

10. What mass of a 12.5%(m/m) solution of methyl alcohol in glycol can be prepared from 145 g of methyl alcohol?

11. Calculate the %(m/v) concentration of the following mixtures:

    (a) 15 g NaOH dissolved in sufficient water to make 250. mL of solution.

    (b) 23.5 g KNO_3 dissolved in enough water to prepare 920. mL of solution.

12. What volume in milliliters of a 5.50%(m/v) solution of NaCl in water contains:

(a) 100. g NaCl.

(b) 2.50 g NaCl.

(c) 0.350 mole NaCl. The molar mass of NaCl is 58.4 g.

13. What mass of $CaCl_2$ is in 1,200. mL of a 5.0%(m/v) solution of calcium chloride in water?

14. What mass of $CuSO_4$ is in 2.800 L of a 2.45%(m/v) solution of copper(II) sulfate?

15. What is the molarity of the following solutions?

(a) 0.650 mole of sulfuric acid, H_2SO_4, in 350 mL of solution.

(b) 30.0 g $AlCl_3$ in 1.30 L of solution. The molar mass of $AlCl_3$ is 133.3 g.

(c) 0.800 mole of $Fe(NO_3)_3$ in 1,250 mL of solution.

16. Calculate the number of moles of solute in each of the following:

(a) 200 mL of 0.850 M NaCl.

(b) 2.35 L of 2.15×10^{-2} M $CaCl_2$.

17. Calculate the mass of solute needed to prepare the following solutions:

(a) 2,000. mL of 0.750 M NaOH. The molar mass of NaOH is 40.0 g.

(b) 800. mL of 1.50 M $MgSO_4$. The molar mass of $MgSO_4$ is 120.4 g.

18. What volume of 0.700 M sucrose solution contains 45.0 g of sucrose, $C_{12}H_{22}O_{11}$? The molar mass of sucrose is 342.3 g.

19. What volume of 3.50 M potassium nitrate, KNO_3, contains 1.00 mole of KNO_3?

20. (a) How many milliliters of 6.0 M HCl(*aq*) must be used to prepare 2.000 L of 0.20 M HCl(*aq*)?

(b) What volume of concentrated nitric acid, 15.0 M HNO_3(*aq*), must be used to prepare 6.000 L of 0.250 M nitric acid?

21. How many milliliters of 0.250 M $BaCl_2$ are required to just consume 0.0825 mole of K_2CrO_4? The balanced equation is

$BaCl_2(aq) + K_2CrO_4(aq) \rightarrow BaCrO_4(s) + 2\ KCl(aq)$

22. It required 45.20 mL of 0.105 M NaOH(*aq*) to just exactly neutralize the acid in 25.00 mL of H_2SO_4(*aq*). What was the molarity of the sulfuric acid? The balanced equation is

$2\ NaOH(aq) + H_2SO_4(aq) \rightarrow Na_2SO_4(aq) + 2\ H_2O(l)$

23. How many milliliters of 1.50 M nitric acid, $HNO_3(aq)$, are required to just consume 25.0 g of barium oxide, BaO? The molar mass of BaO is 153.3 g. The balanced equation is

$BaO(s) + 2\ HNO_3(aq) \rightarrow Ba(NO_3)_2(aq) + H_2O(l)$

24. Calculate the molality of the following solutions:

(a) 50.0 g of ethyl alcohol, C_2H_6O, in 500 g of water. The molar mass of ethyl alcohol is 46.07 g.

(b) 1.45 moles of glucose, $C_6H_{12}O_6$, in 1,750 g of water.

25. What are the four colligative properties?

26. (a) Give two examples of an electrolyte and define what the term says about a solute.

(b) Give two examples of a nonelectrolyte and define what this term says about a solute.

27. The freezing point depression constant for water is 1.86°C/m, and the boiling point elevation constant is 0.52°C/m.

(a) What is the freezing point of a 0.85 m solution of urea, a nonelectrolyte, in water?

(b) What is the freezing point of a 0.85 m solution of NaCl, an electrolyte, in water?

(c) What is the molality of an aqueous solution that freezes at −7.5°C?

(d) What is the boiling point of a 2.0 m sucrose solution?

28. What is the osmotic pressure of a 0.45 M solution of glucose in water at 25°C?

29. A 0.50 M solution of glucose is separated by a semipermeable membrane from a 0.50 M solution of NaCl, both at the same temperature. Will water flow from the glucose solution into the NaCl solution or vice versa? Or will there be no net flow of water from one to another.

Answers

1. **Solution—A homogeneous mixture**

Solute—The substance dissolved in the solvent of a solution

Solvent—The substance into which the solute is dissolved in a solution

Miscible—Liquids that are soluble in one another in any proportion, such as alcohol and water

Immiscible—Two liquids that are not soluble in each other, such as oil and water

2. **(a) In a solution of KI, the solute exists as separated potassium and iodide ions, $K^+(aq)$, $I^-(aq)$.**

(b) In a solution of glucose, the solute exists as individual molecules, $C_6H_{12}O_6(aq)$.

3. **The molecules in a polar solvent are dipoles and might also be capable of hydrogen bonding.**

4. **An electrolyte ionizes when dissolved in water; it exists in solutions as ions.** A nonelectrolyte does not ionize when dissolved in water, rather it exists in solution as neutral molecules.

5. **(a) $(NH_4)_2SO_4$ is soluble, rule 1; (b) CaS is not soluble, rule 5; (c) $Ca(OH)_2$ is soluble, rule 7.**

6. **(a) $Hg_2(NO_3)_2$ is soluble, rule 2; (b) K_2S is soluble, rule 1; (c) CaC_2O_4 is not soluble, rule 5.**

7. **(a) 8.27%(m/m)**

$$\%(\text{m/m}) = \frac{62.0 \not{g}\ NH_4Cl}{750. \not{g}\ \text{solution}} \times 100\% = 8.27\%(\text{m/m})$$

(b) 7.66%(m/m)

$$\%(\text{m/m}) = \frac{8.30 \not{g}\ NaI}{108.3 \not{g}\ \text{solution}} \times 100\% = 7.66\%(\text{m/m})$$

(c) 8.34%(m/m)

$$\%(\text{m/m}) = \frac{45.5 \not{g}\ K_2SO_4}{545.5 \not{g}\ \text{solution}} \times 100\% = 8.34\%(\text{m/m})$$

(d) 10.3%(m/m)

mass of NaCl = 0.685 mole × (58.4 g/mole) = 40.0 g NaCl

$$\%(\text{m/m}) = \frac{40.0 \not{g}\ NaCl}{390.0 \not{g}\ \text{solution}} \times 100\% = 10.3\%(\text{m/m})$$

8. **(a) 200 g of solution**

$$\text{mass of solution} = 7.00 \text{ g } \cancel{NiCl_2} \times \left(\frac{100.\text{ g solution}}{3.50 \text{ g } \cancel{NiCl_2}}\right) = 200.\text{ g}$$

(b) 2.86×10^3 g solution

$$\text{mass of solution} = 100.\text{ g } \cancel{NiCl_2} \times \left(\frac{100.\text{ g solution}}{3.50 \text{ g } \cancel{NiCl_2}}\right) = 2.86 \times 10^3\text{ g}$$

(c) 1.85×10^3 g solution

mass of NiCl2 = (0.500 mole NiCl2) × (129.6 g/mole) = 64.8 g

$$\text{mass of solution} = 64.8 \text{ g } \cancel{NiCl_2} \times \left(\frac{100.\text{ g solution}}{3.50 \text{ g } \cancel{NiCl_2}}\right) = 1.85 \times 10^3\text{ g}$$

9. **400 g of solution**

$$\text{mass of solution} = 20.0 \text{ g } \cancel{\text{sugar}} \times \left(\frac{100.\text{ g solution}}{5.00 \text{ g } \cancel{\text{sugar}}}\right) = 400.\text{ g}$$

10. **1.16×10^3 g solution**

$$\text{mass of solution} = 145\ \text{g methyl alcohol} \times \left(\frac{100.\ \text{g solution}}{12.5\ \text{g methyl alcohol}}\right) = 1.16 \times 10^3\ \text{g}$$

11. **(a) 6.0%(m/v) NaOH**

$$\%(\text{m/v}) = \frac{15\ \text{g NaOH}}{250.\ \text{mL solution}} \times 100\% = 6.0\%$$

(b) 2.55%(m/v) KNO_3

$$\%(\text{m/v}) = \frac{23.5\ \text{g KNO}_3}{920.\ \text{mL solution}} \times 100\% = 2.55\%(\text{m/v})$$

12. **(a) 1.82×10^3 mL of solution**

$$\text{volume of solution} = 100.\ \text{g NaCl} \times \left(\frac{100.\ \text{mL solution}}{5.50\ \text{g NaCl}}\right) = 1.82 \times 10^3\ \text{mL}$$

(b) 45.4 mL of solution

$$\text{volume of solution} = 2.50\ \text{g NaCl} \times \left(\frac{100.\ \text{mL solution}}{5.50\ \text{g NaCl}}\right) = 45.5\ \text{mL}$$

(c) 371 mL of solution

mass of NaCl = (0.350 mole) × (58.4 g/mole) = 20.4 g NaCl

$$\text{volume of solution} = 20.4\ \text{g NaCl} \times \left(\frac{100.\ \text{mL solution}}{5.50\ \text{g NaCl}}\right) = 371\ \text{mL}$$

13. **60 g $CaCl_2$**

$$\text{mass of CaCl}_2 = 1{,}200.\ \text{mL} \times \left(\frac{5.0\ \text{g CaCl}_2}{100.\ \text{mL solution}}\right) = 60.\ \text{g}$$

14. **68.6g $CuSO_4$**

2.800 L = 2,800 mL

$$\text{mass of CuSO}_4 = 2{,}800.\ \text{mL} \times \left(\frac{2.45\ \text{g CuSO}_4}{100.\ \text{mL solution}}\right) = 68.6\ \text{g}$$

15. **(a) 1.86 M**

$$M = \frac{0.650\ \text{mole}}{0.350\ \text{L}} = 1.86\ \frac{\text{moles}}{\text{L}} = 1.86\ \text{M}$$

(b) 0.173 M Mole of $AlCl_3$ = 30.0 g × (1.00 mole/133.3 g) = 0.225 mole

$$M = \frac{0.225\ \text{mole}}{1.30\ \text{L}} = 0.173\ \frac{\text{mole}}{\text{L}} = 0.173\ \text{M}$$

(c) 0.640 M

$$M = \frac{0.800\ \text{mole}}{1.250\ \text{L}} = 0.640\ \frac{\text{moles}}{\text{L}} = 0.640\ \text{M}$$

16. **(a) 0.170 mole NaCl**

200 mL = 0.200 L

$$\text{moles NaCl} = 0.200\ \cancel{L} \times \left(\frac{0.850\ \text{mole}}{1.00\ \cancel{L}}\right) = 0.170\ \text{mole NaCl}$$

(b) 5.05×10^{-2} mole $CaCl_2$

$$\text{moles } CaCl_2 = 2.35\ \cancel{L} \times \left(\frac{2.15 \times 10^{-2}\ \text{moles}}{1.00\ \cancel{L}}\right) = 5.05 \times 10^{-2}\ \text{moles } CaCl_2$$

17. **(a) 60.0 g NaOH** 2,000 mL = 2.000 L

$$\text{mass NaOH} = 2.000\ \cancel{L} \times \left(\frac{0.750\ \cancel{\text{mole NaOH}}}{1.00\ \cancel{L}}\right)\left(\frac{40.0\ \text{g NaOH}}{1.00\ \cancel{\text{mole NaOH}}}\right) = 60.0\ \text{g NaOH}$$

(b) 144 g $MgSO_4$ 800 mL = 0.800 L

$$\text{mass } MgSO_4 = 0.800\ \cancel{L} \times \left(\frac{1.50\ \cancel{\text{moles } MgSO_4}}{1.00\ \cancel{L}}\right)\left(\frac{120.4\ \text{g } MgSO_4}{1.00\ \cancel{\text{mole } MgSO_4}}\right) = 144\ \text{g } MgSO_4$$

18. **0.187 L or 187 mL**

Mole of sucrose = 45.0 g × (1.00 mole/342.3 g) = 0.131 mole

$$\text{volume} = 0.131\ \cancel{\text{mole sucrose}} \times \left(\frac{1.00\ \text{L}}{0.700\ \cancel{\text{mole sucrose}}}\right) = 0.187\ \text{L}$$

19. **0.286 L or 286 mL**

$$\text{volume} = 1.00\ \cancel{\text{mole } KNO_3} \times \left(\frac{1.00\ \text{L}}{3.50\ \cancel{\text{moles } KNO_3}}\right) = 0.286\ \text{L}$$

20. **(a) 67 mL** 2.000 L = 2,000. mL

$$V_{conc} = \frac{V_{dil} M_{dil}}{M_{conc}} = \frac{(2{,}000.\ \text{mL})(0.20\ \cancel{M})}{(6.0\ \cancel{M})} = 66.7\ \text{mL} = 67\ \text{mL}$$

(b) 100. mL

$$V_{conc} = \frac{V_{dil} M_{dil}}{M_{conc}} = \frac{(6{,}000.\ \text{mL})(0.250\ \cancel{M})}{(15.0\ \cancel{M})} = 100\ \text{mL}$$

21. **330 mL of 0.250 M $BaCl_2$ are required.**

The balanced equation shows that 1 mole of $BaCl_2$ is required to consume 1 mole of K_2CrO_4. Therefore, the number of moles of $BaCl_2$ equals the number of moles of K_2CrO_4, 0.0825 mole.

$$\text{volume } BaCl_2 = 0.0825\ \cancel{\text{mole}} \times \left(\frac{1.00\ \text{L}}{0.250\ \cancel{\text{mole}}}\right) = 0.330\ \text{L} = 330\ \text{mL}$$

22. **The sulfuric acid solution was 0.0952 M.**

Moles of NaOH consumed = 0.04520 L × (0.105 mole/1.00 L) = 0.00475 mole.

From the coefficients in the balanced equation:

$$\text{mole } H_2SO_4 = \left(\frac{1}{2}\right)(0.00475 \text{ mole}) = 0.00238 \text{ mole}$$

$$M = \frac{0.00238 \text{ mole}}{0.02500 \text{ L}} = 0.0952 \text{ M } H_2SO_4$$

23. **217 mL of 1.50 M HNO_3*(aq)* are required.**

Mole of BaO = 25.0 g × (1.00 mole/153.3 g) = 0.163 mole

The coefficients in the balanced equation are used to find the mole of HNO_3*(aq)* required:

$$\text{moles } HNO_3 = \left(\frac{2}{1}\right)(0.163 \text{ mole}) = 0.326 \text{ mole } HNO_3$$

Volume of HNO_3*(aq)* = 0.326 mole × (1.00 L/1.50 mole)=0.217 L = 217 mL

24. **(a) 2.18 m ethyl alcohol** 500 g = 0.500 kg

Mole of ethyl alcohol = 50.0 g × (1 mole/46.07 g) = 1.085 moles = 1.09 moles

$$\text{molality} = \frac{1.09 \text{ moles alcohol}}{0.500 \text{ kg water}} = 2.18 \frac{\text{moles}}{\text{kg}} = 2.18 \text{ m}$$

(b) 0.829 m glucose 1,750 g = 1.750 kg

$$\text{molality} = \frac{1.45 \text{ moles}}{1.750 \text{ kg}} = 0.829 \frac{\text{moles}}{\text{kg}} = 0.829 \text{ m}$$

25. **The four colligative properties are 1) the depression of the freezing point; 2) the elevation of the boiling point; 3) the lowering of the vapor pressure; and 4) osmotic pressure.**

26. **(a) An electrolyte is a solute that ionizes when dissolved in water.** All soluble ionic compounds, such as NaCl or $MgSO_4$, and all bases and acids, such as NaOH, H_2SO_4, are electrolytes in water.

(b) A nonelectrolyte is a solute that does not ionize when dissolved in water. Soluble, molecular compounds like the sugars and alcohols are nonelectrolytes.

27. **(a) The urea solution freezes at −1.6°C.**

$$\Delta T_{FP} = mK_f = 0.85 \not{m} \times \left(\frac{1.86^\circ C}{1.00 \not{m}}\right) = 1.58^\circ C = 1.6^\circ C$$

$$FP_{solution} = FP_{water} - \Delta T_{FP} = 0.00^\circ C - 1.6^\circ C = -1.6^\circ C$$

(b) The NaCl solution freezes at −3.2°C. The freezing point of the 0.85 m NaCl solution will be depressed twice as much as the 0.85 m urea solution. Each unit of NaCl ionizes in solution forming two ions. The molality of solute particles in 0.85 m NaCl is 2 × 0.85 m or 1.7 m. The NaCl solution will freeze at 2 × (−1.6°C) = −3.2°C.

(c) The molality of the solution is 4.0 m. ΔT_{FP} = 7.5°C, therefore:

$$m = \frac{\Delta T_{KP}}{K_f} = 7.5^\circ \not{C} \times \left(\frac{1.00 \text{ m}}{1.86^\circ \not{C}}\right) = 4.03 \text{ m} = 4.0 \text{ m}$$

(d) The solution will boil at 101.0°C.

$$\Delta T_{BP} = mK_b = 2.0\ \cancel{m} \times \left(\frac{0.52^\circ C}{1.0\ \cancel{m}}\right) = 1.04^\circ C = 1.0^\circ C$$

The boiling point of the solution is then 100.00°C + 1.0°C = 101.0°C.

28. **Osmotic pressure equals 11.0 atm.** Glucose is a nonelectrolyte.

$$\Pi = MRT = \left(0.45\frac{\cancel{mole}}{\cancel{L}}\right)\left(0.0821\frac{\cancel{L}-atm}{\cancel{mole}-\cancel{K}}\right)(298\ K) = 11.0\ atm$$

29. **The molarity of solute particles in 0.50 M NaCl is 2 × 0.50 M or 1.0 M because it ionizes to form 0.50 M Na^+ plus 0.50 M Cl^- in solution. The net flow of water will be from the solution of lower solute concentration, 0.50 M glucose, into the solution of greater solute particle concentration, the 0.50 M NaCl solution.**

Supplemental Chapter Problems

Problems

1. Isopropyl alcohol, commonly known as rubbing alcohol, can dissolve in water, and water can dissolve in it in any proportion. How would you describe the solution-forming nature of these liquids?

2. The molar mass of sucrose is about 342 g and that of water is about 18 g. This means a molecule of sucrose is about 19 times heavier than a molecule of water. Yet, sucrose is very soluble in water. Why?

3. What is meant by the term "solubility" of a compound?

4. What is the general relationship between the charges on the ions of an ionic compound and its solubility in water?

5. Which of these ionic compounds should be soluble in water: (a) $BaSO_4$; (b) Na_3PO_4; (c) $Zn(OH)_2$?

6. Which of these ionic compounds should be soluble in water: (a) $Mg(C_2H_3O_2)_2$; (b) PbI_2; (c) FeO?

7. What is the %(m/m) concentration of these solutions?

   (a) 290 g of glucose in 2,350 g of solution.

   (b) 650 g of KCl in 5.500 kg of water.

8. What mass of a 5.45%(m/m) solution of $FeCl_3$ contains:

   (a) 35.0 g of $FeCl_3$.

   (b) 0.800 mole $FeCl_3$. The molar mass of $FeCl_3$ is 162.2 g.

9. (a) What mass of a 10.0%(m/m) solution can be prepared from 75.0 g of Na_2CO_3?

(b) What mass of a 0.910%(m/m) solution can be prepared from 5.00 g of NaCl?

10. (a) Calculate the %(m/v) of a solution prepared by dissolving 3.50 g of NaCl in enough water to form 15.0 mL of solution.

(b) Calculate the %(m/v) of a solution prepared by dissolving 0.650 mole of $MgSO_4$ in enough water to form 1.85 L of solution. The molar mass of magnesium sulfate is 120.4 g.

11. (a) What volume in milliliters of a 5.50%(m/v) solution of NaCl contains 2.50 g of NaCl?

(b) What volume in liters of a 5.50%(m/v) solution of NaCl contains 0.810 mole of NaCl? The molar mass of NaCl is 58.4 g.

12. (a) What mass of sodium hydroxide, NaOH, is in 2,500. mL of 7.75%(m/v) NaOH solution?

(b) What mass of glucose is in 0.450 L of 0.95%(m/v) glucose solution?

13. (a) What is the molarity, M, of a solution that contains 1.25 moles of KNO_3 in 1.80 L of solution?

(b) What is the molarity of a solution in which 500.0 g of NaOH are dissolved in 7.50 L of solution? The molar mass of NaOH is 40.0 g.

14. (a) What volume of a 2.00 M solution of iron(III) nitrate contains 30.0 g of $Fe(NO_3)_3$? The molar mass of $Fe(NO_3)_3$ is 241.9 g.

(b) What volume of 0.450 M hydrochloric acid contains 0.065 mole of HCl*(aq)*?

15. (a) How many moles of HCl*(aq)* are in 50.0 mL of 0.1250 M hydrochloric acid?

(b) How many grams of $BaCl_2$ are in 375 mL of 0.500 M barium chloride solution? The molar mass of $BaCl_2$ is 208.2g.

16. What mass of potassium chloride, KCl, is needed to prepare 500. mL of 2.80 M KCl solution? The molar mass of KCl is 74.6 g.

17. (a) What volume of 10.0 M HCl*(aq)* must be used to prepare 450 mL of 2.00 M HCl*(aq)*?

(b) What volume of 9.50 M $CaCl_2$ must be used to prepare 3.500 L of 0.25 M $CaCl_2(aq)$?

18. A 200.0 mL volume of 0.150 M $AgNO_3$ was added to an excess of phosphoric acid solution. What mass of Ag_3PO_4 is produced? The molar mass of Ag_3PO_4 is 418.7 g. The balanced equation is

$$3\ AgNO_3(aq) + H_3PO_4(aq) \rightarrow Ag_3PO_4(s) + 3\ HNO_3(aq)$$

19. What volume of 0.125 M KCl*(aq)* is required to just completely react with 50.0 mL of 0.250 M $Pb(NO_3)_2(aq)$? The balanced equation is

$$2\ KCl(aq) + Pb(NO_3)_2(aq) \rightarrow PbCl_2(s) + 2\ KNO_3(aq)$$

20. What volume of 0.250 M HCl(*aq*) is required to just completely consume 8.50 g of calcium carbonate, $CaCO_3$? The molar mass of $CaCO_3$ is 100.1 g. The balanced equation is

$$CaCO_3(s) + 2\ HCl(aq) \rightarrow CaCl_2(aq) + CO_2(g) + H_2O(l)$$

21. What is the molality of a solution that contains 34.6 g methyl alcohol, CH_4O, in 500. g of water? The molar mass of methyl alcohol is 32.0 g.

22. For water: $K_f = 1.86°C/m$ and $K_b = 0.52°C/m$. What is the freezing point and boiling point of a solution of 100. g of ethylene glycol (antifreeze, a nonelectrolyte) in 500. g of water. The molar mass of ethylene glycol, $C_2H_6O_2$, is 62.1 g.

23. How many grams of ethylene glycol must be added to 2,200 g of water to prepare a solution that will not freeze at −20.0°C? $K_f = 1.86°C/m$, and the molar mass of ethylene glycol is 62.1 g.

24. What is the osmotic pressure of a 4.50×10^{-2} M solution of sucrose at 25.0°C? What would be the osmotic pressure of this solution at 90.0°C? R = 0.0821 L-atm/mole-K.

25. What is the osmotic pressure of pure water at 25°C? In terms of the osmotic pressures of two solutions, in what direction does the water flow in osmosis?

Answers

1. **The two liquids are miscible.** (page 357)

2. **Sucrose, like other sugars, can engage in multiple hydrogen bonds with water, allowing it to be accommodated and held in solution by the solvent.** (page 373)

3. **The solubility of a compound is the mass of that compound that can dissolve in a certain mass or volume of solvent, usually 100 g or 100 mL of solvent.** The solution is saturated with solute. (page 358)

4. **As a general rule, the larger the charges on the cation and anion of an ionic compound, the less likely will the compound be soluble in water, especially if both ions carry multiple charges.** (page 360)

5. **Na_3PO_4 is soluble, rule 5.** (page 360)

6. **$Mg(C_2H_3O_2)_2$ is soluble, rule 2.** (page 360)

7. **(a) 12.3%(m/m); (b) 10.6%(m/m).** (page 362)

8. **(a) 642 g solution; (b) 2.38×10^3 g solution.** (page 362)

9. **(a) 750 g of solution; (b) 549 g of solution.** (page 362)

10. **(a) 23.3%(m/v); (b) 4.23%(m/v).** (page 363)

11. **(a) 45.5 mL; (b) 860 mL.** (page 363)

12. **(a) 194 g NaOH; (b) 4.3 g glucose.** (page 363)

13. **(a) 0.694 M; (b) 1.67 M.** (page 366)

14. **(a) 62.0 mL; (b) 144 mL.** (page 366)

15. **(a) 0.00625 mole HCl; (b) 39.0 g $BaCl_2$.** (page 366)

16. **104 g of KCl.** (page 367)

17. **(a) 90.0 mL of 10.0 M HCl(*aq*); (b) 0.0921 L or 92.1 mL of 9.50 M $CaCl_2$.** (page 367)

18. **4.19 g Ag_3PO_4.** (page 370)

19. **200. mL of 0.125 M KCl are required.** (page 370)

20. **680 mL of 0.250 HCl(*aq*) are required.** (page 370)

21. **2.16 m methyl alcohol.** (page 370)

22. **ΔT_{FP} = 6.0°C; the freezing point of the solution is −6.0°C. ΔT_{BP} = 1.7°C; the boiling point of the solution is 101.7°C.** (page 373)

23. **ΔT_{FP} = 20.0°C, requiring the molality of the solution to be 10.8 m.** 23.8 moles of ethylene glycol are required, which is 1.48×10^3 g of ethylene glycol. (page 374)

24. **At 25°C, π = 1.10 atm. At 90°C, π = 1.34 atm.** (page 378)

25. **The osmotic pressure of pure water (or any pure solvent) is zero.** In osmosis, water passes through the membrane from a solution of lower osmotic pressure into one of higher osmotic pressure. (page 378)

Chapter 14

Acids, Bases, and Neutralization

Compounds have been classified as acids or bases based on their characteristic properties since the earliest days of chemistry. **Acids** have a sour taste (Latin, *acidus* meaning sour), they react with many metals to form hydrogen gas, $H_2(g)$, they change the color of litmus, a vegetable dye, from blue to red, and acids react with bases, a process called neutralization. **Bases** have a bitter taste, and their solutions feel slippery (soap solutions are bitter and slippery). Behaving just the opposite of acids, bases turn red litmus blue, and they neutralize acids. A **neutralization** reaction is one between an acid and a base that "neutralizes" the properties of both reactants. Acids and bases are linked together because of their abilities to destroy, simultaneously, the properties of each other. You cannot study acids without also studying bases; they are two sides of the same coin. The common acids and bases were first presented in Chapter 4, and several of those are again listed below. Notice, the formulas of all the acids and bases indicate they are in aqueous solution, though several exist as pure compounds outside of solutions.

The Common Acids	***The Common Bases***
$HF(aq)$ – hydrofluoric acid	LiOH – lithium hydroxide
$HCl(aq)$ – hydrochloric acid	NaOH – sodium hydroxide
$HBr(aq)$ – hydrobromic acid	KOH – potassium hydroxide
$HI(aq)$ – hydroiodic acid	$Ca(OH)_2$ – calcium hydroxide
$HNO_3(aq)$ – nitric acid	$Ba(OH)_2$ – barium hydroxide
$HNO_2(aq)$ – nitrous acid	$NH_3(aq)$ – aqueous ammonia
$H_2CO_3(aq)$ – carbonic acid	
$H_2SO_4(aq)$ – sulfuric acid	
$H_3PO_4(aq)$ – phosphoric acid	
$H_3BO_3(aq)$ – boric acid	
$HC_2H_3O_2(aq)$ – acetic acid	

Note that aqueous ammonia, $NH_3(aq)$, is sometimes called ammonium hydroxide, $NH_4OH(aq)$, although there is no evidence that a compound with this formula exists in solution.

The Arrhenius Concept of Acids and Bases

Though acids and bases had been known for many years, the question remained, "chemically, what makes a substance behave as an acid or behave as a base?" The first successful answer to that question was put forth in 1884 by the Swedish chemist **Svante Arrhenius** (1859-1927), who defined acids, bases and neutralization in terms of definite chemical species.

- **Acid** – Any substance that produces **hydrogen ion, H^+**, when dissolved in water.
- **Base** – Any substance that produces **hydroxide ion, OH^-**, when dissolved in water.
- **Neutralization** – The reaction of an acid and base to produce a salt and water.

The ionization of nitric acid and hydrochloric acid show they are typical Arrhenius acids:

- Nitric acid in water: $HNO_3(aq) \rightarrow H^+(aq) + NO_{3-}{}^{(aq)}$
- Hydrochloric acid in water: $HCl(aq) \rightarrow H^+(aq) + Cl^-(aq)$

The ionization of sodium hydroxide and calcium hydroxide show they are typical Arrhenius bases:

- Sodium hydroxide in water: $NaOH(aq) \rightarrow Na^+(aq) + OH^-(aq)$
- Calcium hydroxide in water: $Ca(OH)_2(aq) \rightarrow Ca^{2+}(aq) + 2\ OH^-(aq)$

All acids and bases are electrolytes in water, that is, they form ions in solution, and their solutions conduct electricity.

The Hydronium Ion

During the time of Arrhenius, the idea of a free hydrogen ion, H^+, (The hydrogen ion is the nucleus of a hydrogen atom, a proton, and hydrogen ions are often called protons for this reason.) existing in water did not seem unreasonable, but in time it became clear that the hydrogen ion is intimately associated with water and exists in solution as the **hydronium ion**, H_3O^+. The symbol of the hydronium ion is the combination of the formula of water, H_2O, and a hydrogen ion, H^+, $H_2O + H^+ = H_3O^+$. As an acid ionizes the hydrogen ion immediately bonds to a water molecule through a pair of nonbonding electrons on the oxygen atom of water. **All hydrogen ions in water exist as hydronium ions.** The hydronium ion can be represented in equations as H_3O^+, $H_3O^+(aq)$ or simply as $H^+(aq)$, which represents the hydrogen ion as it exists in water. Hydroxide ions do not exist as unique ions in water. Hydroxide ions are simply surrounded by water molecules in the same way as that all ions are surrounded by water molecules. They are represented in equations as OH^- or $OH^-(aq)$.

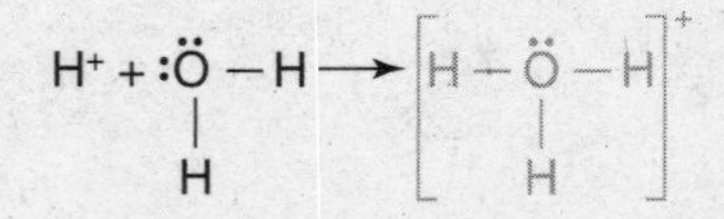

The hydrogen ion bonds to water forming the hydronium ion.

The ionization of nitric acid in water can be rewritten to show the hydrogen ion as the hydronium ion:

$$HNO_3(aq) + H_2O(l) \rightarrow H_3O^+(aq) + NO_3^-(aq)$$

It is now best to update *the Arrhenius definition of an acid*:

> **Acid** – Any substance that produces **hydronium ion, H_3O^+**, when dissolved in water.

The Bronsted-Lowry Concept

The Arrhenius concept of acids and bases was a tremendous advance in the understanding of these compounds, but is it limited to aqueous solutions, and a lot of chemistry takes place out of water. In 1923 a Danish chemist, **Johannes Bronsted** (1879-1947), and an English chemist, **Thomas Lowry** (1874–1936), proposed a more general way to describe acids and bases centered on the ability of a species to donate or accept a proton, H^+. It was not limited to aqueous solutions. Here is how they defined acids, bases and neutralization:

- **Acid** – any species (molecule or ion) that can donate a proton, H^+, to another species.
- **Base** – any species (molecule or ion) than can accept a proton.
- **Neutralization** – the transfer of a proton from an acid to a base.

The Bronsted-Lowry concept describes many more processes as acid-base reactions than does the Arrhenius concept. For example, the ionization of hydrogen chloride gas as it dissolves in water, $HCl(g) \rightarrow HCl(aq)$, can be described as an acid-base process as the proton from HCl is transferred to water.

$$\underset{\text{acid}}{HCl(aq)} + H_2O(l) \rightarrow H_3O^+(aq) + \underset{\text{base}}{Cl^-(aq)}$$

In water, $HCl(aq)$, acts as a Bronsted-Lowry acid donating a proton to water, $H_2O(l)$, which accepts the proton and acts as a Bronsted-Lowry base. The solvent, water, is more than just a solvent, it is part of a neutralization.

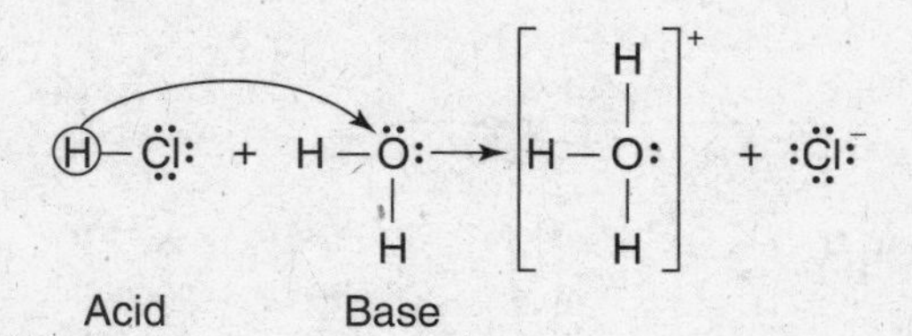

Proton donation can also occur in the gas phase as $HCl(g)$ reacts with $NH_3(g)$ to form solid ammonium chloride.

$$\underset{\text{acid}}{HCl(g)} + NH_3(g) \rightarrow \underset{\text{base}}{(NH_4^+, Cl^-)} \rightarrow NH_4Cl(s)$$

Hydrogen chloride, the acid, donates a proton to ammonia, the base, which becomes the ammonium ion, NH_4^+.

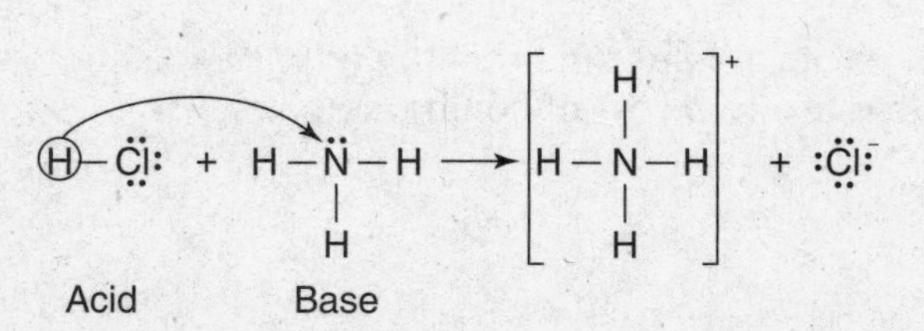

All compounds that are acids and bases in the Arrhenius concept are also acids and bases in the Bronsted-Lowry scheme. Ammonia gas, $NH_3(g)$, is very soluble in water, quickly engaging in an equilibrium that produces hydroxide ion making the solution basic. In doing so, water acts as a Bronsted-Lowry acid, donating a proton to ammonia, leaving the remainder of the water molecule, the hydroxide ion, behind.

$$\underset{\text{base}}{NH_3(aq)} + \underset{\text{acid}}{H_2O(l)} \rightleftharpoons NH_4^+(aq) + OH^-(aq)$$

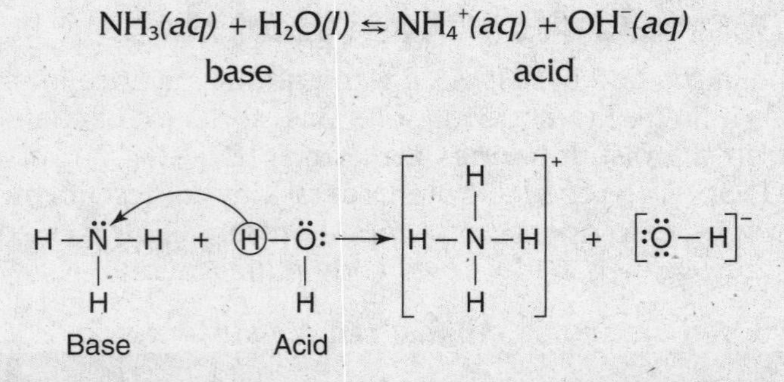

Clearly, ammonia is a Bronsted-Lowry base, it accepts a proton, but it is also a substance that forms hydroxide ion when dissolved in water, so it qualifies as an Arrhenius base too. The equation given earlier showing $HCl(aq)$ in water shows $HCl(aq)$ to be an acid in both the Arrhenius and Bronsted-Lowry concepts.

The Bronsted-Lowry concept provides yet an additional way to view acid-base reactions using the language of conjugate pairs of acids and bases. **Conjugate acid-base pairs** are two species that differ by a single proton, the acid of the pair has the proton, the base does not. For example, HCl is an acid, Cl^- is its conjugate base (the acid, HCl, minus one H^+). *Every acid has a conjugate base*. The nitrate ion, NO_3^-, is the conjugate base of nitric acid, HNO_3.The same conjugate relationship holds for bases. In the previous equation, ammonia, NH_3, was the base that accepted a proton to form the ammonium ion, NH_4^+, which is its conjugate acid, (the base, NH_3, plus one H^+). *Every base has a conjugate acid.* The hydronium ion, H_3O^+, is the conjugate acid of the base, water, H_2O. In any Bronsted-Lowry acid-base reaction there are two sets of conjugate pairs. Let's look again at the equation describing ammonia as a base in water:

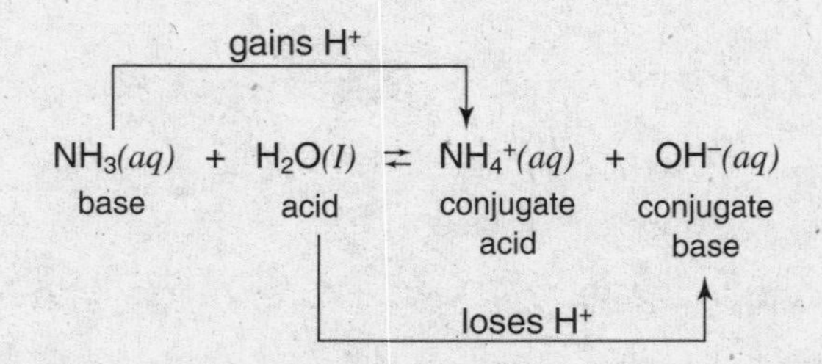

Ammonia (the base) accepts a proton from water (the acid) to form the ammonium ion (the conjugate acid of the base, ammonia) and hydroxide ion (the conjugate base of the acid, water). Another way of looking at the equation is that both sides have an acid and a base. In the forward reaction (→) ammonia is the base, and water is the acid. In the reverse reaction (←), ammonium ion, NH_4^+, is the acid, and hydroxide ion, OH^-, is the base. Each acid in the equation is related to a base on the opposite side

Acetic acid, $HC_2H_3O_2(aq)$, exists in water in equilibrium with acetate ion, $C_2H_3O_2^-(aq)$, and the hydronium ion. Again, there are two sets of conjugate pairs.

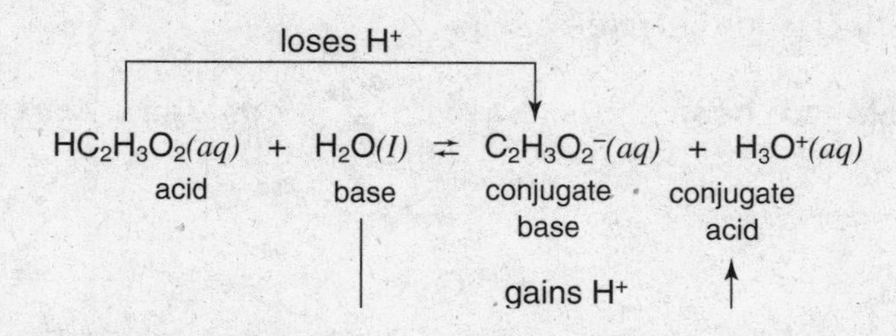

The language "acid and its conjugate base" or "base and its conjugate acid" may sound confusing, but this is the important thing to remember: Conjugate pairs are an acid and a base that differ by only one proton, H^+. The acid of the pair has that proton and the base does not. In equations it is customary to place those species termed "acid" and "base" on the left-hand side and the "conjugate acid" and "conjugate base" on the "right-hand side" Several conjugate pairs are listed in the following table.

Conjugate Pairs of Acids and Bases

Acid	*Conjugate Base*	*Base*	*Conjugate Acid*
H_2O	OH^-	OH^-	H_2O
NH_4^+	NH_3	NH_3	NH_{4+}
HCl	Cl^-	S^{2-}	HS^-
H_2SO_4	HSO_4^-	NO_3^-	HNO_3
$HC_2H_3O_2$	$C_2H_3O_2^-$	H_2O	H_3O^+

Example Problems

These problems have both answers and solutions given.

1. What are the Arrhenius definitions of an acid, base and neutralization?

   Answer: An acid is any substance that produces hydronium ion, H_3O^+, when dissolved in water. A base is any substance that produces hydroxide ion, OH^-, when dissolved in water, and neutralization is the reaction between an acid and base to produce water.

2. In terms of the Bronsted-Lowry concept, define an acid, a base and neutralization.

   Answer: An acid is a proton donor, a base a proton acceptor and neutralization is the transfer of a proton from an acid to a base.

3. Identify the conjugate pairs in the following equation that shows the equilibrium of nitrous acid, a weak acid, in water.

   $HNO_2(aq) + H_2O(l) \leftrightharpoons H_3O^+(aq) + NO_2^-(aq)$

   Answer: HNO_2 is the acid, NO_2^- is its conjugate base. H_2O is the base and the hydronium ion, H_3O^+ is its conjugate acid.

4. List the missing species in the acid-base pairs:

acid	conjugate base	base	conjugate acid
HBr	________	________	H_2O
________	SO_4^{2-}	H_2PO_{4-}	________

Answer: There is only one proton difference between conjugate pairs.

acid	conjugate base	base	conjugate acid
HBr	Br^-	OH^-	H_2O
HSO_4^-	SO_4^{2-}	$H_2PO_4^-$	H_3PO_4

Work Problems

Use these problems for additional practice.

1. Using words and a single chemical equation, show how nitric acid, HNO_3*(aq)*, is classed as an acid in both the Arrhenius and Bronsted-Lowry Concepts.

2. Identify the conjugate pairs in the following equation describing methylamine, CH_3NH_2, in water.

$$CH_3NH_2(aq) + H_2O(l) \rightleftarrows CH_3NH_3^+(aq) + OH^-(aq)$$

Worked Solutions

1. **Nitric acid ionizes in water to produce hydrogen ion, which exists in water as the hydronium ion. This makes it an Arrhenius acid.** The equation showing the formation of the hydronium ion, H_3O^+, shows nitric acid as a proton donor, making it an acid in the Bronsted-Lowry concept.

$$HNO_3(aq) + H_2O(l) \rightarrow H_3O^+(aq) + NO_3^-(aq)$$

2. **CH_3NH_2 is a base accepting a proton to form its conjugate acid, $CH_3NH_3^+$.** Water in donating the proton serves as the Bronsted-Lowry acid leaving hydroxide ion, its conjugate base.

The Strength of Acids and Bases

All acids and bases are electrolytes, that is, they form ions when dissolved in water. But, not all acids and bases form ions to the same degree. Some acids ionize completely (100%) in solution; others ionize only slightly. The same is true for bases; some ionize completely (100%), some only slightly in water. Acids and bases that are completely ionized in solution are called strong acids and strong bases; they are all **strong electrolytes**. Those that ionize only slightly are called weak acids or weak bases; they are **weak electrolytes**. Let's begin by looking at the acids.

Strong and weak acids

Strong acids are completely ionized in solution. They are strong electrolytes and exist in solution entirely as ions. Five of the common acids listed at the beginning of the chapter are strong acids, and equations describing their behavior in solution show complete ionization:

Hydrochloric acid	$HCl(aq) \rightarrow H^+(aq) + Cl^-(aq)$
Hydrobromic acid	$HBr(aq) \rightarrow H^+(aq) + Br^-(aq)$
Hydroiodic acid	$HI(aq) \rightarrow H^+(aq) + I^-(aq)$
Nitric acid	$HNO_3(aq) \rightarrow H^+(aq) + NO_3^-(aq)$
Sulfuric acid	$H_2SO_4(aq) \rightarrow H^+(aq) + HSO_4^-(aq)$

With the exception of sulfuric acid that has two ionizable hydrogens, the H_3O^+ molarity in solutions of these strong acids is the same as the molarity of the acid itself. A 0.10 M solution of $HNO_3(aq)$ is 0.10 M in H_3O^+. Most of the remaining acids are classed as weak acids.

A **weak acid** ionizes only slightly in solution, perhaps only to a few percent, and in solution between the molecular acid is in equilibrium with its ions. There are thousands of known weak acids and five of the more common ones are listed below with the equation for their equilibrium in solution. The equilibrium constant for the ionization of weak acids is symbolized K_a, and is called the acid-ionization constant.

Hydrofluoric acid	$HF(aq) \leftrightharpoons H^+(aq) + F^-(aq)$	8.2% ionized
Acetic acid	$HC_2H_3O_2(aq) \leftrightharpoons H^+(aq) + C_2H_3O_2^-(aq)$	1.3% ionized
Nitrous acid	$HNO_2(aq) \leftrightharpoons H^+(aq) + NO_2^-(aq)$	6.7% ionized
Carbonic acid	$H_2CO_3(aq) \leftrightharpoons H^+(aq) + HCO_3^-(aq)$	0.21% ionized
Hydrocyanic acid	$HCN(aq) \leftrightharpoons H^+(aq) + CN^-(aq)$	0.0020% ionized

If a weak acid is 1% ionized it means 1% of the molecules (1 of every 100) exist as ions in equilibrium with the molecular acid. Of course, the molecular acid (not ionized) will be the most abundant form of any weak acid in solution. Hydrochloric acid, $HCl(aq)$, a strong acid, and is 100% ionized in solution as are all strong acids.

The concentration of hydronium ion in solutions of weak acids is always much lower than the concentration of the acid itself. In the following table the hydronium ion concentrations in 0.10 M solutions of these same acids are compared. Notice how much lower the H_3O^+ concentrations are compared to the concentrations of the acids themselves. Unlike the weak acids, the H_3O^+ molarity in $HCl(aq)$ is the same as the molarity of the acid. It is completely ionized.

Weak Acids—%-Ionization and H_3O^+ in 0.10 M Acid Solutions

Acid	*%-Ionization*	*H_3O^+ Molarity*
0.10 M $HF(aq)$	8.2 %	0.0082 M
0.10 M $HNO_2(aq)$	6.7 %	0.0067 M
0.10 M $HC_2H_3O_2(aq)$	1.3 %	0.0013 M
0.10 M $H_2CO_3(aq)$	0.21 %	0.00021
0.10 M $HCN(aq)$	0.0070 %	0.0000070 M
0.10 M $HCl(aq)$ a strong acid, no K_a	100 %	0.10 M

Most of the acids discussed up to now have been **monoprotic acids**, having only *one* ionizable hydrogen in each acid molecule be it a strong or weak acid. Sulfuric acid, $H_2SO_4(aq)$, is a **diprotic acid**, an acid with *two* ionizable hydrogens. It is listed as a strong acid because the first hydrogen ionizes completely in solution forming the hydronium ion and the hydrogensulfate ion, HSO_4^-.

$$H_2SO_4(aq) \rightarrow H^+(aq) + HSO_4^-(aq)$$

The second hydrogen is part of the hydrogensulfate ion, HSO_4^-, which behaves like a *weak* acid, ionizing to a much smaller extent and existing in equilibrium with hydronium and sulfate ions.

$$HSO_4^-(aq) \leftrightharpoons H^+(aq) + SO_4^{2-}(aq)$$

A little less than 10% of HSO_4^- is ionized in 0.10 M sulfuric acid. Yet, sulfuric acid is classed as a strong acid because the *first* hydrogen is completely ionized. The hydronium ion molarity in 0.10 M $H_2SO_4(aq)$ is just a bit higher than the concentration of the acid itself, about 0.11M.

Strong and weak bases

All **strong bases** are *hydroxide compounds* of the Group IA metals (Li, Na, K, etc.) and the heavier Group IIA metals (Ca and Ba) that ionize completely in water. Three common strong bases are listed below showing how their ionization produces hydroxide ion in solution.

Sodium hydroxide	$NaOH(aq) \rightarrow Na^+(aq) + OH^-(aq)$
Potassium hydroxide	$KOH(aq) \rightarrow K^+(aq) + OH^-(aq)$
Calcium hydroxide	$Ca(OH)_2(aq) \rightarrow Ca^{2+}(aq) + 2\ OH^-(aq)$

Because strong bases are completely ionized, the molarity of hydroxide ion in sodium and potassium hydroxide solutions, the **monohydroxy bases**, is the same as the molarity of the base itself. The OH^- molarity in 0.10 M NaOH is 0.10 M. In the case of $Ca(OH)_2$, a **dihydroxy base**, the OH^- molarity is exactly two times the molarity of $Ca(OH)_2$. Not all hydroxide compounds function well as bases because of their low solubility in water. Aluminum hydroxide, $Al(OH)_3$, and magnesium hydroxide, $Mg(OH)_2$ can both neutralize acids (they are used in several antacids) but neither is very soluble and cannot used to prepare solutions.

Weak bases form hydroxide ions by reacting with water, removing H^+ from water leaving OH^- in solution. They are weak electrolytes and only a small fraction of base molecules produce hydroxide ion at any instant. Reactions like this in which water is a reactant are called **hydrolysis** reactions. The most common weak bases are the nitrogen bases, neutral molecules containing a nitrogen atom that can bond to the H^+ removed from water. Aqueous ammonia, a solution of ammonia gas dissolved in water, is the most common nitrogen base. A small fraction of the ammonia molecules react with water and exist in equilibrium with ammonium and hydroxide ions, $NH_4^+(aq)$ and $OH^-(aq)$.

$$NH_3(aq) + H_2O(l) \leftrightharpoons NH_4^+(aq) + OH^-(aq)$$

Three weak bases are listed below showing their equilibria in water that generates hydroxide ion.

Ammonia	$NH_3(aq) + H_2O(l) \leftrightharpoons NH_4^+(aq) + OH^-(aq)$
Pyridine	$C_5H_5N(aq) + H_2O(l) \leftrightharpoons C_5H_5NH^+(aq) + OH^-(aq)$
Methylamine	$CH_3NH_2(aq) + H_2O(l) \leftrightharpoons CH_3NH_3^+(aq) + OH^-(aq)$

In the following table the molarity of OH^- in 0.10 M solutions of these three weak bases is shown. Unlike the weak bases, the concentration of OH^- is the same as the molarity of NaOH(*aq*) itself since it is 100% ionized in solution.

Weak Bases—%-Ionization and OH^- M in 0.10 M Base Solutions

Base	*%-Ionization*	*OH^- Molarity*
0.10 M NH_3(*aq*)	1.3 %	0.0013 M
0.10 M C_5H_5N(*aq*)	0.013 %	0.000013 M
0.10 M CH_3NH_2(*aq*)	6.6 %	0.0066 M
0.10 M NaOH(*aq*)	100 %	0.10 M

Do not confuse the terms concentrated and dilute with strong and weak; they concern two different things. Concentrated and dilute refer to the concentration of an acid or base, while strong and weak refer to the ability of an acid or base to ionize in solution. There are concentrated solutions of weak acids and dilute solutions of strong bases, and so forth.

Example Problems

These problems have both answers and solutions given.

1. What is a strong acid? What is a strong base?

   Answer: A strong acid is one that ionizes completely in water to produce hydronium ions. A strong base is one that ionizes completely in water to produce hydroxide ions.

2. Which solution in each pair is more acidic, that is, has the higher concentration of hydronium ion?

   (a) 0.050 M HNO_3 or 0.050 M HNO_2

   (b) 0.050 M HCl or 0.075 M HNO_3

   Answer: (a) 0.050 M HNO_3 is a solution of a strong acid and has the greater H_3O^+ concentration. (b) 0.075 M HNO_3 is more concentrated than the other strong acid and has the greater H_3O^+ concentration.

Work Problems

Use these problems for additional practice.

1. What is a weak acid? What is a weak base?

2. Write the equilibrium equation for the ionization of ammonia in water.

3. Which solution in each pair has the higher OH^- concentration?

   (a) 0.050 M $Ca(OH)_2$ or 0.050 M KOH

   (b) 0.050 M NaOH or 0.50 M C_5H_5N

Worked Solutions

1. **A weak acid is one that ionizes to a small extent in water producing a hydronium ion molarity much smaller than the molarity of the acid. A weak base is one that reacts with water and ionizes to a small extent producing a hydroxide ion molarity much smaller than the molarity of the base.**

2. **$NH_3(aq) + H_2O(l) \rightleftharpoons NH_4^+(aq) + OH^-(aq)$**

3. **(a) 0.050 M $Ca(OH)_2$ has the greater OH^- concentration since it is a dihydroxy base compared to the same concentration of a monohydroxy base.**

   (b) 0.050 M NaOH is a strong base and has the greater OH^- concentration compared to the same concentration of the weak base pyridine, C_5H_5N.

Neutralization and Net-Ionic Equations

If H_3O^+ ions are responsible for the properties of acids, and OH^- responsible for the properties of bases, neutralization reactions should show how acids and bases destroy the properties of each other. The products of an acid-base neutralization reaction are a salt and water.

$$\underset{\text{the acid}}{HCl(aq)} + \underset{\text{the base}}{KOH(aq)} \rightarrow \underset{\text{the salt}}{KCl(aq)} + \underset{\text{the water}}{H_2O(l)}$$

To a chemist, the term **salt** refers to the ionic product of a neutralization reaction, the positive ion coming from the base (K^+ from KOH), and the negative ion coming from the acid (Cl^- from HCl). Potassium chloride, KCl, is the salt produced in the neutralization of hydrochloric acid and potassium hydroxide. Sodium nitrate, $NaNO_3$, is the salt produced in the neutralization of nitric acid, HNO_3, with sodium hydroxide, NaOH. The other product, water, is the covalent product of neutralization, formed by the combination of H^+ from the acid, and OH^-, from, the base. Notice now the ions responsible for the properties of acids, $H^+(aq)$, and the properties of bases, $OH^-(aq)$, combine to form neutral water, $H_2O(l)$. This is how acids and bases destroy the properties of each other simultaneously.

$$H^+(aq) + OH^-(aq) \rightarrow H_2O(l)$$

The formation of water is the core, or heart, of neutralization, a fact revealed in the net-ionic equation for the reaction. Because net-ionic equations reveal the actual reactions taking place, it would be useful to learn more about them. Net-ionic equations are developed in a three-step process that begins with the formula equation, then onto the ionic equation and ending with the net-ionic equation. Let's derive the net-ionic equation for a typical neutralization reaction. We need to start with a formula equation. **Formula equations** use neutral formulas or symbols for all reactants and products. The balanced formula equation for the neutralization of nitric acid with sodium hydroxide is:

Formula equation: $HNO_3(aq) + NaOH(aq) \rightarrow NaNO_3(aq) + H_2O(l)$

The formula equation shows that neutralization is a double-replacement reaction (Chapter 7), but there is more that can be shown about this reaction. Nitric acid, sodium hydroxide and sodium nitrate exist in solution as ions – they ionize in water. An **ionic equation** will separate the neutral formulas into the ions they form in solution. Water, being a molecular compound, is *not* ionized and is retained as a molecule.

Ionic equation: $H^+(aq) + NO_3^-(aq) + Na^+(aq) + OH^-(aq) \rightarrow Na^+(aq) + NO_3^-(aq) + H_2O(l)$

the acid the base the salt the water

The ionic equation tells even more about the neutralization, showing the actual way each compound exists in solution, and this leads to an interesting observation: Some ions don't seem to be involved in the reaction. Both the nitrate ion, $NO_3^-(aq)$, and the sodium ion, $Na^+(aq)$ remain unchanged as reactants become products. They are regarded as *spectator ions*, not directly involved in the reaction though present in solution. Removing the spectator ions by canceling them from both sides of the equation leaves only those species involved in the actual reaction. The **net-ionic equation** is what remains after removal of the spectator ions and it shows only those ions that take part in the neutralization reaction. Crossing out the spectator ions

$$H^+(aq) + \cancel{NO_3}^-(aq) + \cancel{Na}^+(aq) + OH^-(aq) \rightarrow \cancel{Na}^+(aq) + \cancel{NO_3}^-(aq) + H_2O(l)$$

leaves the net-ionic equation.

Net-ionic equation: $H^+(aq) + OH^-(aq) \rightarrow H_2O(l)$

The net-ionic equation clearly shows that the core of neutralization is the combining of those ions that define acids and bases to form neutral water.

When deriving net-ionic equations, be certain to start with a *balanced* formula equation. That way, you will end up with a balanced net-ionic equation. The formula, ionic and net-ionic equations for the reaction between hydrochloric and calcium hydroxide appear below. The net-ionic equation shows that acid-base neutralization here is the same reaction seen in the nitric acid-sodium hydroxide reaction.

Formula equation: $2\ HCl(aq) + Ca(OH)_2(aq) \rightarrow CaCl_2(aq) + 2\ H_2O(l)$

Ionic equation: $2\ H^+(aq) + \cancel{2\ Cl}^-(aq) + \cancel{Ca}^{2+}(aq) + 2\ OH^-(aq) \rightarrow \cancel{Ca}^{2+}(aq) + \cancel{2\ Cl}^-(aq) + 2\ H_2O(l)$

Net-Ionic equation: $2\ H^+(aq) + 2\ OH^-(aq) \rightarrow 2\ H_2O(l)$

The net-ionic equation is simplified by dividing all coefficients by 2: $H^+(aq) + OH^-(aq) \rightarrow H_2O(l)$

Not all acids are completely ionized when converting formula equations to ionic equations; they are kept together as molecules because only a small fraction of their molecules actually ionize in solution. These are the "weak acids" which were discussed previously in this chapter. We will be concerned with these three: acetic acid, $HC_2H_3O_2(aq)$, nitrous acid, $HNO_2(aq)$ and hydrofluoric acid, $HF(aq)$. The formula, ionic and net-ionic equations for the reaction of acetic acid with sodium hydroxide are shown below.

Formula equation: $HC_2H_3O_2(aq) + NaOH(aq) \rightarrow NaC_2H_3O_2(aq) + H_2O(l)$

Ionic equation: $HC_2H_3O_2(aq) + \cancel{Na}^+(aq) + OH^-(aq) \rightarrow \cancel{Na}^+(aq) + C_2H_3O_2^-(aq) + H_2O(l)$

Net-Ionic equation: $HC_2H_3O_2(aq) + OH^-(aq) \rightarrow C_2H_3O_2^-(aq) + H_2O(l)$

The net-ionic equation again shows the formation of water from OH^- and, in this case, the hydrogen ion from the acetic acid molecule.

Example Problems

These problems have both answers and solutions given.

1. Salts are produced in neutralization reactions. Using both formulas and names, identify the acid and base that reacted to produce these salts: (a) KNO_3, (b) $CaSO_4$, (c) Na_3PO_4

   Answer: The positive ion, the cation, comes from the base, the negative ion, the anion, comes from the acid. (a) potassium hydroxide, KOH, and nitric acid, $HNO_3(aq)$, (b) calcium hydroxide, $Ca(OH)_2$, and sulfuric acid, $H_2SO_4(aq)$, (c) sodium hydroxide, NaOH and phosphoric acid, $H_3PO_4(aq)$.

2. Write the formula, ionic and net-ionic equations for the neutralization of hydrochloric acid with potassium hydroxide: $HCl(aq) + KOH(aq) \rightarrow KCl(aq) + H_2O(l)$

   Answer:

   Formula equation: $HCl(aq) + KOH(aq) \rightarrow KCl(aq) + H_2O(l)$

   Ionic equation: $H^+(aq) + Cl^-(aq) + K^+(aq) + OH^-(aq) \rightarrow K^+(aq) + Cl^-(aq) + H_2O(l)$

   Net-ionic equation: $H^+(aq) + OH^-(aq) \rightarrow H_2O(l)$

3. Write the formula, ionic and net-ionic equations for the neutralization of nitrous acid, HNO_2, with barium hydroxide, $Ba(OH)_2$: $2\ HNO_2(aq) + Ba(OH)_2(aq) \rightarrow Ba(NO_2)_2(aq) + 2\ H_2O(l)$

   Answer:

   Formula equation: $2\ HNO_2(aq) + Ba(OH)_2(aq) \rightarrow Ba(NO_2)_2(aq) + 2\ H_2O(l)$

   Ionic equation: $2\ HNO_2(aq) + Ba^{2+}(aq) + 2\ OH^-(aq) \rightarrow Ba^{2+}(aq) + 2\ NO_2^-(aq) + 2\ H_2O(l)$

   Net-ionic equation: $2\ HNO_2(aq) + 2\ OH^-(aq) \rightarrow 2\ NO_2^-(aq) + 2\ H_2O(l)$

4. Complete and balance the following neutralization reactions:

   (a) $LiOH(aq) + HBr(aq) \rightarrow$

   (b) $Ca(OH)_2(aq) + HF(aq) \rightarrow$

   Answer: The neutralizations will produce water and a salt formed from the remaining ions.

   (a) $LiOH(aq) + HBr(aq) \rightarrow LiBr(aq) + H_2O(l)$

   (b) $Ca(OH)_2(aq) + 2\ HF(aq) \rightarrow CaF_2(aq) + 2\ H_2O$

Work Problems

Use these problems for additional practice.

1. Using both formulas and names, identify the acid and base that produced the following salts in neutralization reactions: (a) $BaCl_2$, (b) K_2SO_4, (c) NaBr

2. Write the formula, ionic and net-ionic equations for the neutralization of sulfuric acid with sodium hydroxide: $H_2SO_4(aq) + 2\ NaOH(aq) \rightarrow Na_2SO_4(aq) + 2\ H_2O(l)$

3. Write the formula, ionic and net-ionic equations for the neutralization of hydrofluoric acid, HF, with potassium hydroxide, KOH: $HF(aq) + KOH(aq) \rightarrow KF(aq) + H_2O(l)$

4. Complete and balance the following neutralization reactions.

(a) $H_2SO_4(aq) + NaOH(aq) \rightarrow$

(b) $HC_2H_3O_2(aq) + Ba(OH)_2(aq) \rightarrow$

Worked Solutions

1. **(a) Barium hydroxide, $Ba(OH)_2$ and hydrochloric acid, $HCl(aq)$, (b) potassium hydroxide, KOH, and sulfuric acid, $H_2SO_4(aq)$, (c) sodium hydroxide, NaOH, and hydrobromic acid, $HBr(aq)$.**

2. **Formula equation: $H_2SO_4(aq) + 2\ NaOH(aq) \rightarrow Na_2SO_4(aq) + 2\ H_2O(l)$**

Ionic equation: $2\ H^+(aq) + SO_4^{2-}(aq) + 2\ Na^+(aq) + 2\ OH^-(aq) \rightarrow 2\ Na^+(aq) + SO_4^{2-}(aq) + 2\ H_2O(l)$

Net-ionic equation: $2\ H^+(aq) + 2\ OH^-(aq) \rightarrow 2\ H_2O(l)$

Dividing the coefficients by 2: $H^+(aq) + OH^-(aq) \rightarrow H_2O(l)$

3. **Formula equation: $HF(aq) + KOH(aq) \rightarrow KF(aq) + H_2O(l)$**

Ionic equation: $HF(aq) + K^+(aq) + OH^-(aq) \rightarrow K^+(aq) + F^-(aq) + H_2O(l)$

Net-ionic equation: $HF(aq) + OH^-(aq) \rightarrow F^-(aq) + H_2O(l)$

4. **(a) $H_2SO_4(aq) + 2\ NaOH(aq) \rightarrow Na_2SO_4(aq) + 2\ H_2O(l)$**

(b) $HC_2H_3O_2(aq) + Ba(OH)_2(aq) \rightarrow Ba(C_2H_3O_2)_2(aq) + 2\ H_2O(l)$

Chapter Problems and Answers

Problems

1. List at least three properties of acids and three properties of bases.

2. In the Arrhenius concept of acids and bases, what species are produced by acids and bases in water? What is formed from these species in neutralization?

3. Complete and balance the following equations describing neutralization reactions.

(a) $HNO_3(aq) + LiOH(aq) \rightarrow$

(b) $Ba(OH)_2(aq) + HC_2H_3O_2(aq) \rightarrow$

4. List three ways the hydronium ion can be symbolized.

5. Write the ionic and net-ionic equations for the following neutralization reactions.

   (a) $2\ HCl(aq) + Sr(OH)_2(aq) \rightarrow SrCl_2(aq) + 2\ H_2O(l)$

   (b) $HC_2H_3O_2(aq) + NaOH(aq) \rightarrow NaC_2H_3O_2(aq) + H_2O(l)$

6. Define an acid, base and neutralization in the Bronsted-Lowry concept of acids and bases.

7. Why is ammonia, NH_3, considered to be a base in both in the Arrhenius and Bronsted-Lowry concepts of acids and bases?

8. Why is HCl(aq) considered to be an acid in both in the Arrhenius and Bronsted-Lowry concepts of acids and bases?

9. Identify the acid, base, conjugate base and conjugate acid in the following equations:

   (a) $HNO_2(aq) + H_2O(l) \leftrightharpoons H_3O^+(aq) + OH^-(aq)$

   (b) $HBr(g) + NH_3(g) \rightarrow NH_4^+, Br^-(s)$

   (c) $H_2O(l) + H_2O(l) \leftrightharpoons H_3O^+(aq) + OH^-(aq)$

10. Write the formula for the missing species.

acid	conjugate base	base	conjugate acid
HNO3	________	S^{2-}	________
________	HSO_4^-	F^-	________

11. What is the principal difference between weak and strong acids?

12. Write the ionization equations for acetic acid, $HC_2H_3O_2(aq)$, and Hydrocyanic acid, $HCN(aq)$, both weak acids.

13. In terms of composition and behavior, what are the differences between weak and strong bases?

14. Write the equations showing (a) ammonia, NH_3, and (b) pyridine, C_5H_5N, acting as weak bases in water.

Answers

1. Acids have a sour taste, react with many metals to produce hydrogen gas, turn blue litmus red and neutralize bases. Bases taste bitter, their solutions feel slippery, they turn red litmus blue and they neutralize acids. Both are electrolytes, they ionize in solution.

2. An Arrhenius acid produces hydronium ion, H_3O^+, in water and an Arrhenius base produces hydroxide ion, OH^-, in water. When acids and bases neutralize one another these ions combine to form water.

3. (a) $HNO_3(aq) + LiOH(aq) \rightarrow LiNO_3(aq) + H_2O(l)$

   (b) $Ba(OH)_2(aq) + 2\ HC_2H_3O_2(aq) \rightarrow Ba(C_2H_3O_2)_2(aq) + 2\ H_2O(l)$

4. The hydronium ion can be written, H_3O^+, $H_3O^+(aq)$, or $H^+(aq)$.

5. (a) Ionic: $2\ H^+(aq) + 2\ Cl^-(aq) + Sr^{2+}(aq) + 2\ OH^-(aq) \rightarrow Sr^{2+}(aq) + 2\ Cl^-(aq) + 2\ H_2O(l)$

   Net-ionic: $2\ H^+(aq) + 2\ OH^+(aq) \rightarrow 2\ H_2O(l)$ (simplify by dividing all coefficients by 2)

   (b) Acetic acid is a weak acid and is written as the *molecule* in ionic and net-ionic equations.

   Ionic: $HC_2H_3O_2(aq) + Na^+(aq) + OH^-(aq) \rightarrow Na^+(aq) + C_2H_3O_2^-(aq) + H_2O(l)$

   Net-ionic: $HC_2H_3O_2(aq) + OH^-(aq) \rightarrow C_2H_3O_2^-(aq) + H_2O(l)$

6. In the Bronsted-Lowry concept, an acid is a proton donor, a base a proton acceptor and neutralization is the donation of one proton by an acid to a base.

7. When ammonia is dissolved in water, it can accept a proton from water to form the ammonium ion, and hydroxide ion. Ammonia is a base in the Arrhenius concept because it produces hydroxide ion in water, and it is a base in the Bronsted-Lowry concept because it is a proton acceptor.

   $NH_3(aq) + H_2O(l) \leftrightharpoons NH_4^+(aq) + OH^-(aq)$

8. $HCl(aq)$ ionizes completely in water, producing hydronium ion, H_3O^+, as it donates a proton to water. It is an Arrhenius acid because it produces hydronium ion and a Bronsted-Lowry acid because it is a proton donor.

   $HCl(aq) + H_2O(l) \rightarrow H_3O^+(aq) + Cl^-(aq)$

9. (a) $HNO_2(aq) + H_2O(l) \leftrightharpoons H_3O^+(aq) + OH^-(aq)$

   acid base conjugate acid conjugate base

   (b) $HBr(g) + NH_3(g) \rightarrow NH_4^+, Br^-(s)$

   acid base conjugate acid conjugate base

   (c) $H_2O(l) + H_2O(l) \leftrightharpoons H_3O^+(aq) + OH^-(aq)$

   acid base conjugate acid conjugate base

   Water is acting as both a proton donor and a proton acceptor in this last reaction, so one water, either one, acts as an acid and the other as a base.

10.

acid	**conjugate base**	**base**	**conjugate acid**
HNO3	NO_3^-	S_2^-	HS^-
H_2SO_4	HSO_4^-	F^-	HF

11. A strong acid ionizes completely in solution; it exists completely as ions in solution. A weak acid ionizes only slightly and exists as an equilibrium between the molecular acid and the ions it forms in solution.

12. $HC_2H_3O_2(aq) \leftrightharpoons H^+(aq) + C_2H_3O_2^-(aq)$ and $HCN(aq) \leftrightharpoons H^+(aq) + CN^-(aq)$

13. Strong bases are all composed of a cation and hydroxide ion. They are strong electrolytes. Weak bases react with water to make hydroxide ion, accepting a proton from water leaving free hydroxide ion in solution. They are weak electrolytes and exist in solution as an equilibrium between the molecular base and the ions it forms upon reaction with water.

14. (a) $NH_3(aq) + H_2O(l) \leftrightharpoons NH_4^+(aq) + OH^-(aq)$

(b) $C_5H_5N(aq) + H_2O(l) \leftrightharpoons C_5H_5NH^+(aq) + OH^-(aq)$

Supplemental Chapter Problems

Problems

1. Twelve statements concerning acids and bases follow. If the statement is true, indicate that it is true. If it is false, rewrite the statement to make it true.

(a) All Arrhenius acids produce hydronium ions in water.

(b) All substances that are acids in the Arrhenius concept are also acids in the Bronsted-Lowry concept.

(c) A dilute acid solution and a solution of a weak acid mean the same thing.

(d) The conjugate acid of sulfuric acid, H_2SO_4, is the hydrogensulfate ion, HSO_4^-.

(e) Neutralization in the Bronsted-Lowry concept is the transfer of a proton from an acid to a base.

(f) The base used in a neutralization reaction that produced the salt, $LiNO_3$, was LiOH.

(g) Acetic acid is not ionized in an ionic equation because it is a weak acid.

(h) The acetate ion, $C_2H_3O_2^-$, is the conjugate base of a weak acid.

(i) The net-ionic equation for the neutralization of nitric acid with calcium hydroxide does not show any nitrate ions or calcium ions.

(j) Water can be a Bronsted-Lowry base but never a Bronsted-Lowry acid.

(k) The hydroxide ion molarity in a 0.5 M solution of KOH is 0.5 M. In a 0.5 M solution of NH_3 it is less than 0.5 M.

(i) A salt is the ionic product of an acid-base neutralization.

Answers

1. (a) True (page 394)

   (b) True (page 395)

   (c) False. Dilute is a concentration term; weak is a "degree of ionization" term. It is certainly possible to have a concentrated solution of a weak acid. (page 399)

   (d) False. The conjugate acid of H_2SO_4 would have one more H^+ than H_2SO_4. HSO_4^- is the conjugate base of sulfuric acid. (page 396)

   (e) True (page 395)

   (f) True (page 402)

   (g) True (page 399)

   (h) True (page 400)

   (i) True (page 402)

   (j) False. Water can act as a Bronsted-Lowry acid or base under most all conditions. (page 402)

   (k) True (page 402)

   (l) True (page 403)

Glossary

The number(s) in parentheses indicate the chapter(s) where the term is important.

acid (4, 14) *See Arrhenius acid and Bronsted-Lowry acid.*

acid, strong (14) An acid (a source of hydrogen ions, H^+, that is completely ionized in solution.

acid, weak (14) An acid that is only partially ionized in solution.

alkali metals (2) The metals in Group IA on the periodic table (Li, Na, K, Rb, Cs).

alkaline earth metals (2) The metals in Group IIA on the periodic table (Be, Mg, Ca, Sr, Ba).

anion (2) An ion that bears a negative charge, for example, Cl^- and SO_4^{2-}.

aqueous solution (13) A solution (homogeneous mixture) in which water is the solvent.

Arrhenius acid (14) Any substance that produces hydrogen ions (hydronium ions) when dissolved in water.

Arrhenius base (14) Any substance that produces hydroxide ions when dissolved in water.

Arrhenius concept (14) An acid-base concept in which an acid produces hydrogen ion (hydronium ion) and a base produces hydroxide ion when dissolved in water. Neutralization combines hydronium and hydroxide ions to form water.

atmosphere, atm (11) A unit of pressure equal to 760 mmHg or 1.013225×10^5 Pascal.

atomic mass (3) The mass of an atom in atomic mass units (amu).

atomic mass unit, amu (3, 5) Exactly one-twelfth the mass of one C–12 atom, 1.66×10^{-24} g.

atomic number (3) The number of protons in the nucleus and the number of electrons about the nucleus of a neutral atom.

atomic orbital (9) A region of space about a nucleus where electrons in that orbital are most likely to be found. An orbital can contain a maximum of two electrons.

atomic weight (3) *See molar mass.*

Aufbau principle (9) The principle followed to construct ground-state electron configurations of atoms and monatomic ions.

Avogadro's law (11) The volume of a gas is directly proportional to the number of moles of gas as long as pressure and temperature remain constant.

Avogadro's number (5) The number of particles in one mole, 6.022×10^{23}. The number of atoms in exactly 12 grams of C–12.

balanced equation (7) A chemical equation that obeys the Law of Conservation of Mass. The same number of atoms of each element appear on both sides of the equation.

base (4, 14) *See Arrhenius base and Bronsted-Lowry base.*

base, strong (14) A base (a source of hydroxide ions, OH^-) that is completely ionized in solution.

base, weak (14) A molecular base that, by reaction with water, is only partially ionized in solution.

binary compound (4) One composed of only two elements.

boiling-point elevation, ΔTBP (13) The increase in the boiling point of a solvent caused by the presence of a solute. The boiling point elevation is a colligative property of solution.

bond (10) An attractive force between atoms or between ions of opposite charge.

bond angle (10) The angle of arc between two adjacent bonds in a covalent species.

bonding pair (10) A pair of electrons shared between two atoms linking them together.

bond length (10) The distance between the nuclei of two bonded atoms.

Boyle's law (11) The volume of a gas varies inversely with the pressure applied to the gas if temperature and the amount of gas remain constant.

Bronsted-Lowry acid (14) Any species that can donate a proton, H^+.

Bronsted-Lowry base (14) Any species than can accept a proton, H^+.

Bronsted-Lowry concept (14) An acid-base concept that defines an acid as a proton donor and a base as a proton acceptor. Neutralization is the transfer of a proton from an acid to a base.

cation (2) An ion that bears a positive charge, for example Na^+, NH_4^+, and Al^{3+}.

Celsius temperature scale (1) A scale of temperature on which water normally freezes at 0°C and boils at 100°C.

Charles' law (11) The volume of a gas varies directly with its Kelvin temperature when pressure and the amount of gas remain constant.

chemical change (2) A chemical reaction in which one or more substances are formed from one or more different substances. The Law of Conservation of Mass is obeyed.

chemical equation (7) A qualitative way of describing a chemical reaction using formulas of compounds and symbols of elements for reactants and products.

chemical formula (2, 4) An array of symbols and subscript numbers to indicate the composition of a compound. $C_6H_{12}O_6$ is the formula of the compound glucose.

chemical nomenclature (2, 4) The use of sets of rules to develop names of chemical compounds.

chemical property (2) A property of an element or compound that is learned from its participation in chemical changes.

chemical reaction (2) *See chemical change.*

chemical symbol (2) A one or two-letter abbreviation of the name of an element.

chemistry (1) The study of matter and the changes it undergoes.

colligative property (13) A property of solutions that depends only on the concentration of solute particles, such as boiling point elevation, freezing point depression, and osmotic pressure.

combination reaction (7) A chemical change in which two or more substances combine to form a new substance. Also known as a synthesis reaction.

combustion (7) A reaction in which a substance burns in oxygen.

compound (2, 4) A pure substance in which two or more elements are combined in a definite mass ratio, for example, atom ratio.

concentration (13) A measure of the amount of one substance in a given amount (mass or volume) of another. Molarity and percent are common units of concentration.

conjugate acid (14) In the Bronsted-Lowry concept, the species formed when a base accepts a H^+. The conjugate acid of the base ammonia, NH_3, is the ammonium ion, NH_4^+.

conjugate base (14) In the Bronsted-Lowry concept, the species formed when an acid donates a H^+. The conjugate base of the acid, HCl, is the chloride ion, Cl^-.

conversion factor (1) A factor usually written as a fraction that is used to convert a measured quantity from one unit to another, such as a measured length from inches to meters or a quantity of a substance from mass to moles.

covalent bond (10) A bond between two atoms caused by the sharing of one or more pair(s) of electrons between them.

covalent compound (10) A compound containing only covalent bonds.

Dalton's law (11) The total pressure exerted by a mixture of gases equals the sum of the partial pressures of each gas in the mixture.

decomposition reaction (7) A chemical change in which a compound is broken down to simpler substances.

density (1) The mass of a sample of matter divided by its volume.

diatomic molecule (2) A molecule composed of two atoms.

dipole-dipole force (12) An attractive force between polar molecules.

double bond (10) A covalent bond in which two pairs of electrons are shared.

double-replacement reaction (7) A reaction between ionic compounds in which the cations and anions exchange partners with the formation of a precipitate, water, or a gas.

electrolyte (13) A substance that forms a solution that conducts electricity when dissolved in water. The formation of ions by the substance allows electrical conduction.

electron, e– (3) The negatively charged subatomic particle found outside the nucleus of an atom.

electron configuration (9) The pattern of electrons occupying the orbitals of an atom or ion. The lowest energy configuration is called a ground-state configuration.

electronegativity (10, 12) A measure of the attraction an atom has for the electrons it shares with another atom. Generally, metals have lower and nonmetals higher electronegativities.

electronic structure (9) The arrangement of electrons in orbitals about a nucleus that is shown by the electronic configuration of an atom or ion.

element (2) A pure substance that cannot be broken down to simpler substances by chemical means. Elements are the building blocks of compounds.

empirical formula (6) Also known as a simple formula. The smallest whole number ratio of atoms of each element that comprise a compound. The formulas of ionic compounds are empirical formulas.

endothermic reaction (7) A chemical reaction that absorbs heat as it proceeds.

excited state (9) Any electronic configuration that is not the ground-state configuration.

exothermic reaction (7) A chemical reaction that produces heat as it proceeds.

Fahrenheit temperature scale (1) A scale of temperature in which water normally freezes at 32°F and boils at 212°F.

formula unit (2, 4) The set of ions that is identical to the empirical formula of an ionic compound.

freezing-point depression, ΔTFP (13) The lowering of the freezing point of a solvent caused by the concentration of solute particles in the solution.

gas constant, R (11) The constant of the ideal gas law, R = 0.0821 L-atm/mole-K.

ground-state configuration (9) The lowest energy electronic configuration of an atom or ion.

group (2) A vertical column of elements in the periodic table.

halogens (2) The elements in Group VIIA (F, Cl, Br, I).

heterogeneous mixture (2, 13) A mixture in which the components remain physically separate to such a degree that they can be seen as separate particles with the naked eye.

homogeneous mixture (2, 13) A mixture in which the components are intimately mixed on the molecular level making it impossible to see the component parts with the naked eye.

Hund's rule (9) Electrons will fill orbitals of the same energy (those in p- or d-subshells) singly before any pairing of electrons occurs.

hydrate (4, 5) A compound containing water molecules in its composition.

hydrogen bond (12) An attractive link formed by a hydrogen atom bridging two very electronegative atoms, most always O, N, or F.

hydronium ion (14) The H_3O^+ ion.

ideal gas law (11) PV = nRT

intermolecular force (12) A force of attraction between molecules: London force, dipole-dipole force, and hydrogen bonds are intermolecular forces.

ion (2, 3, 4) Any chemical species that bears an electrical charge, such as, NO_3^-, K^+, or I^-.

ionic bond (10) The net attraction of ions of opposite charge in the crystal of an ionic compound.

ionic compound (10) A crystalline compound composed of ions of opposite charge.

ionic equation (14) A chemical equation that represents the reactants and products as the ions they form in solution: $H^+(aq) + Cl^-(aq) + Na^+(aq) + OH^-(aq) \rightarrow Na^+(aq) + Cl^-(aq) + H_2O(l)$.

ionization (2) The gain or loss of electrons by a neutral species to form an ion. The formation of ions as an acid, base, or salt is dissolved in water.

ionization energy (9) The minimum energy required to remove an electron from an atom.

isotopes (3) Atoms with the same atomic number (same element) but different mass numbers.

Joule (1) A unit of energy in the International System of Units (SI).

Kelvin scale (1) A temperature scale starting at absolute zero (0 K) with the freezing point of water at 273.15 K and the boiling point of water at 373.15 K.

kinetic energy (1) The energy possessed by a body by virtue of its motion.

kinetic molecular theory (11) The theory that describes the properties of an ideal gas in which molecules are in constant, random, straight-line motion colliding with each other and the wall of its container.

Law of Conservation of Energy (1) Energy can be neither created nor destroyed.

Law of Conservation of Mass (7) In chemical changes, mass (matter) is neither created nor destroyed.

Law of Definite Composition (2) All samples of a given pure compound contain the same elements in the same mass ratio (same number of atoms).

Lewis structure (10) A diagram of connected atoms showing the distribution of all valence electrons among the atoms in a covalent species.

Lewis symbol (9) An element symbol surrounded by dots that represent its valence electrons.

limiting reactant (8) The reactant that determines the amount of product formed in a reaction.

London force (12) An intermolecular force arising from the formation of instantaneous dipoles between atoms or molecules that are close together.

main-group elements (2) Any of the elements in the A-groups in the periodic table.

mass (1) The amount of matter in a sample expressed in units of mass (g, kg, mg).

mass number (3) The sum of the number of protons and neutrons in a nucleus.

metal (2) A substance that conducts electricity, has a luster, and is likely to be malleable and ductile. Metals commonly for positive ions.

metalloid (2) An element that shares both metallic and nonmetallic properties. They are the elements along the stair-step on the periodic table that separates metals from nonmetals.

mixture (2) A sample of matter composed of two or more pure substances in a ratio that is not fixed. The substances can be separated by physical means, for example, filtering, solubility, and so on.

mmHg (11) A unit of pressure taken from the height of the mercury column in a mercury barometer; 1 mmHg = 1 torr, and 760 mmHg = 1 atmosphere (atm).

molarity (13) A unit of solution concentration equal to the number of moles of solute divided by the volume of the solution in liters. Molarity is symbolized with a capital M.

molar mass (5, 8) The mass in grams (or kilograms) of one mole of atoms of an element, one mole of molecules of a molecular compound, or one mole of formula units of an ionic compound. One molar mass of sodium is 22.99 g of Na; one molar mass of water is 18.016 g of H_2O.

molar volume of a gas (11) The volume of one mole of gas at STP, 22.4 L.

mole (5) 6.022×10^{23} units, the number of atoms in exactly 12 grams of C–12.

molecular compound (2, 10) A covalent compound that exists as molecules.

molecular formula (6) An array of symbols and subscripts showing the kind and number of atoms of all the elements in a molecule. $C_6H_{12}O_6$ is the molecular formula of glucose.

molecular mass (5) The mass of a molecule in atomic mass units (amu).

molecule (2) The smallest particle of a molecular compound.

mole ratio (7) The stoichiometric ratio given by the coefficients of two species in a balanced chemical equation. The ratio is usually written as a conversion factor.

net-ionic equation (14) An equation showing only those ions directly involved in the chemical reaction.

neutralization reaction (14) The reaction between an acid and a base.

neutron, n (3) A subatomic particle in the nucleus that is electrically neutral.

noble gas (2) The elements in Group VIIIA in the periodic table (He, Ne, Ar, Kr, Xe, Rn).

nonbonding pair (10) A pair of valence electrons not involved in bonding.

nonelectrolyte (13) Any substance that does not form ions when dissolved in water.

nucleus (3) The massive, small, positively charged particle at the center of the atom.

octet rule (10) Atoms in chemical reactions, other than hydrogen, tend to lose, gain, or share the necessary number of electrons to achieve an octet (8) of electrons in their valence shells.

orbital (9) A region of space where there is a high probability of finding an electron; s-orbitals are spherical and p-orbitals are dumbbell shaped. An orbital can hold a maximum of two electrons.

osmosis (13) The net movement of solvent across a semipermeable membrane from a region of higher solvent concentration (lower solute concentration) into one of lower solvent concentration (higher solute concentration).

osmotic pressure, π (13) The pressure that must be applied to stop the movement of solvent across a semipermeable membrane during osmosis.

oxidation (7) The loss of one or more electrons by a species.

oxyacid (4) An acid that contains oxygen, for example, HNO_3, $HClO_4$, H_2CO_3.

paired electrons (9) Two electrons with opposite spins (↑↓). Two electrons in the same orbital or in a covalent bond are paired electrons.

partial charge (10) Small charges, $\delta-$, $\delta+$, arising from the unsymmetrical motion of electrons in an atom or molecule.

Pauli principle (9) The maximum of two electrons in the same orbital must have opposite spins.

percent composition, %(m/m) (6) The mass of one component in a sample expressed as a percent of the mass of the entire sample.

percent yield (8) The percent of the calculated or theoretical yield of a chemical reaction actually obtained.

period (2) The horizontal rows of elements in the periodic table.

physical property (2) A property of a substance determined without changing its identity.

polar-covalent bond (10) One in which the electron pair(s) are shared unequally between the bonded atoms.

polar molecule (10, 12, 13) A molecule that exhibits a permanent dipole.

polyatomic ion (4) An ion containing two or more atoms.

potential energy (1) The energy possessed by a sample by virtue of its position (water behind a dam), condition (a wound clock spring), or composition (gun powder).

principal quantum number, n (9) A whole number, 1 or greater, identifying the principal energy level (shell) of an electron.

product (7) A substance produced in a chemical reaction.

properties (2) The identifying characteristics of a substance, for example, color, density, boiling point, and so on.

proton, p+ (3) The positively charged subatomic particle in the nucleus of an atom.

reactant (7) A substance consumed in a chemical reaction.

redox reaction (7) One in which electrons are transferred from one reactant to another.

reduction (7) The gain of one or more electrons by a species.

resonance (10) The use of two or more Lewis structures to better represent a species.

salt (7, 14) The ionic product of an acid-base neutralization reaction.

significant figures (1) The digits in a measurement up to and including the first uncertain digit.

single bond (10) A single pair of electrons shared between two atoms.

single-replacement reaction (7) A reaction in which one element replaces another in a compound.

solute (13) The substance dissolved in the solvent of a solution.

solution (2, 13) A homogeneous mixture.

solvent (13) The medium in which the solute is dissolved to form a solution.

specific heat (1) The amount of heat needed to raise the temperature of 1 gram of a substance 1°C.

state of matter (2) The physical condition of a sample, solid, liquid, or gas.

stock number in nomenclature (4) A Roman numeral indicating the charge of a monatomic ion.

stoichiometric coefficients (7) The coefficients in a balanced equation.

stoichiometry (8) The quantitative relationships of reactants and products in chemical reactions.

STP (11) Standard temperature and pressure for gases, 0°C and 1 atm of pressure.

subshell (9) A subdivision of a principal shell: s-subshell, p-subshell, d-subshell.

substance (2) A pure sample of matter. Elements and compounds are substances.

temperature (1) A measure of hotness or coldness of a sample of matter.

torr (11) A measure of pressure named in honor of Torricelli; 1 torr = 1 mmHg.

transition metal (2, 9) The metals in the B-Groups in the center of the periodic table.

triple bond (10) Three pairs of electrons shared between two atoms.

valence electrons (10) The outermost electrons of an atom, those involved in chemical changes and bonding.

volume (1) The amount of space occupied by a sample of matter.

VSEPR theory (10) Valence shell electron pair repulsion theory, which predicts the three-dimensional shape of molecules and ions based on the arrangement of electron pairs (nonbonding pairs and bonds) about the central atom.

weak electrolyte (14) A molecular substance that is only partially ionized in solution.

Customized Full-Length Exam

1. Round off 5,345,662 to three significant figures and express the answer in scientific notation.

   Answer: 5.35×10^6

   If you answered **correctly,** go to problem 2.
   If you answered **incorrectly,** review "Significant Figures and Rounding Off Numbers," page 13.

2. 0.458 mg = _____ kg

   Answer: 4.58×10^{-7} kg

   If you answered **correctly,** go to problem 3.
   If you answered **incorrectly,** review "Metric-Metric Conversions," page 22.

3. 8.94 L = _____ qt

   Answer: 9.45 qt

   If you answered **correctly,** go to problem 4.
   If you answered **incorrectly,** review "English-Metric Conversions," page 24.

4. The density of mercury is 13.6 g/mL. What mass of mercury has a volume of 150. mL?

   Answer: 2.04×10^3 g

   If you answered **correctly,** go to problem 5.
   If you answered **incorrectly,** review "Density," page 26.

5. 60.0°C = _____ °F = _____ K

   Answer: 140°F and 333 K

   If you answered **correctly,** go to problem 6.
   If you answered **incorrectly,** review "Measuring Temperature," page 28.

6. What amount of heat is required to raise the temperature of 125 g of water from 23.0°C to 66.0°C? The specific heat of water is 4.184 J/g°C.

   Answer: 2.45×10^4 Joule

   If you answered **correctly,** go to problem 7.
   If you answered **incorrectly,** review "Energy and the Measurement of Heat," page 30.

7. Is the fact that gold is impervious to attack by hydrochloric acid a physical or chemical property of gold?

Answer: Chemical property

If you answered **correctly,** go to problem 8.
If you answered **incorrectly,** review "Properties and Changes," page 41.

8. Which element — O, Se, P, or Cl — has chemical properties similar to bromine?

Answer: Cl, chlorine

If you answered **correctly,** go to problem 9.
If you answered **incorrectly,** review "The Periodic Table of the Elements," page 44.

9. What are the correct names for P_2O_5 and CCl_4?

Answer: Diphosphorus pentoxide (or diphosphorus pentaoxide) and carbon tetrachloride

If you answered **correctly,** go to problem 10.
If you answered **incorrectly,** review "Naming Molecular Compounds," page 52.

10. How many protons, neutrons, and electrons are in one atom of element U-238? The atomic number or uranium is 92.

Answer: 92 protons, 92 electrons, and 146 neutrons

If you answered **correctly,** go to problem 11.
If you answered **incorrectly,** review "The Atomic Number and Mass Number," page 65.

11. A hypothetical element is a mixture of two isotopes. The natural abundance and atomic mass of the isotopes are: Isotope-1 40.00% – 19.95 amu; Isotope-2 60.00% – 21.89 amu. What is the mass-average atomic mass of this element?

Answer: 21.11 amu

If you answered **correctly,** go to problem 12.
If you answered **incorrectly,** review "Isotopes of the Elements," page 68, and "The Atomic Mass (Atomic Weight) of the Elements," page 70.

12. Correctly name the following compounds: (a) $FeSO_4$, (b) Na_3PO_4, and (c) CaC_2O_4

Answer: (a) iron(II) sulfate, (b) sodium phosphate, (c) calcium oxalate

If you answered **correctly,** go to problem 13.
If you answered **incorrectly,** review "The Polyatomic Ions," page 92, and "Naming Ionic Compounds with Polyatomic Ions," page 96.

13. Correctly name the following common acids and bases: (a) $HC_2H_3O_2(aq)$, (b) $LiOH(aq)$, (c) $NH_3(aq)$, and (d) $H_2SO_4(aq)$

Answer: (a) acetic acid, (b) lithium hydroxide, (c) aqueous ammonia, and (d) sulfuric acid

If you answered **correctly,** go to problem 14.
If you answered **incorrectly,** review "The Common Acids and Bases," page 98.

14. How many iron atoms are in 0.0450 g of iron? The molar mass of iron is 55.85 g.

Answer: 4.85×10^{20} atoms

If you answered **correctly,** go to problem 15.
If you answered **incorrectly,** review "Elements and the Mole—Molar Mass," page 111.

15. 150. grams of N_2O_5 are _____ moles of N_2O_5 and contain _____ nitrogen atoms. The molar mass of N is 14.01 g and that for O is 16.00 g.

Answer: 1.39 moles and 1.67×10^{24} atoms of N

If you answered **correctly,** go to problem 16.
If you answered **incorrectly,** review "Compounds and the Mole," page 118.

16. What mass of C_4H_{10} contains 0.800 mole of carbon atoms? The molar mass of C_4H_{10} is 58.12 g.

Answer: 11.6 g

If you answered **correctly,** go to problem 17.
If you answered **incorrectly,** review "Compounds: Moles and Counting Molecules," page 123.

17. What is the percent copper by mass in $CuSO_4$? The molar masses are: Cu = 63.55 g, S = 32.07 g, O = 16.00 g.

Answer: 39.81 % Cu

If you answered **correctly,** go to problem 18.
If you answered **incorrectly,** review "Calculating the Percent Composition of Compounds," page 135.

18. A molecular compound of aluminum and chlorine exists in the gas phase. It is 20.24 % aluminum and 79.76 % chlorine. The molar mass of the compound is approximately 266 g. What is the molecular formula of this compound? The molar masses are: Al = 26.98 g, Cl = 35.45 g.

Answer: Al_2Cl_6

If you answered **correctly,** go to problem 19.
If you answered **incorrectly,** review "Calculating Molecular Formulas," page 146.

19. (a) Only a balanced equation obeys the Law of _____________. (b) When this equation is correctly balanced, the sum of all the coefficients equals _____.

$Fe_3O_4(s) + H_2(g) \rightarrow Fe(s) + H_2O(g)$

Answer: (a) Conservation of mass (b) 12

If you answered **correctly,** go to problem 20.
If you answered **incorrectly,** review "Balancing Chemical Equations," page 160.

20. Of the four classes of chemical reactions—synthesis, decomposition, single-replacement, and double-replacement—which would be the appropriate class for the reaction of hydrochloric acid with sodium hydroxide?

$HCl(aq) + NaOH(aq) \rightarrow NaCl(aq) + H_2O(l)$

Answer: Double-replacement

If you answered **correctly,** go to problem 21.
If you answered **incorrectly,** review "Types of Chemical Reactions," page 165.

21. Complete the following balanced equation with the correct product, including its coefficient?

$2\ Fe_2O_3(s) + 6\ C(s) \rightarrow 4\ Fe(s) + \_\_\_\_$

Answer: $6\ CO(g)$

If you answered **correctly,** go to problem 22.
If you answered **incorrectly,** review "Predicting Reactants and Products in Replacement Equations," page 173.

22. How many grams of oxygen gas, $O_2(g)$, are consumed in the combustion of 100. g of $C_4H_{10}(g)$? The molar masses are: C = 12.01 g, O = 16.00 g, H = 1.008 g.

$2\ C_4H_{10}(g) + 13\ O_2(g) \rightarrow 8\ CO_2(g) + 10\ H_2O(l)$

Answer: 358 g O_2

If you answered **correctly,** go to problem 23.
If you answered **incorrectly,** review "Mass-to-Mass Conversions," page 201.

23. A 50.0 g mass of sulfur is burned in 65.0 g of $O_2(g)$ to form $SO_3(g)$. How many grams of sulfur trioxide form? The molar masses are: S = 32.07 g, O = 16.00 g.

$2\ S(s) + 3\ O_2(g) \rightarrow 2\ SO_3(g)$

Answer: 108 g SO_3; O_2 is the limiting reactant.

If you answered **correctly,** go to problem 24.
If you answered **incorrectly,** review "The Limiting Reactant," page 204.

24. An 8.45 g piece of pure gold(III) oxide, Au_2O_3, was heated with an excess of carbon to obtain gold metal. Exactly 6.59 g of gold metal was obtained. What is the percent yield of the reaction? The molar masses are: Au = 197.0 g, O = 16.00 g, C = 12.01 g.

$Au_2O_3(s) + 3\ C(s) \rightarrow 2\ Au(s) + 3\ CO(g)$

Answer: 87.5 % yield

If you answered **correctly,** go to problem 25.
If you answered **incorrectly,** review "Percent Yield of Reactions," page 209.

25. How much heat is produced in the combustion of 100. g of methane gas, $CH_4(g)$? The molar masses are: C = 12.01 g, H = 1.008 g, O = 16.00 g.

$$CH_4(g) + 2\,O_2(g) \rightarrow CO_2(g) + 2\,H_2O(l) \quad \Delta H = -890 \text{ kJ}$$

Answer: 5.55×10^3 kJ

If you answered **correctly,** go to problem 26.
If you answered **incorrectly,** review "Standard Temperature and Pressure," page 305.

26. (a) Write the correct ground state electronic configuration for manganese, Mn, atomic number 25. (b) How many unpaired electrons are in the ground state electronic configuration of manganese?

Answer: (a) $1s^22s^22p^63s^23p^64s^23d^5$ (b) 5 unpaired electrons

If you answered **correctly,** go to problem 27.
If you answered **incorrectly,** review "The Electronic Configuration of Atoms," page 234.

27. Arrange these three species in order of *decreasing* radius: Fe^{3+}, Fe, Fe^{2+}.

Answer: Fe > Fe^{2+} > Fe^{3+}

If you answered **correctly,** go to problem 28.
If you answered **incorrectly,** review "Properties of Atoms and the Periodic Table," page 246.

28. What is the expected formula of a covalent compound formed between phosphorus and hydrogen that uses one atom of phosphorus? (b) How many bonding electron pairs and how many nonbonding electron pairs are in one molecule of this compound?

Answer: (a) PH_3 (b) 3 bonding pairs and 1 nonbonding pair

If you answered **correctly,** go to problem 29.
If you answered **incorrectly,** review "Covalent Bonding—Covalent Compounds," page 265.

29. Which of the following species is expected to display "resonance": (a) CO_2, (b) NO_2^+, (c) SO_3^{2-}?

Answer: (c) SO_3^{2-}

If you answered **correctly,** go to problem 30.
If you answered **incorrectly,** review "Lewis Structures," page 274.

30. The presence of four electron domains about a central atom can lead to three distinct molecular shapes. What are they?

Answer: Bent (3 atoms, H_2O), trigonal pyramidal (4 atoms, NH_3), and tetrahedral (5 atoms, CH_4)

If you answered **correctly,** go to problem 31.
If you answered **incorrectly,** review "VSEPR Theory—Predicting Molecular Shape," page 280.

31. A sample of nitrogen gas has a volume of 9.35 L at 550 mmHg and 100°C. What volume will it occupy at 1.00 atm and 27.0°C?

Answer: 5.44 L

If you answered **correctly,** go to problem 32.
If you answered **incorrectly,** review "The Combined Gas Law," page 305.

32. How many moles of water vapor occupy 250. L at 500.°C and 22.0 atm of pressure?

Answer: 86.7 moles

If you answered **correctly,** go to problem 33.
If you answered **incorrectly,** review "The Ideal Gas Law," page 307.

33. How many liters of $O_2(g)$ are required to just react with 100. L of $C_3H_6(g)$, both volumes being measured at the same temperature and pressure. The balanced equation is:

$$2\ C_3H_6(g) + 9\ O_2(g) \rightarrow 6\ CO_2(g) + 6\ H_2O(l)$$

Answer: 450. L

If you answered **correctly,** go to problem 34.
If you answered **incorrectly,** review "Stoichiometry of Reactions Involving Gases," page 315.

34. The pressure exerted by a gas is the result of ______.

Answer: molecular collisions with the surface of the container that holds it

If you answered **correctly,** go to problem 35.
If you answered **incorrectly,** review "The Kinetic-Molecular Theory of Gases," page 319.

35. What kind of intermolecular forces are present in liquid ammonia, $NH_3(l)$?

Answer: All three are present: London forces, dipole-dipole forces, and hydrogen-bonding.

If you answered **correctly,** go to problem 36.
If you answered **incorrectly,** review "The Intermolecular Forces of Attraction," page 334.

36. Which molecule can exhibit the greatest London dispersion force: (a) CH_4, (b) CS_2, (c) I_2?

Answer: I_2; it has the largest molar mass.

If you answered **correctly,** go to problem 37.
If you answered **incorrectly,** review "The Intermolecular Forces of Attraction," page 334.

37. How many grams of glucose are in 250. g of a 0.500 %(m/m) glucose solution?

Answer: 1.25 g

If you answered **correctly,** go to problem 38.
If you answered **incorrectly,** review "Mass/Mass Percent," page 362.

38. What is the molarity of a solution of calcium chloride, $CaCl_2$, that contains 35.0 g of calcium chloride in 745 mL of solution? The molar masses are: Ca = 40.08 g, Cl = 35.45 g.

Answer: 0.423 M

If you answered **correctly,** go to problem 39.
If you answered **incorrectly,** review "Molarity," page 366.

39. What volume of 6.00 M $CuSO_4$ solution is needed to prepare 120. mL of 0.175 M $CuSO_4$ solution?

Answer: 3.50 mL

If you answered **correctly,** go to problem 40.
If you answered **incorrectly,** review "Dilution of Molar Solutions," page 367.

40. Which aqueous solution will have the highest boiling point: (a) 0.10 M glucose (b) 0.10 M NaCl (c) 0.10 M $CaCl_2$?

Answer: 0.10 M $CaCl_2$

If you answered **correctly,** go to problem 41.
If you answered **incorrectly,** review "Properties of Solutions," page 373.

41. In terms of the Bronsted-Lowry concept of acids and bases, identify the acid, base, conjugate acid, and conjugate base in the following equation:

$HNO_3(aq) + H_2O(l) \rightarrow H_3O^+(aq) + NO_3^-(aq)$

Answer: Acid = HNO_3, base = H_2O, conjugate acid = H_3O^+, and conjugate base = NO_3^-.

If you answered **correctly,** go to problem 46.
If you answered **incorrectly,** review "The Bronsted-Lowry Concept," page 395.

Index

A

B

C

D

E

F

G

H

I

J

K

L

N

O

P

Q

R

S

T

U

V

W

Z